CLASSICAL AND MODERN ENGINEERING METHODS IN FLUID FLOW AND HEAT TRANSFER

CLASSICAL AND MODERN ENGINEERING METHODS IN FLUID FLOW AND HEAT TRANSFER

An Introduction for Engineers and Students

ABRAM DORFMAN

MP MOMENTUM PRESS

MOMENTUM PRESS, LLC, NEW YORK

First published by Momentum Press®, LLC
222 East 46th Street, New York, NY 10017
www.momentumpress.net

ISBN-13: 978-1-60650-269-3 (paperback)
ISBN-10: 1-60650-269-7 (paperback)
ISBN-13: 978-1-60650-271-6 (e-book)
ISBN-10: 1-60650-271-9 (e-book)

DOI: 10.5643/9781606502716

Cover design by Jonathan Pennell
Cover image: Basic equipment of scholarly created classical and modern methods of contemporary results of employing those methods
Interior design by Exeter Premedia Services Private Ltd.
Chennai, India

10 9 8 7 6 5 4 3 2 1

Printed in the United States of America

To my great-grandchildren David and Emily Fridman with love and best wishes

CONTENTS

LIST OF FIGURES   XV

LIST OF EXAMPLES   XIX

NOMENCLATURE   XXIII

PREFACE   XXVII

ACKNOWLEDGMENT   XXXV

ABOUT THE AUTHOR   XXXVII

PART I CLASSICAL METHODS IN FLUID FLOW AND HEAT TRANSFER   1

1 METHODS IN HEAT TRANSFER OF SOLIDS   3

1.1 Historical Notes   3

1.2 Heat Conduction Equation and Problem Formulation   4

  1.2.1 Cartesian Coordinates   4

  1.2.2 Orthogonal Curvilinear Coordinates   6
   1.2.2.1 Cylindrical coordinates ($\xi_1 = r$, $\xi_2 = \gamma$, $\xi_3 = z$)   6
   1.2.2.2 Spherical coordinates ($\xi_1 = r$, $\xi_2 = \gamma$, $\xi_3 = \phi$)   7
   1.2.2.3 Elliptical cylindrical coordinates ($\xi_1 = u$, $\xi_2 = v$, $\xi_3 = z$)   7

  1.2.3 Universal Function for Heat Flux on an Arbitrary Nonisothermal Surface   7

  1.2.4 Initial, Boundary, and Conjugate Conditions   8

  EXERCISES 1.1–1.12   10–11

1.3 Solution Using Error Integral   11

  1.3.1 An Infinite Solid or Thin, Laterally Insulated Rod   12

  1.3.2 A Semi-Infinite Solid or Thin, Laterally Insulated Rod   13

1.4 Duhamel's Method   15

  1.4.1 Duhamel Integral Derivation   15

  1.4.2 Time-Dependent Surface Temperature   17

  EXERCISES 1.13–1.27   18–20

1.5 Method of Separation Variables   20

　　1.5.1 General Approach, Homogeneous, and Inhomogeneous Problems   20

　　1.5.2 One-Dimensional Unsteady Problems   21

　　1.5.3 Orthogonality of Eigenfunctions   24

　　EXERCISES 1.28–1.43   27–28

　　1.5.4 Two-Dimensional Steady Problems   28

1.6 Integral Transforms   30

　　1.6.1 Fourier Transform   30

　　1.6.2 Laplace Transform   33

1.7 Green's Function Method   36

EXERCISES 1.44–1.60   38–39

2 METHODS IN LAMINAR FLUID FLOW AND HEAT TRANSFER   **41**

2.1 A Brief History   41

2.2 Navier–Stokes, Energy, and Mass Transfer Equations   43

　　2.2.1 Two Types of Transport Mechanism. Analogy Between Transfer Processes 44

　　2.2.2 Different Forms of Navier–Stokes, Energy, and Diffusion Equations   45
　　　　2.2.2.1 Vector form   45
　　　　2.2.2.2 Einstein and other index notation   47
　　　　2.2.2.3 Vorticity form of the Navier–Stokes equation   48
　　　　2.2.2.4 Stream function form of the Navier–Stokes equation   48
　　　　2.2.2.5 Irrotational inviscid two-dimensional flows   49
　　　　2.2.2.6 Curvilinear orthogonal coordinates   51
　　　　EXERCISES 2.1–2.24   55–56

2.3 Initial and Boundary Conditions   56

　　2.3.1 Navier–Stokes Equations   56

　　2.3.2 Specific Issues of the Energy Equation   57

2.4 Exact Solutions of Navier–Stokes and Energy Equations   58

　　2.4.1 Two Stokes Problems   58

　　2.4.2 Solutions of Three Other Unsteady Problems   59

　　2.4.3 Steady Flow in Channels and in a Circular Tube   59

　　2.4.4 Stagnation Point Flow (Hiemenz Flow)   61

　　2.4.5 Other Exact Solutions   61

　　2.4.6 Some Exact Solutions of the Energy Equation   62
　　　　2.4.6.1 Couette flow in a channel with heated walls   62
　　　　2.4.6.2 Adiabatic wall temperature   62
　　　　2.4.6.3 Temperature distributions in channels and in a tube   63

2.5 Cases of Small and Large Reynolds and Peclet Numbers   64

    2.5.1 Creeping Approximation (Small Reynolds and Peclet Numbers)   65
        2.5.1.1 Stokes flow past a sphere   65
        2.5.1.2 Oseen's approximation   67
        2.5.1.3 Heat transfer from the sphere in the stokes flow   68

    2.5.2 Boundary-Layer Approximation (Large Reynolds and Peclet Numbers)   68
        2.5.2.1 Derivation of boundary-layer equations   69
        2.5.2.2 Prandtl–Mises and Görtler transformations   72
        2.5.2.3 Theory of similarity and dimensionless numbers   74
        2.5.2.4 Boundary-layer equations of higher order   76
        EXERCISES 2.25–2.65   77–80

2.6 Exact Solutions of the Boundary-Layer Equations   80

    2.6.1 Flow and Heat Transfer on an Isothermal Semi-Infinite Flat Plate
        (Blasius and Pohlhausen Solutions)   80

    2.6.2 Self-Similar Flows in Dynamic and Thermal Boundary Layers   84

    2.6.3 Solutions in the Power Series Form   86

    2.6.4 Flow in the Case of Potential Velocity $U(x) = U_0 - ax^n$ (Howarth Flow)   87

    2.6.5 Fluid Flows Interaction   87
        2.6.5.1 Flow in the wake of a body   87
        2.6.5.2 Two-dimensional jet   88
        2.6.5.3 Mixing layer of two parallel streams   89

    2.6.6 Flow in Straight and Convergent Channels   90

    2.6.7 Solutions of Second-Order Boundary-Layer Equations   91

    2.6.8 Solutions of the Thermal Boundary-Layer Equation   91

    EXERCISES 2.66–2.88   92–93

2.7 Approximate Methods in the Boundary-Layer Theory   93

    2.7.1 Karman-Pohlhausen Integral Method   93
        2.7.1.1 Friction and heat transfer on a flat plate   95
        2.7.1.2 Flows with pressure gradients   97

    2.7.2 Linearization of the Momentum Boundary-Layer Equation   99
        2.7.2.1 Flow at the outer edge of the boundary layer   99
        2.7.2.2 Universal function for the skin friction coefficient   100

    2.7.3 Thermal Boundary-Layer Equations for Limiting Prandtl Numbers   103

2.8 Natural Convection   104

EXERCISES 2.89–2.17   106–108

3. METHODS IN TURBULENT FLUID FLOW AND HEAT TRANSFER   111

3.1 Transition from Laminar to Turbulent Flow   111

    3.1.1 Basic Characteristics   111

    3.1.2 The Problem of Laminar Flow Stability   112

3.2 Reynolds-Averaged Navier–Stokes Equation   114

   3.2.1 Some Physical Aspects   114

   3.2.2 Reynolds Averaging   115

   3.2.3 Reynolds Equations and Reynolds Stresses   116

3.3 Algebraic Models   117

   3.3.1 Prandtl's Mixing-Length Hypothesis   117

   3.3.2 Modern Structure of Velocity Profile in Turbulent Boundary Layer   118

   EXERCISES 3.1–3.22   121–122

   3.3.3 Mellor–Gibson Model [9, 10, 13, 18]   122

   3.3.4 Cebeci-Smith Model [13]   124

   3.3.5 Baldwin–Lomax Model [18]   125

   3.3.6 Application of the Algebraic Models   125
      3.3.6.1 The far wake   125
      3.3.6.2 The two-dimensional jet   127
      3.3.6.3 Mixing layer of two parallel streams   127
      3.3.6.4 Flows in channel and pipe   128
      3.3.6.5 The boundary-layer flows   129
      3.3.6.6 Heat transfer from an isothermal surface   129
      3.3.6.7 The effect of the turbulent Prandtl number   130

   3.3.7 The 1/2 Equation Model   131

   3.3.8 Applicability of the Algebraic Models   132

   EXERCISES 3.23–3.40   133

3.4 One-Equation and Two-Equation Models   134

   3.4.1 Turbulence Kinetic Energy Equation   134

   3.4.2 One-Equation Models   135

   3.4.3 Two-Equation Models   136
      3.4.3.1 The $k - \omega$ model   136
      3.4.3.2 The $k - \varepsilon$ model   138
      3.4.3.3 The other turbulence models   138

   3.4.4 Applicability of the One-Equation and Two-Equation Models   138

3.5 Integral Methods   139

   EXERCISES 3.41–3.56   142–143

PART II MODERN CONJUGATE METHODS IN HEAT TRANSFER AND FLUID FLOW   **145**

Introduction   145

   Concept of Conjugation   145

   Why and When are Conjugate Methods Required?   146

4 CONJUGATE HEAT TRANSFER PROBLEM AS A CONDUCTION PROBLEM   **149**

4.1 Formulation of Conjugate Heat Transfer Problem   149

4.2 Universal Function for Laminar Fluid Flow   150

    4.2.1 Universal Function for Heat Flux in Self-Similar Flows as an Exact Solution of a Thermal Boundary-Layer Equation   150

    4.2.2 Universal Function for Heat Flux in Arbitrary Pressure Gradient Flow   154

    4.2.3 Integral Universal Function for Heat Flux in Arbitrary Pressure Gradient Flow   156

    4.2.4 Examples of Applications of Universal Functions for Heart Flux   158

    EXERCISES 4.1–4.32   163–165

    4.2.5 Universal Function for a Temperature Head   165

    4.2.6 Universal Function for Unsteady Heat Flux in Self-Similar Flow   169

    4.2.7 Universal Function for Heat Flux in Compressible Fluid Flow   171

    4.2.8 Universal Function for Heat Flux for a Moving Continuous Sheet   172

    4.2.9 Universal Function for Power-Law Non-Newtonian Fluids   174

    4.2.10 Universal Function for the Recovery Factor   175

    4.2.11 Universal Function for an Axisymmetric Body   176

    EXERCISES 4.33–4.50   176–177

4.3 Universal Functions for Turbulent Flow   177

4.4 Reducing a Conjugate Problem to a Conduction Problem   182

    4.4.1 Universal Function as a General Boundary Condition   182

    4.4.2 Estimation of Errors Caused by Boundary Condition of the Third Kind   182

    4.4.3 Equivalent Conduction Problem with the Combined Boundary Condition   184

    4.4.4 Equivalent Conduction Problem for Unsteady Heat Transfer   185

    EXERCISES 4.51–4.61   186–187

5 GENERAL PROPERTIES OF NONISOTHERMAL AND CONJUGATE HEAT TRANSFER   **189**

5.1 Effect of Temperature Head Distribution: Temperature Head Decreasing-Basic Reason for Low Heat Transfer Rate   189

    5.1.1 Effect of the Temperature Head Gradient   190

    5.1.2 Effect of Flow Regime   191

    5.1.3 Effect of Pressure Gradient   192

5.2 Biot Number-A Measure of Problem Conjugation   192

5.3 Gradient Analogy   193

5.4 Heat Flux Inversion   195

5.5 Zero Heat Transfer Surfaces   197

5.6 Examples of Optimizing Heat Transfer in Flow Over Bodies   199

EXERCISES 5.1–5.30   206–208

6 **CONJUGATE HEAT TRANSFER IN FLOW PAST PLATES.**
   CHARTS FOR SOLVING CONJUGATE HEAT TRANSFER PROBLEMS   **209**

6.1 Temperature Singularities on the Solid–Fluid Interface   209

   6.1.1 Basic Equations   210

   6.1.2 Singularity Types   211
      6.1.2.1. Laminar Flow at the Stagnation Point   211
      6.1.2.2. Laminar Flow at Zero-Pressure Gradient   212
      6.1.2.3. Turbulent Flow at Zero-Pressure Gradient   212
      6.1.2.4. Laminar Gradient Flow with Power-Law
         Free-Stream Velocity cx^m 212
      6.1.2.5. Asymmetric Laminar-Turbulent Flow   212

6.2 Charts for Solving Conjugate Heat Transfer   212

   6.2.1 Charts Development   212

   6.2.2 Using Charts   215

   EXERCISES 6.1–6.17   220–221

6. 3 Applicability of Charts and One-Dimensional Approach   221

   6.3.1 Refining and Estimating Accuracy of the Charts Data   221

   6.3.2 Applicability of Thermally Thin Body Assumption   221

   6.3.3 Applicability of the One-Dimensional
       Approach and Two-Dimensional Effects   222

6.4 Conjugate Heat Transfer in Flow Past Plates   225

EXERCISES 6.18–6.31   232

Conclusion of Heat Transfer Investigation (Chapters 4–6)   233

Should any Heat Transfer Problem be Considered as a Conjugate?   233

7 **PERISTALTIC MOTION AS A CONJUGATE PROBLEM:**
   MOTION IN CHANNELS WITH FLEXIBLE WALLS   **235**

7.1 What is the Peristaltic Motion Like?   235

7.2 Formulation of the Conjugate Problem   236

7.3 Early Works   237

7.4 Semi-Conjugate Solutions   238

7.5 Conjugate Solutions   243

EXERCISES 7.1–7.24   247–248

PART III NUMERICAL METHODS IN FLUID FLOW AND HEAT TRANSFER   **249**

8 CLASSICAL NUMERICAL METHODS IN FLUID FLOW AND HEAT TRANSFER   **251**

8.1 Way Analytical or Numerical Methods?   251

8.2 Approximate Methods for Solving Differential Equations   253

8.3 Some Features of Computing Flow and Heat Transfer Characteristics   258

    8.3.1 Control-Volume Finite-Difference Method   259
        8.3.1.1 Computing pressure and velocity   259
        8.3.1.2 Computing convection-diffusion terms   260
        8.3.1.3 False diffusion   261

    8.3.2 Control-Volume Finite-Element Method   262

8.4 Numerial Methods of Conjugation   263

EXERCISES 8.1–8.27   265–267

9 MODERN NUMERICAL METHODS IN TURBULENCE   **269**

9.1 Introduction   269

9.2 Direct Numerical Simulation   270

9.3 Large Eddy Simulation   271

9.4 Detached Eddy Simulation   272

9.5 Chaos Theory   273

9.6 Concluding Remarks   274

EXERCISES 9.1–9.12   274–275

PART IV APPLICATIONS IN ENGINEERING, BIOLOGY, AND MEDICINE   **277**

10 HEAT TRANSFER IN THERMAL AND COOLING SYSTEMS   **279**

10.1 Heat Exchangers and Pipes   279

    10.1.1 Pipes and Channels   279

    10.1.2 Heat Exchangers and Finned Surfaces   285

10.2 Cooling Systems   294

    10.2.1 Electronic Packages   294

    10.2.2 Turbine Blades and Rocket   297

    10.2.3 Nuclear Reactor   300

10.3 Energy Systems   303

11 HEAT AND MASS TRANSFER IN TECHNOLOGY PROCESSES   **309**

   11.1 Multiphase and Phase-Changing Processes   309

   11.2 Manufacturing Processes Simulation   314

   11.3 Draing Technology   321

   11.4 Food Processing   330

12 FLUID FLOW AND HEAT TRANSFER IN BIOLOGY AND CLINICAL MEDICINE   **337**

   12.1 Blood Flow in Normal and Pathologic Vessels   337

   12.2 Peristaltic Flow in Disordered Human Organs   347

   12.3 Biologic Transport Processes   351

CONCLUSION   **359**

APPENDIX   **363**

PIONEERS—CONTRIBUTORS   **369**

AUTHOR INDEX   **373**

INDEX   **377**

LIST OF FIGURES

Figure 1.1. Approximation of arbitrary dependence by step function.   16

Figure 1.2. Graphical solution of the equation $\tan \mu = \mu/\text{Bi}$.   25

Figure 2.1. Flow past a wage. At the leading edge, the potential velocity is $U = Cx^m$.   51

Figure 4.1. Coefficient $g_1\left(\text{Pr}, \beta\right)$ of universal function (4.1) for laminar boundary layer. Asymptotes $1 - \text{Pr} = 0, 6 - \text{Pr} \to \infty$; β: 2-1 (stagnation point), 3-0.5 (favorable pressure gradient), 4-0 (zero pressure gradient), 5 (−0.16) (preseparation pressure gradient).   153

Figure 4.2. Coefficients $g_k\left(\text{Pr}\right)$ of universal function (4.1) for laminar boundary layer: $1 - (-g_2), 2 - g_3, 3 - (-g_4)$, ○ - numerical integration   153

Figure 4.3. Dependence of exponents C_1 and C_2 on the Prandtl number and β on the laminar boundary layer: 1- $\text{Pr} = 0$, 2- $\beta = 1$, 3- 0, 4- (−0.16).   157

Figure 4.4. Scheme of heat exchange between two fluids through a thin wall.   158

Figure 4.5. Coefficient $\hat{g}_1$ as a function of the Prandtl number.   159

Figure 4.6. Effect of conjugation in heat exchange between two fluids through thin wall in laminar flow: $1 - \text{Pr} - 0, 2 - 0.01, 3 - 0.1, 4 - \text{Pr} \to \infty$,   160

Figure 4.7. Effect of nonisothermicity for linear temperature head distribution on a plate: $1 - \theta_{we}/\theta_{wi} = -1 - 2, 2 - 1.75, 3 - 1.5, 4 - 1.25, 5 - 1.1, 6 - 1.0,$
$7 - 0.9, 8 - 0.8, 9 - 0.7, 10 - 0.6, 11 - 0.5, 12 - 0.4, 13 - 0.3$ 161

Figure 4.8. Heat transfer from a cylinder with linear temperature head in transverse flow of air ($\text{Pr} = 0.7$). $1 - \theta_{we}/\theta_{wi} = 2, 2 - 1.5, 3 - 1.25, 4 - 1.0, 5 - 0.75, 6 - 0.5$ 162

Figure 4.9. Coefficients $h_k\left(\text{Pr}, \beta\right)$ of universal function (4.26) for the laminar boundary layer. Asymptotes $1 - \text{Pr} = 0, \beta:2 - 1, 3 - 0, 4 - (-0.16)$ 166

Figure 4.10. Two cases of heat transfer from the cylinder in transverse flow.   168

Figure 4.11. Heat transfer on a plate and a cylinder in transverse air flow after heat flux jump.   168

Figure 4.12. Coefficients $g_{ki}\left(i \neq 0\right)$ as functions of z for the zero pressure gradient and $\text{Pr} = 1$.   170

Figure 4.13. Unsteady heat transfer on a plate with time linearly variation temperature head: ___ $q_w/h_* a_0$, − − − χ_t.   170

Figure 4.14. Schematic pattern of the boundary layer on a moving continuous sheet for symmetric $(U_1 = U_2, T_{\infty 1} = T_{\infty 2})$ and asymmetric flows.   172

Figure 4.15. Coefficient $g_0/\text{Pr}^{1/2}$ as a function of Prandtl number and a ratio $\varepsilon = U_\infty/U_w$ for a moving isothermal continuous sheet: $1 - \varepsilon = 1, 2 - 0.8, 3 - 0.5, 4 - 0.3, 5 - 0.1, 6 - 0, 7 - (-0.05)$.   173

Figure 4.16. Coefficients g_k as a function of Prandtl number and a ratio $\varepsilon = U_\infty/U_w$ for a moving continuous sheet: 1- streamlined plate.   173

Figure 4.17. Heat transfer from nonisothermal cylinder in transverse flow of power-law non-Newtonian fluid: ___ $\theta_{we}/\theta_{wi} = 1$, − − − $\theta_{we}/\theta_{wi} = 1 + x/D$, − − − − − $\theta_{we}/\theta_{wi} = 1 - x/D$ (notations in Example 4.2) $\text{Pr} = 1,000$.   175

Figure 4.18. Coefficient g_1 for the turbulent boundary layer: ___ $\text{Re}_{\delta_1} = 10^3$, _ _ _ 10^5, _ _ _ _ _ _ _   178

Figure 4.19. Coefficients g_k for the turbulent boundary layer:
$g_2 : 1 - \beta = -0.3, 2 - \beta = 0, 3 - \beta = 1, 4 - g_3, 5 - g_4$ ___ $\mathrm{Re}_{\delta_1} = 10^3$,
___ 10^5, _____   179

Figure 4.20. Exponent C_1 for the turbulent boundary layer: $1\text{-}\mathrm{Re}_{\delta_1} = 10^3, 2 - 10^5, 3 - 10^9$.   179

Figure 4.21. Exponent C_2 for the turbulent boundary layer: $1\text{-} \mathrm{Re}_{\delta_1} = 10^3, 2 - 10^5, 3 - 10^9$.   180

Figure 4.22. Coefficient h_1 for the turbulent boundary layer: ___ $\mathrm{Re}_{\delta_1} = 10^3$, ___ 10^5,
_ _ _ _ _10^9.   180

Figure 4.23. Coefficients $h_k, k = 2, 3, 4$ for the turbulent boundary layer ___
$\mathrm{Re}_{\delta_1} = 10^3$, ___ 10^5, _____10^9.   181

Figure 4.24. Comparison of calculated and experimental [7] data for stepwise
temperature head.   181

Figure 5.1. Nonisothermicity coefficient for different flow regimes. _ _ _ _ _ $\mathrm{Re}_{\delta_1} = 10^3$,
_ _ _ _ Re_{δ_1} ——— Re_{δ} 191

Figure 5.2. Deformation of the excess temperature profile for the linear decreasing
temperature head $\theta_w/\theta_{wi} = 1 - \bar{x}$, with $\bar{x} = K(x/L)$ and $\mathrm{Pr} = 0.7$, for laminar
(a) and turbulent $\mathrm{Re}_{\delta_1} = 10^3$ (b) boundary layer.   196

Figure 5.3. Absolute value of difference between maximum temperature heads under
linear decreasing and increasing heat fluxes: 1-laminar wall jet, 2-laminar
flow, 3 and 4-turbulent flow $\mathrm{Re} = 10^8$ and $\mathrm{Re} = 5 \cdot 10^5$, 5-stagnation point   200

Figure 5.4. Different distributions of the temperature head (solid curves) and the
corresponding heat fluxes distributions (dashed curves) providing the same
total heat removed from surface a) stagnation point
$a_2/\theta_{w.\max} : 1 - 0.4, \ 2 - 1.0, \ 3 - 1.88$ b) jet wall: $1 - 0, \ 2 - 1.0, \ 3 - 3, \ 4 - 4$.   204

Figure 6.1. Universal functions ϑ_1 and ϑ_2 for laminar flow, $\mathrm{Pr} > 0.5$, —— ϑ_1, - - - - ϑ_2,
$1 - \vartheta/\exp z, 2 - \vartheta'/\exp z$.   213

Figure 6.2. Universal functions ϑ_1 and ϑ_2 for turbulent flow: $\mathrm{Pr} = 0.7$, $\mathrm{Re} = 10^6 \ldots 10^7$,
—— ϑ_1, - - - ϑ_2, $1 - \vartheta/\exp(3z/4), 2 - \vartheta'/\exp(3z/4)$,   214

Figure 6.3. Universal functions ϑ_3 and ϑ_4 for laminar flow. $\mathrm{Pr} > 0.5$, —— ϑ_3, - - - - ϑ_4,
$1 - \vartheta/\exp(3z/4), 2 - \vartheta'/\exp(3z/4), 3 - \vartheta''/\exp(3z/4)$.   214

Figure 6.4. Universal functions ϑ_3 and ϑ_4 for turbulent flow. $\mathrm{Pr} = 0.7$, $\mathrm{Re} = 10^6 \ldots 10^7$, ——
$\vartheta_3, - - - \vartheta_4, 1 - \vartheta_3 / 2, 2 - \vartheta_3', 3 - \vartheta_3'', 4 - [-\vartheta_4/\exp(3z/4)], 5 - [-\vartheta'_4/\exp(3z/4)]$,
$6 - [-\vartheta_4''/\exp(3z/4)]$.   215

Figure 6.5. Heat transfer characteristics for a plate streamlined on the one side by
turbulent flow.   217

Figure 6.6. Comparison of one- and two-dimensional solutions for the heat exchange
between two fluids separated by a plate: (1) thin plate, (2) second
approximation for a thick plate according to equation (6.27), (3) final,
two-dimensional result for a thick plate according to equation (6.29).   223

Figure 6.7. Initial length of a plate as a function of a ratio of thermal resistances of the
plate and fluids: (1) turbulent flow and (2) laminar flow.   223

Figure 6.8. Dependence of the coefficient a_0 in equations (6.29) and (6.30) on the
ratio of thermal resistances of the plate and fluids: (1) turbulent flow and
(2) laminar flow.   224

Figure 6.9. (a and b) Heat transfer characteristics for the plate heated from one end: (a) local
characteristics $\mathrm{Bi}_{*L} = 1.4$, I _____ first case, II- - - - second case,
$1 - 2\bar{q}_w, 2 - \theta/\theta_h, 3 - \chi_t$; (b) ratio of total heat fluxes removed from plate
(1) turbulent flow and (2) laminar flow.   226

Figure 6.10. Heat transfer characteristics for the plate streamlined on the one side by
laminar (a) and turbulent (b) flows _____ χ_t, - - - - $-\theta/\theta_0$,
$1 - \bar{q}_0 = 10, 2 - \bar{q}_0 = 0, 3 - \bar{q}_0 = -2$.   227

Figure 6.11. Variation of temperature head and nonisothermicity coefficient along
the plate streamlined on both sides by turbulent flow $\sigma_{\mathrm{Bi}} = 0.5$, $\theta_0 = 1$, $\bar{q}_0 = -2$,
_____ χ_t, - - - - θ, (1, 2) different sides of a plate.   228

Figure 6.12. Heat transfer characteristics for the plate with inner heat sources streamlined
by turbulent flow. _____χ_t, - - - - θ, (1, 2) different sides of a plate, $\sigma_{Bi} = 0.5$,
(3) one side streamlined plate ($\sigma_{Bi} = 0$).   229

Figure 6.13. Variation along the plate of the heat transfer characteristics for two countercurrent
flows (a) temperature head, (b) heat transfer coefficient, - - - - for the case of an
isothermal plate.   230

Figure 6.14. Scheme of temperature profile deformation in the heat transfer process between
countercurrent fluids.   231

Figure 7.1. Scheme of the two-dimensional channel of a semi-conjugate model.   236

Figure 10.1. Scheme of a transverse flow over a finned surface.   292

Figure 11.1. Temperature of the plate surface in symmetrical flow: I) $b = 0.042$, Pr = 5.5;
1) $\phi = 0$. 2) $\phi = 0.8$, 3) $\phi = 0$, $g_1 = 0$, $\varepsilon = 0$, II) $b = 8.51$, Pr = 6.1;
4) $\phi = 0$, 5) $\phi = 0$, $\varepsilon = 0$, 6) $\phi = 0.8$, 7) $\phi = 0$, $\varepsilon = 0$, $g_1 = 0$.   321

LIST OF EXAMPLES

Example 1.1. General solution for solid with initial temperature $f(x)$.   12

Example 1.2. A solid or a rod initially at constant temperature, $T_i = const.$ 13

Example 1.3. The region $x > 0$ initially at T_i, and the remainder $x < 0$ at zero.   13

Example 1.4. The region $-l < x < l$ initially at T_i, and the remainder $|x| > l$ at zero: 13

Example 1.5. General solution for arbitrary initial $f(x)$ and zero surface temperatures.   13

Example 1.6. Constant initial T_i and surface T_w temperatures (boundary condition of the first kind) (Exercise 1.17).   14

Example 1.7. Constant initial temperature T_i and constant heat flux q_w at the surface $x = 0$ (boundary condition of the second kind).   14

Example 1.8. Constant initial T_i and surrounding T_∞ temperatures. Heat flux at $x = 0$ is defined by boundary condition of the third kind $q_w = h[T_\infty - T(0,x)]$.   14

Example 1.9. A semi-infinite solid or thin rod with particular time-dependent temperatures.   17

Example 1.10. A thin, laterally insulated rod of length L initially at the temperature $T_i(x)$ and constant ends temperatures T_0 at $x = 0$ and T_L at $x = L$.   21

Example 1.11. A thin rod laterally insulated of length L initially at temperature $T_i(x)$, at temperatures T_0 at the end $x = 0$ and with insulated $(\partial T / \partial x = 0)$ end $x = L$.   22

Example 1.12. A thin, laterally insulated rod of length L initially at temperature $T_i(x)$, at temperatures $T_0(t)$ at the end $x = 0$ and with insulated end $x = L$.   23

Example 1.13. A thin, laterally insulated rod of length L initially at $T_i(x)$, with T_0 at $x = 0$ and heat transfer at $x = L$ into surrounding at temperature T_∞.   24

Example 1.14. A thin, laterally insulated circular plate or infinite long cylinder of radius R initially at $T_i(\bar{r})$ and constant boundary temperature T_0 at $\bar{r} = 1$, where $\bar{r} = r / R$.   26

Example 1.15. Two-dimensional rectangular sheet in the xy plane of length a, height b and prescribed sides temperatures: left $T(0,y) = \varphi_1(y)$, right $T(a,y) = \varphi_2(y)$, lower $T(x,0) = \varphi_3(x)$, and upper $T(x,b) = \varphi_4(x)$ (Dirichlet problem).   29

Example 1.16. Two-dimensional rectangular sheet in the xy plane of length a, height b and prescribed sides boundary conditions: left $T(0,y) = 0$, right $T(a,y) = 0$, lower $T(x,0) - (\partial T / \partial y)_{y=0} = 0$, and upper $T(x,b) = \varphi(x)$ (mixed boundary conditions).   29

Example 1.17. An infinite solid or thin, laterally insulated rod initially at $T_i(x)$.   31

Example 1.18. Two-dimensional infinite sheet in the xy plane of semi-infinite height (half plane) initially at $T_i(x)$.   32

Example 1.19. A semi-infinite solid or thin, laterally insulated rod initially at zero temperature and time-dependent surface temperature $T(0,t) = \phi(t)$.   34

Example 1.20. A thin, literally insulated rod of length L initially at zero temperature with insulated end at $x = 0$ and constant temperature T_L at $x = L$.   34

Example 1.21. A one-dimensional solid at initial temperature $T_i(x)$ with a space-time dependent source $q_v(x,t)$ and boundary conditions: $\partial T / \partial x = 0$ at $x = 0$ and time-dependent temperature $\phi(t)$ at $x = L$.   36

Example 2.1. Deriving the expressions for the divergence and continuum equation.   46

Example 2.2. Deriving the expression for the substantial derivative.   46

Example 2.3.   Applying Einstein and other index notations.   47

Example 2.4.   Determining streamlines in a two-dimensional flow field.   49

Example 2.5.   Potential flow along the wedge of angle $\pi\beta$ (Fig. 2.1) (Exercise 2.17).   50

Example 2.6.   Deriving the del. operator (nabla) for a scalar function in curvilinear orthogonal coordinates.   52

Example 2.7.   Derivation of the Navier–Stokes equation in cylindrical coordinates.   53

Example 2.8.   Boundary conditions in two Dirichlet problem solutions: 58

Example 2.9.   Determining the skin friction.   81

Example 2.10.   The boundary-layer thickness estimation.   82

Example 2.11.   Estimation of the thermal boundary-layer thickness.   83

Example 2.12.   Estimation of an adiabatic wall temperature.   83

Example 2.13.   Estimation of heat fluxes.   84

Example 2.14.   Effect of thermal boundary conditions.   85

Example 2.15.   Friction on a flat plate.   95

Example 2.16.   Heat transfer from the nonisothermal plate [22].   96

Example 2.17.   Preseparation flow.   98

Example 2.18.   Self-similar flows.   101

Example 2.19.   Linear potential velocity distribution $U = U_0 - ax$ (Sec.2.6.4).   102

Example 2.20.   Transverse flow around a cylinder at sinusoidal velocity distribution.   102

Example 2.21.   Deriving an influence function.   103

Example 2.22.   Free convection on the vertical plate.   105

Example 2.23.   Free convection and radiation from horizontal fin array [31].   106

Example 2.24.   Stability of fluid between two horizontal plates.   106

Example 4.1.   Heat exchange process between two fluids.   158

Example 4.2.   Linear temperature head on the plate with a zero pressure gradient.   161

Example 4.3.   Cylinder with linear temperature head in transverse flow.   162

Example 4.4.   Heat transfer from cylinder at $q_w = const.$ and $T_w = const.$ in transverse flow.   167

Example 4.5.   Heat transfer after heat flux jump on a plate and cylinder in transverse flow.   168

Example 4.6.   Unsteady heat transfer on a plate with time-linear temperature head.   170

Example 4.7.   Heat transfer from a cylinder in transverse non-Newtonian fluid flow.   174

Example 4.8.   Effect of mechanical energy dissipation.   175

Example 4.9.   Comparison computed and measured data for turbulent heat transfer.   181

Example 4.10.   Heat exchange process between two fluids (Fig. 4.4).   183

Example 4.11.   Heat transfer from a thermally thin plate heated from one end.   183

Example 4.12.   A thermally treated polymer continuous sheet at temperature T_0 is extruded from a die and passed at velocity U_w through a bath with water (Pr = 6.1) at temperature T_∞ (see scheme in Fig. 4.14).   184

Example 5.1.   Effects of nonisothermicity in turbulent and laminar flows.   191

Example 5.2.   There are several heat sources or sinks, for instance, electronic components with linear varying strengths. How should these be arranged on a plate so that the maximum temperature of the plate would be minimal?   199

Example 5.3.   Suppose the maximum allowable surface temperature is given. Find the mode of change of the temperature head at which the quantity of the heat removed (or supplied) from the surface is maximum.   201

Example 5.4.   It is required to remove (supply) from the plate a heat flux Q_w. Find a heat flux pattern that minimizes the maximum plate temperature.   204

Example 6.1.   A steel plate of a length $L = 0.25$ m and a thickness $\Delta = 0.01$ m is in the air flow of velocity 3 m/s The left end of the plate is isolated; the temperature of the other end is at T_{wL}. The air temperature is 300 K. Calculate the heat transfer characteristics.   215

Example 6.2.   A copper plate of length 0.5 m and 0.02 m in thickness is streamlined on the one side by air at temperature 313 K with velocity 30 m/s Another side of the plate is isolated. The temperatures of ends are maintained at $T_{w0} = 593$ K and $T_{wL} = 293$ K. Find local temperature and heat flux distributions.   216

Example 6.3. Consider the same problem as in Example 6.2 for an aluminum plate of length 0.3 m and thickness of 0.002 m streamlined by a flow of air of a velocity 250 m/s on an altitude 20 km. Air temperature is $T_\infty = 223$ K, and kinematic viscosity is $v = 1.65 \times 10^4$ m²/s. The front end is at stagnation temperature $T_{\infty 0} = 254$ K, and the other is at $T_{wL} = 323$ K.   217

Example 6.4. Air flows (Re $= 5 \times 10^4$) over the one side of a thin $(\Delta/L = 1/600)$ radiating plate $(\lambda/\lambda_w = 0.135 \times 10^{-4})$ with uniform internal heat sources $(\overline{q}_v = 5.1)$. Another side of the plate is isolated. The front end is at the free stream temperature T_∞. The radiation is taken into account by parameter $N = \sigma \varepsilon T_\infty^3 / \lambda_w \Delta = 0.07$, where σ and ε are the Stefan–Boltzmann constant and emissivity.   218

Example 6.5. Plate heated from one end in a symmetrical flow.   225

Example 6.6. A plate streamlined on the one side and isolated on another.   227

Example 6.7. A plate streamlined on both sides by turbulent flow and heated from leading edge.   228

Example 6.8. A plate with inner heat sources   228

Example 6.9. Heat exchange process between two countercurrent fluids.   229

Example 7.1. Peristaltic motion in the two-dimensional channel at low Reynolds number and long wavelengths (linear model) [1, 2].   237

Example 7.2. Peristaltic motion in the two-dimensional channel at a finite Reynolds number and moderate amplitude (nonlinear model) [3].   238

Example 7.3. Peristaltic motion in the two-dimensional channel at a moderate Reynolds number (numerical solution) [4].   240

Example 7.4. Peristaltic motion in closed rectangular container (perturbation solution) [6].   241

Example 7.5. Peristaltic flow in closed cavity (perturbation and numerical solutions) [7].   242

Example 7.6. Two-dimensional peristaltic flow induced by sinusoidal waves [8].   243

Example 7.7. Steaming flows in a channel with elastic vibrating walls [10].   244

Example 8.1. Consider a simple conduction problem for a plane wall governed by one-dimensional equation and ordinary boundary conditions: 255

Example 10.1a. Pipe with fully developed laminar flow heated symmetrically at the outer surface by uniform heat flux [1].   279

Example 10.2a. *A parallel plate duct with outer wall subjected to convection with environment and turbulent flow at varying periodically inlet temperature [2].   280

Example 10.3n. Horizontal channel heated from below by $q_w = const.$ full developed laminar flow [3] (recall: this Dirihlet problem we considered in Example 2.8).   282

Example 10.4n. Heat transfer between the forced convective flow inside and the natural convective flow on the outside of the vertical pipe [4].   283

Example 10.5a. Heat transfer between two fluids separated by a thin wall flowing concurrently or countercurrently in the double pipe: Conjugate Graetz problem [6].   285

Example 10.6a/n. Laminar flow in a double-pipe heat exchanger [7].   285

Example 10.7n. Microchannel heat sink as an element of heat exchanger [10].   288

Example 10.8 *n. Horizontal fin array [15].   289

Example 10.9a. Flat finned surface in a transverse flow [17].   291

Example 10.10n. Horizontal channel with protruding heat sources [19].   294

Example 10.11n. Elements and units of electronic systems [20].   296

Example 10.12n. Turbine blades with radial cooling channels [24].   297

Example 10.13n. Solid propellant rocket [25].   298

Example 10.14a. Emergency loss of coolant in the nuclear reactor [28].   300

Example 10.15n. Diesel engine piston temperature field [31].   303

Example 10.16n. Solar energy storage unit [32].   304

Example 11.1n. Solidification in the enclosed region [1].   309

Example 11.2 *n. Concrete at evaluated temperatures [4, 5].   311

Example 11.3 *n. Czochralski crystal growth process [6–8].   313

Example 11.4n. Continuous wires casting [12].   314

Example 11.5 *n.   Optical fiber coating process [13].   316
Example 11.6n.   Twin-screw extruder simulation [16].   318
Example 11.7a/n   Films and fibers production [17].   319
Example 11.8 *n   A brick drying [19].   321
Example 11.9n   Drying a rectangular wood board [21].   324
Example 11.10n   Drying of porous materials [23].   325
Example 11.11a/n   Drying of pulled continuous materials in the initial period [25].   326
Example 11.12n   Freeze drying of slab-shaped food [26].   330
Example 11.13n   Food and polymer flow through extrusion dies [27].   333
Example 12.1n   Arterial stenoses modeling [1].   337
Example 12.2n   Blood flow though series stenoses [2].   338
Example 12.3a   Blood flow in artery with multi-stenosis under magnetic field [4].   340
Example 12.4a   Simulation of blood flow in small vessels [5].   343
Example 12.5a/n   Simulation of the blood flow during electromagnetic hyperthermia [6].   345
Example 12.6a   Particle motion in peristaltic flow with application to the ureter [8].   347
Example 12.7a   Simulation of chyme flow during gastrointestinal endoscopy [10].   348
Example 12.8a   Simulation of bile flow in a duct with stones [11].   350
Example 12.9a   Modeling transport processes in the cerebral perivascular space [13].   351
Example 12.10a/n   Simulation of macromolecules transport in tumors [15].   352
Example 12.11n   Simulation of embryo transport [8].   355
Example 12.12a   Modeling the bioheat transfer in human tissues [20].   356

Nomenclature

$Bi = \dfrac{h\Delta}{\lambda_w}$ Biot number

$Br = \dfrac{\lambda\Delta}{\lambda_w L} Pr^m Re^n$ Brun number

C_1, C_2 exponents of influence functions in integral forms of universal functions for heat flux and temperature head

$C_f = \dfrac{\tau_w}{\rho U_\infty^2}$ friction coefficient

$2C_f/St$ Reynolds analogy coefficient

C, C_p specific heat and specific heat at constant pressure J/kg K

$\hat{c} = \rho c \Delta$ thermal capacity J/m^{2}K

D, D_h diameter and hydraulic diameter m

D_m diffusion coefficient, m^2/s

$Ec = \dfrac{U^2}{c_p \theta_w}$ Eckert number

$f(\xi / x)$ influence function of the unheated zone at temperature jump

$f_q(\xi / x)$ influence function of the unheated zone at heat flux jump

$Fo = \dfrac{\alpha t}{L^2}$ Fourier number

$Gr = \dfrac{\beta \theta_w g L^2}{v^2}$ Grashof number

g_k, h_k coefficients of series in differential forms of universal functions for temperature head and heat flux

g gravitational acceleration, m/s^2

h, h_m heat and mass transfer coefficients, W/ m^{2}K

k specific heat ratio or turbulent energy

K_τ, K_q constants in power rheology laws for non-Newtonian fluids

$Kn = \dfrac{l}{D_h}$ Knudsen number

l body length or mixing length or free path, m

L characteristic length, m

$Le = \dfrac{D_m}{\alpha}$ Levis number

$$\text{Ls} = \frac{\lambda_w h}{\rho_w^2 c_w^2 U_w^2 \Delta}$$ Leidenfrost number

$$\text{Lu} \frac{\rho c}{\rho_w c_w}$$ Luikov number

$$\text{M} = \frac{U}{U_{sd}}$$ Mach number

M moisture content, kg/kg

n, s exponents in the power rheology law for non-Newtonian fluids

$$\text{Nu} = \frac{hL}{\lambda}, \ \text{Nu} = \frac{hL^{s+1}}{K_q U^s}$$ Nusselt number for Newtonian and non-Newtonian fluids

p pressure, Pa

$$\text{Pe} = \frac{UL}{\alpha}$$ Peclet number

$$\text{Pr} = \frac{v}{\alpha}, \ \text{Pr} = \frac{\rho c_p U^{1-s} L^{1+s}}{K_q}$$ Prandtl number for Newtonian and non-Newtonian fluids

q, q_v heat flux, W/m^2, and volumetric heat source, W/m^3

r/s exponent in the heat transfer coefficient expression for an isothermal surface

$$\text{Ra} = \frac{\beta \theta_w g L^3}{v \alpha}$$ Rayleigh number

$$\text{Re} = \frac{UL}{v}, \ \text{Re} \frac{\rho U^{2-n} L^n}{K_\tau}$$ Reynolds number for Newtonian and non-Newtonian fluids

$$\text{Sc} = \frac{v}{D_m}$$ Schmidt number

$$\text{Sh} = \frac{h_m}{\rho c_p D_m}$$ Sherwood number

$$\text{St} = \frac{h}{\rho c_p U}$$ Stanton number

$$Sk = \frac{4\sigma T_\infty^4 L}{\lambda_\infty}$$ Starks number

$$Ste = \frac{c_p \Delta T}{\Lambda}$$ Stephan number

t time, s

T Temperature, K

u, v, w velocity components; u, v functions in integrating by parts

U velocity on the outer edge of a boundary layer

U_e velocity on the outer edge of a turbulent boundary layer

$u_\tau = \sqrt{\tau_w / \rho}$ friction velocity

$u^+ = u/u_\tau, y^+ = yu_\tau/v$ variables in wall law

x, y, z	coordinates

Greek symbols

α	thermal diffusivity
β	dimensionless pressure gradient in self-similar solutions and turbulent equilibrium boundary layer or volumetric thermal expansion coefficient, 1/K
$\chi_t = \dfrac{h}{h_*}, \chi_p = \dfrac{h_m}{h_{m*}}$	nonisothermicity and nonisobaricity coefficients
$\chi_f = \dfrac{C_f}{C_{f*}}$	nonisotachicity coefficient
$\delta, \delta_1, \delta_2$	boundary layer thicknesses, m
δ, δ_{ij}	delta function and Kronecker delta
Δ	body or wall thickness, m
κ	Karman constant
λ	thermal conductivity, W/mK
Λ	latent heat, J/kg or λ_s / λ
μ	viscosity, kg/s m
ν	kinematic viscosity, m^2/s
ξ	unheated zone length, m
$\theta = T - T_\infty,$	temperature excess
$\theta_w = T_w - T_\infty$	temperature head
ρ	density, kg/m^3
σ	Stefan–Boltzmann constant, W/m^{2}K^4
τ	shear stress, N/m^2
Φ, φ	Prandtl–Mises–Görtler variables
ψ	stream function, m^2/s
ω	specific dissipation rate

Some of these symbols are also used in different ways as it is indicated in each case.

Subscripts

av	average
ad	adiabatic
as	asymptotic
bl	bulk
e	end or effective
i	initial; inside
L	at $x = L$
m	mass average, or mean value, or moisture
o	outside
sd	sound

Superscripts

$+, -$	from both sides of the interface
$+, ++$	wall law and other variables for a turbulent boundary layer

Overscores

$\bar{o}, \tilde{o}$	dimensionless, or transformed or auxiliary

q	constant heat flux
t	thermal
tb	turbulent
w	fluid–solid interface
ξ	after jump
∞	far from solid
$*$	isothermal

PREFACE

To teach is to touch life forever

This book presents systematically contemporary theoretical methods in fluid flow and heat transfer. These methods were developed over time since the science creation and are advanced nowadays as well. They constitute a current powerful set of classical and modern means for investigating natural phenomena and industry processes. A reader who possessed such methods gained a background which engineer or physicist need to understand and mathematically interpret the research problems in the fields of their work.

Although the investigation process is interesting and enjoyable, studying methods of such kind that are full of mathematics is a difficult task demanding much time, patience, and persistence. That is the author's motivation, who worked whole life in this field, for writing a book to share his experience with next generations.

This book is intended for engineers and students. Due to that a special care is given to balance between strictness and comprehensibility of the text. Such compromise is achieved due to strict formulation of the problems on the one hand and detailed explanation of definitions and terminology on the other hand. For example, considering the Nevier–Stokes equation, both types Dirichlet and Neumann boundary condition are discussed, but it is explained with examples why even more simple Dirichlet formulation is difficult to realize and why it is easier to solve similar problems in the case of boundary-layer equation. At the same time, to help a reader, many forms (vector, vorticity, etc.) of Navier–Stokes equation and different notation (Einstein, Kronecker, etc.) are considered.

Other features of the writing style are comments and exercises, which author uses to divide the equation derivation with a reader. For the part of considered problems, the way of solution is explained in the text, and the results are given; however, the mathematical realization is left for the reader in exercise. Such a manner offers to a person a choice or be satisfied only by results or to improve his or her expertise. It is indicated in the text when the exercises of this type should be performed. The same is shown for majority of other exercises, consisting discussion of considered problems, and to help a reader to find a particular exercise, the exercises locations are given in the contents. An option is also given for taking some details, such as computing or experimental data or solutions of more complicated problems, which are not necessary for basic understanding. This is done by specifying the page or figure in corresponding reference, such as, for example, notice Figure 3.7 in [3] or Figure 20.4 in [1] after equation (3.12) in Chapter 3. Such a citation may be ignored without losing any awareness, but at the same time, it provides a possibility to get additional information for an interested person. To reduce the number of sources, in some chapters for those citations one book is chosen, for instance, in Chapters 2 (laminar flow), 4 (conjugate heat transfer), or 8 (numerical methods), Schlichting's monograph [1,1968], author's book [1], or book by Patankar [2], respectively, are used. In those cases, often only page or figure is given since the source is specified once or twice in the text.

The above-mentioned comments are used to clear up special terms, such as singularity, order of magnitude, flow stability, or nonisothermicity factor. Historical notes, such as the fact that the boundary-layer theory was not applied for the first 25 years or the Prandtl's note about his mixing-length concept, are also present via comments. Since the examples and exercises have a different level of sophistication, to help the reader to orientate, the more complex exercises, including application examples and references, are marked by an asterisk (*).

The book begins with short historical notes, showing where the classical methods came from and whose shoulder are we standing on (see epigraph to historical notes and list of cited pioneers-contributors after Appendix). It closes by discussing the question "How complicated a mathematical model should be?" where we review some basic points which should be kept in mind as a researcher chooses an existing or creates a new model for studying. Four parts and 12 chapters incorporate the *Classical methods in fluid flow and heat transfer* (Chapters 1–3), *Modern analytical methods in heat transfer and fluid flow* (Chapters 4–7), *Numerical methods including recently developed direct simulation approaches* (Chapters 8 and 9), and *Application in engineering, biology and medicine* (Chapters 10–12). More than 100 examples show the applicability of the investigating methods in different areas from aerospace to food processing, and more than 400 exercises offer opportunities to improve the skills of using these methods.

Each chapter covers a group of specific methods. The first chapter consists of heat transfer in solid, considering methods of conduction investigation. It starts with different forms of a conduction equation and presents the technique of transforming the equations from Cartesian to orthogonal curvilinear coordinates using Lame coefficients. Boundary conditions of four types and their connection with the Biot number and the relevant form of a conduction equation are discussed. Different methods of the conduction equation solution with illustrating examples are analyzed. Beginning with a simple specific method applying error integral, this analysis proceeds with more general methods of separation variables, Duhamel's integral, and then presents the Fourier and Laplace integral transforms and Green function method. Indeed these methods are general mathematical approaches being used in many other areas. Due to that, some fundamental issues, such as eigenvalue problems and Sturm–Liouville conditions or features of homogeneous and inhomogeneous equations and boundary conditions, are considered as well.

Two other chapters of the first part of the book contain methods in fluid flow and heat transfer. Here, the basic attention is paid to the Navier–Stokes equation, which is a fundamental theoretical core of momentum, heat, and substance transfer processes. Laminar and turbulent flows and heat transfer are studied in Chapters 2 and 3, respectively. Various forms (vector, vorticity, streamlined, and irrotational) in Cartesian and curvilinear coordinates of the Navier–Stokes equation as well as different notation (Einstein, Kronecker, and Levi-Civita) applied in application are considered at the beginning of the second chapter. Discussion of Dirichlet and Neumann problems for the Navier–Stokes equation explains the difficulties arising under boundary conditions formulation in these problems. Two particular solutions are reviewed showing how such boundary conditions are formulated in practice.

These examples along with known 13 simple exact solutions of Navier–Stokes equation give an understanding of complexity of this basic equation and the importance of its even simple exact solutions. The comparison of those issues with some heat transfer problems reveals the similarity and specific features of relevant solutions of energy equation.

Next, the similarity theory, the set of dimensionless numbers important in flow and heat transfer, and two cases of small and high Reynolds and Peclet numbers are considered. In the last cases, the Navier–Stokes equation simplifies, resulting in creeping and boundary-layer equations. The basic exact solutions such as a creeping flow around a sphere or Blasius and Pohlhausen solutions for boundary-layer flow and heat transfer on a plate are analyzed, underlining the essential differences between the reduced and full Navier–Stokes equations due to which the solution of shortcutting equations becomes less complicated. Nevertheless, the complexity remains enough strong so that approximate methods widely used before computer time and even now employed, at least first, due to their simplicity and physical clearness of solutions are still of interest. The Karman–Pohlhausen integral method for boundary-layer problems is described in details along with several examples, demonstrating the technique of solution. Two examples of other approximate approach based on linearization of the boundary-layer equation are reviewed as well.

The final section of Chapter 2 presents basic of natural (or free) convection, occurring due to density difference in contrast to above-discussed forced convection, which exists owing to the driving external force. Several examples show the intrinsic properties of this phenomenon, such as (i) coupled flow and heat transfer characteristics, depending on Grashof or Rayleigh numbers instead of Reynolds and Peclet numbers in the case of forced convection; (ii) small velocities and heat transfer rates requiring usually the radiation effects taken into account; and (iii) Rayleigh–Benard stability problems. Two examples are considered to illustrate the difference between free convection on vertical and horizontal plates. Whereas in the first case there is a dominant flow direction and the problem may be considered as boundary-layer type, the second problem requires consideration of the Navier–Stokes and full-energy equations.

Chapter 2 gives a reader the theoretical background and methods in laminar flows, but the main practical means one finds in Chapter 3 consisting theories of turbulence since majority of essential problems occur in fluids (water, air, and oil) with small viscosity and hence at high Reynolds numbers. The chapter begins with notes about the transition of laminar in turbulent flow and values of critical Reynolds number. The problem of laminar flow stability is briefly considered to explain a reader the concept of transition as a process of transforming a stable laminar in unstable turbulent flow. The method of small disturbances is described giving understanding of the Orr–Sommerfeld equation and the neutral curves, determining the conditions of stability.

Theories of turbulence are considered in a chronological order as they were developed beginning from the suggested by Reynolds derivation of averaged Navier–Stokes equation (RANS). It is shown how the additional Reynolds stresses appear in the process of averaging and how it is connected with that problem of closure. The reviewing of turbulence models starts with Prandtl's mixing-length hypothesis, which is followed by three other most popular algebraic models (Mellor–Gibson, Cebeci–Smith, and Baldwin–Lomax) based on the modern structure of the equilibrium turbulent boundary layer. The basic idea of Wilcox perturbation analysis of the features of the equilibrium turbulent boundary layer is introduced briefly. Several examples of algebraic models application in fluid flow and heat transfer are discussed as well. The one- and two-equations turbulence models are described with more details, especially the $k - \varepsilon$ and $k - \omega$ models, which are the main practical tools in studying and performing calculation of turbulent flows. In the final section, the integral methods similar to those in the laminar boundary layer but in contrast grounded on experimentally obtained velocity profiles are described.

The second part of the book consists of four chapters presenting modern analytical conjugate methods. The principles, general properties, and the methods of the problem solution in conjugate heat transfer (Chapters 4–6) and in peristaltic fluid flow treated as a conjugate problem (Chapter 7) are outlined. A short introduction "Why and when conjugate methods are required?" opens this part. In the first chapter of this part (Chapter 4), the derivation of universal functions describing basic dependences in heat transfer for arbitrary nonisothermal surfaces are given. These dependences are universal because they present expressions between the heat fluxes and temperature heads, which satisfied the thermal boundary-layer equation and are independent of particular boundary conditions. The universal functions are obtained as results of exact or accurate approximate solutions in two forms: in the differential form in the series of consecutive derivatives of the temperature head and in the integral form with the influence function of the unheated zone. Such duality provides high accurate computation results using a differential form when the series converges fast and applying the integral form otherwise. The universal functions are obtained for steady and unsteady laminar incompressible flows, steady compressible flows, moving continuous sheet, power law non-Newtonian flows, and for recovery factor as well as for steady turbulent flows. The inverse universal function is also obtained. In this case, the universal function determine the temperature head in similar two forms of series and as Duhamel integral which correspond to given heat flux distribution. It is shown how universal functions are used for solving conjugate heat transfer problems, and several examples of relevant solutions are present.

The next two chapters contain results of universal functions application. In Chapter 5, the general properties of nonisothermal and conjugate heat transfer are formulated, which are obtained from universal functions analysis. Effect of different factors (temperature head variation, pressure gradient, flow regime, and Biot number) on the heat transfer intensity is studied. It is proved that the temperature head decreasing in flow direction or in time is the basic reason of low heat transfer rate. A so-called gradient analogy, which stands for similar effects of external velocity gradient on friction coefficient and temperature head gradient on heat transfer coefficient, and phenomenon of heat flux inversion analogous to the flow separation are investigated. The final section consists of the some examples of heat transfer optimization.

In Chapter 6 on the basis of the universal functions, the charts for solving simple conjugate problem are created. Temperature singularities in temperature head distribution at the leading edge of a plate are studied, showing that those depend on behavior of the isothermal heat transfer coefficient. The technique of problem solution is presented in details, and examples, including numerical data, are considered. The conjugate heat transfer in flows past plates in different situations is investigated, and some conclusions are formulated. The applicability of issues gained for a plate to more general types of flow is analyzed. The results obtained in Chapters 4–6 are summarized by final discussion "Should any conjugate problem be considered as a conjugate?"

Chapter 7 represents peristaltic flow that is the second inherently conjugate phenomenon considered in the book. Since the peristaltic motion occurs as a result of the interaction between elastic wall and flow inside a channel, we treat this phenomenon as a conjugate problem similar to conjugate heat transfer, consisting of two domains, wall, and fluid, coupled on their interface. It is noted that despite similarity of both problems, the peristaltic flow is a much complicated task due to nonlinearity of the equation for fluid and flexibility of the wall. Due to that, the majority of published solutions are not fully conjugated; rather those are semi-conjugate solutions, considering the motion of a flexible wall as given and, hence, taking into account the only effect of the wall on the fluid flow but ignoring the backward effect of the fluid flow on the wall movement.

Different approaches of studying peristaltic flows are reviewed in historical order from earlier to recent publications. The discussion goes along with typical examples. The most of analytical solutions are obtained using assumptions of low Reynolds number and long wavelength in the form of perturbation series, taking into account one or two first terms up to second power of small parameter. More accurate solutions are numerical but also basically semi-conjugate. Some authors used an approximate approach to take into account the backward effect of flow on the wall movement, assuming that progressive wave propagating along the wall is sinusoidal and given. Few publications regard this effect more accurate, and one analytical of those solutions is considered to show the complexity of such problems.

Part III contains the analysis of numerical methods in two chapters. In Chapter 8, the classical numerical methods are outlined, and the modern special methods in turbulence are presented in Chapter 9. At the beginning of Chapter 8, the mutual importance of both analytical and numerical methods is discussed. Different finite-difference and finite-element methods are described and compared from the same viewpoint using the general weighted residual approach. The difficulties in computing the convection-diffusion terms in the Navier–Stokes equation, velocities, pressure, and methods of resolving these problems are discussed. The central-difference and upwind schemes as well as some others are compared to show their advantages and shortcomings. The SEMPLE and SEMPLER software are described. The existing numerical approaches in conjugate heat transfer for the subdomain solution conjugation, such as the one large program for simultaneously solving both equations for fluid and body, iterative scheme, and the method based on superposition principle, are analyzed. Some modern numerical approaches improving the convergence of iteration, such as multigrid and meshless techniques, are reviewed.

Chapter 9 represents the new numerical methods of solving the exact Navier–Stokes equations without averaging developed in the last 50 years. The basic value of such methods is that they may be considered like experimental data, providing insight into physics of turbulence increasing our understanding of its nature. Three methods of this type are described: direct numerical simulation (DNS), large eddy simulation (LES), and detached eddy simulation (DES). Features of these methods are analyzed, showing that because turbulence contains countless number of eddies of different sizes, such methods require huge computer recourses. Due to that, each method uses a special restriction to be in line with possibilities of nowadays computers. In DNS, the exact Navier–Stokes equations are solved for all range of scales of eddies, but the possible solutions are confined to moderate Reynolds numbers. Two other methods achieved the restriction by computing differently the large and small eddies. Estimations of required number of grid points and time steps calculation for DNS are given, indicating that only simple problems with relatively small Reynolds numbers can be solved by this method nowadays. To increase the limited Reynolds number, in the LES, only large eddies carrying the majority of the energy are computed directly using the DNS approach, whereas the nearly isotropic small eddies domain is treated applying Reynolds-average models. A special method knowing as filtering is used for realizing the process of separation of eddies of different sizes. Description of the filtration procedure is presented, illustrating simple examples helping to understand this process.

The most recently suggested DES method is the promising improved LES approach. Its application showed successful modeling of flows past bluff bodies and vehicles of natural size at the real values of Reynolds number and a manageable cost. This was achieved due to combining the average Navier–Stokes approach (RANS) for near-wall region, which requires very fine grids with LES for other part of computational domain. It is clarified how a special facility called blending function works to distinguish between RANS and LES for treatment in corresponding regions. Some results of modeling of the real objects, such as the sphere in sub–and super–critical flows and flows around aircrafts, obtained by this method are shortly described.

In a short special paragraph, the notion of chaos theory is explained, and its possible future application for studying turbulence is discussed. In the final remarks, some scientific prognoses of direct simulation methods are cited and references for further reading are given.

The fourth part of the book presents applications of modern conjugate methods. Three chapters incorporate examples in engineering (Chapters 10 and 11) and in biology and medicine (Chapter12). Solutions of conjugate problems in thermal and cooling systems are reviewed in Chapters 10. In Chapter 11, the results of conjugate heat and mass transfer investigation in technology processes are outlined. Sixteen examples united in three sections (Heat exchangers and pipe, Cooling systems, and Energy systems) in Chapter 10 introduce solutions of conjugate problems of different purposes including:

- Graetz problem: heat transfer between two fluids in double pipe (Example 10.5)
- Microchannel as element of highly effective heat exchanger (Example 10.7)
- Electronic packages cooling (Examples 10.10 and 10.11)
- Turbine blades cooling (Example 10.12)
- Protection of the reentry rocket by a thermal shield (Example 10.13)
- Emergency loss of coolant in a nuclear reactor (Example 10.14)
- Heat transfer in solar energy storage unit (Example 10.16).

Chapter 11 consists of 13 examples of conjugate investigations described in four sections (Multiphase and phase-changing processes, Manufacturing processes simulation, Drying processes, and Food processing) including, in particular, the following:

- Solidification in the enclosed region (Example 11.1)
- Concrete at evaluated temperatures (Example 11.2)
- Czochralski crystal growth process (Example 11.3)
- Continuous wires casting (Example 11.4)
- Optical fiber coating process (Example 11.5)
- Twin-screw extruder simulation (Example 11.6)
- Drying of pulled continuous materials (Example 11.11)
- Freeze drying of slab-shaped food (Example 11.12)

In Chapter12, are presented twelve examples of studying flows and heat transfer processes in human organs published in the last 50 years. Although due to complexity of the problem, these solutions are obtained using assumptions; the practical and fundamental importance of such results is outstanding because they may be considered like experimental data providing a new way of getting information of human organs performance. Not only characteristics measurable before, such as blood pressure or urine flow rate, but such data as, for example, stresses in blood vessel walls or pattern of separation or reattachment flows at the arterial stenoses, which are difficult or impossible obtain in laboratory, could be gained.Normal and disorder conditions of organs operation as well as some medical treatment procedures are considered in three sections (Blood flow in normal and pathologic vessels, Peristaltic flow in disordered human organs and Biologic transport processes) including among others the following:

- Blood flow though series stenoses (Example 12.2)
- Simulation the blood flow during electromagnetic hyperthermia (Example 12.5)
- Particle motion in peristaltic flow modeling a stone in ureter (Example 12.6)
- Simulation of chyme flow during gastrointestinal endoscopy (Example 12.7)
- Simulation of bile flow in a duct with stones (Example 12.8)
- Simulation of macromolecules transport in tumors (Example 12.10)
- Modeling the bioheat transfer in human tissues (Example 12.12)

The author wrote this book trying to share with you his experience, interest, thirst of knowledge, and hopes that has been sufficiently successful in meeting this goal. So now good luck to you in studying and research.

Abram Dorfman
abram_dorfman@hotmail.com
Ann Arbor, October 2012

Key Words/Terms

fluid flow, heat transfer, classical methods, modern methods, analytical, numerical, exact, approximation, laminar/ turbulent flows, laminar flow stability theory, solid heat transfer, strong comprehension presentation, conjugate heat transfer, criteria conjugate/common approach, conjugate/semi-conjugate peristaltic flow,78 fluid flow/heat transfer examples, 41 application conjugate engineering/biology/ medicine solutions, 417 fluid flow/heat transfer exercises, 109 mathematical/ historical comments, methods errors, criteria selection methods, Navier–Stokes equation forms, average Navier–Stokes equation (RANS), boundary-layer theory, creeping flows, boundary conditions, Derichlet/ Neumann problems, algebraic/$k - \varepsilon$ / $k - \omega$ turbulence models, classical numerical methods as weighted residuals, special numerical calculation difficulties, numerical conjugation approaches, modern numerical turbulence simulation DNS, LES, DES.

ACKNOWLEDGMENTS

Although only my name is indicated on the book cover, I, in fact, have real coauthors, people whose contributions are vital to existence of this book. My dear doctors, who made it possible for me to write this second book in the last five years; Professor R. Prager, who gave me new blood vessels after a heart attack in 2000; Professor M. Grossman and nurses C. Martinez and L. McCrumb from his team, who took care of my heart since that time; Dr. S. Saxe and lovely Dr. Jill Bixler, who gave me back my vision; Dr. A. Courey, who helped me breathe normally; Dr. J. Wolf, who assisted me with my urine problems; and my nice dentist Dr. Shyroze Rehemtulla and Dr. S. Gradwohl, who has been taking care of my family for 18 years since we came to Ann Arbor.

My sincere thanks to Professor Massoud Kaviany from the University of Michigan for his interest in my work, advice, and discussions. I received significant help from the staff of the University of Michigan Libraries, especially from Alena Verdiyan (Art and Engineering Library) and from Karen Mike (Interlibrary Loan). I extend my great appreciation to my young friend, senior student of University of Michigan Zach Renner, who prepared figures for this book.

Our close friends Dr. Brian Schapiro and Dr. Stephen Saxe, and their lovely wives Margot and Kim, shared with us our joys and sorrows, and I use this opportunity to express my gratitude and love to them.

As usual, I have greatly benefited from my family, and I appreciate it very much. I offer most of my thanks and love to my dear wife Sofia, who for 63 years has been beside me with her care and encouragement, making it possible for me to study, to research, and to write.

About the Author

Abram S. Dorfman, PhD, was born in 1923 in Kiev, Ukraine, in the former Soviet Union. He graduated from Moscow Institute of Aviation in 1946 as an engineer of aviation technology. From 1946 to 1947, he worked in Central Institute of Aviation Motors (ZIAM) in Moscow. From 1947 to 1990, Dr. Dorfman studied fluid mechanics and heat transfer at the Institute of Thermophysics of Ukrainian Academy of Science in Kiev: first, as a junior scientist from 1947 to 1959; then, as a senior scientist from 1959 to 1978; and finally, as a leading scientist from 1978 to1990. He earned a PhD with a thesis entitled Theoretical and Experimental Investigation of Supersonic Flows in Nozzles in 1952. In 1978, he received a Doctor of Science degree, which was the highest scientific degree in the Soviet Union, with a thesis and a book *Heat Transfer in Flows around the Nonisothermal Bodies*. From 1978 to 1990, he was an associate editor of *Promyshlennaya Teploteknika,* which was published in English as *Applied Thermal Science* (Wiley). Dr. Dorfman was an adviser to graduate students for many years.

In 1990, he emigrated to the United States and continued his research as a visiting professor at the University of Michigan in Ann Arbor (since 1996). During this period, he published several papers in leading American journals and a book titled *Conjugate Problems in Convective Heat Transfer*, 2010, Taylor & Francis. He is listed in *Who's Who in America 2007.*

Dr. Dorfman has published more than 140 papers and three books in fluid mechanics and heat transfer in Russian (mostly) and in English. More than 50 of his papers published in Russian have been translated and are also available in English. Since 1965, he has been systematically studying conjugate heat transfer.

PART I

CLASSICAL METHODS IN FLUID FLOW AND HEAT TRANSFER

CHAPTER 1

Methods in Heat Transfer of Solids

1.1 Historical Notes

If I have seen a little further it is by standing on the shoulders of Giants.
—Isaac Newton in a letter to Robert Hooke, in 1676

There is a tradition in the similarity theory, hydrodynamics and heat transfer to name dimensionless numbers after pioneers who first formulated laws, methods, equations, and terms. Since these principles determine scientific and educational progress, the names of the dimensionless numbers give an understanding of the history of science considering the works and time of pioneers.

The basic laws and equations of modern hydrodynamics and heat transfer were formulated by great scholars of eightieth and ninetieth centuries. In particular, a fundamental equation of conduction in solids was given by French physicist and mathematician Jean Fourier (1768–1830). He followed Biot (1774–1863) who several years before published an analysis of heat transfer between a heated rod and surrounding.

At that time, two different mechanical schools prevailed [1]. One group of scholars following Isaac Newton (1642–1727) believed that any physical system can be described knowing forces and momenta at each point or particle. At the same time, scientists following Gottfried Leibniz (1746–1816) represented another avenue determining a physical system in terms of work and action. Although both approaches were related to each other, there were some specific differences. In particular, Newton's group to which Biot belonged believed in action at a distance only when important are differences of parameters rather than corresponding gradients. Using temperature gradients instead of temperature differences applied in the Biot study was one of the significant features of Fourier's approach, which led him to successful formulation of the general conduction equation. To solve the conduction equation, Fourier applied trigonometric series establishing a new chapter in mathematics now known as the Fourier series. These ideas led him later to discover Fourier integral currently widely used for the solution of differential equation. Nevertheless, the first manuscript of Fourier submitted in 1807 to French Academy was rejected by reviewers, and only in 1822, Fourier published an extended version of his first presentation of these outstanding results.

Another gifted scholar, an Irish mathematician, and physicist George Stokes (1819–1903) devised an equation now known as the Navier–Stokes equation, which is the base of contemporary hydro- and aerodynamics. This step was a great achievement leading from equations describing the motion of an inviscid, ideal fluid given by Leonard Euler (1707–1783) to equation taking into account viscosity of one of the most important properties of real fluid. The first attempt to add viscosity forces to Euler's equations was made by French engineer and physicist Claude-Louis Navier (1785–1836). He determined friction forces using attraction and repulsion effects between neighboring molecules. Later, the equation of viscous fluid motion was rederived in 1828 by Cauchy (1789–1857), in 1829 by Poisson (1781–1840), in 1843 by Saint-Venant (1797–1889), and finally by Stokes in 1845 who gave the derivation that basically is in use to date for more than 150 years. Stokes also gave first two simple solutions of this nonlinear, very complicated equation known as Stokes' problems.

The next pioneering approach introduced in 1895 by English physicist Osborne Reynolds (1842–1912) is the procedure of averaging the Navier–Stokes equation to get statistical description of turbulent motion. Such a procedure is necessary because the turbulent flow consists of random various fluctuations of parameters. Now, Reynolds's method is the standard framework for the RANS (Reynolds-Averaged Navier–Stokes) equation, which is a usual tool for studying turbulent flows. The other famous work of Reynolds published in 1883 presented the results of experimental investigation of flow in channels. Reynolds first used the dimensionless number now called by his name to determine its critical value when character of flow changes from laminar to turbulent structure. Although this dimensionless combination of parameters was introduced by Stokes in 1851, it was named by Reynolds because he intensively used and popularized it in his works [2].

Thus, history shows that Fourier conduction equation, Nervier–Stokes, and RANS equations for velocity and similar energy equations for temperature, which are basic mathematical expressions of modern hydrodynamics and heat transfer, were formulated by pioneers 150–200 years ago.

Significant contribution in heat transfer and mechanics was also made by other pioneers [3]. For example, Jean Peclet (1793–1857) and Franz Grashof (1826–1893) taught heat transfer for many years and published books, which presented the heat transfer development at their time; Leo Graetz (1856–1941) contributed by studying heat conduction, radiation, and electricity and is known as author of the Graetz problem of heat transfer in a duct. Wilhelm Nusselt (1882–1957) proposed in 1915 the dimensionless number, now called by his name. Lord Rayleigh (1842–1919) contributed in almost the entire physics, including wave theory, hydrodynamics, optics, and elasticity. Thomas Stanton's works (1865–1931) covered fluid flow and heat transfer; in particular, he improved the air-cooling engines for aircraft. Ernst Schmidt (1892–1975) is known as the author of principle of analogy between heat and mass transfer. Names of many other pioneers whose results are used in this book are listed at the end before the subject index.

1.2 Heat Conduction Equation and Problem Formulation

1.2.1 Cartesian Coordinates

In general, the conduction equation describes the transient heat transfer and in the case of an isotropic, homogeneous solid has the form

$$\rho c \frac{\partial T}{\partial t} = \frac{\partial}{\partial x}\left(\lambda_w \frac{\partial T}{\partial x}\right) + \frac{\partial}{\partial y}\left(\lambda_w \frac{\partial T}{\partial y}\right) + \frac{\partial}{\partial z}\left(\lambda_w \frac{\partial T}{\partial z}\right) + q_v. \qquad (1.1)$$

This equation is a balance of the instant heat-rate changes at any solid point. The first three terms on the right-hand side present the spatial changes at a given point in x, y, z directions, respectively, and the last term determines the volumetric rate of energy generated by source inside a solid. Sum of these rates of heat causes the instant change of solid internal storage energy determined by expression on the left-hand side.

In the case of constant thermal conductivity, equation (1.1) simplifies. This equation may be written in different forms two of which are

$$\frac{\partial T}{\partial t} = \alpha \left(\frac{\partial^2 T}{\partial x^2} + \frac{\partial^2 T}{\partial y^2} + \frac{\partial^2 T}{\partial z^2} \right) + \frac{q_v}{\rho c}, \qquad \frac{\partial T}{\partial t} = \alpha \nabla^2 T + \frac{q_v}{\rho c} \qquad (1.2)$$

$$\nabla^2 = \frac{\partial^2}{\partial x^2} + \frac{\partial^2}{\partial y^2} + \frac{\partial^2}{\partial z^2}. \qquad (1.3)$$

Here, ∇^2 is the Laplace operator, and $\nabla^2 T$ is called the Laplacian of T. Equation (1.2) simplifies further for the following two cases: for the one- or two-dimensional heat transfer and for the steady-state regime when the derivative on the left becomes zero:

$$\frac{\partial T}{\partial t} = \alpha \frac{\partial^2 T}{\partial x^2} + \frac{q_v}{\rho c}, \qquad \frac{\partial T}{\partial t} = \alpha \left(\frac{\partial^2 T}{\partial x^2} + \frac{\partial^2 T}{\partial y^2} \right) + \frac{q_v}{\rho c} \qquad \frac{\partial^2 T}{\partial x^2} + \frac{\partial^2 T}{\partial y^2} + \frac{\partial^2 T}{\partial z^2} + \frac{q_v}{\lambda_w} = 0. \quad (1.4)$$

The last equation is often used in the forms known as Laplace's and Poisson's equations:

$$\nabla^2 T = \frac{\partial^2 T}{\partial x^2} + \frac{\partial^2 T}{\partial y^2} + \frac{\partial^2 T}{\partial z^2} = 0, \qquad \nabla^2 T = \frac{\partial^2 T}{\partial x^2} + \frac{\partial^2 T}{\partial y^2} + \frac{\partial^2 T}{\partial z^2} = -\frac{q_v}{\lambda_w}. \qquad (1.5)$$

There are two cases when the additional equation simplifications are possible. If the body is thin $(\Delta/L \ll 1)$ and its thermal resistance is of the same order or greater than that of coolant $(\mathrm{Bi} = h\Delta/\lambda_w \geq 1)$, the longitudinal conductivity is negligibly small in comparison with the transverse one. Thus, the two-dimensional problem approximately transforms in one-dimensional $(\partial T/\partial x \ll \partial T/\partial y)$, and in the case of steady-state condition, second equation (1.4) reduces to the simple ordinary differential equation, which can be easily integrated. The equation obtained in this way shows that the temperature distribution across thickness of such a thin body without internal sources is close to linear $(dT/dy = const.)$. An example where such situation, in practice, exists is the heat transfer between a metal coolant and a metal or not metal body (Exercise 1.3).

In an opposite case, for the pair metal body–metal or not metal coolant $(\mathrm{Bi} \leq 1)$, the body can be considered as thermally thin, which means that its thermal resistance is small in comparison with that of coolant. Due to the high conductivity, the body temperature at any instance is practically uniform. In such a case, the simplification is achieved using the temperature averaged across the body thickness. Integrating two-dimensional equation (1.4) across thickness and taking into account two following expressions yields a relation for temperature of a thermally thin body (Exercise 1.4):

$$\int_0^\Delta \frac{\partial T}{\partial t} dy = \frac{\partial}{\partial t} \int_0^\Delta T dy = \Delta \frac{\partial T_{av}}{\partial t}, \qquad \int_0^\Delta \frac{\partial^2 T}{\partial y^2} dy = \left. \frac{\partial T}{\partial y} \right|_{y=\Delta} - \left. \frac{\partial T}{\partial y} \right|_{y=0} = \frac{1}{\lambda_w} (q_{w1} + q_{w2}), \quad (1.6)$$

$$\frac{1}{\alpha} \frac{\partial T_{av}}{\partial t} - \frac{\partial^2 T_{av}}{\partial x^2} + \frac{q_{w1} + q_{w2}}{\lambda_w \Delta} - \frac{q_{v,av}}{\lambda_w} = 0. \qquad (1.7)$$

Here, q_{w1} and q_{w2} are heat fluxes rate on the body surfaces.

Presented equations can be transformed in dimensionless form scaling coordinates by L and using variable $\theta = (T - T_R)/(T_w - T_R)$, where T_R is the reference temperature. After transformation, the one- and three-dimensional equations take the form

$$\frac{\partial \theta}{\partial \text{Fo}} = \frac{\partial^2 \theta}{\partial \overline{x}^2} + \overline{q}_v, \qquad \frac{\partial \theta}{\partial \text{Fo}} = \frac{\partial^2 \theta}{\partial \overline{x}^2} + l_y \frac{\partial^2 \theta}{\partial \overline{y}^2} + l_z \frac{\partial^2 \theta}{\partial \overline{z}^2} + \overline{q}_v \qquad (1.8)$$

Here, $\text{Fo} = ta / L^2$ is the Fourier number, $l_y = L_y / L$ and $l_z = L_z / L$ are dimensionless height and width of a solid, respectively, and $\overline{q}_v = q_v L^2 / \lambda_w (T_w - T_R)$ is the dimensionless source (Exercise 1.6).

1.2.2 Orthogonal Curvilinear Coordinates

To transform equations given in Cartesian coordinates in the other orthogonal coordinate system, one may use the Lame coefficients. If the new orthogonal coordinates are

$$\xi_1 = \xi_1(x, y, z), \qquad \xi_2 = \xi_2(x, y, z), \qquad \xi_3 = \xi_3(x, y, z), \qquad (1.9)$$

the Lame coefficients l_n are determined by expression [see equations (2.28) and (2.35)]:

$$l_n^2 = \left(\frac{\partial x}{\partial \xi_n}\right)^2 + \left(\frac{\partial y}{\partial \xi_n}\right)^2 + \left(\frac{\partial z}{\partial \xi_n}\right)^2. \qquad (1.10)$$

Then, three-dimensional equation (1.2) $(q_v = 0)$ in new variables takes the form [4]

$$\frac{\partial T}{\partial t} = \frac{1}{l_1 l_2 l_3} a \left[\frac{\partial}{\partial \xi_1}\left(\frac{l_2 l_3}{l_1} \frac{\partial T}{\partial \xi_1}\right) + \frac{\partial}{\partial \xi_2}\left(\frac{l_1 l_3}{l_2} \frac{\partial T}{\partial \xi_2}\right) + \frac{\partial}{\partial \xi_3}\left(\frac{l_1 l_2}{l_3} \frac{\partial T}{\partial \xi_3}\right) \right]. \qquad (1.11)$$

1.2.2.1 Cylindrical Coordinates $(\xi_1 = r, \xi_2 = \gamma, \xi_3 = z)$

$$x = r\cos\gamma, \qquad y = r\sin\gamma, \qquad z = z \qquad (1.12)$$

According to (1.10), one gets

$$l_1^2 = \cos^2\gamma + \sin^2\gamma = 1, \quad l_2^2 = r^2\cos^2\gamma + r^2\sin^2\gamma = r^2, \quad l_3^2 = 1. \qquad (1.13)$$

Substituting these results into equation (1.11) gives

$$\frac{\partial T}{\partial t} = \frac{a}{r}\left[\frac{\partial}{\partial r}\left(r\frac{\partial T}{\partial r}\right) + \frac{\partial}{\partial \gamma}\left(\frac{1}{r}\frac{\partial T}{\partial \gamma}\right) + \frac{\partial}{\partial z}\left(r\frac{\partial T}{\partial z}\right)\right] \qquad (1.14)$$

which finally is present as

$$\frac{\partial T}{\partial t} = a \left[\frac{\partial^2 T}{\partial r^2} + \frac{1}{r} \frac{\partial T}{\partial r} + \frac{1}{r^2} \frac{\partial^2 T}{\partial \gamma^2} + \frac{\partial^2 T}{\partial z^2} \right] \qquad (1.15)$$

1.2.2.2 Spherical Coordinates ($\xi_1 = r$, $\xi_2 = \gamma$, $\xi_3 = \phi$)

$$x = r \sin\gamma \cos\varphi, \qquad y = r \sin\gamma \sin\varphi, \qquad z = r \cos\gamma \qquad (1.16)$$

$$l_1 = 1, \qquad l_2 = r, \qquad l_3 = r \sin\gamma \qquad (1.17)$$

$$\frac{\partial T}{\partial t} = \frac{\alpha}{r^2} \left[\frac{\partial}{\partial r} \left(r^2 \frac{\partial T}{\partial r} \right) + \frac{1}{\sin\gamma} \frac{\partial}{\partial \gamma} \left(\sin\gamma \frac{\partial T}{\partial \gamma} \right) + \frac{1}{\sin^2\gamma} \frac{\partial^2 T}{\partial \phi^2} \right]. \qquad (1.18)$$

1.2.2.3 Elliptical Cylindrical Coordinates ($\xi_1 = u$, $\xi_2 = v$, $\xi_3 = z$)

$$x = chuchv, \qquad y = shushv \qquad (1.19)$$

$$l_1^2 = sh^2uch^2v + ch^2ush^2v, \qquad l_2^2 = ch^2ush^2v + sh^2uch^2v, \qquad l_3^2 = 1. \qquad (1.20)$$

Since in this case, Lame coefficients are equal to each other and

$$sh^2uch^2v + ch^2ush^2v = sh^2u + \sin^2 v, \qquad (1.21)$$

equation (1.11) becomes (Exercise 1.7)

$$\frac{1}{a} \frac{\partial T}{\partial t} = \frac{1}{sh^2u + \sin^2 v} \left(\frac{\partial^2 T}{\partial u^2} + \frac{\partial^2 T}{\partial v^2} \right) + \frac{\partial^2 T}{\partial z^2}. \qquad (1.22)$$

1.2.3 Universal Function for Heat Flux on an Arbitrary Nonisothermal Surface

The conjugate formulation stands for considering a problem with variable unknown conditions on the interface of the interaction elements of a system (Introduction to Part II). Therefore, universal functions providing general relations independent of a specific problem [5] are of fundamental importance for understanding and applying conjugate methods (Chapter 4). In the case of heat transfer, such a universal relation between heat flux on the interface and temperature head $\theta_w = (T_w - T_\infty)$ was obtained in the form of series. For a flow with zero pressure gradient past an arbitrary nonisothermal plate, this universal function is (h_* is the heat transfer coefficient for an isothermal surface) [6]

$$q_w = h_* \left(\theta_w + g_1 x \frac{d\theta_w}{dx} + g_2 x^2 \frac{d^2\theta_w}{dx^2} + g_3 x^3 \frac{d^3\theta_w}{dx^3} + \cdots \right) = h_* \left(\theta_w + \sum_{n=1}^{\infty} g_k x^k \frac{d^k\theta_w}{dx^k} \right). \qquad (1.23)$$

Coefficients g_k for laminar and turbulent flows, as well as in other cases, rapidly decrease with increasing numbers, so the series converges fast, and using equation (1.23) with two or three first terms usually yields practically satisfactory results. Moreover, calculation shows (Table 5.1) that for different cases the first coefficient g_1 is three to ten times larger than the second g_2, while the others are negligibly small. The structure of series (1.23) shows that the first term $h_*\theta_w$ determines the heat flux on an isothermal surface, whereas the others specify the nonisothermicity effect. Taking this fact into account and knowing that g_1 is much greater than others, one concludes that the nonisothermicity effect is basically defined by the second term of series (1.23). In what follows, this fact as well as equation (1.23) for heat flux is frequently applied for various purposes. Here, we use universal equation (1.23) for showing the connection between the Biot number, boundary conditions, and relevant type of conduction equation.

1.2.4   Initial, Boundary, and Conjugate Conditions

Since the conduction equation is the first order in time and the second order in space, the solution of the one-dimensional problem depends on one initial and two boundary conditions at two points. In the case of more complicated two- or three-dimensional problem, the solution should satisfy the initial condition and boundary conditions given around a contour of the problem domain. The initial condition specifies the temperature of the system as a function of coordinates at some instant $t = 0$, which is taken to be a beginning of the process, whereas the boundary conditions determine the values of some parameters on the boundaries of the system as the functions of time and position.

In general, there are four kinds of boundary conditions, depending on prescribed parameters on boundaries of the problem domain. The first kind of boundary condition consists of a given boundary temperature, the second one involves specified heat fluxes on the boundary, and the third kind of boundary condition is assigned by Newton's law with known heat transfer coefficient and difference between surface and environmental temperatures. For example, such conditions on the boundary at $x = L$ have the form

$$T\big|_{x=L} = T_w, \quad \text{or} \quad q_w = -\lambda_w \frac{\partial T}{\partial x}\bigg|_{x=L}, \quad \text{or} \quad q_w = h\left(T\big|_{x=L} - T_\infty\right) \tag{1.24}$$

The fourth kind of boundary condition also called as conjugate or coupled conditions is formulated as equalities of temperatures and heat fluxes on the interface calculated from both sides of conjugate subjects and indicated by (+) and (−). For the flow past a plate, such conditions on the interface between the body (+) and flow (−) may be written in the form

$$T^+ = T^-, \quad \lambda_w^+ \frac{\partial T}{\partial y}\bigg|^+ = \lambda^- \frac{\partial T}{\partial y}\bigg|^- \tag{1.25}$$

Consider the energy balance on the interface given by equality (1.25). The heat flux from a body side may be computed using Fourier's law. The heat flux from the fluid side in the general case of nonisothrmal interface may be estimated, as mentioned above, using the second term of series (1.23). Then, equation (1.25) becomes (Exercise 1.8)

$$\lambda_w \frac{\partial T}{\partial y}\bigg|_{y=0} = h_* g_1 x \frac{\partial \theta_w}{\partial x} \quad \text{or} \quad \frac{1}{\text{Bi}} \frac{\partial T}{\partial(y/\Delta)}\bigg|_{y=0} = g_1 x \frac{\partial \theta_w}{\partial x} \quad \text{Bi} = \frac{h_*\Delta}{\lambda_w} \tag{1.26}$$

This equation shows that a product $x(\partial\theta_w/\partial x) = x(\partial T/\partial x)$ of longitudinal gradient of the interface temperature $(\partial T/\partial x)$ and corresponding coordinate x is inversely proportional to the Biot number defined as a ratio of transverse thermal resistances of a body Δ/λ_w and a fluid for the isothermal surface $1/h_*$. If the Biot number is large due to the large body or small fluid resistance, the product $x(\partial T/\partial x)$ is small. This situation takes place in two cases when the transverse thermal resistance is dominant, and two-dimensional heat transfer problem simplifies to one-dimensional. The first case (we consider it later in Sec. 1.3) is that of the semi-infinite solid or the thin, laterally insulated rod when the transverse thermal resistance and the corresponding Biot number are infinite. Consequently, the heat flow is directed longitudinally, and the problem is longwise one-dimensional. In this case according to equation (1.26), product $x(\partial T/\partial x) \to 0$. Nevertheless, at the surface the longitudinal derivative $\partial T/\partial x \to \infty$ because the product $x(\partial T/\partial x)$ goes to zero with semi-infinite order (the plate is semi-infinite), whereas x at the surface is identically zero. The infinite derivative $\partial T/\partial x$ means that the temperature at the surface $x = 0$ changes in a stepwise manner. This result corresponds to the well-known fact that at the surface of a semi-infinite solid in the case of a large Biot number and any boundary condition, the temperature changes suddenly (see Sec. 1.3.2) (Exercise 1.9).

In another case of dominant transverse resistance when the Biot number is large but finite, the product $x(\partial T/\partial x)$ according to equation (1.26) is small and finite as well. As the distance from the leading edge increases, the longitudinal derivative $\partial T/\partial x$ decreases and soon becomes negligible. Therefore, the problem in this case is also close to one-dimensional, but in contrast to the previous case in the transverse direction. This situation corresponds to a generalized, "thin body" model considered in Section 1.2.1 when two-dimensional conduction equation (1.4) reduces to one-dimensional.

The opposite case is when the body resistance is small or the fluid resistance is large, and as a result, the Biot number is small corresponding to practically uniform body temperature. According to equation (1.26), the longitudinal temperature derivative is large in this case, and consequently, the problem is one dimensional in corresponding direction. This is a generalized, "thermally thin body" model, which is also considered in Section 1.2.1 when after temperature averaging across the body thickness, two-dimensional conduction equation (1.4) reduces to simplified version (1.7).

Any other heat transfer problem is characterized by comparable thermal resistances of a solid and a fluid and by Biot number of order of unity. Such problems should be considered using a conjugate approach and boundary conditions of the fourth kind. The same conjugate procedure is required in the two cases considered above (Sec. 1.2.1) when two-dimensional heat transfer equation (1.4) simplifies to one-dimensional. Before the conjugate approach came in use and frequently as well now, conjugate methods are substituted applying approximate relatively simpler conditions of the third kind. Although using the third-kind boundary condition requires empirical information of heat transfer coefficient, there are groups of problems, which could be solved by this approach with a practically satisfactory accuracy. To give a reader an understanding, which method, conjugate or common simple approach, should be used for solving a particular problem is of critical importance (see Conclusion to Part II).

The boundary conditions required to be satisfied and method of problem solving depend on the type of the equation considered. The type of a linear partial differential equation of the second order to which belong the conduction equations is determined by reducing such an equation to its canonical form. It is said that equation in canonical form is of elliptical or hyperbolic type, depending on whether it consists of a sum or difference of two second derivatives. The third type of linear partial differential equations of the second order called parabolic contains only one second derivative. Thus, the one-dimensional conduction equation is a parabolic equation, whereas the two- and three-dimensional Laplace and Poisson equations are of an elliptical type.

The solution of a parabolic one-dimensional equation requires satisfying two boundary conditions at the points and one initial condition at $t = 0$. These boundary conditions may be of any kind. More complicated type of boundary conditions is required when an elliptical equation is considered. In this case, the conditions should be specified around the bounding contour of surface, and some types of boundary conditions are unacceptable (Sec. 1.5.4).

There are different methods of solving the conduction equations. The choice depends upon the type of a problem, basically defined by the heat transfer regime, the solid shape, its dimensionality, and by the initial and boundary conditions. Here, we consider basic, frequently applied methods. The idea and general approach of each method are followed by some particular examples of solutions.

Comment 1.1 The methods reviewed in this chapter are applicable only to linear conduction equations formulated under assumption of constant physical properties. Other, more complicated examples of conduction equation solutions can be found in the last chapters presenting applications of conjugate methods.

EXERCISES

1.1 What represents physically each term in equation (1.1) and in which cases this equation simplifies in equations (1.2)–(1.5)?

1.2 What is the difference between Laplace operator and Laplacian in application? Which is more general and why?

1.3 Why Biot number can be used in order to explain when the two-dimensional general conduction equation (1.4) simplifies?

1.4 Repeat and explain details of derivation of equation (1.7) from equation (1.4).

1.5 Explain physically the difference between terms "thin body" and "thermally thin body"

1.6 What benefits gives equation (1.8) in dimensionless form? How does such benefits relate to series (1.23)?

1.7 Calculate Lame coefficients for following revolution bodies coordinate system

$$x = f\left(\xi_1,\xi_2\right)\cos\xi_3, \qquad y = f\left(\xi_1,\xi_2\right)\sin\xi_3, \qquad z = z\left(\xi_1,\xi_2\right)$$

where ξ_1,ξ_2,ξ_3 are arbitrary curvilinear coordinates and x, y, z are Cartesian orthogonal coordinates.

1.8 Derive relation between Biot number and temperature head gradient and consider limiting cases. Gave physical interpretation of mathematical results.

1.9 Explain why in the case of semi-infinite plate and $Bi \to \infty$ the derivative $\partial T/\partial x \to \infty$ at $x = 0$ despite the product $x\left(\partial T/\partial x\right) \to 0$. Hint: recall that there are infinite values of different order.

1.10 What kinds of linear partial differential equations of the second order exist? What type is the conduction equation?

1.11 How are determined the number and type of boundary conditions required for conduction equation?

1.12 A thin laterally isolated rod of finite length $a < x < b$ is initially at a constant temperature T_i. The heat transfer coefficient at the front end is h and the other end is thermally isolated. The surrounding temperature is T_∞. Formulate the heat transfer problem. Explain what type of equation and boundary conditions should be used.

1.3 Solution Using Error Integral

Error integral is often used in the form of two tabulated functions (Table A1)

$$erf\, z = \frac{2}{\sqrt{\pi}} \int_0^z \exp\left(-\xi^2\right) d\xi \qquad erfc\, z = 1 - erf\, z, \tag{1.27}$$

having the following properties [4]:

$$erf(0) = 0 \quad erf(\infty) = 1 \quad erf(-z) = -erf(z) \quad erfc(0) = 1 \quad erfc(\infty) = 0. \tag{1.28}$$

Equation (1.27) satisfied the unsteady one-dimensional homogeneous (without source) conduction equation (1.4) and boundary conditions pertaining to infinite and semi-infinite solids. The same equation and boundary conditions describe heat transfer in the infinite and semi-infinite thin rods with lateral insulated surface because in both cases transverse resistance is infinite, and the heat flows only in the longitudinal direction (Sec. 1.2.4).

To see that equation (1.27) give a solution of these problems, consider the dimensionless form of one-dimensional equation (1.8). The dimensionless temperature θ in this equation depends on two dimensionless variables $\bar{x} = x/l$ and Fo $= a\, t/l^2$. In the case of an infinite or semi-infinite body, these variables cannot be used since there is no finite body length. It follows from dimensionless variables that in the case considered, one may use only three-dimensional variables a, x, and t. Since units of these variables are m²/s, m, and s, respectively, there is only one dimensionless combination $x/\sqrt{a\, t}$ on which should be depended the dimensionless temperature θ. Hence, the independent variable in equation (1.27) should be proportional to this expression $\xi = cx/\sqrt{a\, t}$, where c is a constant. Since temperature θ depends only on one variable ξ, the partial conduction equation reduces to an ordinary differential equation, and corresponding derivatives take the form

$$\frac{\partial \theta}{\partial t} = \frac{d\theta}{d\xi} \frac{\partial \xi}{\partial t} = -\frac{d\theta}{d\xi} \frac{cx}{t^{3/2} a^{1/2}} \tag{1.29}$$

$$\frac{\partial^2 \theta}{\partial x^2} = \frac{\partial}{\partial x}\left(\frac{d\theta}{d\xi} \frac{\partial \xi}{\partial x}\right) = \frac{\partial}{\partial \xi}\left(\frac{d\theta}{d\xi} \frac{c}{(a\, t)^{1/2}}\right)\frac{c}{(a\, t)^{1/2}} = \frac{d^2 \theta}{d\xi^2} \frac{c^2}{a\, t}. \tag{1.30}$$

Then, the conduction equation becomes

$$-\frac{d\theta}{d\xi} \frac{x}{t^{3/2} a^{1/2}} = a \frac{d^2 \theta}{d\xi^2} \frac{c}{a\, t} \quad \text{or} \quad -\xi \frac{d\theta}{d\xi} = c^2 \frac{d^2 \theta}{d\xi^2}. \tag{1.31}$$

Equation (1.31) is a second-order ordinary differential equation, which do not contains explicitly the dependent variable θ. Such an equation may be reduced to the first-order ordinary differential equation by substituting a new variable $\vartheta = d\theta/d\xi$. Since $d^2\theta/d\xi^2 = d\vartheta/d\xi$, from the last equation, one gets the following expression (Exercise 1.13):

$$-\xi\vartheta = c^2 \frac{d\vartheta}{d\xi}, \qquad -\xi d\xi = c^2 \frac{d\vartheta}{\vartheta}, \qquad \vartheta = \frac{d\theta}{d\xi} = c_1 \exp\left(-\frac{\xi^2}{2c^2}\right). \tag{1.32}$$

Finally, after integrating and taking $c = \sqrt{2}$, the dimensionless temperature becomes

$$\frac{T - T_R}{T_w - T_R} = c_1 \int_0^z \exp(-\xi^2) d\xi + c_2 \qquad z = \frac{x}{2\sqrt{at}}. \tag{1.33}$$

Constants c_1 and c_2 are defined applying boundary conditions.

Temperature profiles (1.33) depend only on one variable z. This means that profiles for different times and positions are similar, so profiles with the same variable z and different t and x coincide forming one curve. Such variables and corresponding solutions called similar significantly simplified the investigation by reducing partial differential equations to ordinary ones.

Solutions of the unsteady one-dimensional equation for infinite and semi-infinite solids under different boundary conditions obtained applying equation (1.27) are presented in the book by Carslaw and Jaeger [4]. Some solutions adopted from [4] are discussed below.

1.3.1 An Infinite Solid or Thin, Laterally Insulated Rod

Example 1.1 General solution for solid with initial temperature $f(x)$.

Since the spatial domain in this problem is infinite $-\infty < x < \infty$, there are no boundary conditions and only initial temperature specifies this problem. In general, the initial condition $T(x,0) = f(x)$ and dimension-dependent variable $T(x,t)$ are used because there are no other temperatures to determine the dimensionless variable. The solution is obtained by determining the constants in equation (1.33). It follows from the symmetry property of x-domain that $c_2 = 0$. The other constant is defined using the initial condition. Substituting this condition into (1.33) and taking into account that $z \to \infty$ as $t \to 0$ leads us to an equation, which after comparison with (1.27) gives the other constant

$$f(x) = c_1 \int_0^\infty \exp\left(-\xi^2\right) d\xi = c_1 \frac{\sqrt{\pi}}{2} \qquad c_1 = \frac{2}{\sqrt{\pi}} f(x). \tag{1.34}$$

Integrating the last equation (1.32) using this result and replacing the integral with one infinite limit by integral with both infinite limits, one gets

$$T(x,t) = \frac{2}{\sqrt{\pi}} \int_0^\infty f(x) \exp(-\xi^2) d\xi = \frac{1}{\sqrt{\pi}} \int_{-\infty}^\infty f(x) \exp(-\xi^2) d\xi, \tag{1.35}$$

where $\xi = x/2\sqrt{at}$. Introducing a new variable $\xi = (x - \zeta)/2\sqrt{at}$, where ζ (we use the same notation ξ) is the dummy variable, yields the solution in the form [4] (Exercise 1.14)

$$T(x,t) = \frac{1}{2\sqrt{\pi a t}} \int_{-\infty}^{\infty} f(\xi) \exp\left[-(x-\xi)^2/4a t\right] d\xi. \tag{1.36}$$

Example 1.2 A solid or a rod initially at constant temperature, $T_i = const$.

This simplest case does not exist because the condition $T \to 0$ at $x \to \pm\infty$, which follows from equation (1.36), cannot be satisfied.

Example 1.3 The region $x > 0$ initially at T_i, and the remainder $x < 0$ at zero.

This is Heaviside's step function. From equations (1.33) and (1.27), we have $c_1 + c_2 = T_i$ at $x \to \infty$, and $-c_1 + c_2 = 0$ at $x \to -\infty$. Then, $c_1 = c_2 = T_i/2$ and

$$T(x,t) = \frac{T_i}{2}\left[1 + erf\left(\frac{x}{2\sqrt{\alpha t}}\right)\right]. \tag{1.37}$$

Comment 1.2 This equation gives a discontinuity at point $x = 0$, $T(0,t) = T_i/2$ as it should be (Exercise 1.15).

Example 1.4 The region $-l < x < l$ initially at T_i, and the remainder $|x| > l$ at zero:

$$T(x,t) = \frac{T_i}{2}\left[erf\left(\frac{l-x}{2\sqrt{a t}}\right) + erf\left(\frac{l+x}{2\sqrt{a t}}\right)\right]. \tag{1.38}$$

Check the boundary conditions satisfaction: (i) symmetry condition, $c_2 = 0$, and $T(l,t) = T(-l,t)$; (ii) at $x = 0$, $T(0,t) = T_i erf\left(l/2\sqrt{a t}\right)$, so at $l \to \pm\infty$ and at $l = 0$, we have $T(\pm\infty,t) = T_i$ and $T(0,t) = 0$, respectively, as it should be from the point of physics; (iii) at points of discontinuity $x = \pm l$, equation (1.38) gives $T(\pm l,t) = (T_i/2) erf\left(2l/\sqrt{a t}\right)$, and, hence as it should be in both limiting cases $T(\pm l,t) = T_i/2$ and $T(\pm l,t) = 0$ when $l \to \pm\infty$ and at $l = 0$, respectively.

1.3.2 A Semi-Infinite Solid or Thin, Laterally Insulated Rod

The semi-infinite body is a subject bounded by the plane surface $x = 0$ and extended in infinite. Since in that case there are two, initial T_i and surface T_w temperatures, the dimensionless temperature $\theta = (T - T_i)/(T_w - T_i)$ may be used.

Example 1.5 General solution for arbitrary initial $f(x)$ and zero surface temperatures.

Solution for this case may be derived using equation (1.36) for an infinite body. Assume that the body is continued in another side from $x = 0$ to $(-\xi)$ with temperature $[-f(\xi)]$. In such a case of a symmetrical body, the surface $x = 0$ remains at the same zero surface temperature, and from equation (1.36), we have [4]

$$T(x,t) = \frac{1}{2\sqrt{\pi a t}}\left\{\int_{0}^{\infty} f(\xi)\exp\left[-(x-\xi)^2/4a t\right]d\xi + \int_{-\infty}^{0} -f(-\xi)\exp\left[-(x-\xi)^2/4a t\right]d\xi\right\}$$

$$= \int_{0}^{\infty} f(\xi)\left\{\exp\left[-(x-\xi)^2/4a t\right] - \exp\left[-(x+\xi)^2/4a t\right]\right\}d\xi. \tag{1.39}$$

In the last expression, the variable $(-\xi)$ in the second integral is replaced by ξ since the limits of integration are changed.

Example 1.6 Constant initial T_i and surface T_w temperatures (boundary condition of the first kind) (Exercise 1.17).

Solution for this case may be obtained from general formula (1.39). However, it is easy to determine the constants in equation (1.33) applying known conditions $\theta = 1$ at $x = z = 0$ and $\theta = 0$ at $t = 0$ and $z \to \infty$. These give $c_2 = 1$ and $c_1 = -1$, so we have

$$\frac{T(x,t)-T_i}{T_w-T_i} = 1 - erf\left(\frac{x}{2\sqrt{at}}\right) = ertc\left(\frac{x}{2\sqrt{at}}\right) = \frac{2}{\sqrt{\pi}} \int_{x/2\sqrt{at}}^{\infty} \exp\left(-\xi^2\right) d\xi. \tag{1.40}$$

Example 1.7 Constant initial temperature T_i and constant heat flux q_w at the surface $x = 0$ (boundary condition of the second kind).

In this case, there is no surface temperature in the boundary conditions. Therefore, the scale $(T_w - T_i)$ for dimensionless temperature θ cannot be used. To find a new appropriate scale, note that there are two new quantities: heat flux at the surface q_w and the thermal conductivity λ_w from Fourier's law $q = -\lambda_w (dT/dn)$ (n is a normal to the surface). Since both units W/m^2 for q_w and W/mK for λ_w contain Watt, it is reasonable to consider quotient q_w/λ_w unit of which is K/m. Then, it becomes clear that combination of $q_w x/\lambda_w$ or $q_w \sqrt{at}/\lambda_w$ is a proper scale for temperature.

To find a solution of the problem in question, consider a subsidiary function $(1/\sqrt{\pi})$ $\exp(-z^2) - zerfcz$ obtained by integration of equation (1.27) by parts putting $u = erfc\xi$, $dv = d\xi$ [4]. One may see that this function in variables indicated above

$$\theta = \frac{\lambda_w(T-T_i)}{q_w 2\sqrt{at}} = \frac{1}{\sqrt{\pi}} \exp\left(-\frac{x^2}{4at}\right) - \frac{x}{2\sqrt{at}} erfc\frac{x}{2\sqrt{at}} \tag{1.41}$$

satisfies the boundary conditions of the problem: $\theta = 0$ at $t = 0$ or $x \to \infty$ and $q_w = -\lambda_w (\partial\theta/\partial x)_{x=0}$ at $x = 0$.

Example 1.8 Constant initial T_i and surrounding T_∞ temperatures. Heat flux at $x = 0$ is defined by boundary condition of the third kind $q_w = h[T_\infty - T(0,x)]$.

Let $f = \theta - (\lambda_w/h)(d\theta/dx)$, where $\theta = (T_\infty - T)/(T_\infty - T_i)$. Function f should satisfy the conduction equation and given boundary conditions

$$\frac{\partial f}{\partial t} = a\frac{\partial^2 f}{\partial x^2}, \qquad f = \theta = 1 \text{ at } t = 0, \ f = 0 \text{ at } x = 0. \tag{1.42}$$

The solution of this problem gives error equation (1.27):

$$f(x,t) = erf\left(\frac{x}{2\sqrt{at}}\right) = \frac{2}{\sqrt{\pi}} \int_0^{x/2\sqrt{at}} \exp\left(-\xi^2\right) d\xi. \tag{1.43}$$

On the other hand, the first expression for f may be considered as the first-order linear ordinary differential equation for θ. Rewriting this expression in the common form and integrating it using a standard procedure yield the relation

$$\frac{d\theta}{dx} - \frac{h}{\lambda_w}\theta = -\frac{h}{\lambda_w}f(x,t), \quad \theta = C\exp\left(\frac{hx}{\lambda_w}\right) - \frac{h}{\lambda_w}\exp\left(\frac{hx}{\lambda_w}\right)\int_{\infty}^{x}f(\xi,t)\exp\left(-\frac{h\xi}{\lambda_w}\right)d\xi. \quad (1.44)$$

Comment 1.3 Dimensionless temperature θ depends on x and t. Nevertheless, in the relation determining function f and in differential equation (1.44), time is considered as a parameter defining some fixed time, and equation (1.44) is solved as an ordinary differential equation.

Since the temperature must be finite, it is necessary to take $C = 0$. Then, substituting equation (1.43) into equation (1.44) and performing integration lead us to the problem solution

$$\frac{T_\infty - T(x,t)}{T_\infty - T_i} = erf\left(\frac{x}{2\sqrt{at}}\right) + \exp\left(\frac{hx}{\lambda_w} + \frac{h^2 a t}{\lambda_w^2}\right)erfc\left(\frac{x}{2\sqrt{at}} + \frac{h\sqrt{at}}{\lambda_w}\right). \quad (1.45)$$

Subtracting unit from both sides gives another form of solution [4] (Exercise 1.18):

$$\frac{T(x,t) - T_i}{T_\infty - T_i} = erfc\left(\frac{x}{2\sqrt{at}}\right) - \exp\left(\frac{hx}{\lambda_w} + \frac{h^2 a t}{\lambda_w^2}\right)erfc\left(\frac{x}{2\sqrt{at}} + \frac{h\sqrt{at}}{\lambda_w}\right). \quad (1.46)$$

1.4 Duhamel's Method

Duhamel's principle is an approach of presenting the solution of a problem depending on a given arbitrary function in terms of the solution of the similar problem with corresponding constant parameters.

1.4.1 Duhamel Integral Derivation

Let the function determines a problem at hand is $F(t)$, and the known solution of a similar, simpler problem is $f(x,t)$. For example, the first function is a given time-dependent surface temperature of a semi-infinite solid, and the second one is the simple solution for semi-infinite solid with constant surface temperature. If the time variable t is divided in small intervals Δt (Fig. 1.1), the corresponding small values of function $F(t)$ are defined as $\Delta F = F'(t)\Delta t$.

In each small interval, the solution of the problem at hand may be approximated by solution $f(x,t)$ for the case of constant function $F(t)$. Then, the sought solution for the first interval is given by a product of the simple solution $f(x\ t)$ and the value $F(0)$ (which is $\Delta F(0)$) of the given function $F(t)$ at beginning when $t = 0$, i.e., $f(x,t)F(0)$. For the next interval, the simple solution approximates function $F(t)$ starting from time, which is less by Δt. Thus, for the second interval, the time is $t - \Delta t$, instead of t for the first one. Therefore, the corresponding simple solution for this case is $f(x, t - \tau_1)$, where $\tau_1 = \Delta t$ is the time lag in the

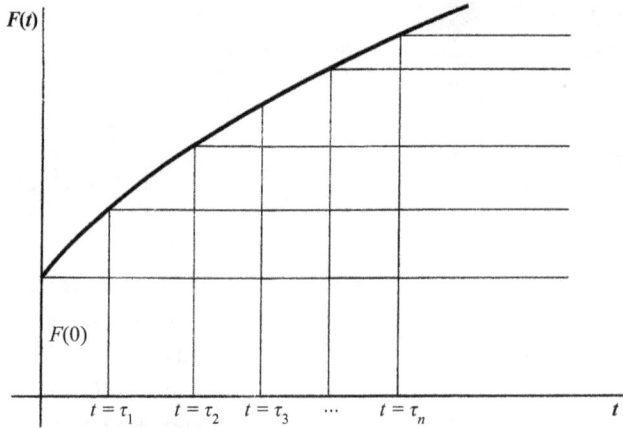

Figure 1.1. Approximation of arbitrary dependence by step function.

second interval. The solution of the problem at hand for this interval is $f\left(x, t - \tau_1\right) \Delta F\left(\tau_1\right) = f\left(x, t - \tau_1\right) F'\left(\tau_1\right) \Delta \tau_1$, where the small variation of function $\Delta F\left(t\right) = F'\left(\tau_1\right) \Delta \tau_1$ is determined for time τ_1 when the second interval begins. Similarly, for the third interval, the solution is $f\left(x, t - \tau_2\right) \Delta F\left(\tau_2\right) = f\left(x, t - \tau_2\right) F'\left(\tau_2\right) \Delta \tau_2$, where τ_2 is the time lag in the third interval, and so on. The summation of overall results obtained for small intervals gives the approximate solution, which in the limit when $\Delta \tau \to 0$ becomes Duhamel's integral presenting the solution of the problem

$$T\left(x.t\right) = f\left(x, t\right) F\left(0\right) + \sum_{k=1}^{n} f\left(x, t - \tau_k\right) F'\left(\tau_k\right) \Delta \tau_k$$

$$T\left(x, t\right) = f\left(x, t\right) F\left(0\right) + \int_{0}^{t} f\left(x, t - \tau\right) F'\left(\tau\right) d\tau.$$

(1.47)

Putting $u = f(x, t - \tau)$, $dv = F'\left(\tau\right) d\tau$, and then $du = f'(x, t - \tau) d\tau$, $v = F\left(\tau\right)$, one transforms equation (1.47) by integration by parts

$$\int_{0}^{t} f(x, t - \tau) F'(\tau) d\tau = f(x, t - \tau) F(\tau)\Big|_{0}^{t} + \int_{0}^{t} f'(x, t - \tau) F(\tau) d\tau.$$

(1.48)

Substituting equation (1.48) into equation (1.47), yields two other forms of Duhamel's integral with not zero and zero initial temperature, respectively (Exercise 1.25):

$$T(x, t) = f\left(x, 0\right) F\left(t\right) + \int_{0}^{t} f'(x, t - \tau) F(\tau) d\tau \qquad T\left(x, t\right) = \int_{0}^{t} f'(x, t - \tau) F(\tau) d\tau.$$

(1.49)

1.4.2 Time-Dependent Surface Temperature

If the initial temperature of a semi-infinite solid or thin, laterally insulated rod is zero and its surface temperature is a function of time $\phi(t)$, the solution may be deduced by Duhamel's integral using result (1.40) obtained for the constant temperature. In such a case, the simple solution $f(x,t)$ is defined by equation (1.40) at $T_i = 0$, while $F(t)$ is a given function $\phi(t)$. Therefore, applying Duhamel's equation (1.49), one gets

$$T(x,t) = \int_0^t \phi(\tau) \frac{d}{dt} f(x,t-\tau) d\tau, \quad f(x,t-\tau) = \frac{2}{\sqrt{\pi}} \int_{x/2\sqrt{\alpha(t-\tau)}}^\infty \exp\left(-\xi^2\right) d\xi x \quad (1.50)$$

$$\frac{d}{dt} f(x,t-\tau) = -\frac{2}{\sqrt{\pi}} \exp\left[-\frac{x^2}{4\alpha(t-\tau)}\right] \frac{d}{dt}\left[\frac{x}{2\sqrt{\alpha(t-\tau)}}\right] = \frac{x}{2\sqrt{\pi\alpha(t-\tau)^3}} \exp\left[-\frac{x^2}{4\alpha(t-\tau)}\right].$$

$$(1.51)$$

Substituting this result into equation (1.50) yields (Exercise 1.25)

$$T(x,t) = \frac{x}{2\sqrt{\pi\alpha}} \int_0^t \frac{\phi(\tau)}{(t-\tau)^{3/2}} \exp\left[-\frac{x^2}{4\alpha(t-\tau)}\right] d\tau. \quad (1.52)$$

Applying a new dummy variable and recalling equation (1.40) transforms this solution to the form

$$\zeta = \frac{x}{2\sqrt{\alpha(t-\tau)}}, \quad t-\tau = \frac{x^2}{4\alpha\zeta^2} \quad T(x,t) = \frac{2}{\sqrt{\pi}} \int_{x/2\sqrt{\alpha t}}^\infty \phi\left(t - \frac{x^2}{4\alpha\zeta^2}\right) \exp\left(-\zeta^2\right) d\zeta. \quad (1.53)$$

Equation (1.53) defines temperature in a semi-infinite solid or thin rod initially being at zero temperature.

A solution for such a body with initially $T_i(x)$ and surface $\phi(t)$ temperatures may be obtained as a sum of equation (1.53) and solution (1.39) or (1.40) for a body with zero surface and initial $T_i(x)$ temperature. Such a procedure is possible because the conduction equation with constant physical parameters is a linear equation. For linear equations, the superposition principle is applicable, which states that a sum of particular solutions is a solution as well. We use this principle in Comment 1.1, in Exercise 1.14, and in fact in deriving the Duhamel's integral (Exercise 1.26).

Example 1.9 A semi-infinite solid or thin rod with particular time-dependent temperatures.
(1) Constant temperatures T_0 for time $0 < t < t_c$ and T_1 for time $t > t_c$:

$$T(x,t) = \frac{2}{\sqrt{\pi}} \int_{x/2\sqrt{\alpha t}}^\infty T_0 \exp\left(-\zeta^2\right) d\zeta = T_0 erfc \frac{x}{2\sqrt{\alpha t}}, \quad 0 < t < t_c \quad (1.54)$$

$$T(x,t) = \frac{2T_0}{\sqrt{\pi}} \int_{x/2\sqrt{\alpha t}}^{\infty} \exp\left(-\zeta^2\right) d\zeta + \frac{2T_0}{\sqrt{\pi}} \int_{\infty}^{x/2\sqrt{\alpha t}} \exp\left(-\zeta^2\right) d\zeta + \frac{2T_1}{\sqrt{\pi}} \int_{x/2\sqrt{\alpha(t-t_c)}}^{\infty} \exp\left(-\zeta^2\right) d\zeta$$

$$= T_0 erfc \frac{x}{2\sqrt{\alpha t}} + \left(T_1 - T_0\right) erfc \frac{x}{2\sqrt{\alpha(t-t_c)}}, \quad t > t_c. \tag{1.55}$$

Comment 1.4 In sum (1.55) of three integrals, the first integral is written for the domain $0 < t < t_c$, while the second one gives the contribution of initial temperature T_0 in domain $t > t_c$ and the third integral takes into account a contribution of T_1 (Exercise 1.27).

(2) Linearly varying time-dependent temperature ct, $c = const$.

$$T(x,t) = \frac{2}{\sqrt{\pi}} \int_{x/2\sqrt{\alpha t}}^{\infty} c\left(t - \frac{x^2}{4\alpha\zeta^2}\right) \exp\left(-\zeta^2\right) d\zeta$$

$$= ct\left\{ -\frac{x^2}{2\alpha t\sqrt{\pi}} \int_{x/2\sqrt{\alpha t}}^{\infty} \frac{\exp\left(-\zeta^2\right)}{\zeta^2} d\zeta + erfc \frac{x}{2\sqrt{\alpha t}} \right\}$$

$$= ct\left\{ \left(1 + \frac{x^2}{2\alpha t}\right) erfc \frac{x}{2\sqrt{\alpha t}} - \frac{x}{\sqrt{\pi\alpha t}} \exp\left(-\frac{x^2}{4\alpha t}\right) \right\}. \tag{1.56}$$

Here, the second integral was obtained using a relation that is a result of integrating by parts the error function, putting $u = \zeta \exp\left(-\zeta^2\right)$ and $dv = d\zeta / \zeta$ to get (Exercise 1.26)

$$\int_{z}^{\infty} \exp\left(-\zeta^2\right) d\zeta = \frac{\exp\left(-z^2\right)}{2z} - \frac{1}{2} \int_{z}^{\infty} \frac{\exp\left(-\zeta^2\right)}{\zeta^2} d\zeta. \tag{1.57}$$

(3) Time-dependent temperature varying as $c\sqrt{t}$, $c = const$.
It follows from equation (1.41) that in the case of constant heat flux at the surface of semi-infinite solid, the temperature varies with time as $c\sqrt{t}$. One may check that corresponding function for zero initial temperature

$$T(x,t) = c\sqrt{t}\left[\frac{1}{\sqrt{\pi}} \exp\left(-\frac{x}{2\sqrt{\alpha t}}\right) - \frac{x}{2\sqrt{\alpha t}} erfc \frac{x}{2\sqrt{\alpha t}} \right] \tag{1.58}$$

satisfies the condition $T = c\sqrt{t}$ at $x = 0$ and $T = 0$ at $t = 0$.

EXERCISES

1.13 Explain why heat transfer inside a semi-infinite solid and inside a thin laterally insulated rod may be considered as the same one-dimensional problem. Perform the integration to get the equation (1.33).

1.14 Transform equation (1.35) into the form (1.36). Because one-dimensional equation (1.4) is linear, any particular part of integrand (1.36) should be a solution of this equation as well. Show by differentiating that expression $t^{-1/2} \exp(x^2/4at)$ satisfies one-dimensional equation (1.4).

1.15 Draw graphs of the Heaviside's step function and function $T(x,t)/T_i = f\left(x/2\sqrt{at}\right)$. Explain the term "discontinuity point." To what parameter pertains this term in that case? How does the discontinuity point look on the graph?

1.16 Analyze formula (1.40). How many variables determine the temperature at a particular time and point? Draw a graph. How does the time required to reach a given temperature at a fixed point depend on the distance from the surface.

1.17 Determine for semi-infinite solid using formula (1.40) (i) the temperature gradient by calculating the derivative $\sqrt{\pi at}\,\partial\left[T/(T_w - T_i)\right]/\partial x = f(x/2\sqrt{at})$. Draw a graph. Explain the term "Gradient." What physically defines the temperature gradient? (ii) the rate of cooling by calculating derivative $2\sqrt{\pi at}\,\partial\left[T/(T_w - T_i)\right]/\partial t = F\left(x/2\sqrt{at}\right)$. Draw a graph. Explain why this curve has a maximum.

1.18 Integrate the ordinary differential equation (1.44) using standard procedure (see Advanced Engineering Mathematics) and derive equations (1.45) and (1.46).

1.19 Determine the depth of freezing soil during winter time caused by changing the average air temperature from 12°C to −12°C. Hint: Consider the Earth as a semi-infinite solid.

1.20 The initial temperature T_i of a slab of length L should be reduced during time t on the length Δ from $x = 0$ ($\Delta \ll L$, so the slab may be considered as semi-infinite). Derive the dependence between T_w/T_i and T/T_i where T_w is the coolant temperature and T is temperature at section $x = \Delta$. Plot the graph $T_w/T_i = f(z)$, $z = x/2\sqrt{at}$ for $T/T_i = 1/2$. What limits are for z? What value of z corresponds to $T_w = 0°C$ if $T_i = 1000\,K$? Determine the required coolant temperature for steel and concrete if $\Delta = 0.02\,m$, time $t = 1h, T_i = 500\,K$ and $T = 300\,K$?

1.21 A concrete firewall of thickness Δ protects a room. Determine the dependence between time that such firewall can provide without rising room temperature and high temperature at the other wall side. Calculate and plot data for $\Delta = 10\,cm$ and room temperature $T = 300\,K$. Consider limiting cases $T_w = T_{room}$ and $T_w = \infty$. Treat the wall as a semi-infinite solid. Hint: consider temperature difference between firewall and room (or $T_R = 0$) and determine constants in equation (1.33).

1.22 * Semi-infinite solid initially at temperature T_i is embedded into environment with temperature T_∞, the heat transfer coefficient is h. Using the boundary condition of the third kind and differentiating equation (1.40) to get the heat flux, we have

$$q_w = -\lambda_w \left.\frac{\partial T}{\partial x}\right|_{x=0} = h\left(T_w - T_\infty\right) = \frac{\lambda_w}{\sqrt{\pi at}}\left(T_i - T_w\right)$$

The last two expressions may be considered as an equation determining T_w. However, this equation contains a mistake. Find it.

1.23 The solution of just stated problem is given in Example 1.8 in the forms (1.45) and (1.46). Show that dimensionless temperature depends on two dimensionless parameters $z = x/2\sqrt{at}$ and Biot number $\mathrm{Bi}_x = hx/\lambda_w$. What determines Biot number in this problem? What is the difference between Biot number considered in Section 1.2.3 and this Biot number?

1.24 Solve the problem 1.21 with boundary condition of the third kind using equation (1.46). Such stated problem (see 1.22) corresponds to a situation when the high temperature side of the firewall is subjected to environment with temperature T_∞. Determine time for $\mathrm{Bi}_x = 10$, $T_\infty = 500\,\mathrm{K}$. Hint: plot calculation results for $z = 0.2$ to $z = 1$.

1.25 What is time lag and how is it used in Duhamel's integral derivation? What is time lag in equation (1.47)? What is the connection between time lags in the first and second expressions (1.47)? Transform equation (1.52) in the form (1.53).

1.26 What is the superposition principle? Is that principle applicable to any equation? When it works and why it could not be used in other cases. Show that superposition could not be applied to one-dimensional conduction equation with temperature dependent physical properties. Think: how we use the superposition in Duhamel integral derivation?

1.27 Explain why the expression (1.55) for domain $t > t_c$ consists of three integrals. Obtain the equation (1.57) and use it to get the equation (1.56).

1.5 Method of Separation Variables

The idea of separation variable is to present a solution of a partial differential equation as a product of several parts in which each part depends only on one variable. Such procedure reduces a problem governed by partial differential equation to a set of ordinary differential equations.

1.5.1 General Approach, Homogeneous, and Inhomogeneous Problems

The problem is homogeneous if the right-hand sides of equation and boundary conditions are zero; otherwise, the equation or boundary condition is inhomogeneous. For example, both equations (1.2) are inhomogeneous because in each equation there is a source on the right-hand side. Consider three cases when variable separation is possible for the conduction equation:

(1) General case when the solution of homogeneous conduction equations (1.2) (without source) is possible to be sought as a product of two functions $T = f_1(t) f_2(x_i)$, one depending on time and another on coordinates. Substituting such solution into equation (1.2) using the initial and homogeneous boundary conditions yields

$$f_1(t)\nabla^2 f_2(x_i) = \frac{f_2(x_i)}{a}\frac{\partial f_1(t)}{\partial t} \qquad T(x_i, 0) = F(t) \qquad \lambda_{wi}\frac{\partial f_1(x_i)}{\partial n_i} + h_i f_1(x_i) = 0. \quad (1.59)$$

Here, T is a temperature excess beyond surrounding, x_i are the coordinates, n is the outward normal, and the boundary condition is written in general form of third kind so that at $\lambda_{wi} = 0$ the first-kind or at $h_i = 0$ the second-kind homogeneous boundary condition is obtained. Separation of variables reduces equation (1.59) to the following two differential equations for two- or three-dimensional and one-dimensional heat conduction problem, respectively:

$$\frac{1}{a f_1(t)}\frac{df_1(t)}{dt} = \frac{1}{f_2(x_i)}\nabla^2 f_2(x_i) = -\mu^2 \qquad \frac{1}{a f_1(t)}\frac{df_1(t)}{dt} = \frac{1}{f_2(x)}\frac{d^2 f_2(x)}{dx^2} = -\mu^2. \quad (1.60)$$

These equations are in separated form because the left-hand side of each relation depends only on time, whereas the right-hand side expression depends only on coordinates, and due to this fact, these relations may be equal to each other only being constant. Integrating the left-hand side gives $f_1(t) = C \exp(-\mu^2 t)$. This result tells us that μ^2 should be negative: otherwise, the solution will be infinite at $t \to \infty$, which contradicts the physics. The right-hand side of each equation (1.60) constitutes the second separated equation—an ordinary differential equation in the case of one-dimensional problem and a partial differential equation in the two- or three-dimensional case:

$$\frac{d^2 f_2(x)}{dx^2} + \mu^2 f_2(x) = 0 \qquad \nabla^2 f_2(x_i) + \mu^2 f_2(x_i) = 0. \qquad (1.61)$$

The partial differential equation (1.61) known as the Helmholtz equation may be separated further to ordinary differential equations if the solution may be present as a product of two or three functions, each depending only on one of coordinates (Exercise 1.28).

(2) Problems with time-independent inhomogeneous boundary conditions may be reduced to homogeneous problems if suitable new variables are possible.

(3) The solution of an unsteady inhomogeneous problem with time-dependent sources and/or boundary conditions may be reduced by Duhamel's integral to a simpler, similar problem independent of time sources and boundary conditions [7]. Simplification is achieved by replacing a time-dependent source or boundary condition by a similar time-independent functions in which time variable t is substituted by parameter τ, defining some fixed time lag as in Duhamel's equation (1.49). The solution of such a simplified problem yields function $f(x, t - \tau)$, which is used for Dihamel's integral evaluation giving the final result (Exercise 1.29). Actually, we used this approach in Section 1.4.2 for a semi-infinite body with time-dependent surface temperature. A similar approach is based on Green's function (Sec. 1.7).

Below are considered examples of the problems with separated variables.

1.5.2 One-Dimensional Unsteady Problems

Example 1.10 A thin, laterally insulated rod of length L initially at the temperature $T_i(x)$ and constant ends temperatures T_0 at $x = 0$ and T_L at $x = L$.

Since both boundary conditions are inhomogeneous, new variable $\vartheta(\bar{x}, \text{Fo})$ is introduced, which transforms the problem to homogeneous one. Then, we have

$$\vartheta(\bar{x}, \text{Fo}) = T(\bar{x}, \text{Fo}) - T_0 - (T_L - T_0)\bar{x} \qquad \vartheta = 0 \text{ at } \bar{x} = 0 \quad \vartheta = 0 \text{ at } \bar{x} = 1. \qquad (1.62)$$

The conduction equation in a new variable as well as both boundary conditions is homogeneous, and the separation of variables gives two ordinary differential equations:

$$\frac{\partial \vartheta}{\partial \text{Fo}} - \frac{\partial^2 \vartheta}{\partial \bar{x}^2} = 0, \quad \vartheta = f_1(\text{Fo}) f_2(\bar{x}), \quad \frac{df_1}{d\text{Fo}} + \mu^2 = 0, \quad \frac{d^2 f_2}{d\bar{x}^2} + \mu^2 f_2 = 0 \qquad (1.63)$$

The solution of equations (1.63) yields expression that should satisfy at any time two conditions (1.62) at the ends of the rod indicated above:

$$\vartheta = (C_1 \sin \mu\bar{x} + C_2 \cos \mu\bar{x}) \exp(-\mu^2 \text{Fo}), \qquad \vartheta(0, \text{Fo}) = \vartheta(1, \text{Fo}) = 0. \qquad (1.64)$$

To satisfy the first condition at $\bar{x} = 0$, it is necessary to take $C_2 = 0$. Then, the second condition at $\bar{x} = 1$ can be satisfied only putting $\mu = n\pi$. Substituting these results into equation (1.64) gives the family of partial solutions, sum of which forms a solution of the problem at hand. Coefficients C_n, which are used instead of C_1, are determined from the initial condition at Fo = 0:

$$\vartheta = \sum_{n=1}^{\infty} C_n \sin(n\pi\bar{x})\exp\left(-n^2\pi^2\mathrm{Fo}\right) \qquad \vartheta_i\left(\bar{x},0\right) = \sum_{n=1}^{\infty} C_n \sin(n\pi\bar{x}). \tag{1.65}$$

Applying equation (1.62) and Fourier's series expansion, one obtains coefficients C_n and then after substituting C_n into (1.65) gets the solution of the problem:

$$\sum_{n=1}^{\infty} C_n \sin(n\pi\bar{x}) = T_i\left(\bar{x}\right) - T_0 - \left(T_L - T_0\right)\bar{x} \qquad C_n = 2\int_0^1 T_i\left(\bar{x}\right)\sin(n\pi\bar{x})d\bar{x} - \frac{2}{n\pi}T_0 \tag{1.66}$$

$$T\left(\bar{x},\mathrm{Fo}\right) = T_0 + \left(T_L - T_0\right)\bar{x} + 2\sum_{n=1}^{\infty}\left[\int_0^1 T_i(\bar{x})\sin(n\pi\bar{x})d\bar{x} - \frac{T_0}{n\pi}\right]\sin(n\pi\bar{x})\exp\left(-n^2\pi^2\mathrm{Fo}\right). \tag{1.67}$$

In the case of constant initial temperature, solution (1.67) simplifies to

$$\frac{T\left(\bar{x},\mathrm{Fo}\right) - T_0}{T_L - T_0} = \bar{x} + \frac{T_i - T_0}{T_L - T_0}\frac{2}{\pi}\sum_{n=1}^{\infty}\frac{\sin(n\pi\bar{x})}{n}\exp\left(-n^2\pi^2\mathrm{Fo}\right). \tag{1.68}$$

Example 1.11 A thin rod laterally insulated of length L initially at temperature $T_i(x)$, at temperatures T_0 at the end $x = 0$ and with insulated $(\partial T / \partial x = 0)$ end $x = L$.

A variable similar to (1.62) transforms the boundary condition to the homogeneous

$$\vartheta\left(\bar{x},\mathrm{Fo}\right) = T\left(\bar{x},\mathrm{Fo}\right) - T_0 \qquad \vartheta = 0 \text{ at } \bar{x} = 0 \quad \partial\vartheta/\partial\bar{x} = 0 \text{ at } \bar{x} = 1. \tag{1.69}$$

Solution (1.64) subjected for the second and then for the first condition yields

$$\vartheta = \sum_{n=0}^{\infty} C_n \cos\left[\left(n + \frac{1}{2}\right)\pi\bar{x}\right]\exp\left[-\left(n + \frac{1}{2}\right)^2\pi^2\mathrm{Fo}\right] \qquad \vartheta_i = \sum_{n=1}^{\infty} C_n \cos\left[\left(n + \frac{1}{2}\right)\pi\bar{x}\right]. \tag{1.70}$$

Determining coefficients C_n as in the previous example, one obtains the problem solution:

$$C_n = 2\int_0^1 T_i\left(\bar{x}\right)\cos\left[\left(n + \frac{1}{2}\right)\pi\bar{x}\right]d\bar{x} - \frac{2(-1)^n}{(n+1/2)\pi}T_0 \tag{1.71}$$

$$T(\bar{x},\mathrm{Fo}) = T_0 + 2 \sum_{n=0}^{\infty} \left\{ \int_0^1 T_i(\bar{x}) \cos\left[\left(n+\frac{1}{2}\right)\pi\bar{x}\right] d\bar{x} - \frac{2(-1)^n T_0}{(n+1/2)\pi} \right\}$$

$$\times \cos\left[\left(n+\frac{1}{2}\right)\pi\bar{x}\right] \exp\left[-\left(n+\frac{1}{2}\right)^2 \pi^2 \mathrm{Fo}\right]. \tag{1.72}$$

In the case of constant initial temperature, solution (1.72) becomes (Exercise 1.30)

$$\frac{T(\bar{x},\mathrm{Fo})-T_0}{T_i - T_0} = \frac{2}{\pi} \sum_{n=0}^{\infty} \frac{(-1)^n}{(n+1/2)} \cos\left[\left(n+\frac{1}{2}\right)\pi\bar{x}\right] \exp\left[-\left(n+\frac{1}{2}\right)^2 \pi^2 \mathrm{Fo}\right]. \tag{1.73}$$

Example 1.12 A thin, laterally insulated rod of length L initially at temperature $T_i(x)$, at temperatures $T_0(t)$ at the end $x = 0$ and with insulated end $x = L$.

The solution is presented as a sum of two others: one for a rod with zero initial temperature and given boundary conditions and another for a rod with given initial temperature and zero boundary conditions. The first problem is an unsteady inhomogeneous problem, which may be solved using the Duhamel approach. To evaluate equation (1.49), function $f(x, t-\tau)$ is required. Such a function is obtained using the solution of a similar problem with a time-dependent boundary condition (see Sec. 1.4.2). Putting in (1.73) $T_i(\bar{x}) = 0$ and applying instead of τ another dummy variable $\hat{\mathrm{Fo}}$, one gets the relations for function $f(x, t-\tau)$ and its derivative required for evaluating equation (1.49):

$$f(\bar{x},\mathrm{Fo}-\hat{\mathrm{Fo}}) = -\frac{2}{\pi} \sum_{n=0}^{\infty} \frac{(-1)^n}{(n+1/2)} \cos\left[\left(n+\frac{1}{2}\right)\pi\bar{x}\right] \exp\left[-\left(n+\frac{1}{2}\right)^2 \pi^2 (\mathrm{Fo}-\hat{\mathrm{Fo}})\right] \tag{1.74}$$

$$\frac{\partial f(\bar{x},\mathrm{Fo}-\hat{\mathrm{Fo}})}{\partial \mathrm{Fo}} = 2\pi \sum_{n=0}^{\infty} (-1)^n \left(n+\frac{1}{2}\right) \cos\left[\left(n+\frac{1}{2}\right)\pi\bar{x}\right] \exp\left[-\left(n+\frac{1}{2}\right)^2 \pi^2 (\mathrm{Fo}-\hat{\mathrm{Fo}})\right]. \tag{1.75}$$

The solution of another problem with general initial condition $T_i(x)$ is obtained from solution (1.72) of the same problem putting $T_0 = 0$. Then, substituting equation (1.75) in the Duhamel's equation (1.49) and adding the solution of the second problem lead us to the solution of the problem in question:

$$T(\bar{x},\mathrm{Fo}) = 2 \sum_{n=0}^{\infty} \cos\left[\left(n+\frac{1}{2}\right)\pi\bar{x}\right] \exp\left[-\left(n+\frac{1}{2}\right)^2 \pi^2 \mathrm{Fo}\right] \left\{ (-1)^n \pi \left(n+\frac{1}{2}\right) \right.$$

$$\left. \times \int_0^{\mathrm{Fo}} T_0(\hat{\mathrm{Fo}}) \exp\left[-\left(n+\frac{1}{2}\right)^2 \pi^2 \hat{\mathrm{Fo}}\right] d\hat{\mathrm{Fo}} + \int_0^1 T_i(\bar{x}) \cos\left[\left(n+\frac{1}{2}\right)\pi\bar{x}\right] d\bar{x} \right\}. \tag{1.76}$$

Comment 1.5 The Fourier series is a powerful tool usable in different areas. To better understand some properties of these series, special Exercises 1.31–1.38 are offered.

1.5.3 Orthogonality of Eigenfunctions

Solutions just considered show the significant importance of Fourier's method of determining coefficients C_n in solutions using the initial condition and trigonometric series. It turns out that there are other groups of functions appearing in problem solutions, which may be used in a similar way for defining coefficients by series. The required properties of such a set of functions gives a solution of the Sturm–Liouville problem named after pioneers who studied this subject in thirtieth of nineteenth century.

Consider the following Sturm–Liouville problem in interval $a < x < b$ with boundary conditions at $x = a$ and $x = b$, respectively:

$$\frac{d}{dx}\left[p(x)\frac{y(x,\mu)}{dx}\right] + q(x)y = \mu\, w(x)y, \quad C_1\frac{dy}{dx} + C_2 y = 0, \quad C_3\frac{dy}{dx} + C_4 y = 0. \quad (1.77)$$

This homogeneous problem has a solution only for a set of values $\mu_1 < \mu_2 < \mu_3 \ldots$, which are called eigenvalues of problem (1.77). The corresponding solutions $y_n(x)$ of equation (1.77), which form a set of functions such that

$$I = \int_a^b y_n(x)y_m(x)w(x)dx = 0 \text{ for } \mu_n \neq \mu_m, \qquad I \neq 0 \text{ for } \mu_n = \mu_m, \qquad (1.78)$$

are called orthogonal eigenfunction with respect to weighting function $w(x)$. Or in other words, a set of functions is orthogonal if equation (1.78) is equal to zero for each pair of eigenfunctions $y_n(x)$ and $y_m(x)$ corresponding to eigenvalues μ_n and μ_m except the case of their equalities when $\mu_n = \mu_m, y_n(x) = y_m(x)$.

The orthogonal property makes it possible to present an arbitrary function as a series of the eigenfunctions similar to Fourier's expansion. To show that, multiply both sides of such a series $f(x) = \sum_{n=0}^{\infty} C_n y_n$ by eigen $y_n(x)$ and weighting $w(x)$ functions to obtain

$$\int_a^b f(x)y_n(x)w(x)dx = C_1 \int_a^b y_1(x)y_n(x)w(x)dx + C_2 \int_a^b y_2(x)y_n(x)w(x)dx$$

$$+ \cdots + C_n \int_a^b y_n(x)y_n(x)w(x)dx. \qquad (1.79)$$

Due to orthogonality equation (1.78), all integrals except one containing $y_n^2(x)$ become zero and Euler formula defining series coefficients follows from (1.79):

$$C_n = \frac{\int_a^b f(x)y_n(x)w(x)dx}{\int_a^b y_n^2(x)w(x)dx} \qquad f(x) = \sum_{n=1}^{\infty} C_n y_n(x). \qquad (1.80)$$

It is easy to check that the Fourier series are orthogonal (Exercise 1.39).

Example 1.13 A thin, laterally insulated rod of length L initially at $T_i(x)$, with T_0 at $x = 0$ and heat transfer at $x = L$ into surrounding at temperature T_∞.

To modify the inhomogeneous boundary conditions and find a corresponding new dependent variable, consider an expression containing two unknown functions

$$T = \vartheta + \varphi_1(\bar{x})T_0 + \varphi_2(\bar{x})T_\infty \quad T = T_0 \text{ at } \bar{x} = 0, \quad \frac{\lambda_w}{L}\frac{\partial T}{\partial \bar{x}} = h(T - T_\infty) \text{ at } \bar{x} = 1. \quad (1.81)$$

Substituting this expression into given boundary conditions (1.81), one finds that homogeneous boundary conditions require the following relations: $\varphi_1(0) = 1, \varphi_2(0) = 0,$ $\varphi_1'(1) + \text{Bi}\varphi_1(1) = 0, \varphi_2'(1) + \text{Bi}[\varphi_2(1) - 1] = 0.$ Assuming that $\varphi(\bar{x})$ is a linear function yields values of $\varphi(0)$ and $\varphi(1)$ and finally homogeneous boundary conditions:

$$\vartheta = T - T_0 + B\bar{x}(T_0 - T_\infty) \quad \vartheta = 0 \text{ at } \bar{x} = 0, \quad \frac{\partial \vartheta}{\partial \bar{x}} - \text{Bi}\vartheta = 0 \text{ at } \bar{x} = 1, \text{ Bi} = \frac{hL}{\lambda_w}, B = \frac{\text{Bi}}{1 + \text{Bi}}. \quad (1.82)$$

The conduction equation in new variable ϑ remains homogeneous because the second derivative of an additional term appearing in the transformed equation is zero (Exercise 1.40).

The particular solution of the transformed equation is the same expression (1.64) as in the previous example. Satisfying first boundary condition (1.82) requires to take $C_2 = 0$.

Then, we have $\vartheta = C_1 \sin \mu \bar{x} \exp(-\mu^2 \text{Fo})$. Using other condition (1.82) gives

$$\mu \cos \mu = \text{Bi} \sin \mu \text{ at } \bar{x} = 1 \text{ or } \tan \mu = \mu/\text{Bi} \quad (1.83)$$

This transcendental equation determines an infinite number of roots. Figure 1.2 shows the graphical solution of equation (1.83). It is seen that the roots are spaced unevenly so that the interval between roots grows with n reaching π as $n \to \infty$. Thus, the solution of the problem in question is $\vartheta = \sum_{n=1}^{\infty} C_n \sin(\mu_n \bar{x}) \exp(-\mu_n^2 \text{Fo})$, where coefficients C_n should be determined as usual using series expansion as $\vartheta_i = \sum_{n=1}^{\infty} C_n \sin(\mu_n \bar{x})$ with roots defined by equation (1.83). However, this is not the usual Fourier series in which the roots are spaced evenly being

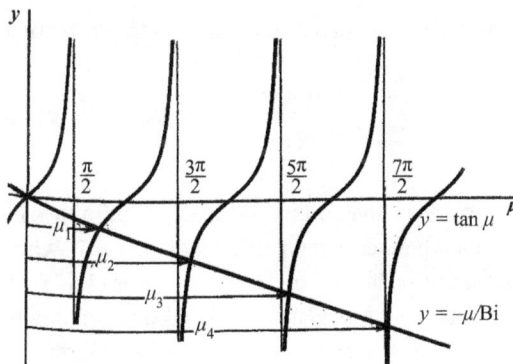

Figure 1.2. Graphical solution of the equation $\tan \mu = \mu/\text{Bi}$.

determined by simple trigonometric equation. In such a case of unusual series, the coefficients can be determined applying formula (1.80) only if the μ_n are eigenvalues of a Sturm–Liouville problem. To check this in some particular case, one should compare the problem in question with the Sturm–Liouville problem formulation.

In this case, equation (1.83) is a solution of ordinary differential equation (1.63) obtained after separation of variables with boundary conditions (1.82): $f'' + \mu^2 f = 0$, $f(0) = 0$, $f'(1) + \text{Bi}f(1) = 0$. Comparing these equations with expressions (1.77) shows that this problem is a special case of the Sturm–Liouville problem in interval $0 < x < 1$ with $p(x) = w(x) = 1, q(x) = 0, C_1 = 0, C_2 = C_3 = 1$ and $C_4 = \text{Bi}$. This implies that μ_n and functions $\sin(\mu_n \overline{x})$ are eigenvalues and eigenfunctions, and hence, the coefficients C_n may be determined using formula (1.80). Taking into account equation (1.82) and performing integration, one gets the problem solution (Exercises 1.41 and 1.42):

$$C_n = \frac{\int_0^1 \vartheta_i(\overline{x}) \sin(\mu_n \overline{x}) d\overline{x}}{\int_0^1 \sin^2(\mu_n \overline{x}) d\overline{x}} = \frac{\int_0^1 \left[T_i(\overline{x}) - T_0 + B\overline{x}(T_0 - T_\infty)\right] \int_0^1 \sin(\mu_n \overline{x}) d\overline{x}}{\int_0^1 \sin^2(\mu_n \overline{x}) d\overline{x}} \tag{1.84}$$

$$T(x, \text{Fo}) = T_0 - B\overline{x}(T_0 - T_e) + 2\sum_{n=1}^{\infty} \frac{1}{\mu_n(\mu_n - \sin \mu_n \cos \mu_n)}$$

$$\times \left[\mu_n^2 \int_0^1 T_i(\overline{x}) \sin(\mu_n \overline{x}) d\overline{x} - T_0 \mu_n (1 - \cos \mu_n) + B(\sin \mu - \mu \cos \mu)(T_0 - T_\infty)\right]$$

$$\times \sin(\mu_n \overline{x}) \exp\left(-\mu_n^2 \text{Fo}\right). \tag{1.85}$$

Some simpler cases follow from this solution. For the case of heat transfer from both ends of a rod into surrounding at T_∞, we obtain, since $T_0 = T_\infty$,

$$T(x, \text{Fo}) = T_\infty + 2\sum_{n=1}^{\infty} \frac{\mu_n \int_0^1 T_i(\overline{x}) \sin(\mu_n \overline{x}) d\overline{x} - T_\infty (1 - \cos \mu_n)}{\mu_n - \sin \mu_n \cos \mu_n} \sin(\mu_n \overline{x}) \exp\left(-\mu_n^2 \text{Fo}\right). \tag{1.86}$$

If, in addition, the initial temperature is constant, this expression becomes

$$\frac{T(x, \text{Fo}) - T_\infty}{T_i - T_\infty} = 2\sum_{n=1}^{\infty} \frac{1 - \cos \mu_n}{\mu_n - \sin \mu_n \cos \mu_n} \sin(\mu_n \overline{x}) \exp\left(-\mu_n^2 \text{Fo}\right). \tag{1.87}$$

Example 1.14 A thin, laterally insulated circular plate or infinite long cylinder of radius R initially at $T_i(\overline{r})$ and constant boundary temperature T_0 at $\overline{r} = 1$, where $\overline{r} = r/R$.

This problem is governed by unsteady conduction equation (1.14) in cylindrical coordinates. Since the plate is thin and initial and boundary conditions depend only on radial coordinate, the problem is one-dimensional, and corresponding equation has a form

$$\frac{\partial T}{\partial t} = \alpha\left(\frac{\partial^2 T}{\partial r^2} + \frac{1}{r}\frac{\partial T}{\partial r}\right) \quad \text{or} \quad \frac{\partial T}{\partial \text{Fo}} = \frac{\partial^2 T}{\partial \overline{r}^2} + \frac{1}{\overline{r}}\frac{\partial T}{\partial \overline{r}} \quad \text{Fo} = \frac{\alpha t}{R^2}. \tag{1.88}$$

Variable $\vartheta = T - T_0$ transforms the problem to homogeneous, and one gets as in the previous examples $f_1(\text{Fo}) = \exp(-\mu^2 \text{Fo})$ and a differential equation $\bar{r} f_2''(\bar{r}) + f_2'(\bar{r}) + \bar{r}\mu^2 f_2(\bar{r}) = 0$, determining a function of coordinate. Substitution $\bar{r} = \xi / \mu$ converts the last equation to Bessel's equation $\xi J'' + J' + \xi J = 0$ which solution is Bessel function of the first kind of order zero $J_0(\mu \bar{r})$. Note that after writing the last equation as $(\bar{r} J')' + \bar{r}\mu^2 J = 0$, it becomes clear that this equation and boundary condition $\vartheta = T - T_0 = J_0 = 0$ at $\bar{r} = 1$ form a Sturm–Liouville problem in interval $0 < \bar{r} < 1$ with $p(\bar{r}) = w(\bar{r}) = \bar{r}$, $q(\bar{r}) = 0$, $C_1 = C_2 = C_3 = 0$, and $C_4 = 1$. Thus, μ_n and $J_0(\mu \bar{r})$ are eigenvalues and eigenfunctions, and the solution of the problem in question is given as follows (Exercise 1.43):

$$T = T_0 + \sum_{n=1}^{\infty} C_n J_0(\mu_n \bar{r}) \exp(-\mu_n^2 \text{Fo}) \qquad C_n = \frac{\int_0^1 \left[T_i(\bar{r}) - T_0 \right] J_0(\mu_n \bar{r}) \bar{r} \, d\bar{r}}{\int_0^1 J_0^2(\mu_n \bar{r}) \bar{r} \, d\bar{r}}. \qquad (1.89)$$

Comment 1.6 Specific nonstandard series satisfying Sturm–Liouville requirements, such as considered above, are known as the generalized Fourier series.

EXERCISES

1.28 What is an idea of separation method? Repeat the analysis of general case of homogeneous conduction equation (1.2). Explain why this method cannot be used for an inhomogeneous problem. Two boundary conditions are given in the form: $x - y = 0, xy = a$. Are they homogeneous?

1.29 How does the Duhamel's approach reduce an inhomogeneous problem to a corresponding homogeneous one?

1.30 Show that equations (1.68) and (1.73) follow from equations (1.67) and (1.72), respectively. Find expressions for temperature and heat flux in the case of constant initial temperature for small values of x and Fourier number.

1.31 Solve a problem from Example 1.11 for $T_i = 400\,\text{K}$ and $T_0 = 300\,\text{K}$ using dimension variables. Check your result comparing it with formula (1.73). Calculate the temperature distribution along the rod for $\text{Fo} = 0.1$ using only the first term of series.

1.32 Investigate the convergence rate of Fourier series in expression (1.73) for small values of Fourier number. Analyze the value of ratio of the second to the first terms of series to show that the maximum error occurs at $x = 0$ and that the error is less than $\approx 5\%$ for $\text{Fo} > 0.04$ Compare your results with numerical data obtained in problem 1.31.

1.33 Analyze the results obtained in previous problem and formulate recommendations for practical computing the Fourier series: (i) When and with what accuracy may be used only the first term neglecting others, (ii) What does it mean that Fourier number is a similarity parameter or universal function? How dose it relate conceptually to approach outlined in Section 1.2.3? Explain how limiting values of dimension values are computed knowing the limits of Fourier number.

1.34 Determine the heat flux from the surface $x = 0$ in Example 1.11 using expression (1.73). Perform similar analysis of Fourier series convergence for the heat flux. Compare results with those obtained in problem 1.32.

1.35 Solve the problem considered in problem 1.21 using equation (1.73). Think: are the boundary conditions at the cold wall side in both cases similar? Compare results and explain why they are different. Which model is more realistic for this problem?

1.36 Investigate analytically how protection time depends on firewall thickness at fixed high temperature in both problems. You will find that the time depends on the wall thickness at fixed temperature identically in both problems. You will also see that this dependence remains as the temperature increases. Nevertheless, the numerical results obtained in previous problem show that difference between protection times in both problems decreases when the temperature grows. What is the reason of this seeming discrepancy?

1.37 Solve the problem considered in Example 1.10 for a long thin slab initially at temperature $T_i(x)$ and symmetrically subjected to temperature T_0. Calculate the temperature distribution across the thickness for $T_i = const$.

1.38 A long thin steel slab of thickness $7\,cm$ is heated to 200°C and then is embedded in a tank with water at 25°C in order to estimate the thermal diffusivity of this alloy. Determine the thermal diffusivity if it is known that in $5\,min$ temperature at the symmetry axis of the slab reduces to 65°C. What type of steels have thermal diffusivity of such order? Hint: use curve $\vartheta = f(Fo)$.

1.39 Give the definition of orthogonal eigenfunction. Define the eigenfunctions and eigenvalue in Fourier series. Give other examples. May any set of functions be considered as eigenfunctions? Check this property for $\cos nx$ in interval $-\pi < x < \pi$.

1.40 Find a new dependent variable required to transform inhomogeneous boundary conditions in homogeneous in Example 1.13 if it is known that both function $\varphi(x)$ are linear. Show that the one-dimensional conduction equation in new dependent variable remains homogeneous.

1.41 Solve the problem from example 1.13 for constant initial and boundary temperatures and check your result with formula (1.87). Derive the expression for heat flux.

1.42 Calculate the first roots of equation (1.83) for Bi = 1 and the heat flux at the end $x = 0$ using result for first nonzero term of series obtained in previous problem. Compare the obtained value with second term of series to estimate the accuracy of neglecting this term. Hint: solve equation $\tan x / x$ by trial and error method.

1.43 The heat flow in laterally insulated rod and in circular plate is one-dimensional in both cases. However, the governing equations are different and consequently the solutions are presented via different functions: trigonometric functions in the first and Bessel function in second cases respectively. Explain physically how these one-dimensional heat flows differ from each other. Think also about difference between elementary trigonometric function and special Bessel function.

1.5.4 Two-Dimensional Steady Problems

The two-dimensional steady heat transfer problems are governed by homogeneous Laplace's and inhomogeneous Poisson's equations (1.5). As is indicated in Section 1.2.4 for solving these equations, the boundary conditions should be specified on each side of the computation domain. Two types of problems depending on the kind of boundary conditions used are usually considered: the Dirichlet problem when the boundary condition of the first kind is used, specifying the temperature on the boundaries of domain, and the Neumann problem in which the second kind of boundary condition in the form of normal temperature derivative on domain sides is given.

The Neumann problem is ill-posed, which physically means that the object reaches thermal equilibrium only when the total heat flow inside the object and corresponding integral defined it become zero:

$$\int_S \frac{\partial T}{\partial n} ds = 0. \tag{1.90}$$

Unless this condition is satisfied, the Neumann problem does not have a solution.

Example 1.15 Two-dimensional rectangular sheet in the xy plane of length a, height b and pre-scribed sides temperatures: left $T(0, y) = \varphi_1(y)$, right $T(a, y) = \varphi_2(y)$, lower $T(x, 0) = \varphi_3(x)$, and upper $T(x, b) = \varphi_4(x)$ (Dirichlet problem).

Substitution of the product $f_1(x) f_2(y)$ in Laplace's equation (1.5) and separation of variables yield two ordinary differential equations and their solutions:

$$\frac{f_1''(x)}{f_1(x)} = -\frac{f_2''(y)}{f_2(y)} = -\mu^2 \qquad f_1''(x) + \mu^2 f_1(x) = 0 \qquad f_2''(y) - \mu^2 f_2(y) = 0 \tag{1.91}$$

$$T(x, y) = [C_1 \sin(\mu x) + C_2 \cos(\mu x)][C_3 sh(\mu y) - C_4 ch(\mu y)]. \tag{1.92}$$

Assuming temperatures along vertical sides being zero $\varphi_1(y) = \varphi_2(y) = 0$ and satisfying zero temperatures conditions at $x = 0$ and $x = a$, we take $C_2 = 0$ and $\mu = n\pi / a$. We should also take $C_4 = 0$ because temperature is zero at $y = 0$ and $y = b$. Satisfying the first of those conditions, we have from equation (1.92) $C_{1n} \sin(n\pi x / a) sh(n\pi y / a)$. To satisfy the third condition of zero at $y = b$, we use superposition principle adding a similar solution $C_{2n} \sin(n\pi x / a) sh[(n\pi / a)(b - y)]$ to the first one. The sum of these two results gives the solution with unknown coefficients C_n,

$$T(x, y) = \sum_{n=1}^{\infty} \left[C_{1n} sh \frac{n\pi}{a} y + C_{2n} sh \frac{n\pi}{a} (b - y) \right] \sin \frac{n\pi}{a} x, \tag{1.93}$$

which are determined using given boundary conditions. Corresponding equations are obtained by putting $y = 0$ and $y = b$ into solution (1.93), yielding formulae

$$\varphi(x) = \sum_{n=1}^{\infty} C_n sh \frac{n\pi b}{a} \sin \frac{n\pi}{a} x \qquad C_n = \frac{2}{ash \frac{n\pi b}{a}} \int_0^a \varphi(x) \sin \frac{n FX \pi}{a} x dx \tag{1.94}$$

valid in both cases so that $\varphi_3(x)$ and $\varphi_4(x)$ are used to define C_{1n} and C_{2n}, respectively.

The other part of the problem should also be considered in the same way, assuming two other boundary conditions being zero: $\varphi_3(x) = \varphi_4(x) = 0$ (Exercise 1.44). The complete solution is found as a sum of results obtained for horizontal and vertical sides.

Example 1.16 Two-dimensional rectangular sheet in the xy plane of length a, height b and prescribed sides boundary conditions: left $T(0, y) = 0$, right $T(a, y) = 0$, lower $T(x, 0) - (\partial T / \partial y)_{y=0} = 0$, and upper $T(x, b) = \varphi(x)$ (mixed boundary conditions).

The particular solution of this problem is the same expression (1.92) that after satisfying conditions $T = 0$ at $x = 0$ and $x = a$ becomes

$$T(x,y) = C_1 \sin\frac{n\pi}{a}x\left(C_3 sh\frac{n\pi}{a}y - C_4 ch\frac{n\pi}{a}y\right). \tag{1.95}$$

Satisfying the condition given for the lower side using this relation leads us to expression

$$C_1 \sin\frac{n\pi}{a}x\left[C_3 sh\frac{n\pi}{a}y - C_4 ch\frac{n\pi}{a}y - \frac{n\pi}{a}\left(C_3 ch\frac{n\pi}{a}y - C_4 sh\frac{n\pi}{a}y\right)\right]_{y=0} = 0, \tag{1.96}$$

which gives $C_4 + \dfrac{n\pi}{a}C_3 = 0$. Substituting that result into equations (1.95) and following summation lead us to the problem solution (Exercise 1.47)

$$T(x,y) = \sum_{n=1}^{\infty} C_n \sin\frac{n\pi}{a}x\left(sh\frac{n\pi}{a}y + \frac{n\pi}{a}ch\frac{n\pi}{a}y\right) \quad C_n = \frac{2\int_0^a \varphi(x)\sin\frac{n\pi}{a}x\,dx}{a\left(sh\frac{n\pi}{a}b + \frac{n\pi}{a}ch\frac{n\pi}{a}b\right)}, \tag{197}$$

where coefficients are found putting $y = b$ and applying the given condition for upper side.

1.6 Integral Transforms

The integral transform technique significantly simplifies differential equations solutions, reducing an ordinary differential equation to an algebraic relation and transforming a partial differential equation to an ordinary differential equation. Here, we consider briefly Fourier and Laplace transforms, the most often used in applications. A systematical usage of integral transform to heat conduction problems may be found in [7].

Although the transformed equations are simpler than originals and usually can be solved readily, the most difficult procedure is to inverse the solution obtained in the subsidiary space to physical variables. Therefore, it is common to use tables of transforms for obtaining inverse solution. Such short tables are given in advanced mathematical courses. More complete tables may be found in the special literature (e.g., [8,9]).

1.6.1 Fourier Transform

The expansion in the Fourier series presents an arbitrary functions as a sum of harmonic oscillations with finite frequencies $n\pi/L = \pi/L, 2\pi/L, 3\pi/L\dots$. As L increases, the distance between frequencies π/L decreases so that the number of terms in series increases. Therefore, in the case of infinite or semi-infinite domain, the difference between frequencies goes to zero, while the number of terms becomes infinite, and in the limit, the Fourier series converts to integral with the continuous frequency spectrum ω:

$$f(x) = \frac{1}{2\pi}\int_{-\infty}^{\infty} C_n \exp(i\omega x)\,d\omega \qquad C_n = \int_{-\infty}^{\infty} f(x)\exp(-i\omega x)\,dx. \tag{1.98}$$

The last integral denoting as $\hat{f}(\omega)$ gives the Fourier transform of function $f(x)$, whereas the first one called the inverse formula returns the function $f(x)$ if coefficients C_n in its integrant is replaced by Fourier transform $\hat{f}(x)$. These two integrals form the Fourier transform approach:

$$\hat{f}(\omega) = f(x)\exp(-i\omega x)dx \qquad f(x) = \frac{1}{2\pi}\int_{-\infty}^{\infty}\hat{f}(\omega)\exp(i\omega x)d\omega. \qquad (1.99)$$

Equations (1.99) become simple for even or odd function $f(x)$, resulting in two pairs of expressions known as cosine and sine Fourier transforms:

$$\hat{f}_C(\omega) = \int_0^{\infty}f(x)\cos\omega x dx \qquad f(x) = \frac{2}{\pi}\int_0^{\infty}\hat{f}_C(\omega)\cos\omega x d\omega \qquad (1.100)$$

$$\hat{f}_S(\omega) = \int_0^{\infty}f(x)\sin\omega x dx \qquad f(x) = \frac{2}{\pi}\int_0^{\infty}\hat{f}_S(\omega)\sin\omega x d\omega. \qquad (1.101)$$

These equations follow from (1.99) after applying formula $\exp(\pm i\omega x) = \cos\omega x \pm i\sin\omega x$. Putting this relation, for example, in the first equation (1.99), we have

$$f(\omega) = \int_{-\infty}^{\infty}f(x)(\cos\omega x - i\sin\omega x)dx = 2\int_0^{\infty}f(x)\cos\omega x dx \qquad (1.102)$$

This result is obtained by taking into account that cosine is even function but sine is odd one. Due to that, the equation in (1.102) containing cosine doubles, whereas the analogous integral with sine vanishes. Performing a similar procedure to other equation (1.99), and comparing results leads us to expressions (1.100) and (1.101) (Exercise 1.48).

Example 1.17 An infinite solid or thin, laterally insulated rod initially at $T_i(x)$.

The Fourier transform of the one-dimensional conduction equation with respect to x yields

$$\alpha\int_{-\infty}^{\infty}\frac{\partial^2 T}{\partial x^2}\exp(-i\omega x)dx = \int_{-\infty}^{\infty}\frac{\partial T}{\partial t}\exp(-i\omega x)dx \qquad \frac{d\hat{T}}{dt} + \alpha\omega^2\hat{T} = 0. \qquad (1.103)$$

The first term in the last equation is obtained using on the right-hand side of equation (1.103) the Leibniz rule of interchanging the integration and differentiation, whereas the second term is derived from the left hand side of integral using double integration by parts putting $u = \exp(-i\omega x)$, $dv = (\partial^2 T / \partial x^2)dx$

$$\int_{-\infty}^{\infty}\frac{\partial^2 T}{\partial x^2}\exp(-i\omega x)dx = \frac{\partial T}{\partial x}\exp(-i\omega x)\Big|_{-\infty}^{\infty} + i\omega\int_{-\infty}^{\infty}\frac{\partial T}{\partial x}\exp(-i\omega x)dx. \qquad (1.104)$$

The first term in this expression vanishes due to usual assumption that the temperature and its derivatives go to zero as $x \to \pm\infty$. Then, repeating the integration by parts of the last integral gives the second term in equation (1.103) (Exercise 1.49).

Since ω in equation (1.103) is a parameter considered as a constant, this equation is a simple ordinary differential equation with constant coefficients, the solution of which is $\hat{T} = C\exp(-\alpha\omega^2 t)$. Determining the constant C using transformed initial temperature

$C = \hat{T}\big|_{t=0} = \hat{T}_i(\omega)$, we get the problem solution in the Fourier space $\hat{T}(\omega, t) = \hat{T}_i(\omega)\exp(-a\omega^2 t)$. To return to physical variables, it is necessary to use a convolution theorem. According to this theorem, the inverse of a product of two transformed functions is given by integral of a product of the inversed functions obtained for each of these functions. Thus, for functions $\hat{T}_i(\omega)$ and $\exp(-\alpha\omega^2 t)$, we should have

$$T(x,t) = \int_{-\infty}^{\infty} T_i(\xi) I(x - \xi, t)\, d\xi, \qquad I(x,t) = \frac{1}{\pi}\int_0^{\infty}\exp(-\alpha\omega^2 t)\cos\omega x\, d\omega, \qquad (1.105)$$

where $I(x,t)$ is an inverted function obtained for $\exp(-\alpha\omega^2 t)$ using cosine transform (1.100) since that function is even with respect to ω. A half of this function is taken into account only because of the relation between Fourier integrals with infinite and semi-infinite limits (see equation 1.102) (Exercise 1.50).

In contrast to the previous standard steps of the Fourier transform, there is no standard technique for second equation (1.105) evaluation. Such a situation when an inverse requires a special approach is typical. In this case, the result may be achieved by expanding cosine in series or by differentiating equation (1.105) with respect to x. In the last case, the derivative of this integral after integration by part ($u = \sin(\omega x), dv = \exp(-\alpha\omega^2 t)\omega\, d\omega$) yields a relation proportional to the second equation (1.105):

$$\frac{dI}{dx} = -\frac{1}{\pi}\int_0^{\infty}\exp(-\alpha\omega^2 t)\sin(\omega x)\omega\, d\omega = -\frac{x}{2\pi\alpha t}\int_0^{\infty}\exp(-\alpha\omega^2 t)\cos\omega x\, dx = -\frac{x}{2\alpha t}I. \qquad (1.106)$$

Thus, the first and the last terms in that expression construct an ordinary differential equation $dI/dx = -(x/2\alpha t)I$, the solution of which gives a result of equation (1.105) evaluation $I = C_1\exp(-x^2/4\alpha t)$. The constant C_1 is defined as a value of integral at $x \to 0$ when $\cos\omega x \to 1$ and integral becomes an error function giving $C_1 = I_{x \to 0} = 1/2\sqrt{\pi\alpha t}\, erf(1)$

Substituting these results into first equation (1.105), one gets the solution, which coincides with equation (1.36) obtained in Example 1.1 (Exercise 1.51):

$$T(x,t) = \frac{1}{2\sqrt{\pi\alpha t}}\int_{-\infty}^{\infty} T_i(\xi)\exp\left[-(x - \xi)^2/4\alpha t\right] d\xi. \qquad (1.107)$$

Comment 1.7 The result equation (1.105) evaluation may be found easier using the tables of Fourier transforms. Here, this calculation is presented to show an example of the inverse, applying a special artificial approach.

Example 1.18 Two-dimensional infinite sheet in the xy plane of semi-infinite height (half plane) initially at $T_i(x)$.

This problem is governed by Laplace equation (1.5), Fourier transform of which with respect to x yields

$$\int_{-\infty}^{\infty}\frac{\partial^2 T}{\partial x^2}\exp(-i\omega x)\, dx + \int_{-\infty}^{\infty}\frac{\partial^2 T}{\partial y^2}\exp(-i\omega x)\, dx = 0 \qquad \frac{d^2\hat{T}}{dy^2} - \omega^2\hat{T} = 0. \qquad (1.108)$$

The derivation of this equation is similar to that of equation (1.103). The first term is the Fourier transform of the second derivative obtained applying double integration by parts (see (1.104)), whereas the second one is a result of changing integration and differentiation in the second term of first equation (1.108) (Example 1.52).

The solution of ordinary differential equation (1.108), satisfying the transformed initial condition $\hat{T}\Big|_{y=0} = \hat{T}_i(\omega)$, is $\hat{T} = \hat{T}(\omega)\exp(-|\omega|y)$. To return to physical variables, the convolution theorem and sine inverse formula (1.101) are used in the way similar to that described in the previous example (Exercise 1.53):

$$T(x,y) = \int_{-\infty}^{\infty} T_i(\xi) I(x-\xi,y)\,d\xi, \qquad I(x,y) = \frac{1}{\pi}\int_0^{\infty}\exp(-|\omega|y)\cos\omega x\,d\omega. \tag{1.109}$$

The inverse of equation (1.109) in contrast to a similar equation (1.105) in the previous problem may be evaluated using a standard technique of integrating. Since the variable in the integral is ω, the coordinates x and y are considering as parameters

$$I(x,y) = \frac{1}{\pi}\int_0^{\infty}\exp(-|\omega|y)\cos\omega x\,d\omega$$

$$= \frac{\exp(-|\omega y|)}{\pi\left(x^2+y^2\right)}(y\cos\omega x + x\sin\omega x)\Bigg|_0^{\infty} = \frac{y}{\pi\left(x^2+y^2\right)}. \tag{1.110}$$

Substituting this result in the first equation (1.109) yields problem solution (Exercise 1.54)

$$T(x,y) = \frac{y}{\pi}\int_{-\infty}^{\infty}\frac{T_i(\xi)}{\left(x-\xi\right)^2+y^3}\,d\xi. \tag{1.111}$$

1.6.2 Laplace Transform

Laplace transform is another widely used integral transform. While the Fourier transform is usually used for infinite variable domains, the Laplace transform is applicable to problems with domains restricted to semi-infinite positive part of numerical axis. Accordingly, basic equations (1.99) contains instead of a complex variable $(-i\omega x)$ in the Fourier integral a variable $(-st)$ with $s > 0$, and the transform expressions are

$$\hat{f}(s) = \int_0^{\infty} f(t)\exp(-st)\,dt \qquad f(t) = \frac{1}{2\pi i}\int_{\gamma-\infty}^{\gamma+\infty}\hat{f}(s)\exp(st)\,ds, \tag{1.112}$$

where functions $f(t)$ and $\hat{f}(s)$ substitute functions $f(x)$ and $\hat{f}(\omega)$ in Fourier integrals, respectively.

It is common to use in Laplace transforms variable t instead of x because time is often independent variable in applications. In fact, for dummy variable in (1.112), it does not matter. The Laplace transform is the most often used integral transform due to better integral convergence with real kernels in comparison with the complex ones, which sometimes result in divergent trigonometric functions as $x \to \pm\infty$. This becomes clear if one recalls the relation $\exp(-i\omega x) = \cos\omega x - i\sin\omega x$. At the same time, the more complicated inverse procedure in the Laplace transform by second formula (1.112) usually overcomes applying tables of transforms.

Example 1.19 A semi-infinite solid or thin, laterally insulated rod initially at zero temperature and time-dependent surface temperature $T(0,t) = \phi(t)$.

Since the domain is semi-infinite and temperature is a function of time, the Laplace transform with respect to t is appropriate. For one-dimensional equation, one gets

$$\alpha \int_0^\infty \frac{\partial^2 T}{\partial x^2} \exp(-st)\,dt = \int_0^\infty \frac{\partial T}{\partial t} \exp(-st)\,dt \qquad \alpha \frac{d^2 \hat{T}}{dx^2} - s\hat{T} = 0. \qquad (1.113)$$

The first term in the last equation is obtained by interchanging the integration and differentiation as in the previous example. The second term in this equation is a Laplace transform of derivative $\partial T/\partial t$. To find this transform, the integral on the right-hand side of first equation (1.113) is integrated by parts putting $u = \exp(-st)$ and $dv = (\partial T/\partial t)dt$, which results in the following relation:

$$\int_0^\infty \frac{\partial T}{\partial t} \exp(-st)\,dt = \exp(-st)T(t)\Big|_0^\infty + s\int_0^\infty T(t)\exp(-st)\,dt = s\hat{T} - T(0). \qquad (1.114)$$

Since the initial temperature is zero, the second term in this sum vanishes, and the first one determines the second term in equation (1.113).

Solving the ordinary differential equation (1.113) and using the common condition of the limited final result $\lim_{x \to \infty} \hat{T}(x,s) = 0$ gives the solution in the Laplace space:

$$\hat{T}(x,s) = C_1 \exp\left(\sqrt{s/\alpha}\,x\right) + C_2 \exp\left(-\sqrt{s/\alpha}\,x\right) \qquad \hat{T}(x,s) = \hat{\phi}(s)\exp\left(-x\sqrt{s/\alpha}\right), \qquad (1.115)$$

where the constant C_2 is defined applying the boundary conditions in the transformed space $T(0,s) = \hat{\phi}(s)$. To inverse the solution (1.115), we use the convolution theorem in the same way as in the derivation of equation (1.105) (Exercise 1.50) and then the table of transforms in order to get the inverse of function $\hat{f}(x,s) = \exp\left(-x\sqrt{s/\alpha}\right)$:

$$T(x,t) = \int_0^t \phi(\tau)f(t-\tau,x)\,d\tau, \qquad f(x,t) = \frac{x\exp\left(-x^2/4\alpha t\right)}{2\sqrt{\pi\alpha}\,t^{3/2}} \qquad (1.116)$$

Substitution of the second relation into the first one leads us to problem solution

$$T(x,t) = \frac{1}{2\sqrt{\pi\alpha}} \int_0^t \frac{\phi(\tau)}{(t-\tau)^{3/2}} \exp\left[-x^2/4\alpha(t-\tau)\right]\,d\tau. \qquad (1.117)$$

As it should be, this result coincides with equation (1.52), obtained applying error integral.

Example 1.20 A thin, literally insulated rod of length L initially at zero temperature with insulated end at $x = 0$ and constant temperature T_L at $x = L$.

We considered analogous problem in Example 1.11. The series obtained there as well as others of such type convergence slowly at small values of Fourier numbers close to $t = 0$. Here, we present an example of solution obtained by Laplace transform in the form of series that fast converges at small times.

In Laplace space, the problem is governed by the same equation (1.113) as in the previous example and following boundary conditions:

$$\alpha \frac{d^2 \hat{T}}{dx^2} - s\hat{T} = 0, \quad \frac{d\hat{T}}{dx} = 0 \text{ at } x = 0, \quad \int_0^\infty T_L \exp(-st) dt = \frac{\hat{T}_L}{s} \text{ at } x = L. \quad (1.118)$$

The solution of this equation after satisfying the boundary conditions gives

$$\hat{T}(x,s) = C_1 sh\sqrt{\frac{s}{\alpha}} x + C_2 ch\sqrt{\frac{s}{\alpha}} x, \quad C_2 ch\sqrt{\frac{s}{\alpha}} L = \frac{\hat{T}_L}{s}, \quad \hat{T}(x,s) = \frac{\hat{T}_L ch\sqrt{s/\alpha} x}{s\, ch\sqrt{s/\alpha} L}. \quad (1.119)$$

The last equation is a solution in the Laplace space obtained satisfying first the condition at $x = 0$ and then determining the constant C_2 using the other condition at $x = L$. One way to inverse solution (1.119) is to express hyperbolic functions in the Taylor series presenting the fraction in the form [4]

$$\hat{T}(x,s) = \frac{\hat{T}_L}{s} \left(e^{ax} + e^{-ax} \right) e^{-aL} \left(1 + e^{-2aL} \right)^{-1} = \frac{\hat{T}_L}{s} \left[e^{-a(L-x)} + e^{-a(L+x)} \right] \sum_{n=0}^\infty (-1)^n e^{-2naL}$$

$$= \frac{\hat{T}_L}{s} \sum_{n=0}^\infty (-1)^n e^{-a[(2n+1)L-x]} + \frac{\hat{T}_L}{s} \sum_{n=0}^\infty (-1)^n e^{-a[(2n+1)L+x]}, \quad a = \sqrt{\frac{s}{\alpha}}. \quad (1.120)$$

Then, using the table of transforms for e^{-ax}/s, we obtain the problem solution

$$\frac{T(x,t)}{T_L} = \sum_{n=0}^\infty (-1)^n \, erfc \frac{(2n+1)L - x}{2\sqrt{\alpha t}} + \sum_{n=0}^\infty (-1)^n \, erfc \frac{(2n+1)L + x}{2\sqrt{\alpha t}}. \quad (1.121)$$

Since this solution converges fast at small times, it is convenient to use this series along with solution (1.73), converging fast at relatively large times (Exercises 1.55 and 1.56).

Comment 1.8 Observe that in some cases, we use hyperbolic function, whereas in others the exponential functions are employed. For instance, the solution of the same equation (1.113) in the last example is presented by the first functions, but in former example, the other way is used. In fact, both expressions are equivalent and only compactness of formulae dictate a choice.

Comment 1.9 In examples 1.17–1.19, we used the convolution theorem. Study this theorem and other basic properties of Fourier and Laplace transforms. Integral transforms are widely used for solving ordinary and partial differential equations in different areas. Advanced Engineering Mathematics courses usually considered the Fourier and Laplace transforms and offer problems for exercises.

1.7   Green's Function Method

The idea of Green's function is similar to Duhamel's principle. This method presents a solution of a given problem in terms of the same type of a simple problem. In creating a Green's function, the simplicity is achieved due to applying homogeneous boundary conditions and Dirac delta function instead of inhomogeneous conditions and space-time dependent heat sources, respectively. Dirac delta function is defined as being zero for all x and being infinite for $x = x_0$ so that integral of it is equal to unity:

$$\delta\left(x - x_0\right) = \begin{cases} 0 & x \neq x_0 \\ \infty & x = x_0 \end{cases} \quad \int_{-\infty}^{\infty} \delta\left(x - x_0\right) dx = 1 \tag{1.122}$$

In formulating Green's function, the space-time dependent heat source is substituted by product of delta functions $\delta\left(x - \xi\right)\delta\left(t - \tau\right)$. Physically, it means that such a product of delta functions determines the temperature at location x and time t produced by an instantaneous source of strength unity at point ξ and time τ.

The solution of the one-dimensional conduction problem for finite domain is given in terms of Green's function $G_{\tau=0}\left(x, t \mid \xi, \tau\right)$ by the following expression [7]:

$$T\left(x, t\right) = \int_0^L G\big|_{\tau=0} T_i\left(\xi\right) d\xi + \frac{\alpha}{\lambda_w} \int_0^t d\tau \int_0^L q_v\left(\xi, \tau\right) G d\xi + \frac{\alpha}{\lambda_w} \int_0^t d\tau \int_0^L \left(Gf\big|_{\xi=0} + Gf\big|_{\xi=L}\right) d\xi. \tag{1.123}$$

This equation is written for Green's function satisfying the boundary conditions $G = 0$ at $t < \tau$ and the general boundary condition $hG - \lambda_w\left(\partial G / \partial n\right) = f\left(x, t\right)$ at $t > \tau$. The first term in equation (1.123) takes into account the initial temperature distribution, the second determines the effect of the source, and the third one defines the contribution of the boundary conditions at $x = 0$ and $x = L$. If the problem in question consists of the boundary condition of the first kind, it corresponds to $\lambda_w = 0$ in the general boundary condition. In that case, in the last integral in equation (1.123), λ_w should be omitted, and the terms Gf should be replaced by $\left(-1/h\right)\left(\partial G / \partial n\right)$.

Example 1.21 A one-dimensional solid at initial temperature $T_i\left(x\right)$ with a space-time dependent source $q_v\left(x, t\right)$ and boundary conditions: $\partial T / \partial x = 0$ at $x = 0$ and time- dependent temperature $\phi\left(t\right)$ at $x = L$.

The problem is governed by equation (1.4) and just indicated initial and boundary conditions:

$$\alpha \frac{\partial^2 T}{\partial x^2} + \frac{\alpha}{\lambda_w} q_v - \frac{\partial T}{\partial t} = 0, \quad T = T_i\left(x\right), t = 0. \quad \frac{\partial T}{\partial x} = 0, x = 0 \quad T = \phi\left(t\right), x = L. \tag{1.124}$$

Green's function is found from the equation similar to equation (1.124) for the problem with homogeneous boundary conditions and product of delta functions instead of the source

$$\alpha \frac{\partial^2 G}{\partial x^2} + \frac{\alpha}{\lambda_w} \delta\left(x - \xi\right)\delta\left(t - \tau\right) = \frac{\partial G}{\partial t}, \quad G = 0 \, t < \tau, \quad \frac{\partial G}{\partial x}\bigg|_{x=0} = 0, \quad G\big|_{x=l} = 0. \tag{1.125}$$

To satisfy the corresponding homogeneous (without delta functions) problem, separation of variables is used, resulting (see (1.64)) according to conditions (1.125) in the following equation:

$$(C_1 \sin \mu_n x + C_2 \cos \mu_n x) \exp\left(-\alpha \mu_n^2 t\right) \quad C_1 = 0, \ \mu_n = \frac{(2n+1)\pi}{2L}, \ n = 0,1,2.... \quad (1.126)$$

In view of this result ($C_2 \neq 0$, $\cos \mu_n x$ is even), to find Green's function, we apply to equation (1.125) the cosine Fourier transform with respect to x:

$$\alpha \int_0^L \frac{\partial^2 G}{\partial x^2} \cos \mu_n x \, dx + \frac{\alpha}{\lambda_w} \int_0^L \delta(x-\xi)\delta(t-\tau)\cos \mu_n x \, dx - \int_0^L \frac{\partial G}{\partial t} \cos \mu_n x \, dx = 0 \quad (1.127)$$

These integrals are evaluated using Green's identity (Exercise 1.57):

$$\int_0^L \left(u \frac{\partial^2 v}{\partial x^2} - v \frac{\partial^2 u}{\partial x^2} \right) dx = u \frac{\partial v}{\partial x}\bigg|_0^L - v \frac{\partial u}{\partial x}\bigg|_0^L, \quad (1.128)$$

for the first integral, the property of delta function for the second, and the rule of interchanging integration and differentiating for the third one. For the first integral, we take in (1.128) $u = \cos \mu_n x$ and $v = G$ to obtain

$$\int_0^L \frac{\partial^2 G}{\partial x^2} \cos \mu_n x \, dx = -\mu_n^2 \int_0^L G \cos \mu x \, dx + \cos \mu_n x \frac{\partial G}{\partial x}\bigg|_0^L + G \sin \mu_n x\big|_0^L = -\mu_n^2 \hat{G}. \quad (1.129)$$

The second and third terms on the right-hand side in this expression vanish due to the fact that μ_n is determined by equation (1.126). For the second equation in (1.127), we have according to (1.122) (Exercise 1.58)

$$\int_0^L \delta(x-\xi) f(x) dx = \begin{cases} f(\xi) & 0 < \xi < L \\ 0 & \xi < 0, \ \xi > L \end{cases},$$

$$\delta(t-\tau) \int_0^L \delta(x-\xi)\cos \mu_n x \, dx = \delta(t-\tau)\cos \mu_n \xi. \quad (1.130)$$

Then, performing the change of integration and differentiating in the third term in equation (1.127) and using (1.129) and (1.130), we obtain an linear ordinary differential equation in the Fourier space and its solution:

$$\frac{d\hat{G}}{dt} + \alpha \mu_n^2 \hat{G} = \delta(t-\tau)\cos \mu_n \xi,$$

$$\hat{G} = \cos \mu_n \xi \exp(-\alpha \mu_n^2 t)\left[C + \int_0^t \exp\left(\alpha \mu_n^2 \tau\right)\delta(t-\tau) d\tau \right]. \quad (1.131)$$

After taking $C = 0$ according to condition (1.125) and using delta function property (1.130), equation (1.131) simplifies giving Green's equation (1.132) in the Fourier space.

The inverse of this relation, which yields Green's function in physical domain, is obtained using the Fourier series since the space domain is finite:

$$\hat{G} = \cos \mu_n \xi \exp\left[-\alpha\mu_n^2 \left(t-\tau\right)\right], \quad G(x,t) = \frac{2}{L}\sum_{n=1}^{\infty}\exp\left[-\alpha\mu_n^2\left(t-\tau\right)\right]\cos \mu_n \xi \cos \mu_n x \quad (1.132)$$

Finally, substitution of Green function into equation (1.123) leads us to the problem solution:

$$T(x,t) = \frac{2}{L}\sum_{n=1}^{\infty}\exp\left(-\alpha\mu_n^2 t\right)\cos \mu_n x \left\{\int_0^L T_i(\xi)\cos \mu_n \xi d\xi + \int_0^t \exp\left(\alpha\mu_n^2 \tau\right)\right.$$

$$\left. \times \left[\frac{\alpha}{\lambda_w}\int_0^L q_v(\xi,\tau)\, \cos \mu_n \xi d\xi + (-1)^n \alpha\mu_n \phi(\tau)\right]d\tau\right\}, \quad (1.133)$$

where μ_n is given by equation (1.126). The last term in equation (1.133) corresponds to the boundary condition of the first kind at the end $x = L$. It is obtained according to the note after equation (1.123) by computing the derivative $\partial G/\partial x$. using equation (1.132) (Exercise 1.59).

Comment 1.10 Note that Green's equation (1.132) contains both types of variables:
x,t and ξ,τ as it should be according to physical interpretation mentioned at the beginning of Section 1.7 (Exercise 1.60).

Comment 1.11 There are usually different methods to solve the same problem. Thus, similar problems to the one considered in the last example by Green's function are solved applying Duhamel's integral (1.4.2) and Laplace transform (1.6.2). The other examples are the solution of some one-dimensional conduction problem, which may be obtained using the integral transform with respect to time and to coordinate as well as two pairs of identical solutions gained by error integral (1.3.1, 1.4.2) and integral transform (1.6.1, 1.6.2). Although identity of solutions obtained by different means seems to be natural, it should be noted that sometimes it is not easy to compare such two results (see for instance, equations (1.117) and (1.133)).

EXERCISES

1.44 Solve the problem 1.15 for the rectangular sheet with temperatures $\varphi_1(y)$ and $\varphi_2(y)$ assuming that $\varphi_3(x) = \varphi_4(x) = 0$.

1.45 Derive expression for temperature distribution $T(x,y)/T_1$ using results obtained in problem 1.44 if $\varphi_1(y) = T_1 = const$ and $\varphi_2(y) = 0$. Calculate the temperature distribution using the first three terms of series for $a = b = 1$, $x = 0.25, 0.5, 0.75$, and $y = 0, 0.1...1$

1.46 Functions satisfying the Laplace equation are known as harmonic functions. One of the properties of such a function is the maximum principle. Study what this principle is like. Think how this principle may be demonstrated using graph obtained in the previous problem.

1.47 Solve the problem 1.16 for the same rectangular sheet with zero temperatures for horizontal sides and prescribed temperatures: constant for left and arbitrary $\varphi(x)$ for right sides.

1.48 Define the difference of even and odd functions. Give examples of such functions. Explain why in expression (1.102), the even function doubles but the odd function vanishes.

1.49 In deriving equation (1.103), we used the Leibniz rule and Fourier transform of the second derivative with respect to x. Recall the Leibniz rule and derive the expression for the second derivative using twice integration by parts, like it is done in equation (1.104).

1.50 According the convolution theorem, the inverse of a product of two Fourier transformed functions $\hat{f}(\omega)$ and $\hat{g}(\omega)$ is given by the integral $\hat{f}(\omega)\hat{g}(\omega) = \int_{\infty}^{\infty} f(\xi)g(x-\xi)d\xi$. Read about this theorem in Advanced Engineering Mathematics and perform some exercises.

1.51 Solve the problem 1.17 using the table of Fourier integral transforms for the evaluation of the second equation (1.105).

1.52 The Laplace equation contains two second derivatives. Explain why by performing the Fourier transform we used different methods for derivatives with respect to longitudinal and transverse coordinates: the integration by parts in the first case and the rule of changing integration and differentiation in second case.

1.53 Explain why the absolute value of ω is used in Example 1.18. Is the function $\exp(-|\omega|y)$ odd or even? Explain your answer.

1.54 Solve the problem 1.18 using the table of Fourier transforms for the evaluation of the second equation (1.109).

1.55 Change in the problem 1.20 the boundary conditions to constant temperature T_0 at $x = 0$ and to insulated end at $x = L$. Solve that problem for those boundary conditions using the Laplace transform. Compare the result with solution (1.73) for zero initial temperature $T_i = 0$. Observe that independent variables are different in both series. Define them. How many variables determine the temperature in each case?

1.56 Calculate the temperature using one and two terms of series obtained in the previous problem for some values of x and small values of time. Compare the domains of convergence of two series applying your results and data from problem 1.32. Formulate the suggestion for practical using both series.

1.57 Prove Green's identity (1.128) used in example 1.21. Hint: apply the product of the first derivatives of functions u and v.

1.58 Study properties of delta function from Advanced Engineering Mathematics to understand the phenomena describing by delta function and its usage in this example. Obtain equation (1.130).

1.59 Derive equation (1.133) following directions given in the text.

1.60 Explain the difference between variables x, t and ξ, τ. Compare ideas of Green's function method and Duhamel's approach.

REFRENCES

[1] Norseman, T. N. (1999). Fourier's heat conduction equation: history, influence and connections. *Reviews of Geophysics 37*, 151–172.

[2] http://en.wikipedia.org/wiki/Reynolds_number

[3] http://www.me.utexas.edu/%7Eme339/history.html

[4] Carslaw, H. S. & Jaeger, J. C. (1986). *Conduction of heat in solids* (2nd ed.). Oxford: Clarendon Press.

[5] Dorfman, A. S. (2011). Universal functions in boundary later theory. *Fundamental Journal of Thermal Science and Engineering* 1, 35–72.

[6] Dorfman, A. S. (2009) *Conjugate problems in convective heat transfer*. Boca Raton: CRC Press Taylor & Francis.

[7] Ozisik, M. N. (1968). *Boundary value problem of heat conduction*. Seranton, Pennsylvania: Int. Textbook Company.

[8] Erdélyi, Arthur, ed. (1954). *Tables of integral transforms*, (1). New York: McGraw-Hill.

[9] Polyanin, A. D., & Manzhirov, A. V. (1998). *Handbook of integral equations*. Boca Raton: CRC Press, ISBN 0-8493-2876-4. doi: http://dx.doi.org/10.1201/9781420050066

CHAPTER 2

Methods in Laminar Fluid Flow and Heat Transfer

2.1 A Brief History

Although current theories of momentum, energy, and substance transfer processes are very close, the basic mathematical equations and methods of solutions used in all transfer theories were developed mainly in the fluid mechanics. As is mentioned in historical notes at the beginning of Chapter 1, the initiation of a modern approach of mathematical describing the fluid motion dates back to the eighteenth century when Euler in 1757 formulated the equations of the motion of the perfect inviscid fluid. Since that time, countless theoretical and experimental studies formed the contemporary fluid mechanics, which is a keystone in our knowledge of all transfer processes. While many researches contributed to the current level of hydrodynamics, the conceptual changes were achieved by great scholars.

Two fundamental results published after Euler's work in the next little more than hundred years—Navier–Stokes equations and Reynolds procedure of statistical averaging Navier–Stokes equations (RANS) for turbulent flows—practically completed classical hydrodynamics basics of our mathematical means for understanding and modeling physical flow processes. However, nonlinear, sophisticated Navier–Stokes and Reynolds equations could not be solved before the computers came into existence. Thus, despite the known equations of the real viscous fluid motion, only the Euler's equations for perfect fluid can be actually used at the end of nineteenth century. Since it was clear that the friction forces arising in the air or water around a moving body are small in comparison with pressure and gravity forces, it was expected that solutions of Euler's equations should be close to the authentic pattern. In fact, the theory of perfect fluid motion leads to satisfactory results in many problems, such as formatting waves, jets, or in determining the pressure distribution around moving body, but this theory fails in predicting the pressure drags. That contradiction between theory and reality is known as a D'Alemberts paradox: the pressure losses in a flow moving around a body are zero.

The great German scholar Ludwig Prandtl (1875–1953) resolved this challenging problem in 1904. In a seven-page article, he explained that to understand the importance of small friction forces, one should divide the flow around a moving body in two regions: the small layer near the body surface where the friction forces are significant and the other part of flow, out of this layer, where the friction effects may be neglected considering the fluid as perfect without friction. Since the velocity on the surface is zero, the fluid velocity changes dramatically across the thin boundary layer from zero at the surface to the velocity value far away from the body. This is the reason why the small in average of whole flow friction forces result in appreciable pressure losses and resistance of the moving bodies. Boundary-layer concept allows significantly simplifying the Navier–Stokes equations taking into account friction forces only in a thin boundary layer. In more than hundred years following Prandtl's article, the boundary-layer theory became a specific part of the fluid mechanics with wide applications in determining the resistance forces of the moving objects from simple plates to ships, aircrafts, and reentry vehicles. Even now despite steady increase in computer resources, the boundary-layer theory is still important due to simplification of governing equations, especially in complicated problems.

Although the Reynolds averaging of Navier–Stokes equations was very important achievement in studying turbulent flows, it did not complete the theory of turbulence because averaging leads to other difficulties known as the closure problem. This problem arises due to additional unknown terms called turbulent stresses that are produced during the process of averaging, leading to an unclosed system of equations. Determining those turbulent stresses requires some hypothesis based on the extra information, which actually resolves the closure problem.

The first closure hypothesis was suggested by Prandtl in 1925. He used the relation of proportionality between the turbulent stress and the velocity gradient similar to that in laminar flow. To determine the coefficient of proportionality (eddy viscosity), he introduced an idea of mixing length and postulated that in the proposed simple model, the mixing length may be considered to be proportional to distance from the surface. Despite the simplicity of Prandtl's hypothesis, his model showed a notable agreement with experimental data and was improved by many researches. Contemporary models of this type known as algebraic turbulence models are still widely used in applications.

Later (1945), Prandtl formulated a more physically grounded model taking into account the dependence of eddy viscosity on the energy of turbulent fluctuations and the proposed differential equation defining the energy of fluctuations. This idea established a class of models, which require (in contrast to algebraic models) a solution of one or more differential equations in addition to basic conservation laws. Models of that type are called one or two equations' turbulence models, depending on the numbers of additional equations. Thus, the original Prandtl's model of this type is a one-equation turbulence model, whereas the algebraic models are a zero-equation turbulence model.

The first two-equation turbulence model was proposed in 1942 by preeminent Russian mathematician Andrey Kolmogorov (1903–1987). This model consists of a differential equation for energy fluctuation k and a similar equation for dissipation energy ω. Such a two-equation model is an accomplished pattern in contrast to one-equation models, which usually require some subsidiary information of a length scale. The other most used in applications two-equations model is the $k - \varepsilon$ model developed by two English scholars Brian Launder and Brian Spalding. Both two-equations models used for determining k and ε or k and ω so-called transport equations similar to that formulated by Kolmogorov. With this in view, several authors modified Prandtl's model by additional appropriate relations. Nevertheless, one-equation models are not widely used for applications.

In modern times, new powerful methods for numerical studying of turbulent flows were developed. The direct numerical simulation (DNS) gives the turbulent flow structure in details using the solution of Navier–Stokes equations without additional assumptions. This is achieved by computing the large and small eddies practically of the whole space and time scales. Since such a numerical modeling requires a huge amount of grid points that depends on the Reynolds number, current DNS applicability is restricted to relatively low Reynolds numbers.

The two other methods of this type are the large-eddy simulation (LES) and detached-eddy simulation (DES). To lessen the amount of grid points, in LES and DES, only large eddies are computed, whereas the small eddies are modeled using the usual wall law approach. Due to that, these methods are applicable to greater but also limited Reynolds numbers. Nevertheless, the results obtained by DNS, LES, and DES are in essence different from those given by turbulence models because they may be considered as specific experimental data obtained for much less cost.

The more detailed historical information one may obtain in [1] and [2].

2.2   Navier–Stokes, Energy, and Mass Transfer Equations

The mathematical description of transfer processes is given by system containing the continuity equation, the Navier–Stokes equations for momentum transfer, and similar equations for energy and mass transfer. These equations are expressions of conservation laws written in terms of velocity components, temperature, and concentration. For a flow of homogeneous incompressible fluid with constant properties, they are as follows:

$$\frac{\partial u}{\partial x}+\frac{\partial v}{\partial y}+\frac{\partial w}{\partial z}=0 \tag{2.1}$$

$$\rho\left(\frac{\partial u}{\partial t}+u\frac{\partial u}{\partial x}+v\frac{\partial u}{\partial y}w\frac{\partial u}{\partial z}\right)=-\frac{\partial p}{\partial x}+\mu\left(\frac{\partial^2 u}{\partial x^2}+\frac{\partial^2 u}{\partial y^2}+\frac{\partial^2 u}{\partial z^2}\right) \tag{2.2}$$

$$\rho\left(\frac{\partial v}{\partial t}+u\frac{\partial v}{\partial x}+v\frac{\partial v}{\partial y}+w\frac{\partial v}{\partial z}\right)=-\frac{\partial p}{\partial y}+\mu\left(\frac{\partial^2 v}{\partial x^2}+\frac{\partial^2 v}{\partial y^2}+\frac{\partial^2 v}{\partial z^2}\right) \tag{2.3}$$

$$\rho\left(\frac{\partial w}{\partial t}+u\frac{\partial w}{\partial x}+v\frac{\partial w}{\partial y}+w\frac{\partial w}{\partial z}\right)=-\frac{\partial p}{\partial z}+\mu\left(\frac{\partial^2 w}{\partial x^2}+\frac{\partial^2 w}{\partial y^2}+\frac{\partial^2 w}{\partial z^2}\right) \tag{2.4}$$

$$\rho c_p\left(\frac{\partial T}{\partial t}+u\frac{\partial T}{\partial x}+v\frac{\partial T}{\partial y}+w\frac{\partial T}{\partial z}\right)=\lambda\left(\frac{\partial^2 T}{\partial x^2}+\frac{\partial^2 T}{\partial y^2}+\frac{\partial^2 T}{\partial z^2}\right)+\mu S \tag{2.5}$$

$$S=2\left[\left(\frac{\partial u}{\partial x}\right)^2+\left(\frac{\partial v}{\partial y}\right)^2+\left(\frac{\partial w}{\partial z}\right)^2\right]+\left(\frac{\partial u}{\partial y}+\frac{\partial v}{\partial x}\right)^2+\left(\frac{\partial u}{\partial z}+\frac{\partial w}{\partial x}\right)^2+\left(\frac{\partial v}{\partial z}+\frac{\partial w}{\partial y}\right)^2$$

$$\rho\left(\frac{\partial C}{\partial t}+u\frac{\partial C}{\partial x}+v\frac{\partial C}{\partial y}+w\frac{\partial C}{\partial z}\right)=\rho D_m\left(\frac{\partial^2 C}{\partial x^2}+\frac{\partial^2 C}{\partial y^2}+\frac{\partial^2 C}{\partial z^2}\right). \tag{2.6}$$

Here, p is a pressure in flow without a hydrostatic term expressing only the difference between the pressure values in resting and moving fluid (Exercise 2.2).

Comment 2.1 The term "Navier–Stokes equations" strictly implies the entire system of equations (2.1)–(2.4), including the continuity equation (2.1). However, for simplicity, the same term is used when only the momentum equations (2.2)–(2.4) are considered.

2.2.1 Two Types of Transport Mechanism. Analogy Between Transfer Processes

Consider a system of equations (2.1)–(2.6) for a steady regime. This system of six equations determines six flow characteristics: three velocity components u, v, w, pressure p, temperature T, and concentration C. The continuity and the three Navier–Stokes equations describe the momentum transfer, whereas the heat and mass transfer are specified by energy and diffusion equations (2.5) and (2.6), respectively. Analysis of these fundamental equations shows that three transfer processes of momentum, heat, and mass are similar being based in essence on the same physical principles.

The major part (without pressure gradients and dissipation function) of each equation is an identical structure being comprised of two analogous groups of terms. Such groups are built of derivatives of a quantity that corresponds to the driving force of a relevant transfer process. In Navier–Stokes equations, these groups are constructed from velocity derivatives because namely the velocity gradient is a driving force in Newton's viscosity law. In the energy and mass transfer equations, analogous groups are made up of the derivatives of temperature and concentration according to the pertinent gradients forming driving forces in Fourier's and in Fick's laws of heat and diffusion, respectively.

Another interpretation of transfer processes similarity may be carried out, considering the transfer phenomena in terms of potential field like gravity or electricity field. In such a case, the velocity, temperature, and concentration are considered as field potentials. Then, driving forces in transfer processes, that is, velocity, temperature, and concentration gradients, are viewed as analogous gradients of altitude in a gravitational field or of voltage in an electrical field.

Two groups of terms in transfer equations mentioned above correspond to two basic mechanisms of transport phenomenon. The group of terms on the right-hand side of each equation corresponds to the molecular transport, whereas the other group on the left-hand side of each equation represents the convective transport. According to that, the structure of each group depends on the nature of transport. In each Navier–Stokes equation, the molecular transport group has the structure of viscous force in conformity with molecular mechanism. Such a group consists of a sum of the second derivatives of relevant velocity component with respect to each coordinate multiplied by viscosity coefficient μ. The molecular transport groups in energy and mass equations have an analogous structure. Each group is composed of a similar product of sums of the second derivatives of temperature or of concentration and corresponding molecular transfer coefficient: thermal conductivity λ or similar molecular diffusion coefficient ρD_m (Exercise 2.4).

Another structure is inherent in the convective transport groups. Since in this case, the momentum, heat, and species are transported by the fluid flow, the mechanism of this process is defined by hydrodynamic laws. Therefore, in each Navier–Stokes equation, the convective group has a structure of inertia force, which in conformity with the second Newton's law is a driving force of the fluid flow. According to that, each of these groups is composed as a product of a mass and acceleration. Since the changes may occur in time and in space, the result is presented as a sum of the four derivatives of velocity times corresponding unit mass: density for time derivative and $\rho u, \rho v$, and ρw for derivatives with respect to x, y, and z. Similarly, the convective groups in energy and mass equations are composed. They consist of the derivatives of temperature or of concentration instead of velocity derivatives and the specific thermal capacity ρc or the mixture density instead of fluid density in the first and the second cases, respectively.

To express mathematically the similarity of transfer processes, we transform system (2.1)–(2.6) to a dimensionless form. Let ϕ is any dependent variable: velocity components, temperature, or concentration transformed by corresponding scale to dimensionless form so that $\phi = (u, v, w)/U, (T - T_\infty)/(T_w - T_\infty)$ or $(C - C_\infty)/(C_w - C_\infty)$, respectively. Introducing as well dimensionless coordinates $\bar{x}, \bar{y}, \bar{z}$ scaled by characteristic length L, we obtain an equation in a dimensionless general form valid for any of the three considering transfer processes:

$$\phi \frac{\partial \phi}{\partial \bar{x}} + \phi \frac{\partial \phi}{\partial \bar{y}} + \phi \frac{\partial \phi}{\partial \bar{z}} = \frac{1}{N} \left(\frac{\partial^2 \phi}{\partial \bar{x}^2} + \frac{\partial^2 \phi}{\partial \bar{y}^2} + \frac{\partial^2 \phi}{\partial \bar{z}^2} \right), \quad N = \frac{UL}{\gamma} \tag{2.7}$$

where γ is the kinematic viscosity ν, thermal diffusivity α, or diffusion coefficient D_m, and dimensionless number $N = \text{Re}$, Pe or $\text{Re}_m = \text{ReSc}$ whereas $\text{Sc} = \nu/D_m$ is the Shmidt number. Equation (2.7) shows that the three transfer processes are so much similar that they may be described by the same equation differing only by dynamic Reynolds, Re, thermal Peclet, Pe, or Reynolds mass, $\text{Re}_m = \text{ReSc}$ characteristic numbers, and in addition by the pressure gradients for the Navier–Stokes and by the dissipation function for energy equations, respectively.

Considering analogy is useful in understanding the mechanism of transfer processes by comparing similar effects. Another advantage of such an idea is the possibility of using results obtained in studying one phenomenon for investigating similar others. In particular, as mentioned above in "Brief history," due to this principle, the basic laws and methods of problem solutions developed in fluid mechanics are used in heat transfer and diffusion theory. In this text, we will often see how such a principle of adaptation works.

Equation (2.7) also indicates that all geometrically similar objects having the same characteristic numbers, Re, Pe, or Re_m, behave similarly if the boundary conditions are identical. This similarity principle first expressed by Reynolds number is important in modeling, especially in experimental investigations because it makes possible to extrapolate dimensionless characteristics obtained on the models to natural objects and systems. For example, the drag coefficient measured on a car or a plane model in a wind tunnel may be used to estimate the resistance force for a real prototype.

2.2.2 Different Forms of Navier–Stokes, Energy, and Diffusion Equations

2.2.2.1 Vector Form

In Section 1.2.1, we use Laplace operator ∇^2 to present conduction equations in vector form (1.5). To present the Navier–Stokes equation in vector form, we need, in addition, the Hamilton operator ∇ called del. or nabla. Using nabla as a vector makes it possible to express the three basic field characteristics: the gradient, the divergence (as a dot product), and the curl (as a cross-product):

$$\nabla = \frac{\partial}{\partial x} \hat{\mathbf{i}} + \frac{\partial}{\partial y} \hat{\mathbf{j}} + \frac{\partial}{\partial z} \hat{\mathbf{k}}, \qquad \text{grad } V = \nabla V = \frac{\partial V}{\partial x} \hat{\mathbf{i}} + \frac{\partial V}{\partial y} \hat{\mathbf{j}} + \frac{\partial V}{\partial z} \hat{\mathbf{k}}$$

$$\text{div} \mathbf{V} = \nabla \cdot \mathbf{V} = \frac{\partial u}{\partial x} + \frac{\partial v}{\partial y} + \frac{\partial w}{\partial z} \tag{2.8}$$

$$\text{curl}\mathbf{V} = \nabla \times \mathbf{V} = \left(\frac{\partial w}{\partial y} - \frac{\partial v}{\partial z}\right)\hat{\mathbf{i}} + \left(\frac{\partial u}{\partial z} - \frac{\partial w}{\partial x}\right)\hat{\mathbf{j}} + \left(\frac{\partial v}{\partial x} - \frac{\partial u}{\partial y}\right)\hat{\mathbf{k}},$$

where $\hat{\mathbf{i}}, \hat{\mathbf{j}},$ and $\hat{\mathbf{k}}$ are outward unit normal vectors in x, y, and z directions. To use these equations, recall that the scalar products of the same unit vectors equal unity, whereas the other scalar products of unit vectors are zero and vice versa: the vector products are equal to zero of the same units but equal to 1 for product of two others, for example, $\hat{\mathbf{i}} \times \hat{\mathbf{j}} = \hat{\mathbf{k}}, \hat{\mathbf{j}} \times \hat{\mathbf{i}} = -\hat{\mathbf{k}}$ and similar others (Exercise 2.5).

Comment 2.2 Nabla is not a usual vector, rather it is a symbolic vector, which simplifies some mathematical operations. Some vector properties are not applicable to nabla (see next example).

Example 2.1 Deriving the expressions for the divergence and continuum equation.
　　Because the divergence is presented as a scalar product (2.8) and the continuity equation (2.1) proved that divergence equals zero, they are determined in vector form respectively as follows:

$$\nabla \cdot \mathbf{V} = \left(\frac{\partial}{\partial x}\hat{\mathbf{i}} + \frac{\partial}{\partial y}\hat{\mathbf{j}} + \frac{\partial}{\partial z}\hat{\mathbf{k}}\right) \cdot \left(u\hat{\mathbf{i}} + v\hat{\mathbf{j}} + w\hat{\mathbf{k}}\right) = \frac{\partial u}{\partial x} + \frac{\partial v}{\partial y} + \frac{\partial w}{\partial z} \qquad \nabla \cdot \mathbf{V} = 0. \qquad (2.9)$$

As an example, when usual vector property is not applicable to nabla, consider the product

$$\mathbf{V} \cdot \nabla = \left(u\hat{\mathbf{i}} + v\hat{\mathbf{j}} + w\hat{\mathbf{k}}\right) \cdot \left(\frac{\partial}{\partial x}\hat{\mathbf{i}} + \frac{\partial}{\partial y}\hat{\mathbf{j}} + \frac{\partial}{\partial z}\hat{\mathbf{k}}\right) = u\frac{\partial}{\partial x} + v\frac{\partial}{\partial y} + w\frac{\partial}{\partial z} \neq \nabla \cdot \mathbf{V}. \qquad (2.10)$$

Thus, the result of dot product with nabla depends on the order, which is in contrast to the property of usual dot product.

Example 2.2 Deriving the expression for the substantial derivative.
　　The sum of time and space derivatives, like that in parentheses on left in each Navier–Stokes equation, determines the time derivative for an observer moving with the flow. Such a derivative is known as the substantial derivative and is used as an operator:

$$\frac{D}{Dt} = \frac{\partial}{\partial t} + u\frac{\partial}{\partial x} + v\frac{\partial}{\partial y} + w\frac{\partial}{\partial z} \qquad \frac{D\mathbf{V}}{Dt} = \frac{\partial\mathbf{V}}{\partial t} + \mathbf{V} \cdot \nabla\mathbf{V}. \qquad (2.11)$$

The spatial part $\mathbf{V} \cdot \nabla\mathbf{V}$ called convective derivative has x, y, and z components. Since they are similar, we show the deriving only for the x-component:

$$[\mathbf{V} \cdot \nabla\mathbf{V}]_x = \left(u\hat{\mathbf{i}} + v\hat{\mathbf{j}} + w\hat{\mathbf{k}}\right) \cdot \left(\frac{\partial u}{\partial x}\hat{\mathbf{i}} + \frac{\partial u}{\partial y}\hat{\mathbf{j}} + \frac{\partial u}{\partial z}\hat{\mathbf{k}}\right) = u\frac{\partial u}{\partial x} + v\frac{\partial u}{\partial y} + w\frac{\partial u}{\partial z}. \qquad (2.12)$$

In terms of field vectors, continuity equation (2.1) and Navier–Stokes equations (2.2)–(2.4) became the compact form:

$$\nabla \cdot \mathbf{V} = 0 \qquad \rho\frac{D\mathbf{V}}{Dt} = \rho\left(\frac{\partial\mathbf{V}}{\partial t} + \mathbf{V} \cdot \nabla\mathbf{V}\right) = -\nabla p + \mu\nabla^2\mathbf{V}. \qquad (2.13)$$

Vector forms of heat and mass transfer equations are similar to the last equation (2.13):

$$\rho c \frac{DT}{Dt} = \rho c \left(\frac{\partial T}{\partial t} + \mathbf{V} \cdot \nabla T \right) = \lambda \nabla^2 T + \mu S \quad \rho \frac{DC}{Dt} = \rho \left(\frac{\partial C}{\partial t} + \mathbf{V} \cdot \nabla C \right) = \rho D_m \nabla^2 C \quad (2.14)$$

2.2.2.2 Einstein and Other Index Notation

$$\frac{\partial V_i}{\partial x_i} = 0 \quad \rho \left(\frac{\partial V_i}{\partial t} + V_j \frac{\partial V_i}{\partial x_j} \right) = -\frac{\partial p}{\partial x_i} + \mu \frac{\partial^2 V_i}{\partial x_j \partial x_j}. \quad (2.15)$$

The idea of index notation known as Einstein's convention came from a sum. Since sum implies a repeated index, one may omit the sign of summation knowing the number of terms. According to Einstein's convention when an variable index appears twice in a single term, it implies that we are summing over all indicated values. In particular, for coordinate components, each index should be repeated three times. Thus, for continuity equation (2.15), we take $i = 1, 2, 3$, and $V_1 = u$, $V_2 = v$, $V_3 = w$, $x_1 = x$, $x_2 = y$, $x_3 = z$. whereas for the Navier–Stokes equation, we should put $i = 1$, $j = 1, 2, 3$, which results in the first equation (2.2). Similar, putting $i = 2$, $i = 3$, and $j = 1, 2, 3$, we obtain second and third Navier–Stokes equations (2.3) and (2.4) (Exercise 2.8).

Other index notations often used are Kronecker delta δ_{ij} and Levi–Civita symbol ε_{ijk}:

$$\delta_{ij} = \begin{cases} 1 & if \quad i = j \\ 0 & if \quad i \neq j \end{cases} \quad \varepsilon_{ijk} = \begin{cases} +1 & if \ (i, j, k) = (1, 2, 3), (3, 1, 2) \, or \, (2, 3, 1) \\ -1 & if \ (i, j, k) = (2, 1, 3), (3, 2, 1) \, or \, (1, 3, 2) \\ 0 & if \ any \ two \ equal \end{cases} \quad (2.16)$$

The first equation (2.16) states that symbol δ_{ij} is 1 if the indices are even ($i = j$) and is zero if they are different ($i \neq j$). The second equation (2.16) indicates that the value of symbol ε_{ijk} is +1 or −1 depending on indices order, or is zero if two of indices are even.

To understand the indices order, start with 1, 2, 3, then move 3 in the front of 1 to get 3, 1, 2 and again move 2 in the front of 3 to get 2, 3, 1. Do the same to obtain the second row of indices. Start once more with 1, 2, 3, put 2 in the front of 1 to get 2, 1, 3, and finally use permutations similar to that just done in the first row to obtain 3, 2, 1 and 1, 3, 2 (Exercise 2.11).

Example 2.3 Applying Einstein and other index notations.

In applying index notation, it is convenient to use instead of unit vectors $\hat{\mathbf{i}}$, $\hat{\mathbf{j}}$, and $\hat{\mathbf{k}}$ unit vectors with numbers, for example, $\hat{\boldsymbol{\delta}}_1, \hat{\boldsymbol{\delta}}_2$, and $\hat{\boldsymbol{\delta}}_3$. Then, the relations between unit vectors [see equation (2.8) and below] may be presented using the index notations as the scalar products $\hat{\boldsymbol{\delta}}_i \cdot \hat{\boldsymbol{\delta}}_j = \delta_{ij}$ and the vector products $\hat{\boldsymbol{\delta}}_i \times \hat{\boldsymbol{\delta}}_j = \varepsilon_{ijk} \hat{\boldsymbol{\delta}}_k$, where $i, j, k = 1, 2, 3$. These notations simplifies the presentation of operators. For instance, the dot and cross-products of two vectors take the following forms:

- The dot product $\mathbf{A} \cdot \mathbf{B} = \sum_{i=1}^{3} A_i B_i = A_{ij} \delta_{ij}$. This simple expression contains nine terms, but only three of them are not zero since only $\delta_{ii} = 1$, $i = 1, 2, 3$.

- The three components of cross-product for $i = 1, 2, 3$ $\mathbf{A} \times \mathbf{B} = \sum_{j=1}^{3} \sum_{k=1}^{3} \varepsilon_{ijk} A_j B_k = \varepsilon_{ijk} A_j B_k$.

In this case, from nine components only two are not zero because only $\varepsilon_{123} = 1$ and

$\varepsilon_{213} = -1$. Two other equations for $i = 2, 3$ are analogous with $\varepsilon_{312} = 1$, $\varepsilon_{321} = -1$ and $\varepsilon_{231} = 1$, $\varepsilon_{132} = -1$.

2.2.2.3 Vorticity Form of the Navier–Stokes equation

This form is usually used for two-dimensional flows for which the stream function ψ may be introduced, and expression (2.8) for curl simplifies to one term:

$$u = \frac{\partial \psi}{\partial y}, \quad v = -\frac{\partial \psi}{\partial x}, \quad \text{curl} = \omega = \left(\frac{\partial v}{\partial x} - \frac{\partial u}{\partial y} \right) = -\left(\frac{\partial^2 \psi}{\partial x^2} + \frac{\partial^2 \psi}{\partial y^2} \right) = -\nabla^2 \psi. \quad (2.17)$$

The stream function introduced in such a way satisfies continuity equation (2.1), and the problem of flow simplifies to the system of Navier–Stokes equations (2.2) and (2.3) that is reduced to one equation as follows. Differentiating equations (2.2) and (2.3) with respect to y and to x respectively yields

$$\frac{\partial^2 u}{\partial t \partial y} + u \frac{\partial^2 u}{\partial x \partial y} + \frac{\partial u}{\partial y} \frac{\partial u}{\partial x} + v \frac{\partial^2 u}{\partial y^2} + \frac{\partial u}{\partial y} \frac{\partial v}{\partial y} = -\frac{1}{\rho} \frac{\partial^2 p}{\partial x \partial y} + v \left(\frac{\partial^3 u}{\partial x^2 \partial y} + \frac{\partial^3 u}{\partial y^3} \right) \quad (2.18)$$

$$\frac{\partial^2 v}{\partial t \partial y} + u \frac{\partial^2 v}{\partial x^2} + \frac{\partial u}{\partial x} \frac{\partial v}{\partial x} + v \frac{\partial^2 v}{\partial x \partial y} + \frac{\partial v}{\partial x} \frac{\partial v}{\partial y} = -\frac{1}{\rho} \frac{\partial^2 p}{\partial y \partial x} + v \left(\frac{\partial^3 v}{\partial x^3} + \frac{\partial^3 v}{\partial x \partial y^2} \right). \quad (2.19)$$

Since mixed derivatives of pressure in both equations are equal, subtracting the first equation from the second eliminates the pressure leading to the Navier–Stokes equation in vorticity form:

$$\frac{\partial \omega}{\partial t} + u \frac{\partial \omega}{\partial x} + v \frac{\partial \omega}{\partial y} = v \left(\frac{\partial^2 \omega}{\partial x^2} + \frac{\partial^2 \omega}{\partial y^2} \right) \quad \text{or} \quad \frac{D\omega}{Dt} = v \nabla^2 \omega. \quad (2.20)$$

The first term of this equation is obtained as a difference between first terms of equations (2.19) and (2.18); the second, third, and the last terms are differences of the second, fourth, and last terms of these equations. Other terms vanish because they can be arranged in the form $\omega \left(\partial u / \partial x + \partial v / \partial y \right)$ where the sum in parentheses is zero (Exercise 2.12).

Equation (2.20) describes the vorticity transport and is useful in studying some general properties of lows. For example, it follows from (2.20) that the vorticity of a fluid flow does not change with time ($D\omega / Dt = 0$). Thus, the initially irrotational inviscid flow remains irrotational in space and time (Sec. 2.2.2.5).

2.2.2.4 Stream Function Form of the Navier–Stokes Equation

Substituting equation (2.17) into equation (2.20) leads to another form of the Navier–Stokes equation containing only stream function as an unknown variable:

$$\frac{\partial \nabla^2 \psi}{\partial t} + \frac{\partial \psi}{\partial y} \frac{\partial \nabla^2 \psi}{\partial x} - \frac{\partial \psi}{\partial x} \frac{\partial \nabla^2 \psi}{\partial y} = v \nabla^4 \psi \quad (2.21)$$

$$\nabla^4 \psi = \nabla^2 (\nabla^2 \psi) = \frac{\partial^4 \psi}{\partial x^4} + 2 \frac{\partial^4 \psi}{\partial x^2 \partial y^2} + \frac{\partial^4 \psi}{\partial y^4}. \quad (2.21)$$

Example 2.4 Determining streamlines in a two-dimensional flow field.

Streamlines are curves plotted in a flow field such that they are tangent to the direction of flow at each point in the flow field. If γ is an angle between velocity components, then $\tan \gamma = v/u = dy/dx$. Thus, the streamline equation is $vdx = udy$ or

$$udy - vdx = 0, \quad \frac{\partial \psi}{\partial x}dx + \frac{\partial \psi}{\partial y}dy = 0, \quad d\psi = 0, \quad \psi = const. \tag{2.23}$$

Here, the second relation is obtained using the stream function definition (2.17), and the other two results follow from the fact that the left-hand side of the second relation determines the exact differential of the stream function. Since ψ is constant along the streamlines, the volume flow rate between two streamlines is determined as difference $\psi_2 - \psi_1$. This follows from the result of integrating the elementary volume rate $d(\dot{m}/\rho) = udy = d\psi$ (Exercise 2.13).

2.2.2.5 Irrotational Inviscid Two-Dimensional Flows

It follows from equation (2.17) that in the case of irrotational flow ($\omega = 0$), the stream function satisfies the Laplace equation. Another useful function satisfying the Laplace equation also exists in this case. This function called potential of a flow field is defined by similar-to-stream-function expression (2.17) $\nabla^2 \varphi = 0$. Then, we have

$$u = \frac{\partial \varphi}{\partial x}, v = \frac{\partial \varphi}{\partial y}, \quad \frac{\partial \varphi}{\partial x} = \frac{\partial \psi}{\partial y}, \frac{\partial \varphi}{\partial y} = -\frac{\partial \psi}{\partial x}. \tag{2.24}$$

The last two relations are known as Cauchy–Riemann conditions for differentiability.

These must be satisfied by any complex function $w(z) = \varphi(x, y) + i\psi(x, y)$ where $z = x + iy$ to be differentiable (analytic). Functions $\psi(x, y)$ and $\varphi(x, y)$ are also harmonic functions since they satisfied the Laplace equation (Exercise 1.46). Hence, such functions describe many physical processes governed by the Laplace equation (e.g., Sec. 1.5.4).

In particular, any analytic function $w(z)$gives functions $\psi(x, y)$ and $\varphi(x, y)$ representing stream functions and velocity potentials of some irrotational inviscid fluid flow. Two families of curves $\psi(x, y) = c_1$ and $\varphi(x, y) = c_2$ constitute the streamlines and equipotential lines, which are orthogonal to each other. To show this, consider the slopes of these curves using results from Example 2.4 and compare them:

$$\left.\frac{dy}{dx}\right|_{\psi=c} = \frac{v}{u} = -\frac{\partial \psi/\partial x}{\partial \psi/\partial y}, \quad \left.\frac{dy}{dx}\right|_{\varphi=c} = -\frac{\partial \varphi/\partial x}{\partial \varphi/\partial y} = \frac{\partial \psi/\partial y}{\partial \psi/\partial x}. \tag{2.25}$$

Here, the last result follows from Cauchy–Riemann conditions (2.24). It is clear that comparing slops are negative reciprocal values, which proves that the streamlines and equipotential lines are orthogonal. The property of these curves is used for graphical presentation and analyzing potential flows. From this property, it also follows that the body surface is one of streamlines (usually taken as $\psi = 0$) because the normal component of the velocity, which coincides with tangent to the equipotential line at the surface is zero.

Comment 2.3 Strongly speaking, the satisfaction of condition (2.24) may not be sufficient, but except some special examples, the functions that we frequently use in applications are analytic.

The characteristics of potential flow may be calculated if the corresponding complex function $w(z)$ (complex potential) is known. Dividing the complex potential in the form $w(z) = w(x+iy) = \varphi(x,y) + i\psi(x,y)$ in real and imaginary parts yields functions $\varphi(x,y)$ and $\psi(x,y)$. Differentiating these functions and using equations (2.24) give the velocity components. The other way to find the velocity components is to compute first the complex velocity as a derivative of complex potential $w(z)$, and then, dividing the result in real and imaginary parts to obtain velocity components according to formula $dw/dz = u + iv$ which follows from expressions (2.24).

The pressure may be determined from the Bernoulli equation using the knowing velocity. Different forms of the Bernoulli equation follow from the Navier–Stokes equations. For the case of a steady, potential flow along the streamline, the Navier–Stokes equation (2.2) simplifies after omitting time-dependent and viscosity terms. In addition, the flow along the streamline may be considered as one-dimensional because in this case, there is only one velocity directed along the tangent to streamline, and the distance changes in the same direction along the streamline as well. Such a simplified equation may be integrated to get the following result (Exercise 2.16):

$$\rho V dV + dp = 0 \qquad \frac{1}{2}\rho V^2 + p = const \qquad \frac{1}{2}\rho\left(u^2 + v^2\right) + p = const. \qquad (2.26)$$

Despite that all real fluids possess viscosity, models based on idealized inviscid fluid give useful and meaningful results for several kinds of problems. At the same time, the assumption of negligible viscosity significantly simplifies the investigation. That is why the potential theory has been extensively developed. Potential theory methods are applicable in studying low-viscosity flows. In particular, such methods are widely used in the boundary-layer theory because potential flows adequately describe the flow outside the boundary layer. In many cases of flow with favorable (negative) pressure gradient, the potential theory description is almost entirely close to reality. For example, the flows through cylindrical and contractive short channels or flows past relatively short plates are cases of this type. The reason of such close results is that in these flows, the viscose forces are small and, what is most important, the flow structures in real and inviscid flows are practically the same. Thus, the potential pattern gives understanding of flow forms, but the energy loses could not be estimated by potential approaches. The situation is completely different in the opposite cases when the pressure gradient is unfavorable (positive) such that the pressure increases in the flow direction as, for example, in diffuser. Flows of this type only at small pressure gradients have continue structure (like in the case of favorable pressure gradient), which becomes destroyed by flow separation as pressure increases. This flow pattern could not be modeled by potential theory. However, until the flow is unseparated, the inviscid flow theory may be used, and even in separated flow, there are parts to which potential methods may be applied [3].

Example 2.5 Potential flow along the wedge of angle $\pi\beta$ (Fig. 2.1) (Exercise 2.17).

The complex potential for the neighborhood of the leading edge is a power function $w = -cz^m$, where $c = const.$ and the exponent is a function of an angle $m = 2/(2-\beta)$. Complex velocity is found by differentiating the complex potential $dw/dz = -cmz^{m-1} = u + iv$. To divide the complex power function in the real and imaginary parts, the formula Moivre is used: $\left[r(\cos\gamma + i\sin\gamma)\right]^m = r^m(\cos m\gamma + i\sin m\gamma)$, where r and γ are polar coordinates. Then, previous relation takes the form $(dw/dz) = -cmr^{m-1}\left[\cos(m-1)\gamma + i\sin(m-1)\gamma\right] = u + iv$.

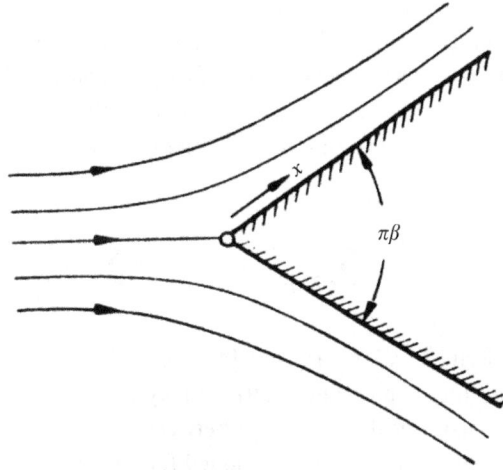

Figure 2.1 Flow past a wage. At the leading edge, the potential velocity is $U = Cx^m$.

From this expression, we find $u = cmr^{m-1}\cos(m-1)\gamma$ and $v = -cmr^{m-1}\sin(m-1)\gamma$. The potential and stream functions are found from equality $\varphi + i\psi = -cr^m(\cos m\gamma + i\sin m\gamma)$ to obtain $\varphi = -cr^m\cos m\gamma$ and $\psi = -cr^m\sin m\gamma$. At the leading edge ($y = 0$), we have (Fig. 2.1)

$$y = \gamma = v = \psi = 0, r = x \quad \text{and} \quad u = c[2/(2-\beta)]x^{\beta/(2-\beta)}. \tag{2.27}$$

2.2.2.6 Curvilinear Orthogonal Coordinates

In Section 1.2.2, we use the Lame coefficients to transform the conduction equation from Cartesian coordinates to orthogonal curvilinear coordinates. Here, we show how the Lame coefficients may be applied to transform the Navier–Stokes equations to curvilinear orthogonal coordinates. Since in this case, the vectors should be transformed, first the unit vectors are determined.

Let the new orthogonal coordinates are $\xi_1(x,y,z)$, $\xi_2(x,y,z)$, and $\xi_3(x,y,z)$. Then, the Cartesian coordinates $x(\xi_1,\xi_2,\xi_3)$, $y(\xi_1,\xi_2,\xi_3)$, and $z(\xi_1,\xi_2,\xi_3)$ are presented some vector field as $\mathbf{R} = x(\xi_1,\xi_2,\xi_3)\hat{\mathbf{i}} + y(\xi_1,\xi_2,\xi_3)\hat{\mathbf{j}} + z(\xi_1,\xi_2,\xi_3)\hat{\mathbf{k}}$ with further denoting unit vectors as $\mathbf{l}_n$ and Lame coefficients l_n, defined by computing a derivative $\partial\mathbf{R}/\partial\xi_n$ [4]:

$$\mathbf{l}_n = \frac{\partial\mathbf{R}}{\partial\xi_n}, \quad l_n = \left|\frac{\partial\mathbf{R}}{\partial\xi_n}\right|, \quad l_n^2 = \left(\frac{\partial x}{\partial\xi_n}\right)^2 + \left(\frac{\partial y}{\partial\xi_n}\right)^2 + \left(\frac{\partial z}{\partial\xi_n}\right)^2 \quad n = 1,2,3. \tag{2.28}$$

This is expression (1.10) that we used to transform the conduction equation to curvilinear coordinates. The expressions for basic operators in curvilinear orthogonal coordinates are formally the same as those in Cartesian ones. However, in this case, the directions of unit vectors depend on the location so that they change from point to point, and hence unit vectors are functions of coordinates. This property should be taken into account in calculations. One way to do it is to express Cartesian units in terms of curvilinear units. Since in this procedure, the units with indices are needed, we use for Cartesian units notation $\hat{\delta}_1, \hat{\delta}_2, \hat{\delta}_3$ instead of $\hat{i}, \hat{j}, \hat{k}$ (Example 2.3). Note that units $\hat{\delta}_i$ in contrast to curvilinear $\hat{e}_n$ are constant as well as usual Cartesian units. We express also the curvilinear units in terms of Cartesian ones using equation (2.28) and vice versa $\hat{\delta}_i$ in terms of $\hat{e}_i$:

$$\hat{e}_n = \frac{l_n}{l_n} = \frac{1}{l_n}\frac{\partial \mathbf{R}}{\partial \xi_n} = \sum_i \frac{1}{l_n}\frac{\partial x_i}{\partial \xi_n}\hat{\delta}_i \qquad \hat{\delta}_n = \sum_i \frac{1}{l_i}\frac{\partial x_n}{\partial \xi_i}\hat{e}_i. \tag{2.29}$$

The last equation may be derived like the first one [4] or by solving for $\hat{\delta}_i$ the system of three linear equations, which the first expression (2.29) actually makes up [5]. The two relations in equation (2.29) look similar, but the difference between these relations becomes clear in using the units with indices. Equation (2.29) are applied for obtaining some basic operators in curvilinear coordinates (Equations 2.19).

Comment 2.4 Some authors used instead of the Lame coefficient the term metric coefficient [6], which is more general [4]. There is also reciprocal to (2.28) definitions of the Lame or metric coefficient [4].

Example 2.6 Deriving the del. operator (nabla) for a scalar function in curvilinear orthogonal coordinates.

Using equation (2.8) for nabla in Cartesian coordinates and the last equation (2.29), one gets a relation, which by applying the chain rule [last equation (2.30)] is transformed to expression for ∇ (Exercise 2.21):

$$\nabla = \sum_n \frac{\partial}{\partial x_n}\hat{\delta}_n = \sum_n \frac{\partial}{\partial x_n}\sum_i \frac{\hat{e}_i}{l_i}\frac{\partial x_n}{\partial \xi_i} = \sum_i \frac{\hat{e}_i}{l_i}\left(\sum_n \frac{\partial x_n}{\partial \xi_i}\frac{\partial}{\partial \xi_n}\right) \frac{\partial}{\partial \xi_i} = \sum_i \frac{\partial x_n}{\partial \xi_i}\frac{\partial}{\partial x_n} \tag{2.30}$$

$$\nabla = \sum_i \frac{\hat{e}_i}{l_i}\frac{\partial}{\partial \xi_i} = \frac{\hat{e}_1}{l_1}\frac{\partial}{\partial \xi_1} + \frac{\hat{e}_2}{l_2}\frac{\partial}{\partial \xi_2} + \frac{\hat{e}_3}{l_3}\frac{\partial}{\partial \xi_3}. \tag{2.31}$$

Employing nabla simplifies obtaining the other operators. The simplest is to get the gradient of the scalar field by inserting some scalar function f in equation (2.31):

$$\nabla f = \sum_i \frac{\hat{e}_i}{l_i}\frac{\partial f}{\partial \xi_i} = \frac{\hat{e}_1}{l_1}\frac{\partial f}{\partial \xi_1} + \frac{\hat{e}_2}{l_2}\frac{\partial f}{\partial \xi_2} + \frac{\hat{e}_3}{l_3}\frac{\partial f}{\partial \xi_3}. \tag{2.32}$$

Deriving the relations for operations containing vectors is similar, but the procedure is lengthy and tedious, especially for the expressions with the derivatives of unit vectors.

For example, the divergence, or del operator of any vector, like gradient of vector field, is determined as

$$\nabla \cdot \mathbf{V} = \sum_i \frac{\hat{e}_i}{l_i}\frac{\partial}{\partial \xi_i}\sum_j \hat{e}_j \mathbf{V}_j = \sum_i \sum_j \frac{\hat{e}_i}{l_i}\left(\hat{e}_j \frac{\partial \mathbf{V}_j}{\partial \xi_i} + \mathbf{V}_j \frac{\partial \hat{e}_j}{\partial \xi_i}\right). \tag{2.33}$$

The expressions of such type containing the products of two vectors or two unit vectors of a curvilinear coordinate system (but not Cartesian coordinates for which the unit vectors are constant) are called dyadic. Those are quantities of the third kind (the two others are scalars and vectors) known as tensors. Tensor is a general term defining the other quantities by order or a degree of tensor. A tensor order depends on the array of numerical values (or indices), which determines the tensor. In this terminology, the scalar is a tensor of zero order, whereas the vector has the order one since the array of three values is necessary to define a vector. The tensor of a second order, for example, dyadic, is defined by nine values.

Due to tedious procedures required for transforming the Navier–Stokes equations to curvilinear coordinates, only relatively simple operators are given here in general form. Expressions (2.34) and (2.35) present such general results for the divergence and for the scalar field Laplace operator:

$$\nabla \cdot \mathbf{V} = \mathrm{div}\mathbf{V} = \frac{1}{l_1 l_2 l_3}\left[\frac{\partial(V_1 l_2 l_3)}{\partial \xi_1} + \frac{\partial(V_2 l_1 l_3)}{\partial \xi_2} + \frac{\partial(V_3 l_1 l_2)}{\partial \xi_3}\right] \qquad (2.34)$$

$$\nabla^2 = \frac{1}{l_1 l_2 l_3}\left[\frac{l_2 l_3}{l_1}\frac{\partial}{\partial \xi_1} + \frac{l_1 l_3}{l_2}\frac{\partial}{\partial \xi_2} + \frac{l_1 l_2}{l_3}\frac{\partial}{\partial \xi_3}\right]. \qquad (2.35)$$

Here, V_n are velocity components in curvilinear coordinates. Recall that the last relation is used for transforming the conduction equation in curvilinear coordinates [Equation (1.11)]. More complicated transforming procedures we consider as examples or exercises for particular coordinate systems.

Example 2.7 Derivation of the Navier–Stokes equation in cylindrical coordinates.

The Cartesian and cylindrical coordinates are related by equations (1.12) $x = r\cos\gamma$, $y = r\sin\gamma$, $z = z$. The inverse procedure yields $r = \sqrt{x^2 + y^2}$, $\gamma = \arctan(y/x)$, $z = z$. The relations between derivatives in both coordinate systems are found using the coordinate definitions and the chain rule (Exercise 2.21):

$$\frac{\partial}{\partial x} = \frac{\partial}{\partial r}\frac{\partial r}{\partial x} + \frac{\partial}{\partial \gamma}\frac{\partial \gamma}{\partial x}, \quad \frac{\partial r}{\partial x} = \cos\gamma, \quad \frac{\partial \gamma}{\partial x} = -\frac{y}{x^2 + y^2}, \quad \frac{\partial}{\partial x} = \frac{\partial}{\partial r}\cos\gamma - \frac{\partial}{\partial \gamma}\frac{\sin\gamma}{r} \qquad (2.36)$$

$$\frac{\partial}{\partial y} = \frac{\partial}{\partial r}\frac{\partial r}{\partial y} + \frac{\partial}{\partial \gamma}\frac{\partial \gamma}{\partial y}, \quad \frac{\partial r}{\partial y} = \sin\gamma, \quad \frac{\partial \gamma}{\partial y} = \frac{x}{x^2 + y^2}, \quad \frac{\partial}{\partial y} = \frac{\partial}{\partial r}\sin\gamma + \frac{\partial}{\partial \gamma}\frac{\cos\gamma}{r}. \qquad (2.37)$$

The Lame coefficients are defined employing equation (2.28), which is the same as (1.10)

$$l_1 = \sqrt{\cos^2\gamma + \sin^2\gamma} = 1, \quad l_2 = \sqrt{r^2(\sin^2\gamma + \cos^2\gamma)} = r, \quad l_3 = 1. \qquad (2.38)$$

Knowing the Lame coefficients, one calculates the nabla for cylindrical coordinates by equation (2.31) and the divergence by equation (2.34), determining the continuity equation (2.40) in cylindrical coordinates from second equation (2.39) (Exercise 2.22):

$$\nabla = \hat{\mathbf{e}}_r \frac{\partial}{\partial r} + \frac{\hat{\mathbf{e}}_\gamma}{r}\frac{\partial}{\partial \gamma} + \hat{\mathbf{e}}_z \frac{\partial}{\partial z} \qquad \nabla \cdot \mathbf{V} = \frac{1}{r}\left[\frac{\partial(rv_r)}{\partial r} + \frac{\partial v_\gamma}{\partial \gamma} + \frac{\partial(rv_z)}{\partial z}\right] \qquad (2.39)$$

$$\frac{\partial v_r}{\partial r} + \frac{v_r}{r} + \frac{1}{r}\frac{\partial v_\gamma}{\partial \gamma} + \frac{\partial v_z}{\partial z} = 0. \tag{2.40}$$

To find the convective derivative (2.12) and Laplacian of velocity for the Navier–Stokes equation, the unit vectors and their derivatives are required. These in terms of Cartesian coordinates as well as the inverse relations are obtained from equations (2.29) and (2.38):

$$\hat{\mathbf{e}}_r = \hat{\boldsymbol{\delta}}_1 \cos\gamma + \hat{\boldsymbol{\delta}}_2 \sin\gamma, \ \ \hat{\mathbf{e}}_\gamma = -\hat{\boldsymbol{\delta}}_1 \sin\gamma + \hat{\boldsymbol{\delta}}_2 \cos\gamma, \ \ \hat{\mathbf{e}}_z = 0 \tag{2.41}$$

$$\hat{\boldsymbol{\delta}}_1 = \hat{\mathbf{e}}_r \cos\gamma - \hat{\mathbf{e}}_\gamma \sin\gamma, \ \ \hat{\boldsymbol{\delta}}_2 = \hat{\mathbf{e}}_r \sin\gamma + \hat{\mathbf{e}}_\gamma \cos\gamma, \ \ \hat{\boldsymbol{\delta}}_3 = 0. \tag{2.42}$$

It follows from expressions (2.41) that the derivatives of all unit vectors with respect to r and z as well as all derivatives of $\hat{\mathbf{e}}_z$ are zero. The two others with respect to γ are as

$$\frac{\partial \hat{\mathbf{e}}_r}{\partial \gamma} = -\hat{\boldsymbol{\delta}}_1 \sin\gamma + \hat{\boldsymbol{\delta}}_2 \cos\gamma = \hat{\mathbf{e}}_\gamma, \ \ \frac{\partial \hat{\mathbf{e}}_\gamma}{\partial \gamma} = -(\hat{\boldsymbol{\delta}}_1 \cos\gamma + \hat{\boldsymbol{\delta}}_2 \sin\gamma) = -\hat{\mathbf{e}}_r. \tag{2.43}$$

The convective derivative is computed applying equation (2.33) for divergence and taking into account that there are only two non-zero derivatives (2.43). The given below first expression (2.44) obtained in this way is modified as follows. After opening the brackets, the result is presented in the form of three parts in conformity with unit vectors $\hat{\mathbf{e}}_r, \hat{\mathbf{e}}_\gamma, \hat{\mathbf{e}}_z$ similar to the Cartesian vector form containing $\mathbf{i}, \mathbf{j}, \mathbf{k}$. Then, applying the Kronecker delta (2.16) to terms inside brackets at each unit vector $\hat{\mathbf{e}}_r, \hat{\mathbf{e}}_\gamma, \hat{\mathbf{e}}_z$, the terms with different unit vectors are neglected giving the second equation (2.44), containing only terms with two identical unit vectors such as $\hat{\mathbf{e}}_r\hat{\mathbf{e}}_r\hat{\mathbf{e}}_r$ or $\hat{\mathbf{e}}_r\hat{\mathbf{e}}_\gamma\hat{\mathbf{e}}_\gamma$:

$$\mathbf{V}\cdot\nabla\mathbf{V} = \left(v_r\hat{\mathbf{e}}_r + v_\gamma\hat{\mathbf{e}}_\gamma + v_z\hat{\mathbf{e}}_z\right)\left[\hat{\mathbf{e}}_r\left(\hat{\mathbf{e}}_r\frac{\partial v_r}{\partial r} + \hat{\mathbf{e}}_\gamma\frac{\partial v_\gamma}{\partial r} + \hat{\mathbf{e}}_z\frac{\partial v_z}{\partial r}\right) + \frac{\hat{\mathbf{e}}_r}{r}\left(\hat{\mathbf{e}}_r\frac{\partial v_r}{\partial \gamma} + \hat{\mathbf{e}}_\gamma\frac{\partial v_\gamma}{\partial \gamma} + \hat{\mathbf{e}}_z\frac{\partial v_z}{\partial \gamma}\right)\right.$$

$$\left. + \hat{\mathbf{e}}_z\left(\hat{\mathbf{e}}_r\frac{\partial v_r}{\partial z} + \hat{\mathbf{e}}_\gamma\frac{\partial v_\gamma}{\partial z} + \hat{\mathbf{e}}_z\frac{\partial v_z}{\partial z}\right) + \hat{\mathbf{e}}_\gamma\hat{\mathbf{e}}_\gamma\frac{v_r}{r} - \hat{\mathbf{e}}_\gamma\hat{\mathbf{e}}_r\frac{v_\gamma}{r}\right] \tag{2.44}$$

$$\mathbf{V}\cdot\nabla\mathbf{V} = \hat{\mathbf{e}}_r\left(v_r\frac{\partial v_r}{\partial r} + \frac{v_\gamma}{r}\frac{\partial v_r}{\partial \gamma} + v_z\frac{\partial v_r}{\partial z} - \frac{v_\gamma^2}{r}\right) + \hat{\mathbf{e}}_\gamma\left(v_r\frac{\partial v_\gamma}{\partial r} + \frac{v_\gamma}{r}\frac{\partial v_\gamma}{\partial \gamma} + v_z\frac{\partial v_\gamma}{\partial z} + \frac{v_r v_\gamma}{r}\right)$$

$$+ \hat{\mathbf{e}}_z\left(v_r\frac{\partial v_z}{\partial r} + \frac{v_\gamma}{r}\frac{\partial v_z}{\partial \gamma} + v_z\frac{\partial v_z}{\partial z}\right).$$

Comment 2.5 One may be confused using the Kronecker delta to product of three unit vectors instead of the Levi–Civita rules (2.16). The answer is that there is no cross-product to which the Levi–Civita rules are applicable.

Adding the time derivatives to each expression in the brackets in (2.44) results in the substantial derivatives, determining the left-hand sides of three Navier–Stokes equations in cylindrical coordinates. The right-hand side of each equation, the Laplacian of the velocity field, may be obtained analogously as a divergence of a velocity gradient$\nabla \cdot \nabla \mathbf{V}$, where ∇ is defined by equation (2.31) and $\nabla \mathbf{V}$ is given by the expression in the second brackets of the first equation (2.44) (Exercise 2.23).

The Navier–Stokes equations in cylindrical and spherical coordinates are given in the Appendix (Exercise 2.24).

Comment 2.6 Elements of vector algebra presented here are not actually for deriving the Navier–Stokes equations in cylindrical coordinates, cited in many books (e.g., [1], [5], [7]); rather, this opportunity is used to provide a reader with some knowledge of the involved methods of vector algebra important for understanding some results of application.

EXERCISES

2.1 What is the D'Alemberts paradox? What is the result of resolving this puzzle?

2.2 Navier–Stokes equations include in addition body forces which usually are gravitational forces. Instead of these here the pressure is defined as that in the flow without hydrostatic term. Think why such a formulation allows us to eliminate gravitational forces from equations in the case of homogeneous flow past body.

2.3 Explain how basic mechanisms of transfer processes differ from each other? How does this difference affects the structure of transfer equations?

2.4 What are a field and a field potential? How does this concept relates to transfer processes? Think about fields and corresponding potentials other than discussed here.

2.5 Why the notations of unit vectors contain the overscores? Are these notation used anywhere without overscores?

2.6 What is nabla? How is it used? Derive the formula (2.8) for curl using nabla.

2.7 What is a substantial derivative? How does it differ from convective derivative?

2.8 Where the idea of Einstein notation comes from? How do these notations work?

2.9 Explain to a friend what are Kronecker delta and Levi–Civita symbol.

2.10 Are there any differences in using unit vectors $\hat{\mathbf{i}}, \hat{\mathbf{j}}, \hat{\mathbf{k}}$ and $\hat{\boldsymbol{\delta}}_1, \hat{\boldsymbol{\delta}}_2, \hat{\boldsymbol{\delta}}_3$?

2.11 Find the mistakes in the relation $\varepsilon_{ijk} = \varepsilon_{kij} = \varepsilon_{jki} = 1$ if there are any.

2.12 Prove that there are groups of terms equals zero in equation obtained by subtracting equations (2.18) and (2.19).

2.13 Why any streamline may be taken as a streamlined surface? Hint: see equation (2.25).

2.14 What is the difference between general and irrotational two-dimensional flows? What an additional equation satisfies an irrotational flow?

2.15 What does the term "analytic function" mean? What conditions must satisfy an analytic function?

2.16 Explain how the complex potential and complex derivative are used for determining the inviscid flow characteristics. Derive the Bernoulli equation from Navier–Stokes equations.

2.17 Find the velocity components and field characteristics (φ and ψ) for potential flow inside right angle knowing that complex potential is $w = z^2$. Read article "Potential flow" from Wikipedia to see other potential flows and streamlines and equipotential lines fields.

2.18 What types of flows may be studied by invuscid flow models? Why these models are applicable to one part of problems but could not be used in others?

2.19 What is the difference between unit vectors in curvilinear and Cartesian coordinates? Explain why the procedures in curvilinear coordinates are more complicated than that in Cartesian ones.

2.20 What is a tensor? Explain how the order of a tensor is determined. What are the relations between tensors of different orders?

2.21 Study or recall the chain rule for calculating the derivatives of the functions, which are depended on two others. Compute the relations between derivatives in spherical and Cartesian coordinates [equations (2.36) and (2.37)].

2.22 Derive the Lame coefficients and divergence for spherical coordinates using the equations (2.28) and (2.34). Compare the continuity equations in spherical and cylindrical coordinates.

2.23 Repeat the derivation of equation (2.44) for convective derivative in cylindrical coordinates.

2.24 List and explain ideas used in different presentations of the Navier–Stokes equations.

2.3 Initial and Boundary Conditions

2.3.1 Navier–Stokes Equations

The Navier–Stokes equation is a nonlinear, elliptic-type partial differential equation of the first order in time and of the second order in space. The solution of such an equation should satisfy one initial condition and boundary conditions specified around the contour of the problem domain. An initial condition does not have any specialties and hence as usual defines the velocity distribution in the system as a function of coordinates at some instant, which is taken to be a beginning of the process (Sec. 1.2.4). In contrast to that, boundary conditions formulation for the Navier–Stokes equation is a challenging task theoretically not solved yet. The major problem consists of the boundary condition of no-slip (zero velocity components) on the surface. This condition was suggested by establishers of the Navier–Stokes equation first by Navier (1827) and Poisson (1831) on the base of considering the intermolecular forces and later by de Saint Venant (1843) and Stokes (1845) using the assumption of linearity between the rate of deformation and stresses.

Today, after more than 150 years, we could not exactly prove that the Navier–Stokes equation with such a boundary condition describes correctly the fluid motion. The reason of this is that due to huge mathematical difficulties, there is not even a single exact analytical solution of the full Navier–Stokes equations. Such a solution is necessary because only comparison of results obtained using the exact solution with corresponding experimental data gives the possibility of achieving the desired proof. Nevertheless, the available information, in particular, the analytical solution of simple problem like flow in a channel or in a tube, analytical solution of other reduced Navier–Stokes equations, the numbers of numerical solution of full Navier–Stokes equations, results in the boundary-layer theory so well agree with experimental data that there are no reasons to be in doubt about validity of the Navier–Stokes equation and no-slip boundary condition [1].

Another difficulty in defining the boundary conditions arises due to the elliptic type of the Navier–Stokes equation. As was mentioned above, the elliptic equation requires the boundary conditions on all boundaries of the problem domain. To understand more specifically arising difficulties, consider a flow through a two-dimensional channel or a circular tube of finite length embedded in a plane-parallel infinite stream. In such a case, the problem domain consists of a channel or tube walls and corresponding entrance and exit sections. Hence, the boundary condition should be designed along the perimeter of that rectangular. This means, that, in particular, for the Dirichlet problem (Sec. 1.5.4), one should specify the velocity along the walls and across the sections at entrance and exit.

Indeed in this case, only zero velocity on the walls are known (in conformity with above discussion) and stream velocity far away from a channel or tube, strictly speaking at $x \to \infty$ before embedded body and at $x \to -\infty$ behind it. Since such a body disturbs the surrounding fluid, the flow becomes disturbed the more, the closer it approaches the channel or tube entrance. Therefore, the velocity distribution across the incoming section differs from the known uniform distribution far away from the embedded body. It is obvious that the velocity distribution across the outgoing section differs from the same uniform velocity profile because it results from processes inside the channel or tube affecting the flow characteristics. The same pattern takes place in the case of external infinite stream past embedded finite size body, such as slab or cylinder, resulting in disturbed velocity profiles at the leading and trailing edges.

It follows from above discussion that the velocity distributions required for the boundary conditions formulation in solving the Navier–Stokes equation are usually unknown a priori. Hence, in such a case, one needs additional information about velocity profiles at the beginning and exit of the streamlined subject (Example 2.8).

2.3.2 Specific Issues of the Energy Equation

Despite that the energy equation is linear, the basic problems with formulation of the boundary conditions for energy equation are the same as just considered difficulties arising in solving the Navier–Stokes equation. There are two major reasons for that: (i) both equations are of the same elliptic-type partial differential equations and (ii) the same disturbance mechanisms of velocity and temperature fields before and behind the body. Therefore, the disturbed temperature profiles at the leading and trailing edges of the embedded object are unknown in advance as well, and additional experimental information is needed to create the boundary conditions.

In contrast to that, the boundary conditions on the channel walls or on the body surface for the energy equation are inherently different from those for the Navier–Stokes equation. To understand the physical reasons of this difference, note that the wall or body surface is an interface of a fluid and solid, which is a border of property changes. Since across this border the viscosity changes from the finite value of fluid to infinite of solid, the fluid velocity changes from some particular value to zero on a solid surface. The changes of thermal characteristics across the interface are different because both conductivities of fluid and of solid are finite. This yields two different unknown temperatures and heat fluxes in a fluid and in a solid. Determining these characteristics, including temperature distribution along the interface, is a complex problem because it can be done only by considering two conjugated solutions obtained for a solid and for a fluid. If the solution for the solid and fluid are known, one uses the conjugate condition (1.25), calculate the required derivatives from both sides of interface and, since the temperatures from both sides of interface are equal, obtains the equation, determining the temperature distribution along the interface.

Unfortunately, the solution of both equations for solid and especially for fluid plus their conjugation is a difficult task, making the whole procedure laborious. That is why even simple conjugate problems began to be considered only in 1960s when the computers came into existence.

Comment 2.7 It was said that since the solid viscosity is infinite, the no-slip boundary condition may be expected. This is not always the case. The low density gas slips on the body surface so that the no-slip boundary condition does not hold. Such an effect occurs when the Knudsen number $\mathrm{Kn} = l/L$ is of order unity or greater. The Knudsen number is defined as a ratio of the molecular mean free path l (average distance that molecule travels between collisions with other moving molecules) to characteristic scale length L.

Examples 2.8 Boundary conditions in two Dirichlet problem solutions:
(1) Horizontal channel heated from below [8].
Two-dimensional Navier–Stokes and energy equations with no-slip and conjugate boundary conditions on the interface are solved. The parabolic velocity profile and ambient temperature at the entrance are assumed and experimentally checked. To take into account the recirculation effects at the exit section, some experimental known data were adopted. The following relations are used: for velocity $\partial u/\partial x = \mathrm{v} = 0$, and for temperature $\theta = T - T_\infty = 0$ if $u < 0$ (inflow in the channel), $\partial \theta/\partial x = 0$ if $u > 0$ (outflow).

(2) Rectangular slab heated from one surface with isolated others [9].
Two-dimensional Navier–Stokes equations in variables $\psi - \omega$ and energy equation in variable $\theta = (T - T_\infty)/(T_0 - T_\infty)$ are solved; T_0 is the temperature of a heated surface. No-slip and conjugate conditions are used. Velocity and temperature distributions before and behind the slab far away from it are considered as uniform. Special investigations were performed to estimate the perturbed values of ψ, ω, and θ close to the slab (Exercises 2.25–2.29).

2.4   Exact Solutions of Navier–Stokes and Energy Equations

There are some problems when the Navier–Stokes equation simplifies, and it becomes possible to obtain the exact solution. The importance of such solutions follows from the above discussion. Here, we consider 13 known exact solutions combining similar solutions in two groups. A large group consists of simple unsteady and steady problems when the nonlinear inertia terms identically vanish.

2.4.1   Two Stokes Problems

As was mentioned in historical notes, Stokes was the first to give the two exact solutions of this type. The first solution presents a flow near a plate suddenly accelerated from rest in its own plane. The second Stokes problem describes the flow near an infinite plate, which harmonic oscillates parallel to itself. Both problems are governed by the same simplified Navier–Stokes equation containing only unsteady and viscous terms. The boundary conditions in the first problem specified the starting plate velocity U_0 and zero velocity far from the plate. Similarly, the plate harmonic oscillations are prescribed by boundary condition for the second problem [last equation (2.45)]:

$$\frac{\partial u}{\partial t} = v\frac{\partial^2 u}{\partial y^2}, \quad t \le 0, u = 0, t > 0, u(0) = U_0, u(\infty) = 0; \quad u(0,t) = U_0\cos nt. \quad (2.45)$$

Differential equation (2.45) is the same as one-dimensional conduction equation (1.4) without a source. Thus, the solution of the first problem is given as $u/U_0 = erfc(y/2\sqrt{vt})$ [equation (1.27)]. The solution of the second problem should satisfy the same equation and two boundary conditions: at $y = 0$, due to the no-slip condition, the flow should follow the last equation (2.45), defining the plate oscillations, and at $y \to \infty$, flow velocity should approach asymptotically zero. As a result, one obtains $u(y,t) = U_0\exp(-\xi)\cos(nt - \xi)$ where $\xi = \sqrt{n/2v}$.

2.4.2 Solutions of Three Other Unsteady Problems

The third exact solution satisfying differential equation (2.45) is the unsteady Couette motion, which is a flow in channel with one moving wall. The solution should satisfy the following boundary conditions: $t \le 0, u = 0, t > 0, u(0) = U_0, u(H) = 0$, where H is the distance between channel walls. The solution is found using the superposition principle and is presented in a form of series of error function (1.27). One more exact solution describing the unsteady motion starting from the rest presents the flow of fluid in an infinite long circular tube. The fluid being at rest suddenly is subjected to pressure gradient and starts to move. As a result, the initial uniform velocity profile deforms asymptotically reaching the parabolic Hagen–Poiseuille profile (2.49). The solution of this problem is given by differential equation involving the Bessel functions.

The more general unsteady exact solution satisfying the Navier–Stokes equation including the inertia term and boundary condition $U(t)$ (2.46) is presented as follows:

$$\frac{\partial u}{\partial t} + v_0\frac{\partial u}{\partial y} = \frac{\partial U}{\partial t} + v\frac{\partial^2 u}{\partial y^2}, \quad U(t) = U_0\left[1 + f(t)\right], \quad u(y,t) = U_0\left[\zeta(y) + g(y,t)\right]. \quad (2.46)$$

Here, $U(t)$ is the velocity of external flow and v_0 is the constant suction velocity. Subsidiary functions in equation (2.46) are specified for external flow in two cases: for damped and undamped oscillations and for step and linear changes of velocity values.

The other part of this group of simple exact solutions represents steady flows.

2.4.3 Steady Flow in Channels and in a Circular Tube

Parallel flows containing only longitudinal velocity component in a plane channel and steady Couette flow are described by a simple differential equation such as equation (2.45) but with a pressure gradient instead of an unsteady term. Such an equation, boundary conditions, and solutions are given below:

$$\frac{dp}{dx} = \mu\frac{d^2 u}{dy^2}, \quad y = \pm H, u = 0, \quad y = 0, u = 0, y = 2H, u = U \quad (2.47)$$

$$u = -\frac{1}{2\mu}\frac{dp}{dx}\left(H^2 - y^2\right) \quad u = \frac{y}{2H}U - \frac{H^2}{\mu}\frac{dp}{dx}\frac{y}{H}\left(1 - \frac{y}{2H}\right). \tag{2.48}$$

Here, $2H$ is the distance between channel walls. The first solution (2.48) determines the parabolic velocity profile in the plane channel, whereas the second one shows how the pressure gradient in Couette flow deforms the linear velocity distribution $u = (y/2H)U$, which exists at a zero pressure gradient. According to the calculation, the decreasing pressure in the flow direction leads to positive velocity profile over the whole channel cross-section, whereas the increasing pressure results in the profile containing negative velocity parts near the unmoving wall, which represents the back flow regions (Exercises 2.30, 2.31).

An important exact solution of the Navier–Stokes equation is the flow in circular tube, which is known as the Hagen–Poiseuille flow. This result was obtained in 1930s of nineteenth century from experimental data, and later it was shown that theoretically found inverse proportionality of resistance coefficient ζ on the Reynolds number well agrees with experimental data (Fig. 5.3 in [1]). The dependence $\zeta\,(\mathrm{Re})$ follows from the velocity distribution $u(r)$, which may be derived using the reduced Navier–Stokes equation in cylindrical coordinates and simple boundary condition $u = 0, r = R$ (Exercises 2.32):

$$\mu\left(\frac{d^2u}{dr^2} + \frac{1}{r}\frac{du}{dr}\right) = \frac{dp}{dx}, \quad u(r) = -\frac{1}{4\mu}\frac{dp}{dx}\left(R^2 - r^2\right), \quad \zeta = -\frac{dp}{dx}\frac{4R}{\rho\bar{u}^2} = \frac{64}{\mathrm{Re}}. \tag{2.49}$$

Here, $\mathrm{Re} = 2\bar{u}R/\nu$ and $\bar{u}$ is the mean velocity.

The one more simple exact solution of the Navier–Stokes equation of this group presents the steady flow between two rotating cylinders moving in different directions. The two limiting cases are considered: (i) inner cylinder is at rest, while the other rotates and (ii) the one cylinder rotates in the infinite flow. A comparison is made to the non-steady exact solution of the Navier–Stokes equation for decaying vortex.

The other four exact solutions considered in this text composed a second group, describing the steady flows by nonlinear Navier–Stokes equations containing both inertia and viscosity terms. Since such equations could not be solved analytically, these partial differential equations are reduced to ordinary differential equations in the form applicable for numerical solution and tabulation. To obtain such forms of ordinary differential equation, the given governing equations are transformed using special variables known as similarity variables. One example of similarity variable we considered in Section 1.3, where it is shown that one-dimensional conduction partial differential equation (1.4) is reduced to ordinary one (1.33) applying variable $z = x/2\sqrt{\alpha t}$. The same variable is used in the solution of the first Stokes problem. Physically, this means that velocity distributions for different times and locations are similar and construct one curve $\theta(z)$ in variables (1.33) or curve $u/U_0 = erfc(z)$ in Stokes problem in variables u/U_0 and z. That is why such variables and others of that type are called similarity variables. Another example of similarity relation is the just-considered dependence (2.49) $\zeta\,(\mathrm{Re})$. This expression is universal because using the similarity variable, the Reynolds number gives universal curve for the resistance coefficient instead of many dependences for each combinations of $\bar{u}, R$, and ν (Exercises 2.33).

2.4.4   Stagnation Point Flow (Hiemenz Flow)

The first from the group of the four we consider the exact solution describing the stagnation flow. The two-dimension flow arrives in a perpendicular direction to a plate and impinges on at the point, which is taken as origin $x = y = 0$. Since the potential flow slips at the wall, it leaves in both directions along the plate so that the velocity components of potential flow are proportional to corresponding coordinates. Then, the Bernoulli's equation (2.26) gives the pressure:

$$U = cx, \quad V = -cy, \quad p_o - p = (1/2)\rho(U^2 + V^2) = (1/2)\rho c^2 (x^2 + y^2). \tag{2.50}$$

It is assumed that the solution of the Navier–Stokes equation has the form $u = xf'(y)$ and $v = -f(y)$. Such a form of solution satisfied the continuity equation (2.1) because from the first equation (2.50) follows that $\psi = xf(y)$. It is easy to check that substituting u and v in the first Navier–Stokes equation (2.2) gives the ordinary differential equation $f'^2 - ff'' = c^2 + \nu f'''$. This nonlinear equation could not be solved analytically. To solve this equation numerically, one should try to eliminate two constants c and ν. Otherwise, the calculation and tabulation of results should be performed for each pair of constants c and ν. One way to find the proper similarity variables is to introduce new constants putting $\eta = c_1 y$ and $\phi = c_2 f$. Equating coefficients in the previous equation after substituting new variables, one gets two algebraic equations $(c_1/c_2)^2 = c^2 = (c_1^3 \nu/c_2)$, obtains constants c_1 and c_2 solving these equations, and finally, gets the similarity variables, desired form of differential equation without constants and proper boundary conditions:

$$\eta = \sqrt{\frac{c}{\nu}} y, \phi = \frac{f}{\sqrt{c\nu}}. \quad \phi''' + \phi\phi'' - \phi'^2 - 1 = 0, \quad \eta = 0, \phi = \phi' = 0, \quad \eta \to \infty, \phi' \to 1. \tag{2.51}$$

Last condition (2.51) is found as follows: $\phi' = d\phi/d\eta = (c_2/c_1)df/dy = (1/c)f'(y)$. Then, taken into account that $u = xf'(y)$ and $U = cx$, we obtain $\phi' = u/U$.

The solution of this problem was obtained in 1911 by Hiemenz in his thesis and is known as the Hiemenz flow. Equation (2.51) was tabulated by Howarth (Table A.2). At $\eta = 2.4$, the ratio $\phi' = u/U = 0.99$, which approximately determines the edge of the boundary layer, where the viscosity effects are significant. Thus, the distance from the wall $\delta = 2.4\sqrt{\nu/c}$ defines the boundary-layer thickness with an accuracy of $\approx 1\%$. It follows from this expression that boundary-layer thickness is proportional to the square root of kinematic viscosity, and therefore the boundary layer is small for many low-viscosity common fluids, in particular, for water, air, and oil (Exercise 2.34).

Comment 2.8 The stagnation flow occurs at blunt nose of any cylindrical body in the neighborhood of stagnation point.

2.4.5   Other Exact Solutions

Some other well-known exact solution of this type are: (i) three-dimensional stagnation flow; (ii) the flow induced by a disk rotating about axis perpendicular to its plane in a fluid initially being at rest; and (iii) flow in convergent and divergent channels. For each case, the similarity variables are given, and partial differential Navier–Stokes equations are reduced to the system of ordinary differential equations in similarity variables. These systems are numerically solved, and results are tabulated. The basic results are as follows: (i) the velocity distributions in two- and three-dimensional stagnation flows differ slightly; (ii) the formula for the turning moment well agrees with experimental data for both laminar and turbulent regimes; and (iii) the velocity distribution in convergent and divergent channels differ markedly showing regions with beck flows in divergent channels (Exercise 2.35).

2.4.6 Some Exact Solutions of the Energy Equation

Since the simple exact solutions of the energy equation are of the same type as just-discussed solution of the Navier–Stokes equation, we consider only several examples to show some special effects observed in thermal problem solutions.

2.4.6.1 Couette Flow in a Channel With Heated Walls

Let the temperatures of resting and moving walls are T_0 and $T_1 > T_0$, respectively. In the absence of the pressure gradient when the velocity distribution (2.48) contains only the first linear term $u = (y/2H)U$, the dissipation function (2.5) simplifies to one term $(du/dy)^2$. In this case, the energy equation simplifies as well because one velocity component (u) depends only on y, whereas the other (v) is zero. As a result, the energy equation reduces to a form containing only one viscous term and one term defining the dissipation function. For the considering Couette flow with a linear velocity profile this equation and its solution satisfying the wall temperatures mentioned above are

$$\lambda \frac{d^2 T}{dy^2} + \mu \left(\frac{du}{dy} \right)^2 = \lambda \frac{d^2 T}{dy^2} + \mu \frac{U^2}{4H^2} = 0, \quad \frac{T - T_0}{T_1 - T_0} = \eta + \eta (1 - \eta) \frac{\mu U^2}{2\lambda (T_1 - T_0)}, \quad (2.52)$$

where $\eta = y/2H$. It follows from this result that fluid cools the more heated moving wall only until $(\mu U^2 / 2\lambda) < (T_1 - T_0)$, with further increase in velocity u, the fluid heats the moving wall despite $T_1 > T_0$. The reason is that the heat generated due to friction exceeds the effect produced by cooling fluid.

2.4.6.2 Adiabatic Wall Temperature

The other specific thermal effect leading to so-called adiabatic wall temperature may be analyzed using the solution of equation (2.52) subjected to the boundary condition for thermally isolated wall. For example, considering an isolated unmoving wall $(dT/dy = 0)$ and the other at constant temperature T_0, we have the following boundary conditions, corresponding solution of differential equation (2.52), and the value of adiabatic temperature T_{ad}:

$$y = 0, \frac{dT}{dy} = 0, \, y = 2H, T = T_0, \quad T(y) - T_0 = \mu \frac{U^2}{2\lambda} \left(1 - \frac{y^2}{4H^2} \right), \quad T_{ad} = T_0 + \mu \frac{U^2}{2\lambda}. \quad (2.53)$$

The last result is obtained from the temperature distribution at $y = 0$, that is, for an isolated wall. This expression shows that adiabatic is the temperature that thermally isolated surface reaches when the whole released friction heat is adopted by fluid because the wall is isolated. Comparing formula (2.53) with the just-obtained condition of the cooling–heating process for a hot wall, one sees that fluid cools the wall until it reaches the adiabatic temperature so that the inequalities $\Delta T_w >$ or $<$ than ΔT_{ad} determine what process, cooling or heating, exists. Here, Δ in general is the difference from reference temperature (in this example T_0), whereas according to (2.53) $\Delta T_{ad} = \mu U^2 / 2\lambda$.

2.4.6.3 Temperature Distributions in Channels and in a Tube

In the case of equal walls temperatures $(T_0 = T_1)$, equation (2.52) gives the symmetrical temperature distribution. Analogous symmetrical temperature distribution is obtained from the same differential equation (2.52) for the Poiseuille flow through a channel with flat walls. Both temperature profiles for Couette and Poiseuille flows are represented by parabolas of the second and of the four degrees, respectively:

$$T(y) - T_0 = \frac{\mu U^2}{4\lambda} \frac{y}{H} \left(1 - \frac{y}{2H}\right), \qquad T(y) - T_0 = \frac{\mu u_m^2}{3\lambda} \left[1 - \left(\frac{y}{H}\right)^4\right]. \tag{2.54}$$

As is mentioned above, the last expression is a result of solution of the same equation (2.52) but with the other second term. This term $\mu (du/dy)^3 = 4\mu (u_m y)^2 / H^4$ that accounts for dissipation heat and corresponds to parabolic velocity distribution (2.48) in the flat channel with Poiseuille flow and maximum velocity on the symmetry axis $u_m = -\left(H^2/2\mu\right)(dp/dx)$.

As the last example of exact solution of this type, we mention the temperature distribution in the converge channel with velocity distribution considered above (Sec. 2.4.5). Although both velocity and temperature distributions are similar at the central part of the channel cross-section, the temperature profile exhibits the boundary-layer effect showing increasing temperature near the walls. This effect occurs owing to energy dissipation and becomes pronounced as the Prandtl number increases.

Comment 2.9 One may be confused by a seeming contradiction: first, it is said that there is no exact solutions of the Navier–Stokes equation, and then some exact solutions are discussed. The answer is that the first statement pertains to solutions of the Dirichlet problem of the full Navier–Stokes equation, whereas the considered exact results are simple solutions of different parts of the Navier–Stokes equation. The solution of the Dirichlet problem requires a whole process: entrance, development, and exit as, for example, some numerical solutions presented in Example 2.8, whereas simple solutions consists only of a part of whole problem as, for instance, Couette or Poiseuille flows, or Stokes solutions. Indeed, we study simple solutions because we could not obtain the exact ones. Nevertheless, there is no doubt of usefulness of exact solutions considered above, and we will return to this question in discussing the relation between numerical and analytical methods (Exercise 2.36).

2.5 Cases of Small and Large Reynolds and Peclet Numbers

The Navier–Stokes equations as well as the energy and mass transfer equations consist of two basic groups of terms representing the molecular and convective transfer mechanisms as was discussed in Section 2.2.1. It can be shown that the two limiting cases of small and large Reynolds or Peclet numbers correspond to situation when one of those groups of terms is negligibly small relative to another. In particular, it follows from the Navier–Stokes equations that the inertia terms determining the rate of convective momentum transfer are proportional to the square of the velocity components, whereas the viscous terms which specifies another molecular part of the transfer process are proportional only to the first power of velocity. Proceeding from this fact, it is easy to understand that in the case of small Reynolds numbers, the inertia terms may be neglected, whereas in the opposite case of large Reynolds numbers, the viscous terms are negligibly small relative to inertia ones. Analogously considering the energy equation, one sees that the convective group of terms contains the velocity components as factors, but the terms presenting heat conduction are independent on velocity. Thus again, the conductive terms are dominated at small Peclet numbers, whereas the convective group may be omitted in this case, and vice versa under large Peclet numbers (Exercise 2.38).

The simplified equations obtained by this approach in contrast to reduced exact equations considered above are approximate because in this case, the terms are omitted not naturally due to physically simplicity of a problem, but are neglected causing some inaccuracy. To estimate the error arising in such a simplification, dimensionless equation (2.7) may be used. Since in this equation the derivatives with respect to three coordinates are similar, we use for an approximate error evaluation the one-dimensional version. Writing one-dimensional equation (2.7) in two forms yields a connection between the values of characteristic number N and order of errors in two limiting cases:

$$N\phi\frac{\partial\phi}{\partial\overline{x}} - \frac{\partial^2\phi}{\partial\overline{x}^2} = 0, \quad \phi\frac{\partial\phi}{\partial\overline{x}} - \frac{1}{N}\frac{\partial^2\phi}{\partial\overline{x}^2} = 0, \quad \phi\frac{\partial\phi}{\partial\overline{x}} \approx \frac{1}{N}, \quad \int_0^\varepsilon \phi d\phi = \int_0^\varepsilon \frac{d\overline{x}}{N}, \quad \varepsilon \approx \frac{1.4}{\sqrt{N}} \quad (2.55)$$

It follows from the first equation that the limiting case when the molecular transport is dominant is achieved when the characteristic number is small $N \ll 1$, resulting in neglected convective derivative. In the other limiting case, the number N should be large in the first equation, or inverse value of it has to be small neglecting the molecular part in the second equation. The last condition may be expressed also in the third form, which on integration gives an approximate error evaluation for large N shown in (2.55). The estimation error obtained by the analysis presented here is on the same order as known more accurate results. In particular, according to Van Dyke [10], the error in evaluation of the friction coefficient obtained by the boundary-layer theory is $1.7/\sqrt{\mathrm{Re}_x}$ (Exercises 2.39 and 2.40).

Comment 2.10 The purpose of this preliminary approximate analysis is to show the significance of the Reynolds and Peclet numbers, defining the structure of simplified transfer equations. We derive and discuss the real equations for small and large dimensionless numbers in the following two sections.

2.5.1 Creeping Approximation (Small Reynolds and Peclet Numbers)

Creeping flow is a very slow motion at small Reynolds number (Re ≪ 1), that is, at small velocity, or small object size, or in very viscous fluid. As mentioned above, in this case, the inertia terms are small, and they may be neglected significantly simplifying the Navier–Stokes equations to the form $\nabla p = \mu \nabla^2 \mathbf{V}$ for a steady flow. This type of flow is also known as Stokes flow since he was the first one who considered a flow of such type. The energy and diffusion equations may be simplified to the Laplace equations analogously neglecting small convective terms.

The two-dimensional creeping flow equations may be transformed to a single equation for the stream function, such as general Navier–Stokes equation (2.21). Neglecting the inertia terms (the left-hand side of this equation), one obtains the streamlined form of the Stokes flow equation. Since in this case, the pressure is defined by Laplace equation, the creeping flow problems are governed by the Laplace and simplified (2.21) equations, while the similar thermal and diffusion problems are governed as mentioned above only by the Laplace equation. Thus, we have

$$\nabla^4 \psi = 0, \quad \nabla^2(\nabla^2 \psi) = \frac{\partial^4 \psi}{\partial x^4} + 2\frac{\partial^4 \psi}{\partial x^2 \partial y^2} + \frac{\partial^4 \psi}{\partial y^4} \quad \nabla^2 p = 0 \quad \nabla^2 T = 0 \quad \nabla^2 C = 0. \quad (2.56)$$

To see that the pressure is a harmonic function in this case, develop a Laplacian by differentiating two-dimensional creeping flow equations and show using the continuity equation that additional terms in the equation obtained equal zero, resulting in the Laplace equation for pressure (Exercises 2.41 and 2.42). Creeping flow is also irrotational because from equations (2.17) and (2.56) follows that $\omega = \nabla^2 \psi = 0$.

The creeping approximation approach has wide applications for small objects from a charging electron to pollution of environment. In nature, this type of flow occurs in the moving bacteria and other microorganisms, in the motile sperm, and in a lava flow. In engineering, the application includes polymer and other high-viscosity substances flows, film production, studying of small size, and macro- and nano-technology systems.

2.5.1.1 Stokes Flow Past a Sphere

This problem of parallel flow past a sphere was as well first considered by Stokes. Since a sphere with origin at a centre is axisymmetric, it is convenient to apply the polar coordinates (r, γ) in the meridional plane $\phi = const.$ of the spherical coordinate system (r, γ, ϕ). Determining the vorticity ω_ϕ and velocity components [see equation (2.59)] in the polar coordinates leads to expression for the stream function:

$$\omega_\varphi = \frac{1}{r}\left[\frac{\partial(r v_r)}{\partial r} - \frac{\partial v_\gamma}{\partial \gamma}\right], \quad \left[\frac{\partial^2 \psi}{\partial r^2} + \frac{\sin \gamma}{r^2}\frac{\partial}{\partial \gamma}\left(\frac{1}{\sin \gamma}\frac{\partial \psi}{\partial \gamma}\right)\right]^2 = 0. \quad (2.57)$$

The first expression (2.57) is found as a $\hat{\mathbf{e}}_\phi$ component of cross-product in the r, γ plane:

$$\omega_\phi = \nabla \times \mathbf{V} = \left(\hat{\mathbf{e}}_\mathbf{r}\frac{\partial}{\partial r} + \frac{\hat{\mathbf{e}}_\gamma}{r}\frac{\partial}{\partial \gamma}\right) \times \left(\hat{\mathbf{e}}_\mathbf{r} v_r + \hat{\mathbf{e}}_\gamma v_\gamma\right) + \hat{\mathbf{e}}_\gamma \hat{\mathbf{e}}_\gamma \frac{v_r}{r} - \hat{\mathbf{e}}_\gamma \hat{\mathbf{e}}_r \frac{v_\gamma}{r}. \quad (2.58)$$

The two additional terms in this relation came from the unit vectors derivations like those in the first equation (2.44). Simple algebra and the Levi–Civita rules (Example 2.3) result in the first expression (2.57) (Exercise 2.43). To obtain the second expression (2.57) for the stream function, one should substitute the velocities v_r and v_γ from (2.59) into first equation (2.57) and then put the result into the Laplace equation $\nabla^2 \omega = 0$ (Exercise 2.44).

Comment 2.11 The equation (2.57) obtained in such a way contains exponent 2 on the brackets that comes from ∇^2. Therefore, this exponent is not usual square power; rather, it means that this equation should be used twice: first putting some expression defining ψ and then substituting the result obtained in the first step.

Equation (2.57) for the stream function governed the problem in question. The boundary conditions should guarantee no slip on a sphere surface ($v_r = v_\gamma = 0$) and provide passing the solution into velocity U of parallel flow far away from the sphere. Proceeding from velocities expressions, these boundary conditions in terms of the stream function are

$$v_r = \frac{1}{r^2 \sin\gamma}\frac{\partial\psi}{\partial\gamma}, v_\gamma = -\frac{1}{r\sin\gamma}\frac{\partial\psi}{\partial r}, \quad r = R, \psi = \frac{\partial\psi}{\partial\gamma} = 0, \quad r \to \infty, \psi = \frac{1}{2}Ur^2\sin^2\gamma \quad (2.59)$$

To get the last relation for the stream function, note that far from the sphere the velocity becomes equal to component $v_\gamma = -U\sin\gamma$ of uniform velocity U. Substituting the second equation (2.59) for v_γ gives the expression $d\psi = Ur\sin^2\gamma dr$, which on integration results in equation (2.59) (Exercise 2.45).

The last boundary condition suggests the form of solution $\psi = f(r)\sin^2\gamma$. The equation defining function $f(r)$ is found by substituting this solution into governing equation (2.57) for the stream function. Applying this equation according to notes from Comment 2.11, we obtain the first equation (2.60) by putting the solution for ψ into equation (2.57), and then, after substitution of the first result into the same expression (2.57), we obtain the differential equation determining function $f(r)$ (Exercise 2.46):

$$\psi = (f'' - 2fr^{-2})\sin^2\gamma \quad r^4\frac{d^4 f}{dr^4} - 4r^2\frac{d^2 f}{dr^2} + 8r\frac{df}{dr} - 8f = 0. \quad (2.60)$$

This is fourth-order ordinary differential equation known as an equidimensional Euler–Cauchy equation. Substituting a power-type solution $f = Cr^m$ yields an equation $m(m-1)(m-2)(m-3) - 4[m-1+8(m-1)] = 0$ defining the values of exponent. It is easy to check that the roots are $4, 2, 1, -1$. In conformity with this, the solution consists of four constants determined using boundary conditions (2.59):

$$f = C_4 r^4 + C_2 r^2 + C_1 r + \frac{C_{-1}}{r}, \quad C_4 r^2 + C_2 + \frac{C_1}{r} + \frac{C_{-1}}{r^3} = \frac{U}{2} \text{ at } r \to \infty \quad (2.61)$$

$$\frac{UR^2}{2} + C_1 R + \frac{C_{-1}}{R} = 0 \quad UR + C_1 - \frac{C_{-1}}{R^2} = 0 \text{ at } r = R. \quad (2.62)$$

The second equation (2.61) coming from the last boundary condition (2.59) required $C_4 = 0, C_2 = U/2$. Then, solution of two equations (2.62) obtained from two other condition (2.59) gives the constants $C_1 = -3UR/4$ and $C_{-1} = UR^3/4$. Using these results, one finds the solution for the stream function and calculates flow characteristics. The velocity components are obtained applying equation (2.59) (Exercise 2.47). Under knowing the velocity components, the pressure distribution is found from equation $\nabla p = \mu \nabla^2 \mathbf{V}$ in spherical coordinates. The total drag and drag coefficient are calculated by integrating the pressure and shear stress distribution around the sphere. The last two performances are as well straightforward as a procedure in Exercise 2.47 for velocities computing, but tiresome. The derivation yields the following expressions for pressure and drag characteristics (for example, [11]):

$$\psi = Ur^2 \sin^2 \gamma \left[\frac{1}{2} - \frac{3}{4} \frac{R}{r} + \frac{1}{4}\left(\frac{R}{r}\right)^3 \right], \quad p - p_\infty = -\frac{3}{2}\frac{\mu U}{R}\cos\gamma, \quad F = 6\pi\mu UR, \quad C_f = \frac{24}{\text{Re}}. \quad (2.63)$$

Note that (i) the pressure reaches minimum and maximum $\mp(3/2)(\mu U/R)$ at $\gamma = \pi/2$ and $\gamma = (3/2)\pi$; (ii) the drag is proportional to the first power of velocity and consists of two parts $F = 2\pi\mu UR + 4\pi\mu UR$, where the first one of 1/3 came from pressure and the other results from shear stress; (iii) the drag coefficient is inversely proportional to Reynolds number $\text{Re} = 2\rho UR/\mu$.

Comment 2.12 The similar problem for a liquid drop in immiscible liquid medium was solved independently by Rybczynski and Hadamard at the beginning of the last century. It was shown that in this case, the drag coefficient (2.63) should be multiplied by factor $(2\mu + 3\mu_w)/(\mu + \mu_w)$, where μ and μ_w are viscosity of medium and drop, respectively [11,12].

2.5.1.2   Oseen's Approximation

As the distance from the surface increases, the accuracy of creeping approximation decreases due to growing flow velocity. Oseen took into account the effect of the inertia using the perturbation approach. Presenting the velocity field as $U + u', v', w'$, where U is the constant velocity far from the sphere and u', v', w' are small-field perturbations, resulting in equation with inertia terms of the first and second orders of magnitude. Neglecting the relatively small second-order quadratic terms leads to a second linearized governing equation, considering inertia effects in the first approximation:

$$\rho U \frac{\partial u'}{\partial x} + v'\frac{\partial u'}{\partial y} + w'\frac{\partial u'}{\partial z} + \frac{dp}{dx} = \mu\nabla^2 u', \quad \rho U \frac{\partial u'}{\partial x} + \nabla p = \mu\nabla^2 u', \quad \nabla \cdot \mathbf{V}' = 0. \quad (2.64)$$

Here, the first equation is the Navier–Stokes equation for u' in Oseen approximation. Two others for v' and w' are similar. The second and third equations (2.64) are respectively linearized Oseen and continuity equations in the vector form. Solution of these two equations under the same boundary conditions $U + u' = v' = w' = 0$ gives the flow pattern that unlike the Stokes one is different in the front and behind the sphere.

The velocity field in the front of the sphere is close to the Stokes pattern, whereas the velocities behind sphere are larger than those in the Stokes case. The Oseen drag coefficient may be obtained by multiplying the Stokes drag coefficient (2.63) by $1+(3/16)\mathrm{Re}$ or by increasing it by $(24/\mathrm{Re})(3/16)\mathrm{Re} = 4.5/\mathrm{Re}^2$. Comparison with experimental data indicates that formula (2.63) is applicable only for $\mathrm{Re} < 1$, while the Oseen approximation gives satisfactory results up to $\mathrm{Re} \approx 5$[1] (Exercise 2.48).

2.5.1.3 Heat Transfer From the Sphere in the Stokes Flow

The heat transfer from the rigid or fluid sphere to infinite environment was considered using different approaches simplifying the solution. Analysis of applying models shows that the majority of studies at low Reynolds numbers are based on the assumption that one of the phase thermal resistances is negligible so that only other phase sphere or environment controls the process [13]. Authors of this article reviewed and compared the results obtained employing both models known as internal and external problems with dominant thermal resistance.

As an example, we consider one of early analytical study [14] of unsteady heat transfer from a sphere initially at temperature T_i to an environment with different temperature. To simplify the problem, it is assumed that the sphere thermal conductivity and fluid Prandtl number are high. Due to the first assumption, the sphere spatial temperature is practically uniform, whereas the second one yields high Peclet number $\mathrm{Pe} = \mathrm{Re}\,\mathrm{Pr}$, despite that the Reynolds number is low. This allows neglecting one of inertia terms in the energy equation for fluid around the sphere. Nevertheless, the analytical solution is still involved and is obtained in several consecutive steps. First, the energy equation for fluid is solved for a stagnation point and then for any location on the sphere, supposing that the sphere temperature is constant in time also. Two methods are employed: the Laplace transform with a numerical inversing procedure (Sec. 1.6.2) and asymptotic series expansion in the Laplace space for large s (small times). Finally, these results for constant sphere temperature are used to get the solution for arbitrary time-dependent sphere temperature. The system of equation consisting the Duhamel integral (Sec. 1.4) and balance heat for the sphere is solved.

The basic conclusions are as follows: (i) the sphere temperature decreases monotonically with time approaching the equilibrium condition at $T_w = T_\infty$; (ii) the rate of temperature decaying depends on the ratio of sphere and fluid capacities increasing as the capacity ratio grows; (iii) for large values of capacity ratio, there is a range of Fourier number when sphere temperature is practically constant; (iv) at large values of capacity ratio and Fourier number, the quasi-steady results may be used with satisfactory accuracy.

Other more recent solutions of the problems of this kind may be found in [15] (Examples 6.27, 7.14 and 7.17).

2.5.2 Boundary-Layer Approximation (Large Reynolds and Peclet Numbers)

The concept of the boundary layer was formulated by great scholar Ludwig Prandtl who was the first to understand why the solution of the Euler equation for a perfect fluid does not show any pressure losses and resistance forces of moving body even in the case of small viscosity fluid (the D'Alamber paradox). In article published in 1904, he explained that the perfect flow model describes satisfactorily the majority part of the real flow field except the small layer adjusting to the surface called the boundary layer. Prandtl indicated that the model describing the real fluid motion should consist of two parts: thin boundary layer where the friction forces are significant and another part of flow where friction effects are negligible and hence a perfect fluid model is applicable. The practical significance of the Prandtl approach is determined by the fact that majority of technical important fluids, including water, air, and oil are low viscosity and conductivity liquids (Exercise 2.50).

2.5.2.1 Derivation of Boundary-layer Equations

We have discussed the physical interpretation of the Prandtl model in the historical notes (Sec. 2.1). Here, we outline the procedure of simplifying the Navier–Stokes and energy equations to obtain the dynamic and thermal boundary-layer equations. This procedure is based on the comparison of the order of magnitude of equations terms. To make the terms comparable, one introduces dimensionless variables scaling each variable by its maximum value. Dividing the longitudinal velocities by the free-stream velocity, U_∞, longitudinal coordinates by body length L, transverse coordinates by thicknesses δ and δ_t for dynamic and thermal boundary layers, respectively, pressure by ρU_∞^2, time by L/U_∞, and temperatures by the temperature head $\theta = T_w - T_\infty$, one ensures that the values of each dimensionless variable and dimensionless derivative do not exceed unity within the boundary layers. The similar scale for transverse velocity may be obtained from dimensionless continuity equation (2.1). Considering two-dimensional flow, taking into account that the order of dimensionless derivatives is unity and knowing that both terms in continuity equation are of the same order, we get $U_\infty/L \sim v/\delta$. From the last relation, the order of magnitude of the transverse velocity is obtained as $v \sim \delta\left(U_\infty/L\right)$ (Exercise 2.51).

We begin from estimating the order of magnitude of boundary-layer thicknesses. For this purpose, we note that the friction forces in the velocity boundary layer as well as conduction heat flux in the thermal boundary layer are significant even in the case of small viscosity and conductivity. This occurs because velocity u and temperature difference $T_w - T$ change dramatically across the boundary layer from zero on the surface to the free-stream velocity on the outer border of the dynamic boundary layer in the first case and to the temperature head on the outer border of the thermal boundary layer in the second case. As a result, the large velocity and temperature gradients develop leading to large shear stresses and conduction heat fluxes according to Newton $\tau = \mu\left(\partial u/\partial y\right)$ and Fourier $q = -\lambda\left(\partial T/\partial y\right)$ laws, respectively.

Proceeding from these facts, we assume that both groups of terms in the Navier–Stokes and energy equations presenting convective and molecular transport (Sec. 2.2.1) are of the same order of magnitude. To compare these groups of terms, we estimate the order of magnitude of each term using dimensionless variables. For inertia terms of the first Navier–Stokes equation (2.2) in the case of two-dimensional flow, we have $\rho\left(\dfrac{\partial u}{\partial t} + u\dfrac{\partial u}{\partial x} + v\dfrac{\partial u}{\partial y}\right) \sim \rho\left(\dfrac{U_\infty}{L/U_\infty} \sim U_\infty\dfrac{U_\infty}{L} \sim \delta\dfrac{U_\infty}{L}\cdot\dfrac{U_\infty}{\delta}\right)$. Thus, all inertia terms are of the same order $\rho\dfrac{U_\infty^2}{L}$. Analogous result follows for convective terms of energy equation having the order $\rho c_p\dfrac{U_\infty\theta}{L}$ (Exercise 2.52). We obtain completely different result on comparing terms responsible for molecular transport. In this case, the orders of magnitude of two comparing terms differ markedly, which is clear from two expressions obtained for viscous and conductive terms of Navier–Stokes and energy equations:

$$\mu\left(\frac{\partial^2 u}{\partial x^2} + \frac{\partial^2 u}{\partial y^2}\right) \sim \mu\left(\frac{U_\infty}{L^2} + \frac{U_\infty}{\delta^2}\right) = \mu\frac{U_\infty}{L^2}\left(1 + \frac{L^2}{\delta^2}\right) \qquad (2.65)$$

$$\lambda\left(\frac{\partial^2 T}{\partial x^2}+\frac{\partial^2 T}{\partial y^2}\right)\sim\lambda\left(\frac{\theta}{L^2}+\frac{\theta}{\delta_t^2}\right)=\lambda\frac{\theta}{L^2}\left(1+\frac{L^2}{\delta_t^2}\right).\tag{2.66}$$

Since in the case of thin boundary layer $(L/\delta)^2 \gg 1$, we get from these equations the first conclusion of the Prandtl model: the one viscous term in the Navier–Stokes equation and the one conduction term in the energy equation are negligibly small and can be omitted.

Using this result, we achieve our goal formulated at the beginning of this procedure, namely estimate the order of magnitude of the boundary-layer thicknesses. Comparing the orders of two groups of terms in the Navier–Stokes and in the energy equations yields two algebraic equations determining the order of both thicknesses:

$$\rho\frac{U_\infty^2}{L}\sim\mu\frac{U_\infty}{L^2}\frac{L^2}{\delta^2},\quad\frac{\delta}{L}\sim\frac{1}{\sqrt{Re}}\qquad\rho c_p\frac{U_\infty\theta}{L}\sim\lambda\frac{\theta}{L^2}\frac{L^2}{\delta_t^2},\quad\frac{\delta_t}{L}\sim\frac{1}{\sqrt{Pe}},\tag{2.67}$$

where $Re = U_\infty L/\nu$ and $Pe = U_\infty L/\alpha$. So the boundary-layer thickness is inversely proportional to the square root of the Reynolds or Peclet number. We get the same result in Section 2.4.4, considering solution of the Navier–Stokes equation for the stagnation point.

The second conclusion of the Prandtl model is obtained comparing two first Navier–Stokes equations (2.2) and (2.3). Both these equations have the same structure so that each term in the first equation corresponds to similar one in the second equation. It is seen that the ratio of the orders of each pair of terms is $v/u \sim \delta/L \sim 1/\sqrt{Re}$. Thus, the order of magnitude of each term in the second equation, except the term $\partial p/\partial y$, is $1/\sqrt{Re}$. Using this result, one finds from the second equation (2.3) that the order of pressure gradient term is the same as the order of other terms in this equation, that is, $\partial p/\partial y \sim 1/\sqrt{Re}$. Hence, since y is of the order $\delta \sim 1/\sqrt{Re}$, the result of the last relation integration shows that the change of the pressure across the boundary layer is of the order $1/Re$. Two conclusions of the Prandtl model follow from this analysis: in the case of high Reynolds number the second Navier–Stokes equation may be neglected, and the pressure may be considered as unchanged across the boundary layer being equal to the pressure in external flow at the outer edge of the boundary layer.

The last conclusion of this analysis is obtained on comparing the order of magnitude of dissipation function terms. Applying the same approach to each term of the second equation (2.5) for the case of two-dimensional flow, one gets that the order of only one term is $(\partial u/\partial y)^2 \sim 1/Re$, whereas the others are much smaller and may be omitted.

The outlined Prandtl approach simplifies the Navier–Stokes and energy equations at high Reynolds and Peclet numbers, resulting in the case of two-dimensional flow in the following system of boundary-layer equations (Exercise 2.53):

$$\frac{\partial u}{\partial x}+\frac{\partial v}{\partial y}=0\tag{2.68}$$

$$\frac{\partial u}{\partial t}+u\frac{\partial u}{\partial x}+v\frac{\partial u}{\partial y}+\frac{1}{\rho}\frac{dp}{dx}-\nu\frac{\partial^2 u}{\partial y^2}=0\tag{2.69}$$

$$\frac{\partial T}{\partial t} + u\frac{\partial T}{\partial x} + v\frac{\partial T}{\partial y} - \alpha\frac{\partial^2 T}{\partial y^2} - \frac{v}{c_p}\left(\frac{\partial u}{\partial y}\right)^2 = 0. \tag{2.70}$$

The pressure is calculated from equation (2.69) written for external flow at the outer edge of the boundary layer. Since the external flow is considered as the frictionless, this equation defines the pressure through the velocity U at the edge of the boundary layer, resulting in two Bernoulli equations for unsteady and steady flows [compare to equation (2.26)]:

$$-\frac{1}{\rho}\frac{\partial p}{\partial x} = \frac{\partial U}{\partial t} + U\frac{\partial U}{\partial x} - \frac{1}{\rho}\frac{dp}{dx} = U\frac{dU}{dx}. \tag{2.71}$$

In equations (2.71), $U = U_\infty$ in the case of flow past a flat plate and hence it is given, whereas in the case of flow past another body shape $U \neq U_\infty$ and should be estimated using the potential flow approach (Sec. 2.2.2.5). In the first case, the external potential flow is a parallel flow with given velocity U_∞. Therefore, at the edge of the boundary layer, the velocity U as well as u-component are known being equal U_∞, whereas the v-component is zero. Due to that, the third term in equation (2.69) vanishes as well as the last one due to the potential feature of the external flow. As a result, equation (2.69) transforms into equations (2.71).

To get the same result in other cases, one uses the natural coordinate system when coordinates x and y are measured along the body surface and along the normal to it, respectively. Since at high Reynolds number the boundary layer is very thin, both components of velocity at the outer edge are very close to those on the body surface. Therefore, in natural coordinates, the component u at the outer edge is practically equal to velocity U, which the flow would have on the surface if it were potential, whereas the v-component in this case would be close to zero. Due to these facts, the pattern at the outer edge of the boundary layer in the general case is similar to that on the flat plate, and the equation (2.71) is valid in general case as well. However, in general, the external flow is nonhomogeneous so that velocity U at the outer edge differs from the uniform velocity U_∞ far away from a body and hence should be calculated using the potential flow theory (Sec. 2.2.2.5, Example 2.5, Exercise 2.54).

The system of boundary-layer equations (2.68)–(2.70) is much simpler than the system of Navier–Stokes and energy equations because this system contains three equations instead of four ones accordingly, including three unknown, velocity components and temperature, while the pressure, which is the fourth unknown, is determined by Bernoulli equations (2.71). Moreover, remarkable simplification arises due to reducing the molecular transport terms with second derivatives from two in each full equation to one in each boundary-layer equation. This modification changes the equation type, transforming the Navier–Stokes and energy elliptic equations into parabolic boundary-layer equations (Sec. 1.2.4). The parabolic equations require relatively simple boundary conditions. In particular, the boundary-layer equations should satisfy only no-slip and conjugate [or regular equation (1.24)] conditions for dynamic (2.69) and thermal (2.70) equations on the body surface and asymptotic conditions far from the body on the outer edge of the boundary layer:

$$y = 0, \quad u = v = 0, \quad T^+ = T^-, \lambda\frac{\partial T}{\partial y}\bigg|^+ = \lambda_w\frac{\partial T}{\partial y}\bigg|^-, \quad y \to \infty \quad u \to U, \quad T \to T_\infty. \tag{2.72}$$

In conformity with this, the velocity and temperature distributions at the entrance and exit domain essential for full equations (Sec. 2.3) are not required for a boundary-layer system. Owing to simpler boundary conditions, it is much easier to obtain solutions of boundary-layer problems than to solve the Navier–Stokes or full-energy equations (Exercise 2.55).

The first solutions of boundary-layer problems were obtained soon after the publication of Prandtl article. Only four years later, Prandtl's student Blasius gave the solution of the boundary-layer problem for a flow past the flat plate, and later on Pohlhausen solved a similar problem for the thermal boundary layer. In the following years up to time when the computers came into existence, the boundary-layer methods were only one essential tool for computational estimating the characteristics of moving in air and water objects and in other different cases from simple devices to reentry rockets.

Comment 2.13 The first 25 years, the boundary-layer theory was developed only by Prandtl and his students and hence about one or two articles were published per year. The situation has changed after Prandtl's lecture at the meeting of the Royal Aeronautical Society in London in 1927. In the following years, the amount of publication in the boundary-layer theory was grown steadily and reached about 100 papers per year in the middle of the last century increasing to almost 300 articles yearly 20 years later [1]. This historical fact shows how long it takes a new idea or new result (even as practical important as the boundary-layer theory) to become widely known and be used.

2.5.2.2 Prandtl–Mises and Görtler Transformations

Prandtl and Mises independently show that boundary-layer equations in the case of steady flow can be transformed to the form closed to the one-dimensional heat transfer equation. Such a transformation is achieved using the stream function and longitudinal coordinate as independent variables and $Z = U^2 - u^2$ and T as unknown. Applying the chain rule (Exercise 2.56) yields the equations and boundary conditions for velocity in new variables, whereas for temperature the same boundary conditions (2.72) remain valid:

$$\frac{\partial Z}{\partial x} = vu \frac{\partial^2 Z}{\partial \psi^2}, \quad \frac{\partial T}{\partial x} = \alpha \frac{\partial}{\partial \psi}\left(u \frac{\partial T}{\partial \psi}\right), \quad \psi = 0 \quad Z = U^2(x), \quad \psi \to \infty \quad Z \to 0. \quad (2.73)$$

The form of both equations (2.73) is similar to that of the one-dimensional conduction equation; however, in contrast to the latter, the first equation (2.73) for velocity is nonlinear because the factor at the second derivative depends on unknown function Z.

Boundary-layer equations in the form (2.73) have a singularity at the surface because the derivatives with respect to ψ become infinite at $\psi = 0$. It is said that a function at some point x is singular if the function is not analytic in that point, that is, if it is not differentiable (Sec. 2.2.2.5). Since such a function does not have some derivatives at a singular point, the function at this point could not be presented by Taylor series. To see that equations (2.73) are singular at point $\psi = 0$, note that the left-hand side of both equations is finite on the surface, whereas on the surface $u = 0$, and consequently the second derivative on the right-hand side of each equation should be infinite. The type of singularity may be estimated by considering analytic solution of boundary equation (2.69) for velocity. Presenting the velocity near the surface by the Taylor series, one gets for stream function

$$u = c_1 y + c_2 y^2 + \cdots, \quad \psi = \int_0^y u \, d\xi = \frac{c_1}{2} y^2 + \frac{c_2}{3} y^3 \cdots, \quad u = b_1 \psi^{1/2} + b_2 \psi + b_3 \psi^{3/2} + \cdots. \quad (2.74)$$

It follows from the second series that near the surface at small values of transverse coordinate, $\psi \approx y^2$ and hence $y \approx \psi^{1/2}$. Applying this result to the first series, one obtains the third series (2.74) for velocity in variable ψ. This series is not Taylor one since it contains fractional exponents and is singular at $\psi = 0$, resulting in the infinite first derivative $\partial u / \partial \psi \approx \psi^{-1/2}$ and in the infinite higher derivatives as well. At the same time, the situation may be changed by introducing a new variable $\zeta = \psi^{1/2}$ transforming the singular expression (2.74) for velocity in the Taylor series with finite derivatives with respect to new variable $u = b_1 \zeta + b_2 \zeta^2 + b_3 \zeta^3 + \cdots$ (Exercises 2.56 and 2.57).

Another form of the boundary-layer equations (2.69) and (2.70) is obtained using Görtler variables that according to the chain rule transform those into following [1,1968]:

$$\Phi = \frac{1}{v} \int_0^x U(\xi) \, d\xi, \quad \eta = \frac{yU}{v\sqrt{2\Phi}}, \quad \varphi = \frac{\psi}{v\sqrt{2\Phi}}, \quad \theta = \frac{T - T_w}{T_\infty - T_w} \quad (2.75)$$

$$\frac{\partial^3 \varphi}{\partial \eta^3} + \varphi \frac{\partial^2 \varphi}{\partial \eta^2} + \beta(\Phi) \left[1 - \left(\frac{\partial \varphi}{\partial \eta} \right)^2 \right] = 2\Phi \left[\frac{\partial^2 \varphi}{\partial \Phi \partial \eta} \frac{\partial \varphi}{\partial \eta} - \frac{\partial \varphi}{\partial \Phi} \frac{\partial^2 \varphi}{\partial \eta^2} \right], \quad \beta(\Phi) = \frac{2}{U^2} \frac{dU}{dx} \int_0^x U(\xi) \, d\xi$$

$$(2.76)$$

$$\frac{1}{\Pr} \frac{\partial^2 \theta}{\partial \eta^2} + \varphi \frac{\partial \theta}{\partial \eta} + \Phi \beta_t(\Phi) \frac{\partial \varphi}{\partial \eta} (1 - \theta) = 2\Phi \left[\frac{\partial \varphi}{\partial \eta} \frac{\partial \theta}{\partial \Phi} - \frac{\partial \varphi}{\partial \Phi} \frac{\partial \theta}{\partial \eta} \right], \quad \beta_t(\Phi) = \frac{4}{U\theta_w} \frac{d\theta_w}{dx} \int_0^x U(\xi) \, d\xi$$

Equations (2.76) are obtained taking Φ and η as independent variables and φ and θ as the unknown dependent variables. Görtler variables may be applied to transform Prandtl–Mises equations (2.73). In this case, variables Φ and φ similar to x and ψ are considered as independent, whereas the dependent ones are the same variable Z as in the Prandtl–Mises equation and temperature excess $\theta = T - T_\infty$. Using the chain rule, one gets the boundary-layer equations in the Prandtl–Mises–Görtler form [15] (Exercise 2.58):

$$2\Phi \frac{\partial Z}{\partial \Phi} - \varphi \frac{\partial Z}{\partial \varphi} - \frac{u}{U} \frac{\partial^2 Z}{\partial \varphi^2} = 0, \quad 2\Phi \frac{\partial \theta}{\partial \Phi} - \varphi \frac{\partial \theta}{\partial \varphi} - \frac{1}{\Pr} \frac{\partial}{\partial \varphi} \left(\frac{u}{U} \frac{\partial \theta}{\partial \varphi} \right) = 0 \quad (2.77)$$

Boundary condition (2.72) for temperature and (2.73) for velocity after small changes, θ for T in the first condition and φ for ψ in the second, remains valid for these equations. The properties of equations (2.77) are the same as those of Prandtl–Mises equations (2.73); in particular, the first equation (2.77) is nonlinear, and on the surface at $\varphi = 0$ both relations have a singularity that may be removed employing variable $\varphi^{1/2}$ (Exercise 2.59).

One of the advantages of the Görtler variables and of equations (2.76) and (2.77) written in Görtler variables is that they are dimensionless. At the same time in some cases, the equations (2.73) and (2.77) in the Prandl–Mises form are more relevant than full equations (2.76) because those are free of transverse velocity component v. In particular, the second equation (2.77) was systematically used in [15] in studying the conjugate heat transfer of arbitrary nonisothermal surfaces to obtain the universal functions (Sec. 1.2.3, Chapter 4).

2.5.2.3 Theory of Similarity and Dimensionless Numbers

Each phenomenon may be described by some system of equations with boundary conditions determining the dependence between dependent unknown variables and given independent variables. For example, the system of equations (2.68)–(2.71) and conduction equation (1.2) with boundary conditions (2.72) describe the conjugate heat transfer of a body past incompressible flow and shows two sets of quantities. The first one is a set of the unknown characteristics of the process, including velocity components u and v, temperature and hear flux of flow T and $q = \lambda (\partial T / \partial y)$ and of body T_s and $q_s = \lambda_s (\partial T_s / \partial y)$, as well as temperature and heat flux T_w and q_w on the interface. Another set of known quantities determining the sought characteristics of process consist of coordinates x and y, time t, physical properties of a fluid, viscosity v, thermal diffusivity α, conductivity λ, and specific heat c_p, and physical properties of a body, λ_s and α_s, as well as of velocities U, U_∞ and temperature T_∞ of the external flow. The pressure p and density ρ containing in equation (2.69) are not included in a set of unknown because they are present only in the pressure gradient term that is defined by velocity U according to equation (2.71).

The above analysis shows that the characteristics of the analyzing process are determined by a number of parameters. Such a situation is typical, and the similarity theory is intended for reducing the number of independent variables by combining parameters defining the process in the dimensionless numbers.

There are at least three areas where the results of similarity theory have crucial importance: (i) designing experimental models for simulating the natural objects and processes; (ii) analytical and numerical simulation of phenomena by mathematical models; (iii) presentation of the results of investigation. Employing dimensionless numbers allows performing the experiments on small size models using different values of velocity, temperature and other parameters according to the same as in nature dimensionless numbers. For example, to keep the same Reynolds number on small model one may use in experiment larger velocity or fluid with lesser viscosity. In theoretical simulation, using dimensionless numbers reduces the required number of experiments. Here the situation is the same as in employing similarity variables (Sec. 2.4.3) when investigating the effect of dimensionless number substitute studying influence of each of parameters comprising the dimensionless number. Moreover, the results obtained in some experiment or in theoretical study for particular object, physical properties and regime may be used as well for any other combination of these parameters with the same values of the basic dimensionless numbers. In the presentation of the studying results, reducing the number of independent variables by employing dimensionless numbers is associated with restricted number of parameters that may by taken into account using tables or graphs.

The proper dimensionless numbers are obtained using dimensionless equations scaled similar to that in evaluating the order of magnitude (Sec. 2.5.2.1). The velocity components are scaled by U_∞, coordinates by L, time by L^2 / α for a fluid and by L^2 / α_s for a body, and temperatures by $(T_{w*} - T_\infty)$, where T_{w*} is the temperature of an isothermal surface.

Comment 2.14 The temperature is usually scaled by $(T_w - T_\infty)$, where T_w is the body surface temperature that is given or may be calculated in the case of a nonconjugate problem when a body and fluid are considered separately. However, the conjugate problem treats the body and flow as a unit so that the temperature as well as the heat flux on the surface are unknown and are usually variable. In such a case, it is reasonable to use for a scale the isothermal temperature.

Returning to the system of equations (2.69)–(2.72) and (1.2), we transform it by indicated above scales remaining for dimensionless variables the same notations. Then, after dividing the first transformed equation by U_∞^2/L, the second by $U_\infty(T_{w*} - T_\infty)/L$ and the transformed equation (1.2) by $\alpha_s(T_{w*} - T_\infty)/L^2$, we obtain the following system:

$$\frac{\alpha}{U_\infty L}\frac{\partial u}{\partial t} + u\frac{\partial u}{\partial x} + v\frac{\partial u}{\partial y} - \frac{\partial U}{\partial t} - U\frac{\partial U}{\partial x} + \frac{g\beta L(T_{w*} - T_\infty)}{U_\infty^2}\frac{T - T_\infty}{T_{w*} - T_\infty} - \frac{v}{U_\infty L}\frac{\partial^2 u}{\partial y^2} = 0 \quad (2.78)$$

$$\frac{\alpha}{U_\infty L}\frac{\partial T}{\partial t} + u\frac{\partial T}{\partial x} + v\frac{\partial T}{\partial y} - \frac{\alpha}{U_\infty L}\frac{\partial^2 T}{\partial y^2} - \frac{vU_\infty}{c_p L(T_{w*} - T_\infty)}\left(\frac{\partial u}{\partial y}\right)^2 = 0 \quad (2.79)$$

$$\frac{\partial T_s}{\partial t} - \frac{\partial^2 T_s}{\partial x^2} - \frac{\partial^2 T_s}{\partial y^2} - \frac{q_v L^2}{\lambda_s(T_{w*} - T_\infty)} = 0 \qquad \frac{q_w L}{\lambda(T_{w*} - T_\infty)} = \left(\frac{\partial T}{\partial y}\right)_{y=0}. \quad (2.80)$$

This system unlike the initial dimensional system does not consist of equation (2.68) because the dimensionless continuity equation remains unchanged without any new results. At the same time, equation (2.78) contains additional sixth term included to take into account the natural heat transfer effects. This type of heat transfer is caused by buoyancy forces that occur due to density difference exciting in the inhomogeneous vertically field of temperature (Sec. 2.8). Another result additional to the initial system is the last expression (2.80). This one is obtained from conjugate condition (2.72) and determines the dimensionless heat flux on the surface.

Equations (2.78)–(2.80) consist of nine groups of parameters defining eight dimensionless numbers named (except the group with a heat source) after pioneers-contributors as Peclet, Prandtl, Grashof, Reynolds, Eckert, Fourier, and Nusselt:

$$\mathrm{Pe} = \frac{U_\infty L}{\alpha} = \mathrm{Re}\,\mathrm{Pr},\ \mathrm{Pr} = \frac{v}{\alpha},\ \frac{g\beta L(T_{w*} - T_\infty)}{U_\infty^2} = \frac{\mathrm{Gr}}{\mathrm{Re}^2},\ \mathrm{Gr} = \frac{g\beta L^3(T_{w*} - T_\infty)}{v^2},\ \mathrm{Re} = \frac{U_\infty L}{v}$$

$$(2.81)$$

$$\frac{vU_\infty}{c_p L(T_{w*} - T_\infty)} = \frac{\mathrm{Ec}}{\mathrm{Re}},\ \mathrm{Ec} = \frac{U_\infty^2}{c_p(T_{w*} - T_\infty)},\ \bar{q}_v\frac{q_v L^2}{\lambda_s(T_{w*} - T_\infty)},\ \mathrm{Fo} = \frac{\alpha t}{L^2},\ \mathrm{Nu} = \frac{q_w L}{\lambda(T_{w*} - T_\infty)}.$$

From a physical point of view these dimensionless numbers define the ratio of order of magnitude of (i) convection to conduction heat rate (Peclet), (ii) momentum to thermal diffusivity (Prandtl), (iii) buoyancy to viscous forces (Grashof), (iv) inertia to viscous forces (Reynolds), (v) kinetic energy to enthalpy (Eckert), (vi) heat source to conduction heat rate (Source dimensionless number), (vii) conduction to energy storage heat rates (Fourier), and (viii) convective to conduction heat transfer rates across the interface (Nusselt). Note also that the Fourier number is in fact a dimensionless time defined through the thermal diffusivity α and α_s for a fluid and a body, respectively.

Six dimensionless numbers (Fourier, Reynolds, Prandtl or Peclet, Grashof, Eckert, and Source number) are independent variables, whereas the Nusselt number is unknown dimensionless heat flux through the interface that should be estimated. Proceeding from these facts, one may see that the system (2.78)–(2.80) and dimensionless continuity equation (2.68) together with boundary conditions and Fourier law determine eight dimensionless variables $(u, v, T, q, T_s, q_s, T_w, Nu)$ in terms of two dimensionless coordinates (x, y) and six dimensionless numbers (Fo, Re, Ec, Gr, Pe or Pr and $\bar{q}_v$). This set of eight dimensionless variables substituted a set of 14 dimension variables $(x, y, t, U_\infty, T_\infty, v, \alpha, \lambda, \alpha_s, \lambda_s, g, \beta, c_p, q_v)$, determining the same eight dimension variables from initial set of equations in dimension variables. Thus, for the case in question, employing dimensionless numbers almost halves the number of independent variables. Reducing a number of variables considerable simplifies the analysis and presentation of investigation results, providing the wide application of similarity theory and dimensionless numbers (Exercise 2.61 and 2.62).

2.5.2.4 Boundary-Layer Equations of Higher Order

In Section 2.5.2.1, the boundary-layer equations are obtained by comparing the order of magnitude of terms of Navier–Stokes and energy equations. The alternative, more general, method consists of considering the solutions of the basic equations in the form of asymptotic perturbation expansions using as a small parameter $\varepsilon = 1/\sqrt{Re}$ or $\varepsilon = 1/\sqrt{Pe}$ for the Navier–Stokes equation or for the energy equation. In this case, the boundary-layer equations of the first approximation derived above as well as the equations of higher order are obtained in the same manner as sequential perturbed equations. The sum of expansions terms without a perturbation parameter gives the first approximation equations; the sum of terms containing parameter ε gives the second approximation equations, and the sum of terms with ε^2 results in the third-order equations, and so on (Exercise 2.63).

As well as in the first method, in the perturbation technique, the boundary-layer equations are found as the simplified dimensionless Navier–Stokes and full-energy equations. The first approximation as well as before may be obtained for a flat plate and then simple generalized using the same equations in Cartesian coordinates but measuring them along the body surface and along the normal to it. However, the higher order boundary-layer equations cannot be found in this way because the surface curvature effects are of the second order being proportional to ε. Therefore, in finding the second and other approximations for a common not flat body, the basic full equations should be considered in curvilinear coordinates "surface/normal." Since such a procedure is involved, we give here the second-order boundary-layer equations only for a flat plate presenting the perturbation approach in a relatively simple way. The general case of higher order boundary-layer equations was considered by Van Dyke [10,16]; some results from these and other works one may find in [1]. In [17], the second-order equations in Prandtl–Mises form, including the thermal boundary-layer equation, are derived.

Since the boundary-layer model consists of two parts, it should be considered two perturbation expansions, the outer one for external flow and the inner expansion for boundary layer itself. Using the dimensionless Navier–Stokes and energy equations

$$u\frac{\partial u}{\partial x}+v\frac{\partial u}{\partial y}+\frac{\partial p}{\partial x}-\frac{1}{Re}\left(\frac{\partial^2 u}{\partial x^2}+\frac{\partial^2 u}{\partial y^2}\right)=0 \quad u\frac{\partial T}{\partial x}+v\frac{\partial T}{\partial y}-\frac{1}{Pe}\left(\frac{\partial^2 T}{\partial x^2}+\frac{\partial^2 T}{\partial y^2}\right)=0 \quad (2.82)$$

and noting that the factors at the viscous and conductive terms are of order $1/Re \sim 1/Pe \sim \varepsilon^2$, one concludes that these terms are of order ε^2 or higher. Consequently, the outer expansion for the first and second approximations up to terms with ε^2 presents the flow of inviscid, perfect fluid, and hence the velocity and temperature of external flow in second approximation remain constant as well as in the first one.

To get the inner expansions, the new variables $\bar{y}=y/\varepsilon$ and $\bar{v}=v/\varepsilon$ should be applied. Such a modification transforms the boundary layer in the y-direction corresponding to $y \to \infty$ and makes it possible to match the boundary layer and external flows. In new variables, the system of equations (2.82) and continuity equation becomes (Exercise 2.64)

$$\frac{\partial u}{\partial x}+\frac{\partial \bar{v}}{\partial \bar{y}}=0, \quad u\frac{\partial u}{\partial x}+\bar{v}\frac{\partial u}{\partial \bar{y}}+\frac{\partial p}{\partial x}-\frac{\partial^2 u}{\partial \bar{y}^2}=0, \quad u\frac{\partial T}{\partial x}+\bar{v}\frac{\partial T}{\partial \bar{y}}-\frac{1}{Pr}\frac{\partial^2 T}{\partial \bar{y}^2}=0. \quad (2.83)$$

Substituting into this system, the perturbation expansions

$$u=u_1+\varepsilon u_2+\cdots \quad v=v_1+\varepsilon v_2+\cdots \quad p=p_1+\varepsilon p_2+\cdots \quad T=T_1+\varepsilon T_2+\cdots. \quad (2.84)$$

and collecting terms according to power of parameter ε result in the boundary-layer equations of first and second order marked by indices 1 and 2 (Exercise 2.65):

$$\frac{\partial u_1}{\partial x}+\frac{\partial \bar{v}_1}{\partial \bar{y}}=0 \qquad \frac{\partial u_2}{\partial x}+\frac{\partial \bar{v}_2}{\partial \bar{y}}=0 \qquad (2.85)$$

$$u_1\frac{\partial u_1}{\partial x}+\bar{v}_1\frac{\partial u_1}{\partial \bar{y}}+\frac{\partial p_1}{\partial x}-\frac{\partial^2 u_1}{\partial \bar{y}^2}=0 \quad u_1\frac{\partial u_2}{\partial x}+u_2\frac{\partial u_1}{\partial x}+\bar{v}_1\frac{\partial u_2}{\partial \bar{y}}+\bar{v}_2\frac{\partial u_1}{\partial \bar{y}}+\frac{\partial p_2}{\partial x}-\frac{\partial^2 u_2}{\partial \bar{y}^2}=0 \quad (2.86)$$

$$u_1\frac{\partial T_1}{\partial x}+\bar{v}_1\frac{\partial T_1}{\partial \bar{y}}-\frac{1}{Pr}\frac{\partial^2 T_1}{\partial \bar{y}^2}=0 \quad u_1\frac{\partial T_2}{\partial x}+u_2\frac{\partial T_1}{\partial x}+\bar{v}_1\frac{\partial T_2}{\partial \bar{y}}+\bar{v}_2\frac{\partial T_1}{\partial \bar{y}}-\frac{1}{Pr}\frac{\partial^2 T_2}{\partial \bar{y}^2}=0. \quad (2.87)$$

The second-order equation (2.86) in contrast to the analogous of the first approximation is linear containing the inertia term in the form $u_1\left(\partial u_2/\partial x\right)$ instead of $u_1\left(\partial u_1/\partial x\right)$.

Therefore, to solve this equation as well as the second-order equation (2.87) for temperature, the solution of the first-order equations (2.85) and (2.86) defining u_1 and v_1 should be known. Besides, the second-order solution of the potential flow is required to get the boundary conditions far from a body that are necessary for matching the second-order external and boundary-layer flows.

EXERCISES

2.25 Study some issues of partial differential equations (PDF) using any *Advanced Mathematics Course* to understand what features of the Navier–Stokes equation basically causes huge difficulties in solving this equation. In particular, what is nonlinearity and why nonlinear equation usually is more difficult to solve than linear.

2.26 Recall what is the Dirichlet problem and explain to somebody what causes the problems with boundary conditions for the Navier-Stokes and energy equations.

2.27 Think about physical causes resulting in slip or no-slip effects on the surface of solid. Study (for example, article "Fluid Mechanics" on Wikipedia) what is molecular mean free path length and how it relates to term continuum medium.

2.28 Why there are no exact analytical solutions of Navier-Stokes and energy equations? Why those solutions are required?

2.29 Explain from physical view-point why the interface temperature distribution usually is not known in advance and why its determining requires the conjugate procedure?

2.30 Calculate and plot the velocity distribution in Couette flow using equation (2.48) in the form $u/U = \eta + P\eta(1-\eta)$ for $\eta = 0-1$. Compare this form with equation (2.48) in text to find relation for η and P.

2.31 Analyze the effect of the pressure gradient using results from Exercise 2.30. What part of velocity pattern shows the back flow? What kind of pressure gradient causes the back flow?

2.32 Derive formula (2.49) for resistance coefficient, and explain to somebody in what sense dependence (2.49) is universal.

2.33 What are similarity variables like? Provide some examples other than mentioned in the text.

2.34 *The problem of stagnation point has the solution in the form $u = xf'(y)$, $v = f(y)$. Prove that in this case: (a) The continuity equation is satisfied and (b) The first Navier–Stokes equation may be reduced to ordinary differential equation (2.51). Hint: use information given after equation (2.50).

2.35 What kinds of exact solutions of the Navier–Stokes equation are there? Compare those to understand when it is reasonable to employ the similarity variables.

2.36 Think about differences between exact solutions of the Dirichlet problem and simple solutions, give some examples, and show how these particular problems differ from each other. Read more about adiabatic wall temperature and channel flows with heat transfer [1, pp.277–280].

2.37 Derive the dimension form of equation (2.7) and discuss with friend or colleague the two kinds of similarities following from this equation.

2.38 Think about limiting cases for small and large dimensionless numbers. Why considering limiting cases are important for transfer processes, especially for the momentum transfer described by Navier–Stokes equation?

2.39 It seems that two formulae differ by factors 1.4 and 1.7 are close to each other. In some cases this is true, however, not always. For example, calculate the percentage difference of results obtained by two formulae in two cases. First estimate the error for Re = 1000, and then find values of the Reynolds number that correspond to error of 5%. Make some conclusions.

2.40 Compare equation (2.7) with system (2.2)–(2.6). What terms are not taken into account in equation (2.7)? Think: what results considered in the text may be affected by this approximate approach. Give examples when these terms should be taken into account.

2.41 Write two-dimensional creeping flow equations in the vector form $\nabla p = \mu \nabla^2 \mathbf{V}$ and show that in this case the pressure satisfies the Laplace equation. Hint: differentiate the first creeping equation with respect to x and the second one with respect to y and sum the results to get an equation containing Laplacian for pressure and some additional terms. Then, (i) twice differentiate the continuity equation with respect to x, (ii) do the same with respect to y, and (iii) sum the results to show that additional to Laplacian terms equal zero giving the Laplace equation for the pressure.

2.42 Do the same first part of hint with equation $\nabla^2 \psi$ to get the second equation (2.56) $\nabla^4 \psi = \nabla^2 (\nabla^2 \psi)$.

2.43 Derive equation (2.58) and show that using algebra and rules for unit vectors leads to first equation (2.57).

2.44 Derive the second equation (2.57) following the directions in the text, put expressions (2.59) for velocities v_r and v_γ into the first equation (2.57), and then substitute the result into Laplace equation $\nabla^2 \omega = 0$.

2.45 Show that equation (2.59) for stream function is the boundary condition at $r \to \infty$ and corresponds to flow velocity far from sphere. Follow the steps given in the text: (i) draw geometric representation between vectors U and v_γ to show that $v_\gamma = U \sin \gamma$, (ii) put in this equation v_γ according to the second equation (2.59), and (iii) solve the obtained equation for $d\psi$ and integrate it.

2.46 Derive the ordinary differential equation (2.60) using twice the last equation (2.57) Hint: find the first result by putting solution $\psi = f(r)\sin^2 \gamma$ in (2.57). Then, put the result of the first step in the same equation (2.57). To do this, differentiate the first result with respect to r twice and substitute it in the second term of equation (2.57).

2.47 Obtain the velocity components v_r and v_γ of creeping flow past sphere using solution (2.63) for stream function and formulae (2.59), defining these components.

2.48 Explain how the Stokes equation was improved by Oseen and how the Oseen's equation is linearized.

2.49 Explain the difference between the internal and external heat transfer problems in creeping flow.

2.50 What is the basic idea of the Prandtl boundary-layer model?

2.51 What is the order of magnitude? Describe how this quantity is used in simplifying the Navier–Stokes and energy equations in the boundary-layer system.

2.52 Show that the order of magnitude of convective terms of energy equation is $U_\infty \theta / L$.

2.53 Formulate basic conclusions of the Prandtl model resulting in system (2.68)–(2.71) of boundary-layer equations. Describe how the flow characteristics, in particular, velocity and temperature gradients, depend on the fact that the order of boundary-layer thickness is inversely proportional to the square root of the Reynolds or Peclet number.

2.54 What is the difference between velocities U and U_∞? When this difference is important?

2.55 Explain what basically makes solution of boundary-layer equations much easier than that of the Navier–Stokes or energy equation.

2.56 Repeat the derivation of equation (2.73) for velocity given in [1, p.157] using the chain rule (Exercise 2.22) and function Z instead of g in [1]. Hint: note that $g = (1/2)\rho Z$.

2.57 Learn more about analytic functions and singularities from any *Advanced Engineering Mathematic* course. See other types of singularity and examples of different singular points.

2.58 Transform Prandtl-Mises equations (2.73) in Görtler variables using the chain rule.

2.59 Explain what does the term "removeable singularity" mean? How could one remove some types of singularities? How does the function in new variable be controlled whether it is regular without singularity?

2.60 Why dimensionless numbers are important in designing the experimental models? If it is known that some experimental data of the heat transfer coefficient are obtained for air, determine to what other gases such data are applicable. Hint: use the table of physical properties.

2.61 Rederive the systems (2.78)–(2.80) and obtain the basic in flow and heat transfer dimensionless numbers. Think which of eight variables are determined by this system including continuity equation and which ones are found from boundary conditions and Fourier law.

2.62 Obtain a dimensionless number following from two-dimensional mass transfer equation (2.6) in the nonconjugate problem, assuming that the surface concentration C_w is known. Compare your result with Peclet and Prandtl numbers. Note that the dimensionless mass number similar to the Prandtl number is named as the Shmidt number (Sec. 2.2.1).

2.63 Describe the method of perturbation series using to derive boundary-layer equations.

2.64 Transform system (2.82) to system (2.83) using special variables. Why these variables are used?

2.65 Derive systems (2.85)–(2.87) of the first and second orders applying perturbation expansions.

2.6 Exact Solutions of the Boundary-Layer Equations

Although both boundary-layer equations are simpler than the Navier–Stokes and full-energy equations, their solutions as well encounter considerable difficulties. The dynamic boundary-layer equation is difficult to solve due to its nonlinearity, whereas the number of exact solutions of even linear energy equation is also restricted by the same problem since these solutions depend on the velocity components. Therefore, only a few exact solutions of both velocity and temperature boundary-layer problems are found. Here, we consider two well-known solutions of boundary-layer problems with similar velocity and temperature distributions in external flow and some examples of exact solutions of flow in the boundary layer for different external velocity distribution as well as two known solutions of thermal boundary layer equation. Other exact solutions of the energy equation obtained applying the superposition principle are discussed later in considering the heat transfer of nonisothermal bodies (Chapter 4).

2.6.1 Flow and Heat Transfer on an Isothermal Semi-Infinite Flat Plate (Blasius and Pohlhausen Solutions)

These problems first solved by the Prandtl approach are governed by steady-state boundary-layer system (2.68)–(2.70) at a zero pressure gradient and simple boundary conditions: $y = 0, u = v = 0, T = T_w, y \rightarrow \infty, u \rightarrow U_\infty, T \rightarrow T_\infty$. Since the plate under consideration is semi-infinite, the solution cannot depend on the specific length so that the similarity variables should exist (Sec. 2.4.3). To find these, we note that it is reasonable to use as independent variable dimensionless transverse coordinate $\eta = y/\delta$, while the dimensionless stream function is convenient to apply as dependent variable since we need to calculate both velocity components. Then, knowing that according to equation (2.67) $\delta/x \sim 1/\sqrt{\mathrm{Re}_x}$, where $\mathrm{Re}_x = U_\infty x/\nu$, and that $u = \partial\psi/\partial y$, we get $\eta = y\sqrt{U_\infty/\nu x}$ and $\psi \sim uy \sim U_\infty \delta \sim \sqrt{\nu x U_\infty}$. Using the last expression as a scale for stream function, one seeks the dimensionless solution for the velocity field in the form $\psi/\sqrt{\nu x U_\infty} = f(\eta)$, whereas it is obvious that the analogous temperature distribution should be sought as $(T - T_\infty)/(T_w - T_\infty) = \theta(\eta)$, where T_w is the given constant temperature.

Substituting the velocity components

$$u = \frac{\partial \psi}{\partial y} = \frac{\partial \psi}{\partial \eta}\frac{\partial \eta}{\partial y} = U_\infty f', \ v = -\frac{\partial \psi}{\partial x} = \sqrt{vxU_\infty} f' \frac{\partial \eta}{\partial x} + \frac{1}{2}\sqrt{\frac{vU_\infty}{x}} f = \frac{1}{2}\sqrt{\frac{vU_\infty}{x}}(\eta f' - f) \quad (2.88)$$

into systems (2.68)–(2.70) for the steady flow transforms it to two ordinary differential equations with following boundary conditions (Exercise 2.66):

$$ff'' + 2f''' = 0, \ \eta = 0, f = f' = 0, \eta \to \infty, f' \to 1, \quad \theta'' + \frac{\text{Pr}}{2}f\theta' = -\text{Pr Ec}f''^2 \quad (2.89)$$

The solution of the first nonlinear equation of third order for stream function was first given by Blasius in 1908. He used two expansions: an inner one at the surface and the outer expansion for external flow matching them at some point inside the boundary layer. Substituting the inner expansion in the form of power series $f_{in} = a_2\eta^2 + a_3\eta^3 + \cdots$ into equation (2.89) gives the coefficients a_n in terms of a_2 (Exercise 2.67). This coefficient cannot be defined like the first two because the third boundary condition is specified at $y \to \infty$. Putting $a_n = C_n a_2$ and setting to zero the sums of terms with equal powers of η yields for inner expansion [1, 1951]:

$$f_{in} = \sum_{n=0}^{\infty}\left(-\frac{1}{2}\right)^n \frac{a_2^{n+1}C_n}{(3n+2)!}\eta^{3n+2} = \frac{a_2}{2}\eta^2 - \frac{a_2^2}{2\cdot 5!}\eta^5 + \frac{11a_2^3}{4\cdot 8!}\eta^8 - \frac{375a_2^4}{8\cdot 11!}\eta^{11} + \cdots. \quad (2.90)$$

The outer asymptotic solution was found as a sum of consequent approximations. The first one follows from the third boundary condition $f' = 1$ by integrating $f_1 = \eta - b_1$, where b_1 is the integration constant and minus shows that the stream function decreases because the third boundary condition is specified at $y \to \infty$. Substituting this result into equation (2.89) for f gives the linear equation $(\eta - b_1)f_2'' + 2f_2''' = 0$ for the second approximation. Double integration of this equation (Exercise 2.68) leads to an expression that being added to the first approximation results in a sum defining the outer solution:

$$f_2'' = b_2\exp\left[-\frac{(\eta - b_1)^2}{4}\right], \ f = f_1 + f_2 = \eta - b_1 + b_2\int_\infty^\eta d\xi\int_\infty^\xi \exp\left[-\frac{(\xi - b_1)^2}{4}\right]d\xi. \quad (2.91)$$

Here, the first expression is obtained after the first integration, and ξ is the dummy variable. Note that as it should be, an outer solution depends on two constants b_1 and b_2, whereas the inner one (2.90) consist only one constant a_2. Blasius determined these constants matching both solution at some suitable point obtaining $a_2 = 0.332$, $b_1 = 1.73$, $b_2 = 0.231$. In the following years, several other solutions of the Blasius equation were published, and finally Howarth gave the numerical solution of this equation, which is quoted in Table A.3. The Blasius velocity distribution $u/U_\infty = f'(\eta)$ excellently agrees with experimental data [1, Fig. 7.9].

Example 2.9 Determining the skin friction.

The local skin friction is found from the Newton law applying the value of second derivative on the surface $f''(0) = a_2 = 0.332$. Then, the total skin friction for two sides of a plate of unit width is obtained by integrating the local values (Exercise 2.69):

$$\frac{c_f}{2} = \frac{\tau}{\rho U_\infty^2} = \frac{v}{U_\infty}\sqrt{\frac{U_\infty}{vx}}f''(0) = \frac{0.332}{\sqrt{\text{Re}_x}}, \ C_f = \frac{2}{\rho U_\infty^2 L}\int_0^L \tau \, dx = \frac{0.664}{L}\int_0^L \frac{dx}{\sqrt{\text{Re}_x}} = \frac{1.328}{\sqrt{\text{Re}_L}} \quad (2.92)$$

Example 2.10 The boundary-layer thickness estimation.

In an asymptotic model, the boundary-layer thickness is considered as infinite, but for practical purpose, the model with a finite thickness is more convenient. In this case, the thickness is usually determined as distance from the surface to the point where the velocity inside the boundary layer is 1% less than that in external flow on the outer edge $u = 0.99U_\infty$ (Sec. 2.4.4). The two other using in application boundary-layer thicknesses are displacement δ_1 and momentum δ_2 thicknesses. These characteristics are strongly determined showing how much the boundary layer displaces the potential flow outward of the outer edge (the first), and how much momentum this displacement thickness takes away from the boundary layer. The local volume of external flow displaced by boundary layer of thickness dy is defined as $U_\infty d\delta_1 = (U_\infty - u)dy$, while the corresponding local amount of momentum taken away is given as $U_\infty d\delta_2 = u(U_\infty - u)dy$. Using these definitions and numerical data from Table A.3, one obtains expressions for three boundary-layer thicknesses (Exercise 2.70):

$$\delta = 5\sqrt{\frac{vx}{U\infty}}, \quad \delta_1 = \int_0^\infty \left(1 - \frac{u}{U_\infty}\right)dy = 1.7208\sqrt{\frac{vx}{U\infty}}, \quad \delta_2 = \int_0^\infty \frac{u}{U_\infty}\left(1 - \frac{u}{U_\infty}\right)dy = 0.664\sqrt{\frac{vx}{U\infty}}. \quad (2.93)$$

The growing boundary layer causes a movement of initially parallel external flow in the y-direction, yielding a transverse velocity at the outer edge. This velocity is estimated using equation (2.88) and Tablet A.3 to get $v = 0.86\sqrt{\frac{U_\infty v}{x}}$ (Exercise 2.71).

Comment 2.15 The Blasius analytical solution is presented here as an example of a method applicable to treat some nonlinear problems. The solution of a nonlinear differential equation is an always challenging task because there is no standard procedure like, for example, in the case of solving an ordinary linear differential equation with constant coefficients. Therefore, several offered exercises associated with series application are useful mathematical practice. Another reason of considering both numerical and analytical methods is to compare they underlying the advantages of each. While the numerical approach is widely applicable and relatively simpler due to the almost standard technique, the analytical methods are much helpful in understanding the physical features and singularities of a studying problem. For instance, the series (2.90) shows that the second and third derivatives of velocity at the surface are zero and the velocity changes linearly in a significant part of the boundary layer. Estimating the second term of series for velocity, one finds that this term is less than 5% if compared to the first linear term up to $\eta = 1.92$ (Exercise 2.72). So the part of about 40% of the velocity distribution across the layer is practically linear. This property is used to simplify heat transfer problems in some cases (Sec. 2.7.3).

The second equation (2.89) is a linear inhomogeneous equation of second order governing the problem of heat transfer on a flat plat. The solution of this problem is presented as a sum $\theta = \theta_1 + (Ec/2)\theta_2$ of a general solution θ_1 of homogeneous equation and a particular solution θ_2 of a full inhomogeneous equation (Sec. 1.5.1). The general solution under boundary conditions $\eta = 0, \theta_1 = 1, \eta \to \infty, \theta_1 = 0$ describes the cooling process, while the particular one subjected to the boundary conditions $\eta = 0, \theta_2' = 0, \eta \to \infty, \theta_2 = 0$ determines the adiabatic temperature (Sec. 2.4.6.2). Both equations can be integrated using a standard technique. The homogeneous equation is first integrated by separation of variables. Then, after modifying the result applying the Blasius equation, second integrating and satisfying the boundary condition, the solution published by Pohlhausen in 1921 is obtained:

$$\theta_1 = C\int_\eta^\infty \exp\left(-\Pr\int\frac{f}{2}d\xi\right)d\xi = C\int_\eta^\infty \exp[\ln(f'')\Pr]d\xi = \int_\eta^\infty [f''(\xi)]^{\Pr}d\xi / \int_0^\infty [f''(\xi)]^{\Pr}d\xi. \quad (2.94)$$

Here, the second expression is obtained using the relation $\int(-f/2)d\xi = \ln f''$ which follows from Blasius equation (2.89) (Exercise 2.73).

Example 2.11 Estimation of the thermal boundary-layer thickness.

It follows from equation (2.94) that the temperature distribution depends on the Prandtl number, and consequently the thermal boundary-layer thickness depends on the Prandtl number as well. In the case of $\Pr = 1$, we have from equation (2.94) that $\theta_1 = 1 - f' = 1 - u/U_\infty$ or $(T_w - T)/(T_w - T_\infty) = u/U_\infty$, which means that temperature and velocity distributions are identical (Reynolds analogy). Thus, for $\Pr = 1$, both dynamic and thermal boundary-layer thicknesses are equal $\delta_t/\delta = 1$. Physical reason of that is that $\Pr = v/\alpha$, and hence for $\Pr = 1$, the viscosity and thermal diffusivity of such fluid are equal, resulting in the same distances from the wall to which the viscose and thermal effects extend. In the case of fluid with $\Pr > 1$, the viscosity influence goes farther into fluid than that of the thermal effect, and because of that, the thickness of the thermal boundary layer is smaller than the velocity one $\delta_t < \delta$, whereas it is obvious that the opposite situation takes place for a fluid with $\Pr < 1$ (Exercise 2.74).

The full equation (2.89) for θ_2 defining the adiabatic temperature is a linear ordinary differential equation of the second order. Since this equation does not contain the independent variables, it can be transformed to a linear equation of the first order by substitution $z = \theta_2'$. Using the well-known general solution of the linear differential equation of the first order, one obtains the solution for θ_2' and, after integrating it and satisfying boundary conditions, gets solution of inhomogeneous problem (Exercise 2.75):

$$\theta_2'' + \frac{\Pr}{2}f\theta_2' = -2\Pr f''^2, \quad \theta_2 = 2\Pr\int_\eta^\infty [f''(\xi)]^{\Pr}\int_0^\xi [f''(\zeta)]^{2-\Pr}d\zeta\, d\xi, \quad \theta = \theta_1 + \frac{1}{2}Ec\theta_2 \quad (2.95)$$

Example 2.12 Estimation of an adiabatic wall temperature.

The adiabatic temperature is a temperature of thermally isolated wall. One example of such a temperature on an unmoving wall in Couette flow is considered in Section 2.4.6.2. In this case, equation (2.95) determines the distribution of θ_2 across the boundary layer, and at $\eta = 0$, it gives the adiabatic wall temperature (Exercise 2.76):

$$\frac{T_{ad} - T_\infty}{T_w - T_\infty} = \frac{Ec}{2}\theta_2(\Pr), \quad \theta_2(\Pr) = 2\Pr\int_0^\infty [f''(\xi)]^{\Pr}\int_0^\xi [f''(\zeta)]^{2-\Pr}d\zeta\, d\xi, \quad \theta_2(1) = 1. \quad (2.96)$$

The last result follows from the previous equation at $\Pr = 1$ because in this case, the expression under integral becomes $2f'(\xi)f''(\xi)d\xi = d[f'(\xi)]^2$, which after integration equals unity (Exercise 2.76). Consequently, according to the first equation (2.96), the adiabatic temperature is $T_{ad} = T_\infty + U_\infty^2/2c_p$ [compared to (2.53)] that means for $\Pr = 1$, the adiabatic temperature is equal to temperature rise due the friction when the velocity changes from U_∞ to zero. Calculation shows that the values of $\theta_2(\Pr)$ are greater for $\Pr > 1$ and less for $\Pr < 1$ than for unity. For moderate and large Prandtl numbers, these values can be estimated using approximation formulae $\theta_2(\text{Pr}) = \text{Pr}^{1.2}$ or $1.9\,\text{Pr}^{1/3}$ for $0.6 < \text{Pr} < 10$ and $\text{Pr} \to \infty$, respectively [1].

Example 2.13 Estimation of heat fluxes.

The heat flux is determined by the value of temperature derivative on the surface. Defining this derivative from the last equations (2.94) and recalling that $\theta_2'(0) = 0$, we calculate it and denote as $\theta_1(\text{Pr})$. Then, knowing that in the problem in question the wall is considered to be at an adiabatic temperature, we determine the heat flux in the form $q_w = \theta_1(\text{Pr})(T_{ad} - T_\infty)$ and after using formula (2.95) for adiabatic temperature obtain the expression for the Nusselt number (Exercise 2.77):

$$-\left(\frac{\partial \theta}{\partial \eta}\right)_{\eta=0} = 0.332^{\text{Pr}}/\int_0^\infty \left[f''(\xi)\right]^{\text{Pr}} d\xi = \theta_1(\text{Pr}),$$

$$Nu_{\mathbf{x}} = \frac{q_w x}{\lambda(T_w - T_\infty)} = \theta_1(\text{Pr})\sqrt{\text{Re}}_x\left[1 - \frac{\text{Ec}}{2}\theta_2(\text{Pr})\right]. \tag{2.97}$$

Some values of $\theta_1(\text{Pr})$ as well as the values of $\theta_2(\text{Pr})$ indicated above are given in [1]:

$$\theta_1(\text{Pr}) = 0.564\,\text{Pr}^{1/2}, 0.332\,\text{Pr}^{1/3} \text{ and } 0.339\,\text{Pr}^{1/3} \text{ for } \text{Pr} \to 0, 0.6 < \text{Pr} < 10 \text{ and } \text{Pr} \to \infty.$$

It follows from the last expression (2.97) that $Nu_x > 0$ if $\text{Ec}\theta_2(\text{Pr}) < 2$ and $Nu_x < 0$ if $\text{Ec}\theta_2(\text{Pr}) > 2$. This means that the fluid cools a plate only until the plate temperature is greater than the adiabatic one $T_w > T_{ad}$, and then the heat flux changes the direction to opposite from the fluid to a plate heating it. Thus, in the case of isolated plate, the cooling process is considerably reduced comparing to that for a plate at given temperature when the fluid cools the plate as long as $T_w > T_\infty$. Such a situation occurs due to the fact that flow heated by friction separated cold fluid from the surface (Sec. 2.4.6.2).

2.6.2 Self-Similar Flows in Dynamic and Thermal Boundary Layers

As is shown in Example 2.5, the velocity distribution of a potential flow over the wedge with opening angle $\pi\beta$ has the form of power function $U = Cx^m$, where $m = \beta/(2-\beta)$. If the temperature head distribution along the wedge surface is also of the power-law type $T_w - T_\infty = C_1 x^{m_1}$, then both boundary-layer partial differential equations for velocity and for temperature head can be reduced to ordinary differential equations, which have self-similar solutions. Such solutions describe flows in which the velocity or temperature head profiles at different locations along a flow are similar so that in similarity variables they coincide forming one profile (Sec. 2.4.3). We considered several problems of that type in Sections 2.4.3 and 2.4.4 as exact solutions of the Navier–Stokes equation. The just-discussed problem of boundary layers on a flat plate is also of this type being a particular case of self-similar problems in question at $m = m_1 = 0$. Therefore, the similarity variables analogous to employed in particular case (Sec. 2.6.1) are applicable in the general problem. Using these variables, one obtains equations similar to system (2.89):

$$f''' + ff'' + \beta(1 - f'^2) = 0, \quad \theta'' + \Pr f\theta' + \frac{2m_1}{m+1}\Pr f'(1-\theta) = 0,$$

$$f = \psi\sqrt{\frac{m+1}{2\nu Cx^{m+1}}}, \theta = \frac{T - T_w}{T_\infty - T_w}, \tag{2.98}$$

with an independent variable $\eta = y\sqrt{(m+1)Cx^{m-1}/2\nu}$.

Comment 2.16 These variables were used by Falkner and Skan who first considered equation (2.98) for self-similar velocity boundary layers. It is seen that in the case of a flat plate ($m = 0$), their variables differ from the Blasius variable by factor $\sqrt{2}$. Therefore, equations (2.89) and (2.98), which should be identical, differ as well by factor 2.

The numerical solution of the Folkner–Skan equation for different values of β was obtained by Hartree [1, Fig. 9.1]. These results show that the behavior of accelerated and decelerated flows is essentially different. In the first case to which correspond flows with a range of exponent from $m = 0, \beta = 0$ (plate) to $m = 1, \beta = 1$ (stagnation point), the pressure is constant or decreases in the flow direction. In such flows, the velocity profiles in the boundary layer are regular without singularities, and the flow has a normal structure. In contrast to that, in the other case to which correspond flows with small range of negative exponent only to $m = -0.091, \beta = -0.199$, the pressure increases, and the velocity profiles first exhibit a point of inflexion and then separation occurs. Thus, the laminar boundary-layer flow is able to achieve only small increasing pressure gradients without separation. This result is in conformity with that obtained in studying flows in divergent channels (Sec. 2.4.5).

The numerical solution of the second equation (2.98) for the thermal boundary layer was performed by several authors. Detailed data are given, for example, in [18]. Employing these results, the basic friction and heat transfer characteristics may be obtained. Here, we present like for a flat-plate friction drag coefficient and asymptotic relations for heat transfer for the stagnation point at $\Pr \to 0, \Pr \approx 1$, and $\Pr \to \infty$, which are important being applicable to the entrance of blunt bodies:

$$\frac{c_f}{2} = \frac{1.233}{\sqrt{\mathrm{Re}_x}}, \quad \frac{Nu_x}{\sqrt{\mathrm{Re}_x}} = 0.791\Pr^{1/2}, \ 0.570\Pr^{1/3}, \ 0.661\Pr^{1/3}. \tag{2.99}$$

Example 2.14 Effect of thermal boundary conditions.

In the case of power-law velocity and temperature head, the heat flux changes along a surface also by power law with exponent $m_2 = m_1 + (m-1)/2$ (Exercise 2.79).

Thus, in the case of boundary condition $q_w = const.$ ($m_1 = 0$), the temperature head changes with exponent $(1-m)/2$. Hence, on a flat plate with constant heat flux, the temperature head rises as $x^{1/2}$, whereas the leading edge of a blunt body ($m = 1$) in this case is isothermal. This example shows how remarkably different can be the effect of boundary conditions. On the surface at a stagnation point, both conditions $q_w = const.$ and $T_w = const.$ result in the same heat transfer coefficient, but for a flat plate with constant heat flax, heat transfer coefficients are greater than those for an isothermal plate by 36% (0.453/0.332) and by 57% (0.885/0.564 for mediate Prandtl numbers and $\Pr \to 0$.

2.6.3 Solutions in the Power Series Form

As early as in 1908, Blasius developed a method of solution of the boundary-layer equations for the case when the external flow velocity around a symmetrical body is presented in the form of power series. Later on, Frössling [19] extended the Blasius approach to solution of heat transfer problem for a symmetrical body when both external flow velocity and temperature head are presented by power series:

$$U(x) = u_1 x + u_3 x^3 + u_5 x^5 + \cdots, \quad T_w - T_\infty = \theta_w(x) = \theta_{w0} + \theta_{w2} x^2 + \theta_{w4} x^4 + \cdots. \quad (2.100)$$

Substitution of the solution presented in a similar form of power series:

$$\psi(x, y) = \psi_1(y)x + \psi_3(y)x^3 + \psi_5(y)x^5 + \cdots$$
$$T - T_\infty = \theta(x, y) = \theta_0(y) + \theta_2(y)x^2 + \theta_4(y)x^4 + \cdots \quad (2.101)$$

into boundary-layer equations results in a system of ordinary differential equations defining the coefficients $\psi_n(y)$ and $\theta_n(y)$ of series (2.101). These coefficients depend on the body shape and temperature distribution [i.e., on coefficients in (2.100)] so that in the Blasius approach, the system of ordinary differential equations should be solved for each specific problem. Howarth showed that using the similarity variable $\eta = y\sqrt{u_1/x}$ converts equations for series coefficients into universal form independent on a particular problem. In such a case, the solution of the system of differential equations gives the universal coefficients of series (2.101) and their derivatives. These data are tabulated (i.e., [1, 1968]) and may be used for practical calculations of velocity and temperature distribution in the boundary layer as well as for the friction and heat transfer coefficients estimation. Corresponding relations are obtained by differentiating equations (2.101):

$$\frac{c_f}{2} = \frac{\sqrt{u_1 \nu}}{U_\infty^2}(1.233u_1 x + 2.898u_2 x^3 + 6.192u_5 x^5 + 16.296u_7 x^7 + \cdots) \quad (2.102)$$

$$Nu_x = \frac{x}{\theta_w}\sqrt{\frac{u_1}{\nu}}\left\{\theta_{w0}\left[0.4959 + 0.4764\frac{u_3}{u_1}x^2 + \frac{u_3}{u_1}\left(1.054 + 0.6678\frac{u_3^2}{u_1 u_3}\right)x^4\right]\right.$$
$$\left. + \theta_{w2}\left[0.852x^2 + 0.6678\frac{u_3}{u_1}x^4 + \cdots\right] + \theta_{w4}\left[1.054x^4 + \cdots\right] + \cdots\right\}, \quad (2.103)$$

where the coefficients in the last equation are estimated for $Pr = 0.7$ (air) (Exercise 2.80).

The basic shortage of this method is the slow convergence of the power series. Due to that, the satisfactory results can be achieved only at relatively small part of a surface close to a stagnation point. To get accurate data far from the leading edge up to a point of separation, one needs a number of universal functions that rapidly increases with each following term of solution series. This situation becomes extremely difficult in the case of an asymmetrical body, which requires the series with much more terms. Nevertheless, for the case of flow over a symmetrical body, the universal functions have been calculated up to terms containing x^{11}, and the velocity profiles on the circular cylinder, including a point of separation, are estimated. However, the result accuracy is much higher for accelerated flows than for the flows with increasing pressure, especially close to the separation point. Thus, according to the numerical solution, the separation point on a circular cylinder is at angel 104.5°, whereas the series method gives 108.8°.

2.6.4   Flow in the Case of Potential Velocity $U(x) = U_0 - ax^n$ (Howarth Flow)

In a simplest case of linear distribution with $n = 1$, this flow can be considered as that in a channel consisting of two sections, an entrance part with constant velocity U_0 and following convergent $(a < 0)$ or divergent $(a > 0)$ section. Howarth treated this problem as a particular case of Blasius series and checked the results obtained in that way by numerical solution. He found the velocity profiles in the boundary layer [1, Fig. 9.8] which in conformity with those in other flows of this type (Sec. 2.4.5, 2.6.2) in the case of decreasing pressure $(a < 0)$ are regular but have the specific form with an inflexion point in the opposite case when the pressure increases. The dimensionless coordinate of the separation point was estimated as $ax/U_0 = 0.12$.

2.6.5   Fluid Flows Interaction

The boundary-layer model is applicable as well to study the interaction between two streams with different velocities. The boundary layer occurs on the interface due to viscosity and grows in the flow direction. We consider three problem of this type.

2.6.5.1   Flow in the Wake of a Body

Behind the trailing edge of a body, the two flows come in contact, and their velocity profiles merged forming one profile of the wake. As the distance from the trailing edge increases, the width of wake grows and the mean velocity decreases. Although close to the edge, the wake characteristic depends markedly on a body shape, at the large distance the velocity distribution in the wake is independent of the shape of the body, besides a scale.

The method of calculation of the asymptotic wake profile was developed by Tollmien. It is based on the momentum equation, defining the momentum loss (the drag), and its linearization using assumption that the velocity difference $u_1 = U_\infty - u$ in the wake is small in comparison with the velocity of the flow far from the body. The linearization means that in some expression, the only term proportional to u_1 is taken into account, whereas all terms containing higher power of u_1 are neglected. Applying for linearization the relation $u = U_\infty - u_1$, we find from equation (2.93) for momentum thickness δ_2 the expression (2.104) for drag per unit mass and unit width. The linearized boundary-layer equation [the second equation (2.104)] is obtained using the same simple relation for u and knowing that the pressure in the wake is constant (Exercise 2.81):

$$D = 2 \int_0^\infty u(U_\infty - u)\, dy = 2U_\infty \int_0^\infty u_1 dy, \quad U_\infty \frac{\partial u_1}{\partial x} = v \frac{\partial^2 u_1}{\partial y^2}, \quad y = 0, \frac{\partial u_1}{\partial y} = 0 \quad y \to \infty, u_1 = 0. \quad (2.104)$$

The first boundary condition (2.104) is the symmetry condition; the second follows from u_1 definition given above.

As an example of a complete solution, we consider a wake behind a flat plate. In this case, the Blasius similarity variable η transforms the partial differential equation (2.104) into an ordinary differential equation. In addition, it is clear from previous examples that the local drag should depend on $(x/L)^n$, whereas the total drag should be independent of this parameter. Therefore, it should be taken $n = -1/2$ because in this case assuming $u_1 = CU_\infty (x/L)^{-1/2} f(\eta)$, we get from first equation (2.104) total drag independent of coordinate (Exercise 2.82) and after substitution this relation for u_1 into boundary-layer equation (2.104) obtain the differential equation for function $f(\eta)$ (Exercise 2.83):

$$D = CU_\infty^2 \sqrt{\frac{\nu L}{U_\infty}} \int_{-\infty}^{\infty} f(\eta)\,d\eta, \quad f'' + \frac{1}{2}(\eta f' + f) = 0, \quad y = 0, f' = 0 \quad y \to \infty, f = 0. \quad (2.105)$$

The solution of this equation yields $f = \exp\left[-\left(1/4\eta^2\right)\right]$. Estimating by this function the integral containing in the first equation (2.105) via an error function (1.27) gives the value $2\sqrt{\pi}$ (Exercise 2.84). Then, comparing the total drag (2.105) with the Blasius formula (2.92), we get the constant C from first equation (2.106) and finally obtain the solution for the velocity difference $u_1 = U_\infty - u$ in the wake behind the flat plate:

$$2\sqrt{\pi} CU_\infty^2 \sqrt{\frac{\nu L}{U_\infty}} = \frac{1.328}{\sqrt{\mathrm{Re}}} U_\infty^2 L, \quad \frac{u_1}{U_\infty} = \frac{0.664}{\sqrt{\pi}}\left(\frac{x}{L}\right)^{-1/2} \exp\left(-\frac{1}{4}\frac{y^2 U_\infty}{\nu x}\right). \quad (2.106)$$

This is an asymptotic velocity distribution taking place far away from a flat plate according to Tollmien estimation at the distance $(x/L) > 3$. The laminar wake is unstable even at relatively low Reynolds numbers [i.e., usually is turbulent (Sec. 3.3.6.1)].

2.6.5.2 Two-Dimensional Jet

A stream coming into surrounding from a slit entrances the fluid around it due to the friction forming a jet. In the model describing this process, it is assumed that the slit is infinitely small, but the initial stream velocity is infinite. These two assumptions provide a finite amount of fluid inside a jet and, consequently, finite momentum of it. If the surrounding is initially at rest, and its pressure is constant, the momentum of a jet remains constant as well during its development. Therefore, as jet spreads in the downstream direction, the entraining amount of fluid and corresponding jet width grow yielding a velocity decreasing according to the constant momentum.

Since the problem does not have any linear scale, there should be used a similarity variable y/δ, like in the previous problem, but with $\delta \sim x^n$, where δ is the jet width.

From these considerations, we have

$$I = \rho \int_{-\infty}^{\infty} u^2\,dy = cons., \quad \psi \sim x^m f\left(\frac{y}{\delta}\right) = x^m f\left(\frac{y}{x^n}\right), \quad u = \frac{\partial \psi}{\partial y} \sim x^m f'\left(\frac{y}{x^n}\right) x^{-n}. \quad (2.107)$$

The exponents n and m one defines knowing that (i) the momentum I as well as velocity depends on y/δ but does not depend on x and (ii) the inertia and viscous terms in equation (2.69) are of the same order of magnitude. The first condition gives $u^2 \sim x^{2m-2n}$ and $d(y/\delta)\delta \sim x^{-n}$, so it should be $2m - n = 0$. From the second condition, it follows the $2m - 2n - 1 = m - 3n$. Thus, $n = 2/3$, $m = 1/3$ (Exercise 2.85). Using these data, we introduce the independent variable η and stream function ψ analogous to Blasius variables, knowing that the former is inversely and the latter is directly proportional to $v^{1/2}$. Then, the velocity components are determined and after substituting the velocity components into the steady-state and zero-pressure gradient boundary-layer equation (2.69), an equation for function $f(\eta)$ is obtained:

$$\eta = \frac{1}{3v^{1/2}} \frac{y}{x^{2/3}}, \quad \psi = v^{1/2} x^{1/3} f(\eta), \quad u = \frac{1}{3x^{1/3}} f'(\eta),$$

$$v = -\frac{v^{1/2}}{3x^{2/3}}(f - \eta f'), \quad f'^2 + ff'' + f''' = 0. \tag{2.108}$$

Comment 2.17 In fact in expression for a similarity variable η, the factor is arbitrary, but only its value 1/3 provides for function $f(\eta)$ an equation (2.108) with unity coefficients (Exercise 2.86).

Boundary conditions for differential equation (2.108) follow from symmetry of the problem $\eta = 0, v = 0, \partial u/\partial y = 0$, which yields $f = f'' = 0$ and from the fact that far from a slit the surrounding is undisturbed $\eta \to \infty, u = 0$ and hence $f' = 0$. To integrate the last equation (2.108), note that the first two terms may be presented as $(ff')'$ so that the first integration of equation (2.108) gives $ff' + f'' = 0$, where the integration constant is taken zero according to conditions at $\eta = 0$. For the second integration, we transform the result of the first integration applying a substitution $f = 2f_1$ to get the form $2f_1 f_1' + f_1'' = 0$ that after integration yields an equation $f_1^2 + f_1' = C^2$. The integration constant is taken as C^2 in the view of f_1^2 for the next integration, resulting in the solution of the problem:

$$\eta = \int_0^\infty \frac{df_1}{C^2 - f_1^2} = \frac{1}{C} \tanh^{-1} \frac{f}{2C}, \frac{f}{2C} = \tanh C\eta, \frac{f'}{2C} = 1 - \tanh^2 \xi, u = \frac{2C^2}{3x^{1/3}}(1 - \tanh^2 \xi) \tag{2.109}$$

Here, the first integral is taken from table, the second expression is obtained by inversing the result of integration, variable $\xi = C\eta$, and velocity is found using equation (2.108) (Exercise 2.87). The constant C is determined through momentum I, which is constant for a particular jet and is determined by excess pressure behind the pressure at slit. Substituting equations (2.109) for velocity and (2.108) for variable η into first equation (2.107), one obtains a corresponding relation $I = (16/9)\rho v^{1/2} C^3$ [1].

2.6.5.3 Mixing Layer of Two Parallel Streams

Two streams with uniform velocities $U_1 > U_2$ coming into surrounding in contact form a mixing layer due to friction between each other. Since the pressure gradient is zero, the steady-state boundary-layer governing equations are the same as in the case of a flat plate. The similarity variables are as well defining by one of velocities, for example, $\eta = y\sqrt{U_1/vx}$ and $\psi = \sqrt{vU_1 x} f(\eta)$. Then, the ordinary differential equation determining function $f(\eta)$ is the Blasius equation (2.89). If one assumes that the stream with velocity U_1 is above the other, the boundary conditions are: $\eta \to \infty, f'(\eta) = 1$ and $\eta \to -\infty, f'(\eta) = U_2/U_1$, where $f'(\eta) = u/U_1$. The problem is nonlinear so that only numerical and asymptotic in series solutions are known [1]. The final similarity profiles depend on ratio U_2/U_1 and because of zero pressure gradient are regular.

2.6.6 Flow in Straight and Convergent Channels

Flows in channels and tubes constitute another type of boundary-layer problems. In contrast to problems of flow past bodies and fluid flow interaction considered above, in which the flow domain in the y-direction is infinite, in this case, the flow is confined by channel or tube walls. Due to that in such a flow, the boundary layer exists only on some entrance part of the walls until the growing boundary layers merged forming a velocity profile which finally asymptotically becomes the parabolic Poiseuille profile (Sec. 2.4.3). If the inlet velocity profile is uniform, at the beginning close to inlet, the boundary layer develops as on a flat plate under almost constant velocity. As the distance from the inlet increases, the boundary-layer thickness increases as well, and the flow becomes divided in two parts, a central flow with uniform velocity and boundary layer on the parallel walls in a channel or around the surface in the circular tube. Since the mean velocity in the boundary layer is less than that at the inlet and the rate of flow is the same through a channel or a tube, the uniform velocity of the central part of flow increases downstream as a result of a compensation of the velocity decreasing in the boundary later at the walls. Thus, the flow pattern analysis shows that in a channel or in a tube, the boundary layer develops under an accelerating external flow existing near the axis until the gradually decreasing width of this area becomes zero due to the growing boundary layer, and a full velocity profile establishes that asymptotically converts in parabolic velocity distribution (Exercise 2.88).

The flow in the inlet length of the two-dimensional channel was considered by Schlichting. He used a method similar to the Blasius approach. Two series expansions in downstream and upstream directions are matched at some point where they both are applicable. The expansion in the downstream direction starts at the inlet and is presented as a series in $x^{1/2}$ since at the beginning the boundary layer grows as on the flat plate. The expansion in upstream direction starts from a parabolic velocity distribution and is presented as a perturbation series using the deviation from the parabolic profile as a small parameter. The inlet length was estimated as $L/H = 0.08\,\mathrm{Re}$, where the Reynolds number is defined by the inlet velocity and half of channel width H. The details may be found in [1]. Note that this is also an example of the solution of nonlinear problem (Comment 2.14).

The boundary layer in the convergent channel is another problem of this type solved exactly. This problem is one of those for which the exact solutions of the Navier–Stokes equation are known (Sec. 2.4.5). This problem is also a particular case of self-similar flows with external velocity $U = -u_1/x$. Since in this case the pressure decreases ($dU/dx > 0$), the boundary layer is regular without separation. Similarity variables lead to the ordinary differential equation solved analytically [1].

2.6.7 Solutions of Second-Order Boundary-Layer Equations

The first attempts to obtain a correction to the Blasius formula for skin friction were unsuccessful due to singularity at the leading edge of the semi-infinite plate. As it follows from the solution of the second-order boundary-layer system, the local skin friction at the leading edge is proportional to $x^{-3/2}$. In this case, the total skin friction cannot be computed since the corresponding integral goes to infinity as $x^{-1/2}$ at $x \to 0$. Imai [20] overcame this difficulty considering the balance of momentum, including the external flow. As a result, he found a correction to Blasius formula (2.92) as $2.326/\mathrm{Re}_L$.

In general, the singularity at the leading edge of the semi-infinite plate arises because the boundary-layer equation is a parabolic differential equation, solutions of which cannot take into account edge effects. To get the proper result, one should consider the problem for a plate of finite length with boundary conditions at the edges. This can be done by solving the Dirichler problem for elliptical Navier–Stokes equation (Sec. 2.3). The result of the numerical solution obtained by that way is well described by formula $2.668/\mathrm{Re}_L^{7/8}$. These two equations as well as Van Dyke relation $1.7/\sqrt{\mathrm{Re}_x}$ give an estimation (with some differences) of the accuracy of the first boundary-layer approximation (Exercise 2.39). Other cases of second-order boundary layer, in particular, the stagnation point flows at blunt-nosed bodies, are also considered by Van Dyke [10] (see also [1, p. 194]).

2.6.8 Solutions of the Thermal Boundary-Layer Equation

The exact solutions of the thermal boundary-layer equation may be constructed using superposition of the known simple solutions since this equation is linear. One well-known example is the article published by Chapman and Rubesin in 1949 [21]. They considered the heat transfer from a nonisothermal plate with polynomial surface temperature distribution. In this case, the solution is given as a sum of similarity solutions for the power-law surface temperature distribution. Another example of exact solutions of this type is the universal function (1.23) composed as a sum of consecutive derivatives of the surface temperature distribution. Such exact solutions describe the heat flux distribution on the surface with given temperature head and is widely used in following chapters in considering nonisotermal and conjugate heat transfer. One way to find such universal functions is to apply a well-known general solution for a plate (Sec. 4.2.3):

$$q_w = h_* \left[\theta_w(0) + \int_0^x f(\xi/x) \frac{d\theta_w}{d\xi} d\xi \right] \qquad f(\xi/x) = \frac{h_\xi}{h_*}. \qquad (2.110)$$

This relation obtained by Duhamel method determines the heat flux on the nonisothermal surface with given arbitrary temperature head distribution $\theta_w(x)$ in zero-pressure gradient flow. As is explained in Section 1.4, the Duhamel method allows one to get a solution of a problem with arbitrary boundary condition $\theta_w(x)$ in terms of some known solution of an analogous simple problem. In this case, as a simple solution is used the influence function $f(\xi/x)$, so called because it describes the effect of the unheated zone on the relative heat transfer coefficient h_ξ/h_*, occurring after a surface temperature jump in a standard problem. In such a problem, the temperature of an initial unheated part of a wall remains at the external flow temperature up to some point $x = \xi$ and at this point suddenly changes to another value resulting in heat transfer whose intensity is determined by influence function (2.110) (Example 2.21).

EXERCISES

2.66 Derive velocity components (2.88) and both Blasius and Pohlhausen equations (2.89).

2.67 Obtain relations for coefficients a_n in terms of a_2 by putting power series into the Blasius equation and collecting terms with equal power of η.

2.68 Derive the first expression (2.91) for f_2'' by first integrating the linearized Blasius equation. Hint: use substitution $z = f''$ and define the integration constant as $-b_1^2/4 + \ln b_2$.

2.69 Obtain the expressions (2.92) for local and total skin friction using the Newton law.

2.70 Derive formulae (2.93) for boundary thicknesses following directions from the text.

2.71 Calculate transverse velocity for different values of η applying equation (2.88) and plot $(v/U_\infty)/\sqrt{v/xU_\infty}$ versus η to see variation of this velocity along a plate. Explain why it grows with distance from the leading edge. Hint: use Table A.3.

2.72 Show that a significant part of velocity distribution in the boundary layer is linear. Compare your result with numerical data from Table A.3 by drawing the graph $u/U = f'(\eta)$. Hint: calculate the ratio of two first terms of series (2.90).

2.73 Obtain Pohlhausen solution (2.94). Hint: (i) solve the homogeneous equation (2.89) using the substitution $z = \theta'$ to get the first equation (2.94); (ii) apply the same procedure to Blasius equation (2.89) putting $z = f''$ to get the relation $\int(-f/2)d\xi = \ln f''$; (iii) use this relation to get the second expression (2.94) for θ; and (iv) modify the last equation employing algebra and satisfy the boundary condition $\theta = 1$ at $\eta = 0$ to find the Pohlhausen solution (2.94).

2.74 Describe how Prandtl number affects the relation of thermal and dynamic boundary-layer thicknesses. Explain physically those results.

2.75 Find the solution (2.95) of inhomogeneous equation (2.95) for θ_2. Hint: (i) use the substitution $z = \theta_2'$ to reduce equation (2.95) to the first-order differential equation; (ii) find in some *Advanced Engineering Mathematics* course the general solution of a linear first-order ordinary differential equation to get a solution for $z = \theta'$; (iii) integrate the result for θ_2' to obtain expression for θ_2; (iv) modify the resulting equation using relation $\int(-f/2)d\xi = \ln f''$ like in Exercise 2.73; (v) satisfy the boundary condition for θ_2 to determine the constants and limits of integrals and obtain the final solution (2.95) for θ_2.

2.76 Show following directions in the text that the second expression (2.96) for $Pr = 1$ equals unity.

2.77 Derive the expression (2.97) for the Nusselt number and analyze the difference between two cases of cooling a plate with prescribed and adiabatic surface temperature.

2.78 Compare characteristics of flows with positive and negative pressure gradient, give examples of each type of flow.

2.79 Show that for the case of power-law velocity and temperature head, the heat flux on the surface changes also according power law with exponent $m_2 = m_1 + (m-1)/2$. Compare effects of different boundary conditions for plate and blunt body. Hint: use formula $Nu_x/\sqrt{Re_x}$ for self-similar flows.

2.80 Prove that for the case of a stagnation point when in equations (2.102) and (2.103) remains only the first term, they become equations (2.99) (may be with slightly different numerical factors). Hint: in this case, the first term in equation (2.102) is $u_1x = U$.

2.81 Show that the linearized steady-state boundary-layer equation (2.69) becomes (2.104).

2.82 Derive relation for drag (2.105) from (2.104) using formula $u_1 = CU_\infty (x/L)^{-1/2} f(\eta)$ to show that assumption $n = -1/2$ provides the expression for total drag independent of coordinate.

2.83 Derive differential equation and boundary condition (2.105) for function $f(\eta)$ by substituting assuming function for u_1 (Exercise 2.82) into boundary-layer equation (2.104). Hint: divide the result after substitution by $CU_\infty^2 (x/L)^{-1/2} x^{-1}$.

2.84 (i) Solve differential equation for function $f(\eta)$ (2.105) and (ii) estimate the value of integral of $f(\eta)$ in equation (2.105) for drag by comparing it with error function (1.27). Hints: (i) note that $(\eta f' + f) = (\eta f)'$, (ii) in integral of function $f(\eta)$ make a substitution $(1/4)\eta^2 = \xi^2$.

2.85 (i) Repeat the derivation of the first equation defining exponents n and m, (ii) estimate the order of the first inertia and viscous terms in boundary-layer equation (2.69) to get the second equation for these exponents, and (iii) determine n and m.

2.86 *Prove that similarity variable η (2.108) only with the factor $1/3$ results in equation (2.108) for $f(\eta)$ with unity coefficients. Hints: (i) use variable η with an arbitrary factor C instead of $1/3$, (ii) derive the velocity components u and v using this variable and stream function in equation (2.108), (iii) use the expressions obtained for u and v to find the terms of boundary-layer equation (2.69) (without $\partial u/\partial t$ and $\partial p/\partial x$) for getting the last equation (2.108) in the form $(1/3)f'^2 + (1/3)ff'' + Cf''' = 0$.

2.87 Obtain yourself expression (2.109) for jet velocity by the same way (as in the text) beginning from integrating the differential equation $f_1^2 + f_1' = C^2$. Explain why the integration constant is taken as C^2.

2.88 Describe the difference between two types of boundary-layer flows: past bodies and inside channels.

2.7 Approximate Methods in the Boundary-Layer Theory

The approximate methods were developed in early time before the computers came into existence, and that time, they were widely employed to solve the boundary-layer problems. Even now, despite the presence of the advanced numerical methods, the approximate solutions are helpful in getting the preliminary simpler result, especially in the case of complex problems. However, it should be noted that such simple solutions giving the approximate numerical results cannot substitute the exact analytical solutions in understanding the physical and/or mathematical singularities of a studying problem.

2.7.1 Karman-Pohlhausen Integral Method

The idea of the integral method consists of reducing the two-dimensional boundary-layer problem to one-dimensional task by integrating the boundary-layer equation in the y-direction. Since the integration results in losing the information across the boundary layer, the solution of the one-dimensional integral equation obtained in such a way may give a data only in the downstream direction. Therefore, to get a solution by an integral method, one needs to find some function satisfying the given boundary conditions for describing the profiles in the boundary layer. The examples below provide the details of this relatively simple procedure.

The momentum and energy integral equations are obtained from the steady-state boundary-layer systems (2.68)–(2.70) simplified by using equation (2.71) and by omitting the last term in equation (2.70). For integrating, we present these equations in the form

$$\frac{\partial(Uu)}{\partial x}+\frac{\partial(Uv)}{\partial y}-u\frac{dU}{dx}=0,\quad \frac{\partial(T_\infty u)}{\partial x}+\frac{\partial(T_\infty v)}{\partial y}=0,\quad \frac{\partial u^2}{\partial x}+\frac{\partial(uv)}{\partial y}-U\frac{dU}{dx}-v\frac{\partial^2 u}{\partial y^2}=0. \quad (2.111)$$

The two first equations are obtained by multiplying the continuity equation (2.68) by U or T_∞ in the first and second case, respectively [the first equation is modified by adding terms $\pm u\left(dU/dx\right)$. The third equation is a sum of the continuity equation multiplied by u and the simplified momentum equation (2.69).

The final form of momentum integral equation (2.114) (the first equation) one gets by integrating across the boundary layer the difference between the first and the third equations (2.111). The similar integration of the difference between energy equation (2.70) and the second equation (2.111) yields the energy integral equation (2.114) (the second equation). These procedures consist of several steps: (i) the first and second differences give the following expressions, respectively

$$\int_0^\infty \frac{\partial}{\partial x}[u(U-u)]\,dy+\int_0^\infty \frac{\partial}{\partial y}[v(U-u)]\,dy+\frac{\partial U}{\partial x}\int_0^\infty (U-u)\,dy=v\int_0^\infty \frac{\partial^2 u}{\partial y^2}\,dy \quad (2.112)$$

$$\int_0^\infty \frac{\partial}{\partial x}[u(T-T_\infty)]\,dy+\int_0^\infty \frac{\partial}{\partial x}[v(T-T_\infty)]\,dy=\alpha\int_0^\infty \frac{\partial^2 T}{\partial y^2}\,dy. \quad (2.113)$$

(ii) these expressions are modified by changing in both first terms the integration and differentiation, and by taking into account that (iii) both second terms vanish because the results of integration in these terms equal zero at limits $y=0$ and $y\rightarrow\infty$ and (iv) the integration in the last terms results in shear stress on the surface in expression (2.112) and in heat flux on the surface in expression (2.113) because both derivatives $\partial u/\partial y$ and $\partial T/\partial y$ are zero at $y\rightarrow\infty$. Finally, the momentum and energy integral equation are presented in terms of displacement δ_1, momentum δ_2, and energy δ_{2t} boundary thicknesses (Exercises 2.89, 2.90)

$$\frac{\partial}{\partial x}(U^2\delta_2)+U\frac{\partial U}{\partial x}\delta_1=\frac{\tau_w}{\rho},\qquad \frac{\partial}{\partial x}(U\theta_w\delta_{2t})=\frac{q_w}{\rho c_p} \quad (2.114)$$

$$\delta_1=\int_0^\infty\left(1-\frac{u}{U}\right)dy,\quad \delta_2=\int_0^\infty\frac{u}{U}\left(1-\frac{u}{U}\right)dy,\quad \delta_{2t}=\int_0^\infty\frac{u}{U}\frac{T-T_\infty}{T_w-T_\infty}\,dy.$$

As is mentioned above, to solve the problem using an integral equation, one needs to approximate the profiles in the boundary layer by some function. Such a function $f\left(y/\delta\right)$ of similarity variable $\eta=y/\delta$ should satisfy some boundary conditions on the surface and at the outer edge of boundary layer. Substitution of the approximate profiles $f(\eta)$ into integral equation gives one-dimensional differential equation determining the boundary-layer thickness $\delta(x)$. Knowing function $\delta(x)$, one computes the profiles and boundary-layer characteristics.

2.7.1.1 Friction and Heat Transfer on a Flat Plate

Example 2.15 Friction on a flat plate.

We consider this simple example using three functions for velocity profiles u/U

$$f_1(\eta) = \eta, \; f_2(\eta) = a_0 + a_1\eta + a_2\eta^2 + a_3\eta^3, \; f_3(\eta) = a_0 + a_1\eta + a_2\eta^2 + a_3\eta^3 + a_4\eta^4 \qquad (2.115)$$

to illustrate the role of approximating functions in accuracy of an integral method.

Equation (2.115) should satisfy some of the following boundary conditions:

$$\eta = 0 \;\; f = 0, \;\; v\frac{\partial^2 u}{\partial y^2} = \frac{dp}{dx}, \; \frac{\partial^2 f}{\partial y^2} = 0 \qquad \eta = 1 \;\; f = 1, \; \frac{\partial f}{\partial y} = \frac{\partial^2 f}{\partial y^2} = 0. \qquad (2.116)$$

The second condition follows from steady boundary equation (2.69) since on the surface $u = v = 0$, and it is seen from this condition that in the case of zero-pressure gradient the third condition is valid. Other conditions (2.116) are obvious.

The first function in equation (2.115) satisfies both first conditions at $\eta = 0$ and $\eta = 1$. The coefficients of two other profiles are determined satisfying the same two conditions at $\eta = 0$ and $\eta = 1$, the third condition at $\eta = 0$ and the zero condition for the first derivative at $\eta = 1$. These four conditions gave the system of algebraic equations determining the coefficients of the second profile. The condition for second derivative at $\eta = 1$ is used in addition to find the coefficients of the third profile. Simple calculation yields (Exercise 2.91) $a_0 = a_2 = 0, a_1 = 3/2, a_3 = -1/2$ for the second profile, $a_0 = a_2 = 0, a_1 = -a_3 = 2, a_4 = 1$ for the third profile. Using these profiles, one calculates displacement and momentum thicknesses (2.114) and shear stress to get the momentum integral equation (2.114). As is described above, this procedure leads to differential equation for $\delta(x)$ and finally results in the friction coefficient and other boundary-layer characteristics. Thus, we have for calculation the following expressions:

$$\frac{\delta_2}{\delta} = \int_0^1 f(1-f)d\eta, \;\; \frac{\tau_w}{\rho} = \frac{v a_1 U_\infty}{\delta}, \;\; \delta\frac{d\delta}{dx} = \frac{\delta a_1 v}{\delta_2 U_\infty}, \;\; \delta = \sqrt{\frac{2a_1\delta v x}{\delta_2 U_\infty}}, \;\; C_f = \sqrt{\frac{2v a_1\delta_2}{\delta x U_\infty}}. \qquad (2.117)$$

The first two expressions are obtained using formulae defining δ_2 and τ_w; the third differential equation is found from momentum integral equation; the fourth relation is a result of solution of this differential equation for δ; the friction coefficient is found from the third formula for τ_w; and a_1 is the first coefficient of any function f in equation (2.115). The numerical data are (Exercise 2.92): $\delta\sqrt{U\infty/vx} = 3.46, 4.64, 5.84$, $C_f\sqrt{U_\infty x/v} = 0.577, 0.646, 0.685$, while exact solution gives for thickness 5 and for friction coefficient 0.664. It is seen that this simple approach gives the satisfactory results with profiles of the third and fourth power polynomials. In particular, the friction coefficient is obtained with accuracy $\pm 3\%$ in both the first and the second cases. Less exactness with difference of 7% and 17% is achieved in the thickness estimation. That is because of the conventional definition of the boundary-layer thickness. This becomes clear if one takes a look on the accuracy of displacement and momentum thicknesses, which are strongly defined. Using data for δ and formula (2.114), we have for displacement thickness $\delta_1\sqrt{U_\infty/vx} = 1.73, 1.74, 1.75$. These values are almost the same as exact data 1.72. The momentum thickness is equal to the friction coefficient, which follows from the momentum equation (2.114) in the case of a zero-pressure gradient $(U = const.)$ (Exercise 2.93).

Comment 2.18 The accuracy of less than 9% achieved in the friction coefficient with the simplest first linear profile is unexpected, especially as compared with that of 31% obtained in boundary thickness estimation. The reason of this lies in the fact that the part of exact profile in the vicinity of the surface determining the friction is practically linear (Comment 2.15).

Example 2.16 Heat transfer from the nonisothermal plate [22].

The same four power polynomial $f_3(\eta)$ (2.115) is used for the temperature profile $f_t(\eta_t)$ because the boundary conditions (2.116) are valid in this case as well (Exercise 2.94). Then, the heat flux on the surface is defined by Fourier formula [the third expression (2.118)], and the energy integral equation (2.114) gives similar to (2.117) differential equation determining the thickness of thermal boundary layer δ_t:

$$f_t = \frac{T - T_w}{T_\infty - T_w}, \quad \eta_t = \frac{y}{\delta_t}, \quad \frac{q_w}{\rho c_p} = \frac{\alpha a_1 \theta_w}{\delta_t}, \quad \delta_t \frac{d}{dx}(\theta_w \delta_{2t}) = \frac{\alpha a_1 \theta_w}{U_\infty}. \tag{2.118}$$

Two cases depending on whether the thermal boundary layer is thinner $(\mathrm{Pr} \geq 1)$ or thicker $(\mathrm{Pr} \to 0)$ than the velocity boundary layer should be considered. Denoting the ratio of the thermal and momentum boundary-layer thicknesses as $\varepsilon = \delta_t / \delta$, applying equations (2.114), (2.115), (2.118) and taking into account that in the case of $\delta_t > \delta$ outside of the velocity boundary layer the function $f_t(\eta_t) = 1$, the following relations are obtained [22, also 15, p.8] (Exercise 2.95):

$$\frac{\delta_{2t}}{\delta_t} = \int_0^1 f_3(\eta)[1 - f_t(\eta_t)]d\eta_t = \frac{2\varepsilon}{15} + \frac{3\varepsilon^3}{140} + \frac{\varepsilon^4}{180}, (\varepsilon \leq 1)$$

$$\tag{2.119}$$

$$\frac{\delta_{2t}}{\delta_t} = \int_0^{1/\varepsilon} f_3(\eta)[1 - f_t(\eta_t)]d\eta_t = \frac{3}{10} - \frac{3}{10\varepsilon} + \frac{2}{15\varepsilon^2} - \frac{3}{140\varepsilon^4} - \frac{1}{180\varepsilon^5}(\varepsilon > 1).$$

Considering the integral energy equation (2.118) and taking into account only first terms of these relations yields two linear ordinary differential equations, solution of which gives thickness of the thermal boundary layer for both cases (Exercises 2.96 and 2.97):

$$\delta\varepsilon\frac{d}{dx}(\delta\varepsilon^2\theta_w) = 15\frac{\alpha\theta_w}{U_\infty}, \quad \frac{\delta_t}{\delta} = \frac{0.871}{\mathrm{Pr}^{1/3}\,\theta_w^{1/2}(x)x^{1/4}}\left[\int_0^x \theta_w^{3/2}(x)x^{-1/4}dx\right]^{1/3} \quad \mathrm{Pr} \geq 1 \tag{2.120}$$

$$\delta\varepsilon\frac{d}{dx}(\delta\varepsilon\theta_w) = \frac{20}{3}\frac{\alpha\theta_w}{U_\infty}, \quad \frac{\delta_t}{\delta} = \frac{0.626}{\mathrm{Pr}^{1/2}\,\theta_w(x)x^{1/2}}\left[\int_0^x \theta_w^2(x)dx\right]^{1/2} \quad \mathrm{Pr} \to 0. \tag{2.121}$$

These formulae are obtained using only the first terms of equation (2.119). It is seen that the following terms in the first equation (2.119) are small compared to the first one even for $\varepsilon \approx 1$, whereas for the second expression, this condition is satisfied only for $\varepsilon \gg 1$. Therefore, formula (2.120) is applicable for $\mathrm{Pr} \geq 1$, whereas the equation (2.121) is appropriate only at $\mathrm{Pr} \to 0$. To estimate the accuracy more specifically, we compare these approximate expressions with self-similar solutions. The Nusselt numbers defined for a power-law temperature head using equations (2.118) for heat flux and formulae (2.120) and (2.121) have the form (Exercise 2.98):

$$Nu_x = 0.358(2m+1)^{1/3} \Pr^{1/3} \text{Re}_x^{1/2}, \quad Nu_x = 0.548(2m+1)^{1/2} \Pr^{1/2} \text{Re}_x^{1/2}. \quad (2.122)$$

Then, for surfaces with $\theta_w = const.\ (m = 0)$ and $q_w = const.\ (m = 1/2)$, one obtains 0.358, 0.548 and 0.452, 0.775 for $\Pr \geq 1$ and for $\Pr \to 0$, respectively. Comparing with exact data for the isothermal plate [equation (2.97)] and for the plate with constant heat flux (Example 2.14) shows that the largest difference, about 12% is in the case of $q_w = const.$ at $\Pr \to 0$.

2.7.1.2 Flows With Pressure Gradients

To consider the effect of the pressure gradient, a parameter Λ based on the second boundary condition (2.116) is employed. Using this parameter, the no-slip condition at $\eta = 0$ and other boundary conditions (2.116) at $\eta = 1$ for defining the coefficients of the four power polynomial (2.115) leads to the velocity profile that takes into account the effect of the pressure gradient (Exercise 2.99):

$$\Lambda = \frac{\partial^2 f}{\partial \eta^2} = \frac{\delta^2}{\nu}\frac{dU}{dx}, f = F_1(\eta) + \frac{\Lambda}{6}F_2(\eta), F_1(\eta) = f_3(\eta), F_2(\eta) = \eta - 3\eta^2 + 3\eta^3 - \eta^4. \quad (2.123)$$

Here, the first relation is found by changing the variables in the second boundary condition (2.116) from u and y to u/U and y/δ, and $f_3(\eta)$ is defined by equation (2.115). Physically parameter Λ may be interpreted as a ratio of pressure to viscous forces. This becomes obvious if one modifies formula (2.123) to the form $\Lambda = (-dp/dx)\delta/\nu(U/\delta)$ multiplying and dividing it by U. According to this expression, the flows with increasing and decreasing pressure gradient are described by negative and positive Λ, respectively. The zero value of Λ corresponds to flow with a zero-pressure gradient. A separation occurs when $\tau_w \sim (\partial f/\partial \eta)_{\eta=0} = a_1 = 0$. It follows from velocity profile (2.123) that this condition is satisfied at $a_1 = 2 + (\Lambda/6) = 0$, that is, at $\Lambda = -12$.

The friction coefficient is obtained by determining boundary-layer thickness from the momentum integral equation by the way that was applied for the flat plate. In this case, the momentum integral equation (2.114) is the complicated nonlinear differential equation that may be solved only numerically. However, an analysis of the numerical data shows that there exists a derivative of momentum thickness function, which is practically linear. Using such linear approximation, one transforms the momentum integral equation to the linear ordinary differential equation and gets a handy formula [1, p. 213]:

$$U\frac{d}{dx}\left(\frac{\delta_2^2}{\nu}\right) = 0.47 - 6\frac{\delta_2^2}{\nu}\frac{dU}{dx}, \frac{d}{dx}\left(\frac{U\delta_2^2}{\nu}\right) + \frac{5}{U}\frac{dU}{dx}\frac{U\delta_2^2}{\nu} = 0.47, \frac{U\delta_2^2}{\nu} = \frac{0.47}{U^5}\int_0^x U^5 dx. \quad (2.124)$$

Here, the first equation is modified by adding the term $(dU/dx)\left(\delta_2^2/\nu\right)$ to both its sides to have the derivative of $U\delta_2^2/\nu$ in the second equation (Exercise 2.100).

Comparison between approximate and exact results shows that the predictions of the integral method are satisfactorily accurate for the accelerated flows with negative-pressure gradients. The accuracy achieved for the flows with increasing pressure, especially close to separation point, is considerable less. Thus, in the case of transverse flow past a circular cylinder with theoretical pressure distribution, both results are practically the same in the range of the angle 0–90° where the pressure decreases, while after a pressure minimum, the data differ more significant so that exact prediction of separation point is 104.5° against 109.5° and 108.8° giving by integral and series methods, respectively. A similar integral method based on self-similar profiles is given in [1, 2000].

Example 2.17 Preseparation flow.

Flows with increasing pressure are important for application. For example, such a flow exits on the suction side of airfoils or in divergent channels, known as diffusers, which are used for rising pressure, for instance, in wind tunnels. However, the flows of that type exit only at small pressure gradients because the separation occurs when the pressure gradient exceeds the particular value. Since the separation results in large energy losses, it is desired to avoid this phenomenon. At the same time, there is no restriction in the pressure value if the process of rising pressure goes at a small gradient. The simplest way to achieve this condition is to increase the length of an airfoil or of a diffuser. Therefore, to have the shortest as possible without separation device, one should design it using close to the separation value of pressure gradient providing flow close to separation point, called preseparation flow (Exercise 2.102).

The characteristics of such a flow may be estimated using the data of above-mentioned numerical solution of integral boundary-layer equation (2.114). Knowing that the value $\Lambda = -12$ corresponds to separation, the flow with $\Lambda = -10$ is considered as preseparation flow. The analysis begins from adapting from data of numerical solution the values of two relations $(dU/dx)(\delta_2^2/v) = -0.1369$ and $Ud(\delta_2^2/v)/dx = 1.523$ corresponding to $\Lambda = -10$. Differentiating with respect to x the first relation presented as $\delta_2^2/v = -0.1369/(dU/dx)$ and comparing the result with the second equation yields expression, which after integration gives the solution of the problem (Exercise 2.101):

$$\frac{UU''}{U'^2} \approx 11, \quad \frac{U'}{U^{11}} = -c_1, \quad \frac{1}{10U^{10}} = c_1 x + c_2, \quad U(x) = \frac{U_0}{(1+10c_1 x)^{1/10}},$$

$$U(x) = U_0 \left(1 + 100 \frac{vx}{U_0 \delta_0^2} \right)^{-0.1}. \tag{2.125}$$

Here, the second and the third equations are obtained as the results of the first and the second integration. The constants of integration are found applying the potential velocity U_0 and boundary-layer thickness δ_0 at the point $x = 0$ of a minimum pressure after which the pressure starts to increase. Putting $U = U_0$ at $x = 0$ in the third equation gives $c_2 = 0.1U^{-0.1}$ and then the fourth expression (2.125) containing c_1 is obtained. The latter constant is found employing formula $\Lambda - U'\delta^2/v = -10$. Defining U' by differentiation of the fourth equation (2.125) and substitution of the result in the relation for Λ leads to expression determining δ, δ_0 and finally gives $c_1 = 10v/U_0\delta_0^2$ and the solution (2.125).

Analysis of pre-separation flow shows that (i) according to first relation (2.125), the separation may be avoided if the function $UU''/U'^2 > 11$, while in the opposite case $UU''/U'^2 < 11$ the separation occurs. This means that to avoid a separation, it should be $U'' > 0$ and as a consequence, the curve $U(x)$ starting from some point $x = 0$ is curved upwards with $U' < 0$ (rising pressure) and (ii) a laminar boundary layer with increasing pressure exists without separation only under very small gradients: as it follows from solution (2.125), the decreasing of potential velocity is proportional to $\approx x^{-0.1}$, which corresponds to increase in the boundary-layer thickness proportionally $\approx x^{0.55}$. Such values only slightly differ from the case of the flat plate [1, p.221].

Another solution of this problem with practically the same results is given in [1, 2000], where the problem of preseparation turbulent flow is also considered. The special form of diffuser with preseparation flow is investigated in [23]. Analogous problem for the zero heat transfer surface is analyzed below in Section 5.5.

2.7.2   Linearization of the Momentum Boundary-Layer Equation

Linearization is an effective method to simplify complex nonlinear problems. Most often, linearization is achieved employing only the first term of a perturbation series presenting the nonlinear part of equation. One example of such an approach we have discussing the Tollmien solution for the wake behind the flat plate (Sec. 2.6.5.1). Several other solutions applying this type of linearization are considered in particular in Chapters 7 and 12.

Another approach of linearization consists of approximation of a nonlinear term by appropriate knowing function. Two examples of this method are present below.

2.7.2.1   Flow at the Outer Edge of the Boundary Layer [24]

The boundary-layer equation is taken in the Prandtl-Mises form (2.73). The linearization is achieved by substituting U for the velocity u at the second derivative on the right-hand side. This may be done since the velocity of flow near the outer edge is close to that of external flow $u \approx U$. Another simplification is obtained using displacement thickness definition (Example 2.10) $U\delta_1 = Udy - udy$. Since $udy = d\psi$, we get $d\psi = Udy - Ud\delta_1$. Integrating this equation shows that near the outer edge $\psi \approx U(y - \delta_1)$. This follows from the fact that the last term in the previous equation after integration becomes to be almost equal to the displacement thickness. Using these results and taking into account that due to $u \approx U$, there may be written $Z = (U-u)(U+u) \approx 2U(U-u)$ one transforms equation (2.73) to first equation (2.126) shown below (Exercise 2.103).

$$\frac{\partial U(U-u)}{\partial \hat{\Phi}} = \frac{\partial^2 U(U-u)}{\partial [U(y-\delta_1)]^2}, \quad z = \frac{U\left(y-\delta_1\right)}{2\sqrt{\hat{\Phi}}}, \quad 1-\frac{u}{U} = \frac{2}{U^2\sqrt{\pi}} \int_0^\infty cU^2 \exp(-\xi^2)d\xi \quad (2.126)$$

This equation is of the type of one-dimensional conduction equation (1.4) without a source q_v. Comparing both equations leads to a conclusion that in this case instead of temperature T, coordinate x and product αt should be considered relations $U(U-u)$, $U(y-\delta_1)$, and variable $\hat{\Phi}$, respectively. Here, $\hat{\Phi} = v^2\Phi$, where Φ is the Görtler variable (2.75). The domain of the transverse variable y or ψ is semi-infinite so that equation (2.126) may be solved by the way described in Section 1.3 where the heat transfer in the semi-infinite laterally insulated rod is considered. Employing formula (1.33) we note that in this case: (i) the variable z is given by the second relation in equation (2.126), (ii) the constant $c_2 = 0$ due to symmetry in the ψ direction, (iii) there is no boundary condition at the surface because the linearization is valid only close to the outer edge of the boundary layer. Therefore, similar to equation (1.34), we get only $f(x) \sim cU^2$, knowing that according to condition (2.73) on a surface $Z = U^2$. Based on these notes and on the fact that U is variable, we found the solution in the form similar to equation (1.40) (Exercise 2.104).

To determine the values of c, author compares profiles (2.126) with self-similar solutions (Sec.2.6.2). Calculation performed for three values of $\beta = -0.199$ (preseparation flow), 0 (flat plate), and 2 (accelerated flow) gives the data $c = 0.492, 0.448, 0.393$ showing that parameter c only slightly depends on β. The reason of that is explained in the next section. Despite that the approximate profiles are found regardless the condition at the surface, these coincide with exact profiles almost across the whole boundary layer, except small parts close to the surface. Thus, the preseparation profile shows agreement from the outer edge to the point with coordinate little less than $y/\delta = 1$, while the two others agree with exact profiles even closer to the surface, to the points with $y/\delta = 0.5$ and $y/\delta \approx 0.3$ for $\beta = 0$ and $\beta = 2$, respectively.

2.7.2.2 Universal Function for the Skin Friction Coefficient [25]

More accurate linearization is achieved substituting the self-similar solution instead the velocity at second derivative on the right-hand side of the Prandtl–Mises equation. In this case, equation (2.77) in Görtler variables (2.75) Φ, φ is used, and the dynamic pressure $P = (\rho/2)(U^2 - u^2)$ is considered as a sought function. Physically, the dynamic pressure is defined as a difference between the total pressures in the external flow $p_{0\infty}(x) = \rho U^2/2$, and at any point in the boundary layer, $p_0(x, y) = \rho u^2/2$. Thus, the problem is governed by equation (2.77) for dynamic pressure P with the linearized right-hand side part and following boundary conditions:

$$2\Phi \frac{\partial P}{\partial \Phi} - \varphi \frac{\partial P}{\partial \varphi} - \omega(\varphi, \beta) \frac{\partial^2 P}{\partial \varphi^2} = 0, \quad \varphi = 0 \; P = \rho U^2/2 = P_w(x), \quad \varphi \to \infty \; P = 0 \quad (2.127)$$

It is shown in Section 4.2.1 that the equation of this type has an universal solution in the form of a series of consecutive derivatives of the boundary condition at the surface, like, for example, series 1.23 for heat flux consisting of consecutive derivatives of temperature head $\theta_w(x)$. Hence, in this case such a solution should consists of the derivatives of the pressure on the surface P_w, determining the skin friction coefficient:

$$c_f = c_{f*}\left(1 + b_1 \frac{\Phi}{P_w} \frac{dP_w}{d\Phi} + b_2 \frac{\Phi^2}{P_w} \frac{d^2 P_w}{d\Phi^2} + b_3 \frac{\Phi^3}{P_w} \frac{d^3 P_w}{d\Phi^3} \cdots\right) = c_{f*}\left(1 + \sum_{k=1}^{\infty} b_k \frac{\Phi^k}{P_w} \frac{d^k P_w}{d\Phi^k}\right). \quad (2.128)$$

Here, $c_{f*} = 0.664\sqrt{\mathrm{Re}_{av}} = 0.664\Phi^{-1/2}$ (* denotes a constant U) is the average for the interval $(0, x)$ skin friction coefficient, $\mathrm{Re}_{av} = U_{av}x/v$ is the Reynolds number defined for average velocity U_{av}, and $P_w = \rho U^2/2$ is the dynamic pressure on the surface, which is equal to that in external flow (pressure does not change across boundary layer).

It is also shown in Section 4.2.2 that the expression like series (2.128) in differential form may be presented in equivalent integral form (2.110):

$$c_f = \frac{c_{f*}}{P_w}\left[P_w(0) + \int_0^\Phi f(\xi/\Phi)\frac{dP_w}{d\xi}d\xi\right], \quad f(\xi/\Phi) = \left[1 - \left(\frac{\xi}{\Phi}\right)^{C_1}\right]^{-C_2}. \quad (2.129)$$

Coefficients b_k as well as exponents C given in Table 2.1 only slightly depend on β. This occurs due to the variable Φ that is defined by integral, taking into account the flow history

Table 2.1 Coefficients b_k and exponents C for universal functions (2.128) and (2.129).

β	b_1	b_2	b_3	b_4	C_1	C_2
0	2.28	−0.30	0.058	−0.0096	0.52	0.57
0.5	1.97	−0.26	0.050	−0.0089	0.50	0.54
1	1.85	−0.24	0.047	−0.0085	0.48	0.52
Average	2.0	−0.25	0.05	−0.009	0.50	0.54

(Exercise 2.105). Physically, this means that the value of Φ at the point x is determined not only by data at point x, but rather by data of whole interval from $x = 0$ to x of a considered point (Comment 4.2). Due to slightly dependence on β, the accuracy of its estimation barely affects the final calculation results (Sec. 4.2.2). Therefore, β may be evaluated intuitively or using some approximate formula, for example, $\beta = 2(1 - \Phi/\mathrm{Re}_x)$. This formula follows from equality of average velocities of given and some self-similar flow (Exercise 2.106).

Employing two expressions (2.128) and (2.129) allows one to provide accurate calculations applying the first one when the series converges enough fast and using the integral formula otherwise. As is seen in Table 2.1, coefficients b_k rapidly decrease with the number so that remaining two or three first terms in equation (2.128) are usually enough to get satisfactory results. In examples given below, the data are presented in the form of ratio $\chi_f = c_f/c_{f*}$ which may be named the nonisotachicity coefficient [similar to nonisothermisity one (Sec. 4.2.2)] because it shows how much the friction in a flow with a variable external velocity is more or less than that in flow with average constant velocity.

Example 2.18 Self-similar flows.

In this case $U^2 \sim x^{2\beta/(2-\beta)}$, $\Phi \sim x^{2/(2-\beta)}$, $P_w \sim U^2 \sim \Phi^\beta$ (Sec.2.6.2). Therefore equation (2.128) gives for $\chi_f = c_f/c_{f*}$ first expression (2.130). This series for not integer β divergences so that the integral equation (2.129) should be applied. Defining the derivative $dP_w/d\Phi = \beta\Phi^{\beta-1}$ and using a new dummy variable $\zeta = \xi/\Phi$ leads to the integral formula (2.129) in the form of the second equation (2.130):

$$\chi_f = 1 + b_1\beta - b_2\beta(\beta-1) + \cdots, \qquad \chi_f = \beta \int_0^1 \left[1 - \zeta^{C_1}\right]^{-C_2} \zeta^{\beta-1} d\zeta. \qquad (2.130)$$

For integer β, the first formula gives $\chi_f = 1$ and $\chi_f = 2.85$ for $\beta = 0$ and $\beta = 1$. Hence, $c_f = 0.664/\sqrt{\mathrm{Re}_x}$ and $c_f = 2.649\sqrt{\mathrm{Re}_x}$, which is 7% larger than the exact value (Exercise 2.107). For non-integer β, the second equation (2.130) yields $\chi_f = 1.56, 2.17, 2.89, 3.42$ for $\beta = 0.2, 0.5, 1.0, 1.6$. These values are in reasonable agreement with data for self-similar flows ([1], Fig. 9.1) (Exercise 2.108).

Comment 2.19 More accurate estimation of exactness of formulae (2.130) may be done using Hartree's numerical data [26]. Nevertheless, the suggested coarse evaluation in Exercise 2.108 shows that the average error of universal functions (2.128) and (2.129) is about 10%, which is common for approximate methods. At the same time, the computing method is simple and due to functions universality does not require a solution of any differential equation.

Example 2.19 Linear potential velocity distribution $U = U_0 - ax$ (Sec.2.6.4).

Using new variables $\hat{x} = ax/U_0$ and $\hat{\Phi} = a\Phi/U_0^2$, we have $\hat{\Phi} = \hat{x} - \hat{x}^2/2, P_w = (\rho U_0^2/2)(1 - 2\hat{\Phi}), dP_w/d\hat{\Phi} = -2(\rho U_0^2/2)$. Then, equations (2.128) and (2.129) become (Exercise 2.109)

$$\chi_f = 1 - \frac{2b_1\hat{\Phi}}{1 - 2\hat{\Phi}} \qquad \chi_f = \frac{1}{1 - 2\hat{\Phi}}\left[1 - 2\int_0^{\hat{\Phi}}\left[1 - \left(\frac{\xi}{\hat{\Phi}}\right)^{C_1}\right]^{-C_2} d\xi\right]. \qquad (2.131)$$

Both relations with average values of $b_1 = 2$ and exponents $C_1 = 0.5$ and $C_2 = 0.54$ lead to the practically same numerical results $\chi_f = 1.33, 1.18, 0.78, 0.53$ for $\hat{x} = -0.1, -0.05, 0.05, 0.1$, which are also in reasonable agreement with Howarth data ([1], Fig. 9.8). To obtain a coordinate of separation, we apply $b_1 = 2.31$ and $C_1 = 0.53$, $C_2 = 0.58$, which are found by extrapolation data from Table 2.1 to the self-similar flow preseparation value $\beta = -0.199$. With these values, the both equations (2.131) give the separation coordinate $\hat{x} = 0.16$ (Exercise 2.110), while the exact data are $\hat{x} = 0.12$.

Comment 2.20 Estimation of the separation coordinate from the first equation (2.131) leads to the quadratic equation $\hat{x} - \hat{x}^2/2 = 1/6.62$. The solution of this equation yields small difference of two much larger values, which usually results in poor accuracy of final data. In such a case, better results are obtained using iteration. Presenting the quadratic expression in the form $\hat{x} = 1/6.52 + \hat{x}^2/2$, we get after neglecting the second term the first approximation $\hat{x} = 1/6.62$. Then, adding the second term, we compute applying obtaining result the second approximation: $1/6.62 + \hat{x}^2/2 = 1/6.62 + (1/6.62)^2/2 = 0.16$.

Example 2.20 Transverse flow around a cylinder at sinusoidal velocity distribution.

The potential velocity is $U = 2U_\infty \sin(x/R) = 2U_\infty \sin\hat{x}$, then $\hat{\Phi} = \Phi/\text{Re} = 1 - \cos\hat{x}$, $P_w = 2\rho U_\infty^2(2\hat{\Phi} - \hat{\Phi}^2)$, where $\text{Re} = 2RU_\infty/\nu$ and the last formula is obtained using the trigonometric relations. To compare the results with known data, we modify equation (2.128) by relations $c_f = \tau_w/\rho U^2$ $c_{f*} = 0.664/\sqrt{\hat{\Phi}\text{Re}}$ and define the derivatives of P_w applying above formula to get the skin friction coefficient in the form

$$\hat{\tau} = \frac{\tau_w}{\rho U_\infty^2}\sqrt{\frac{U_\infty R}{\nu}} = 0.940\,[2(1 + b_1)(1 - \cos\hat{x})^{1/2} - (1 + 2b_1 + 2b_2)(1 - \cos\hat{x})^{3/2}]. \quad (2.132)$$

For the separation point, this equation with $b_1 = 2.31$ and $b_2 = 0.31$ (Example 2.19) yields

$$\cos\hat{x} = \frac{1 - 2b_2}{1 + 2b_1 + 2b_2} \qquad \gamma = 108.9°. \qquad (2.133)$$

This value is in conformity with other approximate prediction 109.5° and 108.8° given by integral and Blasius series methods, respectively, whereas the more accurate result found by numerical solution of boundary-layer equations is 104.5°. The results for shearing stress obtained using equation (2.132) and average values $b_1 = 2$ and $b_2 = -0.25$ (Exercise 2.111) agree with other approximate data as well as with numerical solution on the whole surface, except a small region near separation ([1], Figs. 9.6 and 10.7). Other examples, in particular, the general case of distribution $U = U_0 - ax^n$ (Example 2.19) for odd and even exponents, skin friction, and point of separation for airfoil are considered in [25].

2.7.3 Thermal Boundary-Layer Equations for Limiting Prandtl Numbers

In both limiting cases, for fluids with small or large Prandtl numbers, the thermal boundary-layer equation may be simplified because the Prandtl number determining the ratio of boundary-layer thicknesses (Example 2.11) shows that in the first case the thermal thickness is large, and in the second one, it is small compared to the velocity boundary-layer thickness. Due to that in both cases, the velocity distribution in the thermal boundary layer may be considered as known. When the thermal boundary layer is large, the velocity distribution across it may be considered as the same as in external flow since the velocity layer covers only small part of the thermal layer at the surface so that its rest majority part appears to be in external flow. In the other case, the situation is opposite because the small thermal boundary layer lies on the surface covering only minor part of the large velocity boundary layer. Therefore, in that case, the velocity distribution across the thermal boundary layer may be approximated by linear function using tangent to velocity profile on the surface (Comment 2.14). To understand the physical reasons of these facts, note that the Prandtl number is defined as a ratio of the fluid viscosity to its thermal diffusivity, and, consequently, it characterizes the relation between intensity of momentum and heat transport processes. Large Prandtl numbers indicates that transport of momentum is intensive so that the effect of zero velocity on the wall goes far into fluid forming a thick velocity boundary layer. Similarly, the small Prandtl number means that transport of the heat is strong resulting in a large thermal boundary layer (Exercise 2.112).

There are three situations when the real velocity distribution across the thermal boundary layer is close to the linear. The first one is the case of fluids with large Prandtl numbers just discussed. For example, the high Prandtl numbers are typical for non-Newtonian fluids. The other is the case when the unheated surface length precedes the heated zone so that the thermal boundary layer starts to develop inside the velocity boundary layer. This results in a thin thermal boundary layer inside the thicker velocity layer. A solution of a special case when the step change in temperature follows the unheated zone is usually used as a standard influence function in developing the Duhamel integral (2.110) for heat flux on the arbitrary nonisothermal surface. A similar situation of a thin thermal boundary layer inside a thicker velocity layer takes place in the entrance of a tube or a channel when the thermal boundary layer grows inside the fully developed flow.

Example 2.21 Deriving an influence function.

The influence function is found as a solution of a standard problem of heat transfer after the wall temperature jump (Sec. 2.6.8). In general, the solution of this problem is a challenged task so that only approximate solutions are known. For the simplest case of a plate with zero-pressure gradient and $Pr \geq 1$, the influence function was found by integral method [27,28]. Third power polynomial (2.115) $f_2(\eta)$ was applied for describing the velocity and temperature profiles. This leads to the equations (2.117) and (2.120) for the velocity boundary layer thickness δ and for the ratio of thermal to velocity thicknesses ε with coefficients corresponding to polynomial $f_2(\eta)$. Defining δ from the first equation, one substitutes it in the second equation and after performing differentiation gets an ordinary differential equation for the ratio of thicknesses:

$$\delta \frac{d\delta}{dx} = \frac{140}{13} \frac{v}{U_\infty}, \quad \delta\varepsilon \frac{d(\delta\varepsilon^2)}{dx} = 10 \frac{\alpha}{U_\infty}, \quad \varepsilon^3 + 4\varepsilon^2 x \frac{d\varepsilon}{dx} = \frac{13}{14\,\mathrm{Pr}}. \tag{2.134}$$

Assuming $13/14 \approx 1$ and integrating this differential equation yields an expression for ε containing integration constant. Determining this constant knowing that at $x = \xi$ the ratio of thicknesses is zero gives the solution for ε and finally results in influence function:

$$\varepsilon = \mathrm{Pr}^{1/3} \left[1 - \left(\frac{\xi}{x} \right)^{3/4} \right]^{-1/3}, \qquad f(\xi/x) = \left[1 - \left(\frac{\xi}{x} \right)^{3/4} \right]^{-1/3}. \tag{2.135}$$

The second equation follows from the first one because according to equation: (i) (2.118) the heat flux is inversely proportional to thermal boundary-layer thickness and (ii) (2.110) the influence function is a quotient of values of ε at any $\xi \neq 0$ and $\xi = 0$, which equals $\mathrm{Pr}^{1/3}$ and corresponds to the isothermal surface (Exercises 2.113 and 2.114).

Similar results are known for self-similar flows in both limiting cases

$$f(\xi/x) = \left[1 - \left(\frac{\xi}{x} \right)^{(3/4)(m+1)} \right]^{-1/3}, \mathrm{Pr} \to \infty, \quad f(\xi/x) = \left[1 - \left(\frac{\xi}{x} \right)^{m+1} \right]^{-1/2}, \mathrm{Pr} \to 0. \tag{2.136}$$

The first relation follows from a general solution for arbitrary free-stream velocity distribution obtained by Lighthill [29] for this limiting case. Despite that this relation is found for large Prandtl numbers, the calculation shows that it is reasonable accurate up to $\mathrm{Pr} \approx 1$. The other formula (2.136) pertains to small Prandtl numbers when, as mentioned above, the velocity distribution in the thermal boundary layer may be taken equal to the free-stream velocity $U(x)$. Such an assumption is close to reality for liquid metals, like mercury, for which small Prandtl numbers are inherent. The corresponding thermal boundary-layer equation is of one-dimensional conduction type and may be easily integrated [1, p.289]. In the simplest case of the zero-pressure gradient $(U_\infty = const.)$, the velocity across the thermal boundary layer becomes uniform. This case known as a plug assumption as well as a linear velocity approximation is widely used to simplify heat transfer problems, in particular, in conjugate formulation. Examples and analysis may be found in [15]. See also Section 4.2.1.

2.8 Natural Convection

Natural convection occurs whenever there are density differences in the gravitational field. The above-considered forced convection exists due to external force such as pressure difference, which drives the flow or other external action, for example, moving body resulting in fluid flow around it. In contrast to that in natural convection, the buoyancy driving force is produced naturally by density difference, and that is why this type of heat transfer is called natural or free convection. The density difference arises usually due to temperature gradients and decreases with temperature increasing. The velocities in free convention are small and consequently the heat transfer rate is much smaller than that in forced convection. Since in this case, there is no characteristic velocity, the Grashof number (2.81) or related to it the Rayleigh number Ra = Gr Pr is used instead of the Reynolds number. As well as, in general, in the forced heat transfer, the majority problems of natural convection require solution of the Navier–Stokes and full-energy equations. A small group of problems associated with natural heat transfer on vertical surfaces such as a plate or a vertical circular cylinder may be considered in terms of boundary-layer theory. Those have some analytical solutions (Example 2.22), whereas the others are usually treated numerically (Example 2.23). To understand physical reasons of this difference, note that in free convection, the driving force is of gravitational nature so that flow is directed vertically. Such a flow along the vertical surface forms a typical boundary-layer structure similar to that in the case of longitudinal forced flow along the flat plate. A completely different structure has natural convective flow on the horizontal plate. In that case, the whole heated surface is covered by small rising flows in the form of plums without any dominated direction typical to a boundary-layer structure (Exercises 1.115 and 1.116).

In solutions of Navier–Stokes and energy equations for natural convection problems, some specific issues should be taken into account in addition to general requirements to such solutions (Secs. 2.3 and 2.4). In particular, since the rate of natural convection is small, the radiation is often on the same order so that it should be considered along with convection in such a case (Example 2.23). The other phenomenon significant for some free convection flows is special care of stability (Example 2.24).

Example 2.22 Free convection on the vertical plate.

Solutions of this classical problem, including early attempts and comparison with experimental data, are reviewed in [30]. We consider the solution given by Pohlhausen in 1921 now regarded as classical. A problem is governed by the steady-state boundary-layer system (2.68)–(2.70) without pressure gradient and dissipative terms in second and third equations, respectively, but with additional term $g\beta(T-T_\infty)$ in the second equation for taking into account a fluid thermal expansion. In this problem as well as in the case of the forced heat transfer on a flat plate, employing similarity variables reduces the system of partial differential equations to system of two ordinary differential equations [30]:

$$\eta = c\frac{y}{x^{1/4}}, \zeta = \frac{\psi}{4\nu cx^{3/4}}, c = \left[\frac{g(T-T_\infty)}{4\nu^2 T_\infty}\right]^{1/4}, \zeta''' + 3\zeta\zeta'' - 2\zeta'^2 + \theta = 0, \theta'' + 3\Pr\zeta\theta' = 0, \quad (2.137)$$

with boundary conditions: $\zeta = \zeta' = 0, \theta = 1$ at $\eta = 0$, $\zeta' = \theta = 0$ at $\eta \to \infty$, where $\theta = (T-T_\infty)/(T_w - T_\infty)$. It follows from these expressions that (i) according to similarity variable η both boundary-layer thicknesses are proportional to $x^{1/4}$ (like to $x^{1/2}$ in forced convection); (ii) both velocities on the surface and far away from the plate are zero (unlike that in forced convection); and (iii) differential equations (2.137) unlike others similar [e.g., equations (2.89)] are coupled so that the temperature and velocity depend on each other, and hence both equation should be solved simultaneously. Another difference between both convection types follows from similarity analysis [equations (2.78) and (2.79)], which shows that the heat transfer rate depends on the Grashof and Prandtl numbers instead of the Reynolds and Prandtl numbers in forced convection. Proper formulae are [30]:

$$Nu_x = \frac{3A}{4}(Gr_x \Pr)^{1/4}, \quad A = \left[\frac{2\Pr}{5(1+2\Pr^{1/2}+2\Pr)}\right]^{1/4}, \quad Gr_x = \frac{g\beta x^3(T_w - T_\infty)}{\nu^2}, \quad (2.138)$$

with $A = 0.8\,\mathrm{Pr}^{1/4}$ and $A = 0.670$ for $\mathrm{Pr} \to 0$ and $\mathrm{Pr} \to \infty$. The expression (2.138) for A is a result of approximation of tabulated data obtained by numerical solution of differential equations (2.137).

Example 2.23 Free convection and radiation from horizontal fin array [31].

Here, this complex problem is discussed only on purpose to give a reader some understanding of formulation of such problems without solution and results. More detailed review of this work one may find in [15, p. 346] or in the original paper. The model consist of two long adjacent vertical fins attached to base with constant temperature $T_s > T_\infty$. The problem is governed by the two-dimensional Navier–Stokes equation in vorticity-stream function form (Sec. 2.2.2) and full-energy equation for the fluid, one-dimensional conduction equation for the fins and conjugate boundary conditions (Sec. 2.3.2). The radiation heat transfer flux is included in the energy equation as a source. Since the solution of the Navier–Stokes and energy equations requires a closed domain (Sec. 2.3), the configuration of two fins on the base is considered as a closure with the open top and front and rear sides regarded as imaginary surfaces. The radiation heat flux is calculated as a sum of heat exchanges between all six surfaces: two fins, base, and three open imaginary sides. The system of governing equations is solved numerically.

Example 2.24 Stability of fluid between two horizontal plates.

The stability of fluid located between two long parallel heated plats depends on the temperature gradient. If the temperature of the upper plate is higher than that of the lower one, the temperature gradient is directed along the gravitational force. In such a case, the fluid density increases in the gravitational force direction so that lighter fluid layers are located above heavier part of fluid. This situation is stable, and heat transforms in the gravitational force direction from the upper to lower plate by conduction. In the opposite case, the density decreases in the direction of gravitational force, resulting in situation when the buoyancy force rises the lighter fluid layers cooling those, while the heavier layers are descended by the gravitational force being warmed. This situation is unstable resulting in the circulation pattern. In this case, the Rayleigh–Benard cells are forming if the Rayleigh number $Ra_L = g\beta\left(T_u - T_b\right)L^3/\nu\alpha$ (indices u and b refer to upper and bottom plates, L is the height between plates) exceeds its critical value (Exercise 2.117). This pattern was first observed Benard experimentally in 1900. Rayleigh first studied this phenomenon theoretically in 1916.

Comment 2.21 Natural convection has many applications. In metrology, free convection processes are relevant since the stable and unstable phenomena similar to that between two parallel plats occur in atmosphere and ocean significantly affecting the weather. In the Earth's mantle, convection results from a temperature difference between the warmer inside and the surface. Small heat transfer rates are required for cooling in many engineering systems: electronic devices, thermal pipes, refrigerators, and room radiators.

EXERCISES

2.89 Explain the physical meaning of displacement, momentum, and energy thicknesses (Example 2.10).

2.90 Derive the momentum and energy integral equations (2.114) following the way described in the text.

2.91 Calculate coefficients a_n for profiles (2.115).

2.92 Derive the chain of formulas (2.117) and obtain numerical data for boundary-layer thicknesses and friction coefficient.

2.93 Show that in the case of profile $f(\eta) = \sin[(\pi/2)\eta]$ the momentum thickness is defined as $\delta_2 = \delta(4-\pi)/2\pi$. Calculate boundary-layer thicknesses and friction coefficient.

2.94 Show that the boundary conditions (2.116) are valid for the temperature profile in variables (2.118) $f_t(y/\delta_t)$.

2.95 Obtain expression (2.119) following explanation in the text.

2.96 *Derive second equation (2.120) from first one. Hint: (i) perform differentiation of the expression in parentheses presenting it as $\varepsilon^2(\delta\theta_w)$, (ii) introduce a new variable $z = \varepsilon^3$ to get a linear ordinary differential equation of the first order for variable z, (iii) present this equation in the standard form and note that $\exp\left[\ln(\delta\theta_w)\right] = \delta\theta_w$, (iv) use for δ the value corresponding to profile $f_3(\eta)$.

2.97 Obtain second equation (2.121) from first one. Hint: multiply both sides of the first equation by θ_w and apply new variable $z = \delta\varepsilon\theta_w$, use for δ the value corresponding to profile $f_3(\eta)$.

2.98 Derive equations (2.122) for Nusselt numbers and estimate accuracy of equations (2.120) and (2.121).

2.99 Find the coefficients of the velocity profile (2.123) and functions $F_1(\eta)$ and $F_2(\eta)$ using appropriate boundary conditions.

2.100 Derive the second differential equation (2.124) from the first one and solve this linear equation to get the last formula (2.124) (see hint (iii) in Exercise 2.96).

2.101 Repeat the derivation of the chain of equations (2.125). Hint: for first integration present the first relation in equation (2.125) as $U''/U' = 11U'/U$.

2.102 Explain the difference between pressure and pressure gradient. How works the principle: value of increasing pressure is unrestricted if pressure gradient is enough small?

2.103 Explain why the equation (2.73) is nonlinear, whereas equation (2.126) is linear

2.104 Repeat the derivation of equation (2.126) and its solution.

2.105 What is flow history like? Think about some examples.

2.106 Derive the formula $\beta = 2(1-\Phi/\mathrm{Re}_x)$ for estimating value of β. Hint: Integrate both sides of equation $Cx^{\beta/(2-\beta)} = U$.

2.107 Derive the first expression (2.130) and compute skin friction coefficients for $\beta = 0$ and $\beta = 1$ using formula $c_f = 0.664\Phi^{-1/2}\chi_f$.

2.108 Obtain the second equation (2.130) from integral formula (2.129). Perform the same calculation as in the text and compare the results with data for self-similar solutions given in [1, Fig. 9.1]. Hints: (i) the derivative in integrant of (2.130) should be $dP_w/d\xi = \rho\xi^{\beta-1}$, while the pressure should be substituted as $P_w = \rho\Phi^\beta$; (ii) to compare the results for χ_f, measure an angle of tangent to a corresponding curve at the surface and divide it by 0.664. This quotient should be close to the value of χ_f. For instance, the angle of the curve $\beta = 1.6$ is approximately equals $0.8/0.4 = 2$, then the quotient is $2/0.664 = 3.01$, (iii) for calculate integrals determining χ_f use software, for example, Mathcad.

2.109 Derive equation (2.131) and repeat calculation for the same values of $\hat{x}$. See hints in the previous exercise.

2.110 Calculate the coordinate of separation point using both equations (2.131). Hints: (i) Solve the quadratic equation obtained from the first formula to see the problem of small difference arising in this case. Then perform iteration as explained in Comment 2.20, (ii) in using the second formula, compute χ_f for several values of $\hat{\Phi}$, plot the curve $\chi_f(\hat{\Phi})$ and extrapolate this curve to abscissa axis to find the coordinate where $\chi_f = 0$.

2.111 Obtain expressions (2.132) and (2.133). Calculate and plot dependence $\hat{\tau}_w(\gamma^\circ)$. Compare results with the data from Fig. 10.7 [1]. Determine the angle of the separation point.

2.112 Why the thermal boundary-layer equation may be simplified for fluids with limiting values of the Prandtl number?

2.113 Derive equation (2.134) for ratio of thermal and velocity thicknesses.

2.114 Solve equation (2.134) to obtain the expression (2.135) for influence function. Hint: use a variable $z = \varepsilon^3$ noting that $z' = 3z^2$.

2.115 Compare forced and free convection. Explain why characteristic numbers used in two cases are different.

2.116 Clarify physical reasons why natural convection on the vertical plate is considered as the boundary-layer problem, while free convection on the horizontal plate requires the solution of Navier–Stokes and full-energy equations.

2.117 Describe two situations occur when a fluid is contained between two long heated horizontal plates. Read an article from Wikipedia to see photo and know more about cells (http://en.wikipedia.org/wiki/Convection).

REFERENCES

[1] Schlichting, H. (1951,1968,1979, 2000). *Boundary layer theory*. New York: McGraw-Hill. Citations without year pertains to 1979 edition.

[2] Wilcox, D. C. (1994, 2006). *Turbulence modeling for CFD*. La Canada, California: DCW Industries Inc.

[3] Batchelor, G. K. (1967). *An Introduction to fluid dynamics*. New York: Cambridge University Press.

[4] Piercey, V. I. (2007). *The lame and metric coefficients for curvilinear in coordinates R^3*, http://www.google.com/search?hl=en&source=hp&q=lame+and+metric+coefficients+in+curvilinear+co ordinates&aq=0&aqi=m1&aql=&oq=Lame+and+metric+cefficients+in+curvilinear+coordinates &gs_rfai=CRjvxlOG9TLrmL4_uM5vngLoHAAAAqgQFT9CPcnk

[5] Bird, R. B., Stewart,W. E., & Lightfoot, E.N. (2005). *Transport phenomena*, (2nd ed.). New York: Wiley.

[6] Drew, T. B. (1961). *Handbook of Vector and Polyadic Analysis*. New York: Reinhold.

[7] Munson, B. R., Young, D. F., & Okiishi, T. H. (1990). *Fundamentals of fluid mechanics*, (2nd ed.). New York: Wiley.

[8] Chiu, W. K. S., Richards, C, J., & Jaluria, Y. (2001). Experimental and numerical study of conjugate heat transfer in a horizontal channel heated from below. *ASME Journal of Heat Transfer 123*, 688–697. doi: http://dx.doi.org/10.1115/1.1372316

[9] Vynnycky, M., Kimura, S., Kaneva, K., & Pop, I. (1998). Forced convection heat transfer from a flat plate: the conjugate problem. *International Journal of Heat Mass Transfer 41*, 45–59. doi: http://dx.doi.org/10.1002/aic.690260403

[10] Van Dyke, M. D. (1964). *Perturbation methods in fluid mechanics*. New York: Academic Press.

[11] http://web2.clarkson.edu/projects/crcd/me537/downloads/02_Pastsphere.pdf

[12] Levich, V. G. (1962). *Physicochemical hydrodynamics*. Englewood Cliffs, N.J.: Prentice-Hall, Inc.

[13] Abramzon, B. M., & Borde, I. (1980). Conjugate unsteady heat transfer from a droplet in creeping flow. *AIChE Journal 26*, 536–544. doi: http://dx.doi.org/10.1002/aic.690260403

[14] Konopliv N., & Sparrow E. M. (1972). Unsteady heat transfer for Stokesian flow about a sphere. *Journal of Heat Transfer 94*, 266–272. doi: http://dx.doi.org/10.1115/1.3449926

[15] Dorfman, A. S. (2009) *Conjugate problems in convective heat transfer*. Boca Raton: CRC Press Taylor & Francis.

[16] Van Dyke, M. D. (1962). Higher approximatios in boundary layer theory. *Journal of Fluid Mechnics 14*, 161–177; Part 1 General analysis *14*, 481–485.

[17] Dorfman, A. S. (1973). Second approximation boundary layer equations in Prandtl-Mises variables. *High Temperature 11*, 501–506.

[18] Evans, H. L. (1968) *Laminar boundary layer theory*. Reading, MA: Addison-Wesley Publishing Company.

[19] Fröpssing, N. (1958). Calculation by series expansion of the heat transfer in laminar boundary layer at nonisothermal surfaces. *Arkiv för Fysik 14*, 143–151.

[20] Imai, I. (1957). Second approximation to the laminar boundary layer flow over a flat plate. *Journal of the Aeronautical Sciences 24*,155–156.

[21] Chapman, D., & Rubesin, M. (1949). Temperature and velocity profiles in the compressible laminar boundary layer with arbitrary distribution of surface temperature. *Journal of Aeronautical Sciences 16*, 547–565.

[22] Love, G. (1957). An approximate solution of the laminar heat transfer along a plate with arbitrary distribution of the surface temperature. *Journal of Aeronautical Sciences 24*, 920–921.

[23] Ginevskii, A. S. (1969). *Theory of turbulent jets and wakes,* (in Russian). Mashinostroenie, Moscow.

[24] Betz, A. (1955). Zur Berechnung des Uberganges laminarer Grenzschichten in die Ausstromung. In W. Tollmien and H. Görtler (Ed.), *Fifty years of boundary layer research*, ed., pp. 63–70, Braunschweig.

[25] Dorfman, A. S. (2011). Universal function in boundary layer theory. *Fundamental Journal of thermal science and engineering applications 1*, 35–72.

[26] Hartree, D. R. (1937). On an equation occurring in Falkner and Skan's approximate treatment of the equation of the boundary layer. *Proceedings of the Cambridge Philosophical Society 33*, Part II, 223–239. doi: http://dx.doi.org/10.1017/S0305004100019575

[27] Eckert, E. R.G., & Drake, R. M. (1959). *Heat and mass transfer*. New York: McGraw Hill.

[28] Kays, W. M. (1980). *Convective heat and mass transfer*. New York: McGraw-Hill.

[29] Lighthill, M. L. (1950). Contribution to the theory of heat transfer though a laminar boundary layer. *Proceedings of the Royal Society of London* A *202*, 359–377. doi: http://dx.doi.org/10.1098/rspa.1950.0106

[30] Ede, A. J. (1967). Advances in free convection, *Advances in heat transfer*, vol 4 (pp. 1–64). New York: Academic Press.

[31] Rao, V. D., Naidu, S. V., Rao, B. G., & Sharma, K. V. (2006). Heat transfer from a horizontal fin array by natural convection and radiation–a conjugate analysis. *International Journal of Heat Mass Transfer 49*, 3379–3391. doi: http://dx.doi.org/10.1016/j.ijheatmasstransfer.2006.03.010

CHAPTER 3

METHODS IN TURBULENT FLUID FLOW AND HEAT TRANSFER

3.1 Transition from Laminar to Turbulent Flow

3.1.1 Basic Characteristics

Laminar flow exists only at relatively small Reynolds numbers. As the Reynolds number increases, the laminar regime of flow transients in turbulent flow. The laminar flow is well organized so that it looks like thin parallel layers (lamina in Greek means plate or layer) of fluid moving unmixed along a pipe or a plate. In contrast to that, the mixing process inside a fluid, leading to homogeneous disturbed medium, is one of the basic characteristics of turbulent flow. The patterns of these two regimes were first observed by Reynolds in 19th century, who put the dye inside the flow to make it visible. He was also the first to understand that there exists a universal dimensionless number (now known as critical Reynolds number) at which the transition occurs. The critical Reynolds numbers experimentally determined for flows in a pipe of circular cross-section and past a plate are $\text{Re}_{cr} = \hat{u}D/v = 2300$ and $\text{Re}_{cr} = Ux/v = 3.5 \cdot 10^5 / 10^6$, where $\hat{u}$ is an average velocity in a pipe. Studies show that the value of critical Reynolds number depends on the conditions outside of a pipe or a body, namely, the critical Reynolds number increases as the level of disturbances in the inlet flow for pipe or boundary layer decreases. The critical Reynolds numbers indicated above correspond to usually disturbed environment, whereas in the experiments when a special care was taken to reduce the disturbance in the inlet flow, the flow in a pipe remained laminar without transition up to Reynolds number $40,000$ [1]. At the same time at the Reynolds numbers less than 2000, the flow in a pipe remains laminar, independent of the inlet conditions because in this case, the disturbances are dissipated by viscous fluid.

The transition process occurs not at one specific Reynolds number, rather in some interval of its values from the critical one to a somewhat greater Reynolds number. In this interval, the flow alternately becomes laminar or turbulent being fully laminar at the beginning and fully turbulent at the end of the transition interval. This phenomenon called intermittency is characterized by the intermittency factor γ, which is defined at a particular point of flow as a fraction of the whole time of interval when the flow is turbulent. Thus, the intermittency factor is a function of the point coordinates and of the Reynolds number in the limits of transition interval being equal $\gamma = 0$ and $\gamma = 1$ at the beginning and at the end of this interval. The experiments indicate that the intermittency factor in a pipe at a fixed Reynolds number increases with distance from leading edge and in the direction from the axis to the wall [1]. In the boundary layer in the limits of transition, the intermittence factor also increases toward the wall, becoming constant close to the surface at the distance of about 20% of boundary-layer thickness [2].

As the transition occurs, the flow characteristics change: (i) the local parameters, velocity, and pressure, at each point, become unstable, randomly fluctuating; (ii) the velocity distribution across the cross-section in the pipe and in the boundary layer becomes more uniform due to the mixing process in turbulent flow; (iii) the boundary layer becomes thicker being proportional to $x^{4/5}$ instead of $x^{1/2}$ in laminar flow; (iv) in conformity with this, the skin friction on the flat plate increases becoming proportional to the free stream velocity of power 1.8 instead of power 1.5 for a laminar boundary layer, and similarly, the proportionality of the skin friction to average velocity for the laminar flow in the pipe changes to square dependence of average velocity for turbulent flow; (v) analogous to the flat plate, changes occur with flows past thin bodies of other shape streamlined without or with late downstream separation (for example, past aerofoil): the skin friction increases after transition; (vi) in contrast, the resistance decreases after transition in the case of thick bodies (like sphere or cylinder) at which flow separates relatively far from the trailing edge, creating a significant wake behind the body; this resistance decreases because the turbulent boundary flow separates downstream later than laminar flow, resulting in smaller wake and hence in decrease in the energy loss.

Comment 3.1 The basic properties of laminar and turbulent flows are important to know to distinguish these two regimes as well as to find the way to reduce the energy loss, employing laminar airfoils or turbulators. The goal of the first method applicable to thin bodies is to move the point of separation as far on the surface as possible to provide the laminar boundary layer on the majority part of the surface. This is achieved by designing the suitable shape of the streamlined thin body. The aim of the second approach intended for thick bodies is opposite, consisting of a technique providing the transition as early as possible to get separation far from the trailing edge (the turbulent flow separates farther than laminar flow). Such a technique was developed by Prandtl who used the wire on the streamlined sphere as a turbulator to get an early transition, showing the remarkable decrease of the resistance (Exercise 3.4).

3.1.2 The Problem of Laminar Flow Stability

An idea that the transition is a process of transforming a stable laminar flow into unstable turbulent flow goes back to Reynolds and Rayleigh. However, only in the middle of last century after many theoretical and experimental investigations provided by great scholars, it was proved that such an approach works perfectly. Here, we consider a relatively simple version of this extremely complicated problem to give a reader some insight into basic ideas of the method of small disturbances, which finally led to successes. The stability is estimated via imposing on the flow small disturbances in the form of discrete waves propagating along the flow. The flow is considered as stable or unstable, depending on whether the particular disturbance decays or grows with time. The model is formulated using the following assumptions: (i) the main flow is parallel with a velocity distribution $U(y)$ like a fully developed velocity profile in a tube; (ii) the two-dimensional disturbances imposed on the main flow are small compared with main flow so that only terms containing the first degree of corresponding parameters are important (linear problem); (iii) the parameters of the resulting motion may by presented as a sum $U + u'$ of main (U) and imposed (u') flows; (iv) the velocity distribution of the main flow satisfies the Navier–Stokes equation and is considered as given being defined by Reynolds number, $\mathrm{Re}_m = U_m L / v$, where U_m is the maximum velocity and $L = b$ or $L = \delta$ is the characteristic length for a pipe or boundary layer.

In the frame of these assumptions, the problem consists of three unknown parameters of imposed flow: velocities u', v', and pressure p', depending on coordinates and time. Two-dimensional continuity and unsteady Navier–Stokes equations (2.1)–(2.3) governed this problem. To satisfy the continuity equation, the stream function is applied in the complex form, specifying the disturbance waves via their characteristics: a complex amplitude $\phi(y) = \phi_r + i\phi_i$ and two other parameters α and β. Then by differentiating the stream function, the velocity component are obtained (Exercise 3.5)

$$\psi(x, y, t) = \phi(y)\exp[i(\alpha x - \beta t)], \quad u' = \phi'(y)\exp[i(\alpha x - \beta t)],$$
$$v' = -i\alpha\phi(y)\exp[i(\alpha x - \beta t)] \tag{3.1}$$

Comment 3.2 Here, as it becomes common, the primes are used in two ways: at the velocity components, the primes denote that waves' parameters are small (comparing to mean flow), while at the amplitude, they indicate differentiating.

In equations (3.1) parameter $\alpha = 2\pi/\lambda$ determines the disturbances (λ is the wavelength) and is a real number, whereas another parameter is a complex number $\beta = \beta_r + i\beta_i$, where β_r is the circular frequency. To see the physical meaning of β_i, find the real part (Re is common notation of real part of complex number, do not confuse it with Reynolds number) of x-component velocity $\mathrm{Re}(u') = \exp(\beta_i t)[\phi'_r \cos(\alpha x - \beta_r t) - \phi'_i \sin(\alpha x - \beta_r t)]$. It is clear from this relation that β_i is an amplification parameter determining via the term $\exp(\beta_i t)$ the behavior of the some imposed disturbance showing that it grows at $\beta_i > 0$ or decays at $\beta_i < 0$ (Exercise 3.6). Substituting u' and v' into Navier-Stokes equations (2.2) and (2.3) after eliminating the pressure (Sec. 2.2.2.3) yields the equation for amplitude ϕ

$$(U - c)(\phi'' - \alpha^2\phi) - U''\phi = -\frac{i}{\alpha\,\mathrm{Re}}(\phi''' - 2\alpha^2\phi'' + \alpha^4\phi) \tag{3.2}$$

with zero boundary conditions for amplitude at $y = 0$ and $y \to \infty$. Here, $c = \beta/\alpha = c_r + ic_i$ substitutes β defining growing ($c_i > 0$) or decaying ($c_i < 0$) of a particular disturbance wave. The variables in equation (3.2) are made dimensionless, scaling the velocities by U_m and the lengths by b or δ so that derivatives are taken with respect to y/b or to y/δ, depending on the pipe or plate considered. Equation (3.2) is an ordinary linear differential equation of fourth order, which is known as the Orr–Sommerfeld stability equation that was considered first by Orr and a year later by Sommerfeld at the beginning of the last century. This equation reduces the problem of stability to the eigenvalue problem (Sec. 1.5.3) and is usually written in form (3.2) where the left- and right-hand sides present the inertia and viscous terms, respectively. It is convenient to do so because the stability of inviscid fluid is described only by the left-hand side when the viscous terms become zero.

Equation (3.2) containing four parameters $(\mathrm{Re}, \alpha, c_r, c_i)$ presents the amplitude as the eigenfunction of two eigenvalues, giving two possibilities: (i) for each specified pair of Reynolds number (Re) and wavelength (α), estimate complex eigenvalue $c = c_r + ic_i$ to obtain the circular frequency $\beta_r = c_r \alpha$ and the amplification parameter $c_i > 0$ or $c_i < 0$ and (ii) for each specified pair of Reynolds number and circular frequency β_r, obtain α and again the amplification parameter $c_i > 0$ or $c_i < 0$ to know whether the particular wave grows or decays. A special case is $c_i = 0$, where the eigenvalue data as a function of the Reynolds number give the curve called neutral because it separates the stable values of parameters from unstable ones. For example, the values of α at small Reynolds numbers located on the left from the neutral curve $\alpha(\mathrm{Re})$ are stable, whereas those at greater Re in the right region of the neutral curve are unstable. The point on the neutral curve corresponding to the smallest Reynolds number, which is critical Re_{cr}, is found graphically by vertical tangent to the neutral curve. Note that at this point the derivative $d\alpha / d\mathrm{Re} = 0$ so that at any other point of the neutral curve $\mathrm{Re} > \mathrm{Re}_{cr}$, and hence such defined Re_{cr} is indeed the smallest (Exercise 3.7). For all $\mathrm{Re} < \mathrm{Re}_{cr}$, any disturbance wave decays, whereas for $\mathrm{Re} > \mathrm{Re}_{cr}$, at least, some of the waves grow with time.

In the early studies, simplified equation (3.2) without viscous terms was considered. This simplified equation known as Rayleigh equation gives the critical Reynolds number, which is less than Re_{cr} obtained from the full equation (3.2) and is called the limit of frictionless stability. This means that such a determined critical Reynolds number does not show the end of the transition, but only tells us that instability started, and laminar flow is no more stable, while the transition will come later at the larger critical Reynolds number. Due to that both neutral curves found from simplified and full equations usually are plotted on one graph. However, during a long time, comparison with experimental data showed that both critical Reynolds number obtained from simplified and full Orr-Sommerfeld equations are less than values of critical Reynolds numbers indicating the end of the transition observed in experiments.

As is mentioned above only due to intensive, great scholar investigations, the consistent theoretical and experimental results for critical Reynolds number were finally succeeded. An interested reader may find detailed description of those investigations including historical notes, in [1].

3.2 Reynolds-Averaged Navier–Stokes Equation

3.2.1 Some Physical Aspects

The turbulent flow is an extremely complicated phenomenon. In that case, the velocity and pressure at each point of flow fluctuate randomly, depending on coordinates and time. Despite that the range of the scales of those fluctuations is extremely wide, the unsteady, three-dimensional Navier–Stokes equation, in principle, describes such a pattern. However, it is impossible, at least at present, to get a solution of such a problem due to enormous computer memory and speed required for this procedure. Due to that, the practical methods in turbulence are semi-empirical based on hypothesis like Prandtl's mixing-length or statistically grounded approaches.

The turbulent flow behavior is characterized by interaction between fluctuations of quite different scales of length and time, ranging in length from a largest of the order of the boundary-layer thickness to the extremely small sizes. This process of interaction is usually described in terms of turbulent eddies-small swirling flows of size, corresponding to fluctuation scales. Such eddies provide the energy transport across the boundary layer by cascading manner: the large eddies take the energy from the mean flow transferring it to smaller ones, while the last transforms the energy to minor eddies and so on until the smallest eddies are achieved, where the energy is dissipated in the heat by viscous effects. This energy transferring process leads to great transfer of momentum and mass and yields huge waste of energy and additional turbulent stresses several order larger than corresponding laminar stresses (Exercises 3.8 and 3.9).

Comment 3.3 More detailed analysis of turbulence features, including references to original papers, is given, for example, in [3].

3.2.2 Reynolds Averaging

To apply averaging following Reynolds, the instantaneous turbulence parameters should be expressed as a sum of a mean $U_i(x)$ and a fluctuating $u_i'(x,t)$ component

$$u_i(x,t) = U_i(x) + u_i'(x,t), \quad U_i(x) = \frac{1}{t_2} \int_{t_1}^{t_1+t_2} u_i(x,t)dt, \quad \overline{u_i'} = \frac{1}{t_2} \int_{t_1}^{t_1+t_2} [u_i(x,t) - U(x)]dt = 0 \quad (3.3)$$

These expressions are obtained under the assumption that the averaging time is sufficiently large compared to time scale $t_2 \gg t_1$. This makes sure that the mean velocity $U_i(x)$ is independent of time and hence, according to the first and last equations (3.3), provides the average fluctuation $\overline{u_i'}$ to be zero. The zero result follows more precisely from last equation (3.3) due to the same reason: the difference in the integrant is found from first equation (3.3). Analysis based on comparing the averaging and scales times also shows that the averaging value of time derivative of a turbulence velocity is equal to derivative of the mean flow velocity

$$\overline{\frac{\partial u_i}{\partial t}} = \frac{1}{t_2} \int_{t_1}^{t_1+t_2} \frac{\partial}{\partial t}(U_i + u_i')dt = \frac{U_i(x,t+t_2) - U_i(x,t_2)}{t_2} + \frac{u_i'(x,t+t_2) - u_i'(x,t_2)}{t_2} = \frac{\partial U_i}{\partial t}. \quad (3.4)$$

This is true because the scale of mean flow is large so that the value of t_2 in the first term on the right-hand side is relatively small, resulting in the derivative of a mean flow velocity, whereas the second term vanishes due to the same value of t_2, which is large compared to a very small scale of fluctuation component (Exercise 3.10).

In contrast to zero result (3.3) for average single fluctuation, the average product of two fluctuating values $\overline{u_i u_j}$ differs from the corresponding mean product $U_i U_j$ by average product of fluctuation components (Exercise 3.11):

$$\overline{u_i u_j} = \overline{(U_i + u_i')(U_j + u_j')} = \overline{U_i U_j + U_j u_i' + U_i u_j' + u_i' u_j'} = U_i U_j + \overline{u_i' u_j'}. \quad (3.5)$$

The second and third terms of the central part of this equation vanish due to last equation (3.3). Two quantities are correlated if their product is not zero $\overline{u_i u_j} \neq 0$; otherwise, they are uncorrelated. To understand that the turbulent fluctuation velocities are correlated, consider a simple example of two adjusted flow layers with longitudinal mean velocities $\overline{u_1}$ and $\overline{u_2}$ [1]. Let the velocity of the upper layer is $\overline{u_1} > \overline{u_2}$. If a particle from the lower layer enters the upper layer having transverse velocity v', which we count to be positive, it provides due to inertia a negative change u' of velocity $\bar{u}_1$ because the velocity of coming particle is $\bar{u}_2 < \bar{u}_1$. This results in a negative averaged product $(-\overline{u'v'})$ since $v' > 0$ and $u' < 0$. Similarly, the particle with negative transverse velocity v', which enters the lower layer from the upper layer, gives a positive rise u' to velocity $\bar{u}_2$, leading again to a negative averaged product $(-\overline{u'v'})$ because in this case $v' < 0$ and $u' > 0$. This simple example illustrates that the turbulent fluctuation velocities are inherently correlated (Exercise 3.12).

The principles of turbulent flow averaging considered here were first formulated by Reynolds in 1895.

3.2.3 Reynolds Equations and Reynolds Stresses

The Navier–Stokes equation in the Einstein notation (Sec. 2.2.2.2),

$$\frac{\partial u_i}{\partial x_i} = 0, \quad \rho\left(\frac{\partial u_i}{\partial t} + u_j\frac{\partial u_i}{\partial x_j}\right) = -\frac{\partial p}{\partial x_i} + \mu\frac{\partial^2 u_i}{\partial x_j \partial x_j}, \tag{3.6}$$

after modifying the convection terms to the form similar to equation (2.111) [by first equation (3.7)] takes the form that is convenient for averaging (Exercise 3.13):

$$u_j\frac{\partial u_i}{\partial x_j} = \frac{\partial(u_i u_j)}{\partial x_j} - u_i\frac{\partial u_j}{\partial x_j} = \frac{\partial(u_i u_j)}{\partial x_j}, \quad \rho\frac{\partial u_i}{\partial t} + \rho\frac{\partial(u_i u_j)}{\partial x_j} = -\frac{\partial p}{\partial x_i} + \mu\frac{\partial^2 u_i}{\partial x_j \partial x_j}. \tag{3.7}$$

Employing equations (3.3)–(3.5) yields

$$\frac{\partial U_i}{\partial x_i} = 0, \quad \frac{\partial u_i'}{\partial x_i} = 0, \quad \rho\frac{\partial U_i}{\partial t} + \rho\frac{\partial}{\partial x_j}(U_i U_j + \overline{u_i' u_j'}) = -\frac{\partial P}{\partial x_i} + \mu\frac{\partial^2 U_i}{\partial x_j \partial x_j}. \tag{3.8}$$

First equation (3.8) is found from continuity equation (3.6) (the first one) since averaging fluctuation components are zero, while second equation (3.8) for fluctuation velocities (not averaging) is a result of subtracting first equation (3.8) from continuity equation (3.6) (Exercise 3.14). Returning to usual form of convective term gives a system

$$\frac{\partial U_i}{\partial x_i} = 0, \quad \rho\frac{\partial U_i}{\partial t} + \rho U_j\frac{\partial U_i}{\partial x_j} = -\frac{\partial P}{\partial x_i} + \frac{\partial}{\partial x_j}\left(\mu\frac{\partial U_i}{\partial x_j} - \rho\overline{u_i' u_j'}\right). \tag{3.9}$$

It is seen that Reynolds averaging equations (3.9) are Navier–Stokes equations in which the instantaneous parameters are replaced by mean values, but the additional terms $(-\rho\overline{u_i' u_j'})$ known as Reynolds stresses fundamentally change the situation.

The Reynolds stresses form a symmetric $(\tau_{ij} = \tau_{ji})$ tensor $\tau_{ij} = -\rho\overline{u_i' u_j'}$ defined by six unknown independent components. Thus, the averaging process creates six new unknown components without producing any additional equation. This results in the so-called unclosed system of four equations (3.9), containing 10 instead of usual 4 unknown quantities. To close the system, it is necessary to create the deficient equations or to give other means, determining the unknown Reynolds stresses. In turbulent flow theory, the closure problem is solved by semi-empirical or statistical models. As was mentioned in historical notes (Sec. 2.1), according to current terminology, the turbulence models are classified, depending on a number of differential equations used in addition to continuity and Navier–Stokes equations. In conformity with that, the semi-empirical models based on algebraic (without differential) equations are named as algebraic or zero-equation models. The models grounded on one or two differential equations are called as one- or two-equations models, and so on (Exercises 3.15 and 3.16).

Comment 3.4 Attempts to create additional equations for estimating Reynolds stresses by employing, for example, the method of moments failed. The new equations by the method of moments are produced multiplying an existing equation by some parameter in first, second, third, and so on, powers. Such a procedure results in moment equations of the first and higher orders. However, even the first-order moment equation obtained by averaging the Navier–Stokes equation contains due to its nonlinearity 22 new unknown quantities [3].

3.3 Algebraic Models

Boussinesq was the first who in 1877 tried to estimate the turbulent stresses. Proceeding from analogy, he proposed a relation $\mu_{tb}\left(\partial U/\partial y\right)$ similar to that for shear stress in laminar flow $\mu\left(\partial u/\partial y\right)$. These two very similar looking expressions are in fact different in essence. First of all, the viscosity coefficient μ is a physical property of fluid and usually is known, whereas the eddy-viscosity coefficient μ_{tb} is a flow characteristic, depending practically on the same parameters as the flow itself. In particular, in contrast to viscosity coefficient, the eddy viscosity should depend on velocity characteristics because the viscous forces in laminar flow are proportional to velocity, whereas in turbulent flow these depend on the square of mean velocity. The other essential difference between both flows is the mechanism of transport processes responsible for stresses generation. Whereas in a laminar flow, these transport processes are of the molecular nature, in turbulent, they are provided by eddies motion, which is of macroscopic type since even smaller eddies are many orders of magnitude larger than molecules. So it is clear that an analogy between the Boussnesq and laminar shear stress formulae does not help for calculation of turbulent flow characteristics until the relations between eddy-viscosity coefficient and flow parameters are established.

Today, more than 130 years after Boussinesq, the problem of turbulent flow prediction is still far from its complete solution. Nevertheless, the turbulence models developed during this time enable us to solve more or less accurately the turbulent flow problems. Our understanding of turbulence nature improves also remarkably during this time due to a modern experimental technique and recently developed numerical methods such as computational fluid dynamics (CFD), direct numerical simulation (DNS), large eddy simulation (LES), and detached eddy simulation (DES) (Chapter 9).

We begin discussing the turbulence models from the simplest algebraic ones grounded on the semi-empirical approach, which means that deficit of theoretical reasons of model structure is compensated by experimental data.

3.3.1 Prandtl's Mixing-Length Hypothesis

In 1925, Prandtl gave the first method for estimating the eddy-viscosity coefficient, introducing the mixing-length hypothesis. He considered a simple model assuming that in turbulent flow particles coalesce forming lumps, which moves retaining their momentum some distance l that he named mixing length. To understand basic Prandtl's idea, consider three layers of parallel flow in the x-direction. Let particles at layer with coordinate y have mean velocity $U(y)$. Then, the particles at the layers above and below with coordinates $y+l$ and $y-l$ have the velocities $U(y+l)$ and $U(y-l)$. These particles arriving into the middle layer due to transverse fluctuation $v' < 0$ and $v' > 0$ change the velocity $U(y)$ by $\Delta U_a = U(y+l) - U(y)$ and by $\Delta U_b = U(y) - U(y-l)$ owing to the particles from above and below, respectively. Expansion of the functions $U(y+l)$ and $U(y-l)$ in the Taylor series in the vicinity of $U(y)$ and taking into account only first two terms $U(y) \pm l(dU/dy)$ shows that both changes approximately equal $\Delta U_a = \Delta U_b = l(dU/dy)$ (Exercise 3.17). These quantities might be considered as the turbulent fluctuations since they estimate the small changes of the mean velocity $U(y)$. Averaging the absolute values of these fluctuations arriving to a middle layer from adjacent two layers gives the estimation of the absolute value $|u'|$ of x-component fluctuation in the layer with velocity $U(y)$. Then, employing another Prandtl's assumption that both fluctuation components are proportional to each other yields basic relations of the mixing-length model:

$$\overline{|u'|} = \frac{|\Delta U_a| + |\Delta U_b|}{2} = l \left| \frac{dU}{dy} \right|, \quad \overline{|v'|} = const. \cdot l \left| \frac{dU}{dy} \right|, \quad \tau_{tb} = \mu_{tb} \frac{dU}{dy}, \quad \mu_{tb} = \rho l^2 \left| \frac{dU}{dy} \right|. \quad (3.10)$$

The last expression is obtained comparing formula Boussinesq with relation for turbulent stresses $\tau_{tb} = -\rho u'v'$ gained after applying first two equations (3.10) (Exercise 3.18).

It follows from the preceding discussion that the mixing-length l is a distance, which the particle lump travels in the transverse direction before it affects the mean velocities of adjacent layers of flow. The mixing-length concept is similar to free path notion in the kinetic theory of gases in which the microscopic molecules motion is replaced by macroscopic flow of fluid particles lumps. Prandtl postulated that close to the surface the mixing-length is proportional to a distance from the surface $l = \kappa y$, where $\kappa = 0.41$ is the Karman constant. It is also seen that the Reynolds stresses determined by Boussinesq formula using equation (3.10) for mixing-length vanish at the points with $dU/dy = 0$, that is, at the points of maximum or minimum of velocity profile. This is against the fact that the turbulent mixing exists everywhere in turbulent flow, including these points as well. To fix this shortage, Prandtl suggested another relation $\mu_{tb} = \kappa_1 \rho \delta(x)(U_{max} - U_{min})$ according to which the eddy viscosity is proportional to the maximum velocity difference and width of mixing zone $\delta(x)$ via empirical coefficient κ_1. Time shows that this relation yields satisfactory results basically for free turbulent flows.

Comment 3.5 Prandtl himself counted his mixing-length model as a first approximation. Later, detailed analysis based on comparison with molecular transport process indicated some theoretical shortages of the mixing-length model [3]. Nevertheless, for many years, until the computer came to use, Prandtl's model was actually only one approach for practical calculations and was widely used for studying turbulent flows inside the pipes, around bodies, and in free streams, showing an excellent agreement with experimental data in a number of cases.

3.3.2 Modern Structure of Velocity Profile in Turbulent Boundary Layer

Intensive experimental and theoretical investigations lead in the middle of the last century to detailed analysis of special type of turbulent flows known as equilibrium boundary-layer flows.

Despite that strictly speaking turbulence is never in equilibrium, this term becomes common actually defining the turbulent flows with small changes parameters [3]. Flows of such a type studied, in particular, in [3–7], are characterized by constant dimensionless pressure gradient known as equilibrium parameter

$$\beta = (\delta_1/\tau_w)(dp/dx) \tag{3.11}$$

as well as analogous the self-similar laminar boundary-layers flows. For the case of laminar flow, it is easy to check that β is constant (Exercise 3.19). Although it is not as easy for turbulent flow, this might be shown by analyzing the boundary-layer equations [3]. The equilibrium turbulent flows as well as the self-similar laminar flows possess the similarity property (Sec. 2.6.2), significantly simplifying the study of the turbulence processes.

Modern models adopted these results using the equilibrium velocity profile and corresponding eddy-viscosity distribution across the boundary layer as basis of a model. Such a typical velocity profile consists of three parts: (i) the viscous sublayer (inner part), the relatively small region near the surface with dominant laminar stresses, where the law of the wall holds, (ii) the defect layer (outer part), the major region of the boundary layer with dominant turbulent Reynolds stresses and Clauser's law, located between the viscous sublayer and the free stream region, and (iii) the overlap region of inner and outer layers covered by the log layer where both laws are asymptotically valid. Thus, the equilibrium turbulent velocity profile is basically determined by the wall and defect laws

$$U^+ = f_1(y^+), U^+ = \frac{U}{u_\tau}, y^+ = \frac{yu_\tau}{\nu}, u_\tau = \sqrt{\frac{\tau_w}{\rho}}, \frac{U_e - U}{u_\tau} = f_2\left(\frac{y}{\delta_\tau}\right),$$

$$\delta_\tau = \frac{U_e\delta_1}{u_\tau}, \mu_{tb} = \kappa_2\rho U_e\delta_1. \tag{3.12}$$

Here, δ_1 is the displacement thickness defined by equation (2.93), U_e is the velocity on the outer edge in turbulent flow (because in this case U is used for mean flow velocity), u_τ is the so-called friction velocity because it is defined via shear stress on the surface τ_w, $f_1(y^+)$, and $f_2(y/\delta_\tau)$ are universal functions found from a number of experimental data (e.g., Fig. 3.7 in [3] or Fig. 20.4 in [1]) with a linear part $U^+ = y^+$ of $f_1(y^+)$ close to the surface (Exercise 3.20).

The velocity profile in a log layer is found by considering this region as the limiting case of inner and outer layers. Due to that two assumptions are valid inside the log layer: (i) the inertia terms are small compared to viscous ones and (ii) the turbulent stresses are dominant. Whereas the first affirmation is based on the fact that the inner layer lies sufficiently close to the surface, the second one is true because the outer layer is located enough far from the body. The boundary-layer equation (3.9) shows that in the case of a steady-state flow at a zero pressure gradient and negligible convective terms [see (i)], the sum of laminar and turbulent stresses is constant according to given below first equation (3.13). Integrating this equation (Exercise 3.21) and using Prandtl's formulae (3.10) for turbulent stress together with his assumption of linear function for mixing length $l = \kappa y$ leads to the last expression (3.13):

$$\frac{\partial}{\partial y}\left[(\mu + \mu_{tb})\frac{\partial U}{\partial y}\right] = 0, \quad \mu\frac{\partial U}{\partial y} + \mu_{tb}\frac{\partial U}{\partial y} = const., \quad \mu\frac{\partial U}{\partial y} + \rho\left(\kappa y\frac{\partial U}{\partial y}\right)^2 = const. \tag{3.13}$$

At the surface, the second term in the last equation vanishes telling us that the constant in this equation is equal to stress $\tau_w = \mu(\partial U/\partial y)_w$ on the surface. On the other hand, according to assumption (ii) far from the surface, the first term in this equation may be neglected resulting in first equation (3.14), which on integration leads to third expression (3.14) giving logarithmic velocity profile (Exercise 3.22):

$$\rho\left(\kappa y \frac{\partial U}{\partial y}\right)^2 \approx \tau_w \quad \kappa y \frac{\partial U}{\partial y} \approx \sqrt{\frac{\tau_w}{\rho}} = u_\tau, \quad U = \frac{u_\tau}{\kappa}\ln y + C, \quad U^+ = 2.5\ln y^+ + 5. \quad (3.14)$$

The last equation is obtained using variables (3.12) and empirical data for Karman constant $\kappa = 0.41$ and $C = 5$. From empirical data, it is also known that the log layer is usually located between $y^+ = 30$ and $y = 0.1\delta$ so that the sublayer and defect layer are disposed from $y^+ = 0$ to $y^+ = 30$ and from $y = 0.1\delta$ to $y = \delta$, respectively.

Wilcox [3] showed that the basic features of the equilibrium velocity profile established on empirical basis may be obtained theoretically analyzing solutions of the boundary-layer equation. Following Coles' idea [6] of the above-mentioned inner and outer regions in the turbulent boundary layer, he provided perturbation analysis of pertaining solution using two small parameters ε, inverse Reynolds number and ratio u_τ/U_e from Clauser's relation [last equation (3.12)] [4]. The method of perturbation consists of solution expansion as a power series of small parameter presenting the result in the subsequent order of importance: the basic leading terms without small parameter, the second-order terms proportional to the first power of parameter, the next order terms containing the square parameter, and so on (Sec. 2.5.2.4).

In the perturbation analysis of the equilibrium boundary layer, the technique is similar to that described in Section 2.5.2.4, but in this case, only leading term of the expansion is estimated. The steady-state boundary-layer equation (3.9) is modified using the independent variables according to laws (3.12): $\xi = x/L$, y^+ for the sublayer and $\xi, \eta = y/\delta_\tau$ for the defect layer. Presenting the dependent variables in the form similar to equations (2.84) and estimating the order of magnitude of the all terms containing small parameter gives an expression for a leading term. For example, considering the log layer as a limiting case of sublayer yields the following leading term of the inner expansion:

$$\frac{\partial}{\partial y^+}\left[(1+\nu_0)\frac{\partial \hat{u}_0}{\partial y^+}\right] + o(\varepsilon) = \frac{1}{\mathrm{Re}_{\delta_1}}[\beta + o(\delta_1/L)], \quad (3.15)$$

where $\nu_0(\xi, y^+)$ and $\hat{u}_0(\xi, y^+)$ are the first terms in kinematic eddy-viscosity and friction velocity, sign $o(\varepsilon)$ indicates that the order of magnitude of omitting expression is not greater than ε and β is equilibrium parameter (3.11).

Comment 3.6 Signs $O(\varepsilon)$ and $o(\varepsilon)$ with letter "oh" are used to indicate the order of magnitude. For example, in series $\exp(-x) = 1 - x + x^2/2 - x^3/6 + \cdots$ the order of magnitude of omitted terms "…" is $o(x^3)$ or $O(x^4)$ since the next series term is $x^4/24$. The small "oh" tell us that omitting terms are not greater than the last one, whereas the capital "oh" indicate that they are of the order of the next term of series. We use a simplified explanation for $o(\varepsilon)$. For more strong definition, we refer to [3] or [8].

Returning to initial variables performs equation (3.15) in first equation (3.13) obtained from physical reasons. A detailed analysis shows that the inner solution satisfied the no-slip condition at $y = 0$. At the same time as $y^+ \to \infty$, this solution asymptotically reaches the law of the wall indicating that both velocity and eddy-viscosity grow with y increasing indefinitely, but so that the former goes to infinity logarithmically, whereas the latter does the same linearly. Since the log layer is considered as a limiting case of sublayer, these results of perturbation analysis tell us that in the log layer the convection, pressure gradient, and laminar stresses are negligible just in conformity with two assumptions applied in deriving equations (3.13).

The defect layer is treated in a similar way, but the procedure is more involved. The solution gives some other results, proving the empirical considerations mentioned above. In particular, it is shown that the similarity solution exists only if equilibrium parameter β defined by equation (3.11) is constant. That is exactly the same what was observed experimentally and what characterized the equilibrium turbulent boundary layer. The parameter $\hat{\pi}$ in Coles' composite wall-wake profile and the corresponding friction coefficient are also found from the defect layer solution:

$$U^+ = \frac{1}{\kappa}\ln y^+ + \hat{B} + \frac{2\hat{\pi}}{\kappa}\sin^2\left(\frac{\pi}{2}\frac{y}{\delta}\right), \quad \hat{\pi} = \frac{1}{2}\left(u_0 - \ln\frac{\delta}{\delta_\tau}\right), \quad \sqrt{\frac{c_f}{2}} = \left(\hat{B} + \frac{u_0}{\kappa}\right) + \frac{1}{\kappa}\ln \mathrm{Re}_{\delta_1} \quad (3.16)$$

Here, u_0 is a coefficient determined through the integral of the first-order solution in [3]. The composite profile $U^+(y^+, y/\delta)$ [first equation (3.16)] was established in [6] by analyzing experimental data, showing that the turbulent profile may be composed of wall and wake components (Fig. 3.8 in [3]).

Comment 3.7 We considered briefly only the perturbation analysis of the algebraic models. The second part of this valuable research is devoted to comparative analysis of two-equations models.

EXERCISES

3.1 Describe the transition process from laminar to turbulent flow. What is the critical Reynolds number? What are the limits of its values?

3.2 What is the intermittency? How the intermittency factor is defined? Where is it used?

3.3 Repeat the properties which distinguish the laminar and turbulent regimes of flow.

3.4 Explain why different techniques are used to reduce the resistance of the thin and thick streamlined bodies. What is the physical reason of the resistance decreasing in Prandtl's experiment?

3.5 Describe the basic idea of the method of small disturbances. How many critical Reynolds numbers are there? How they differ?

3.6 Derive the real part of u' velocity component (3.1) using the Euler formulae for complex numbers to understand the role of amplification parameter β_i.

3.7 Draw in coordinates $\alpha(\mathrm{Re})$ any curve of parabolic type exposed by its vertex to α-axis and vertical tangent to this curve. This exercise models the situation of neutral curve and gives an understanding that the Re, which corresponds to vertical tangent is the smallest on the curve of such type.

3.8 What are turbulent eddies? What is the cascading energy transferring process?

3.9 What causes the additional stresses in turbulent flow? Why they are several times larger than the laminar stresses?

3.10 Explain why the average time-derivation of fluctuation flow is equal to corresponding derivation of the mean flow.

3.11 What are physical reasons of the fact that single average fluctuation is zero, while the average product of two fluctuation parameters of turbulent flow is not zero? How this average product differs from the corresponding product of mean velocities?

3.12 Repeat the simple explanation of the turbulent fluctuation velocities correlation.

3.13 Derive first equation (3.7) using Einstein notation. Think why the second term in this equation vanishes. Hint: see equation (2.110).

3.14 Obtain equations (3.8) following explanations given in the text.

3.15 Compare viscosity coefficients for laminar and turbulent flows.

3.16 What is the closure problem? How this problem might be solved?

3.17 Draw the flow scheme to understand Prandtl's explanation of the mixing length. Obtain expressions for fluctuations ΔU using Taylor series.

3.18 Compare formula Boussinesq with relation for turbulent stress using expressions (3.10) to get the formula for eddy-viscosity coefficient μ_{tb}.

3.19 Prove that parameter $\beta = (\delta_1/\tau_w)(dp/dx)$ is constant for laminar boundary layers. Hint: use the free stream velocity distribution $U \sim x^{\beta/(2-\beta)}$.

3.20 Derive the linear velocity distribution $U^+ = y^+$. Hint: use Newton's stress law.

3.21 Show that in the case of a steady-state flow with a zero pressure gradient and negligible convective terms, the sum of laminar and turbulent stresses is constant. Hint: obtain the boundary-layer equation for turbulent flow from equation (3.9) and use it to get first equation (3.13).

3.22 Derive the logarithmic formula (3.14) and show that in variables (3.12), U^+, y^+, it has the form $U^+ = (u_\tau/\kappa)\ln y^+ + C$. Hint: consider $\ln u_\tau$ as a constant quantity.

We introduce three algebraic models: the Mellor–Gibson, the Cebeci–Smith, and the Baldwin–Lomax models widely used (see ISI Web of Knowledge).

Comment 3.8 ISI Web of Knowledge indicates who and when cited published articles.

3.3.3 Mellor–Gibson Model [9,10]

The eddy-viscosity function is constructed using the Laufer [11] measured data for the inner part, the Clauser law (3.12) for the outer part, and the Prandtl mixing-length equations (3.10) for the log layer. The Laufer data for inner part are presented in the dimensionless form for total viscosity $\tilde{\nu}_\Sigma = \nu_\Sigma/\nu = 1 + \nu_{tb}/\nu$ by composed function $\tilde{\nu}_\Sigma(\omega)$ using two variables ω and r [see equation (3.18)], while the second and third equations (3.17) determine the dimensional eddy viscosity in the outer part and in the log layer respectively:

$$\left.\begin{array}{l} 0 < r < 2,\ \omega = r^2 - r,\ \tilde{\nu}_\Sigma(\omega) = 1 \\[2mm] 2 < r < 4.5,\ \omega = 2.75r - 1.5,\ \tilde{\nu}_\Sigma(\omega) = \dfrac{r^2}{\omega} \\[2mm] 4.5 < r < 11,\ \omega = 11,\ \tilde{\nu}_\Sigma(\omega) = \dfrac{r^2}{11} \end{array}\right\} \qquad \nu_{tb} = 0.016 U_e \delta_1 \qquad \nu_{tb} = \kappa^2 y^2 \left|\dfrac{dU}{dy}\right|. \qquad (3.17)$$

The velocity profiles are obtained by integrating steady-state boundary-layer equation (3.9) using equations (3.17). For the inner layer, this gives [9]

$$U^+ = u_v^+ + \frac{2}{\kappa}[(1+\beta_v y^+)^{1/2} - 1] + \frac{1}{\kappa}\ln\left[\frac{4}{\beta_v}\frac{(1+\beta_v y^+)^{1/2} - 1}{(1+\beta_v y^+)^{1/2} + 1}\right], \quad \beta_v = \frac{\beta}{\mathrm{Re}_{\delta_1}\sqrt{c_f/2}}$$

$$u_v^+ = \lim_{\zeta \to 0}\left[\int_\zeta^{y^+}\frac{\sigma(r)}{\kappa^2 y^{+2}}dy^+ - \frac{\ln\zeta}{\kappa}\right], \sigma(r) = \omega - r, \omega = \kappa^2 y^{+2}\frac{dU^+}{dy^+}, r = \kappa y^+(1+\beta_v y^+)^{1/2} \quad (3.18)$$

Far from the wall, at $\omega > 11$, function $\sigma(r)$ vanishes, and the first term in the first equation (3.18) becomes a constant $u_v^+ = D^+(\beta_v)$. Equation (3.18) unlike to the common logarithmic velocity profile takes into account the effect of the pressure gradient through parameter β_v. In the case of zero pressure gradient $(\beta_v = 0)$ close to the surface when $y^+ \ll 1/\beta_v$ equation (3.18) has a limit taking usual simple form $U^+ = (1/\kappa)\ln y^+ + D^+$.

Comment 3.9 Note that "far from the wall" in the first sentence after equation (3.18) and "close to the surface" in the last sentence means in fact the same area. Here, as in the case of the log layer considered above by equation (3.13), the log layer overlaps the inner and outer regions, and words "relatively close" or "relatively far" were omitted.

For a positive pressure gradient, especially close to the point of separation, friction coefficient $c_f \to 0$, and both parameters $\beta \sim \beta_v \to \infty$, so that equations (3.18) become not applicable. For this case, Mellor [9] introduced new variables suitable for large values of β and obtained the velocity profile for the inner part in the other form

$$u^{++} = u_v^{++} + \frac{2}{\kappa}\left[(\beta_v^{-2/3} + y^{++})^{1/2} - \beta_v^{-1/3}\right] + \frac{1}{\kappa\beta_v^{1/3}}\ln\left[\frac{4}{\beta_v}\frac{(\beta_v^{-2/3} + y^{++})^{1/2} - \beta_v^{-1/3}}{(\beta_v^{-2/3} + y^{++})^{1/2} + \beta_v^{-1/3}}\right]$$

$$u_v^{++} = \lim_{\zeta \to 0}\left[\int_\zeta^{y^{++}}\frac{\sigma(r)}{\kappa^2 y^{++2}}dy^{++} - \frac{1}{\kappa\beta_v^{1/3}}\ln\frac{\zeta}{\beta_v^{1/3}}\right], \quad r = \kappa\left(\beta_v^{-2/3} + y^{++}\right)^{1/2}y^{++} \quad (3.19)$$

$$u^{++} = \frac{u}{u_p}, \quad y^{++} = \frac{u_p y}{\nu}, \quad u_p = \left(\frac{\nu}{\rho}\frac{dp}{dx}\right)^{1/3}.$$

Here, u_p is the pressure velocity, so called since it is determined via pressure similar to friction velocity u_τ defined through skin friction stress. Equation (3.19) as well as (3.18) close to the wall has the limit $(2/\kappa)y^{++1/2} + (1/\kappa\beta_v^{1/3})[\ln(4/\beta_v) - 2] + D^+/\beta_v^{1/3}$ when $1/\beta_v \to 0$ and $y^{++} \gg 1/\beta_v$ and $\omega > 11$ (Exercise 3.24). The friction coefficient is given in the usual form (3.20) but with coefficients B dependent on β. These and coefficients D^+ in the presented above logarithmical formulae may be found in Table 4.1 in [12, p. 97]:

$$\left(\frac{2}{C_f}\right)^{1/2} = \frac{1}{\kappa}\ln \mathrm{Re}_{\delta_1} + B, (\beta \le 1) \qquad \left(\frac{2}{C_f\beta}\right)^{1/2} = \frac{1}{\kappa}\ln \mathrm{Re}_{\delta_1} + B/\beta^{1/2}, (\beta \ge 1). \quad (3.20)$$

The outer defect layer profiles are found by integrating the boundary-layer equation for an entire range $-0.5 \leq \beta \leq \infty$ and $\text{Re}_{\delta_1} = U\delta_1/v = 10^3, 10^5, 10^9$. Variables (3.12) $\eta, f'(\eta)$ for small β and similar other $\xi, F'(\xi)$ for large β corresponding to equation (3.19),

$$\eta = \frac{y}{\delta_\tau}, \delta_\tau = \frac{\delta_1 U}{u_\tau}, \frac{U-u}{u_\tau} = f'(\eta) \quad \xi = \eta\beta^{1/2}, u_p = u_\tau\beta^{1/2}, \frac{U-u}{u_p} = F'(\xi), \quad (3.21)$$

yield ordinary differential equations, which were solved numerically. The results show that the data in the range $10^3 < \text{Re}_{\delta_1} < 10^9$ differ by less than 2% from that for $\text{Re}_{\delta_1} = 10^5$. Due to that only the last data were tabulated (Table 1 in [10]). The velocity distribution across the entire boundary layer is obtained by using equations (3.18) or (3.19) for inner and the numerical data from Table 1 in [10] for outer parts. It is shown that inner and outer profiles coincide in the overlap area (log layer region).

3.3.4 Cebeci-Smith Model [13]

The eddy viscosity is built as well of the inner μ_{in} and outer μ_{ou} parts

$$\mu_{\text{in}} = \rho l^2 \left[\left(\frac{\partial U}{\partial y}\right)^2 + \left(\frac{\partial V}{\partial x}\right)^2\right]^{1/2}, l = \kappa y\left[1-\exp\left(-\frac{y^+}{A^+}\right)\right], \quad \mu_{\text{ou}} = 0.0168\rho U_e\delta_1 F_K. \quad (3.22)$$

The first two relations are applicable for $y \leq y_m$, and the third equation is used for $y > y_m$, where y_m is coordinate of the matching point. The last value may be found from equation $\mu_{\text{in}} = \mu_{\text{ou}}$. Since the matching point lies in the log layer, the μ_{in} may be found from first equation (3.22). Neglecting the second term in this equation one obtains the logarithmical formula (3.14), and after it differentiating gets $\mu_{\text{in}} = \rho\kappa^2 y^2 u_\tau / \kappa y = \rho\kappa u_\tau y$. Equating this result to the last equation (3.22) (taking $F_K = 1$, see 3.23) gives $y^+ \approx 0.04\,\text{Re}_{\delta_1}$ (Exercise 3.25).

Second equation (3.22) is Van Driest's formula, which takes into account the damping effect near the surface. This effect occurs because in the vicinity of the wall, the laminar stress is so small that the turbulent fluctuations contribution becomes significant.

Assuming that at the surface $u' \sim y$, we obtain from the continuity equation that $v' \sim y^2$, and hence, near a surface the turbulent stresses are proportional to y^3. At the same time, some authors think that the coefficient of the y^3 term is small so that actually the turbulence decays as y^4 [14], and the same result one gets from Van Driest's formula (3.22) (Exercise3.26). The other functions in equations (3.22) are

$$A^+ = 26\left[1+y\frac{dp/dx}{\rho u_\tau^2}\right]^{-1/2} \quad F_K = \left[1+5.5\left(\frac{y}{\delta}\right)^6\right]^{-1}. \quad (3.23)$$

This function for A^+ differs from the initial $A^+ = 26$ in [15] by relation in brackets, which takes into account the effect of pressure gradient. Another equation (3.23) takes account of the intermittence effect. In Section 3.1.1, we considered the intermittency phenomenon, which occurs in transition just before the flow becomes full turbulent. The same behavior of alternative laminar and turbulent flows was observed experimentally [16,17] close to the border of the free stream area. Function F_K estimates the eddy viscosity decreasing in the outer layer caused by intermittent flow at the outer edge that is partly laminar.

Comment 3.10 Van Driest's formula derived theoretically using Prandtl hypothesis and solution of the second Stokes problem of the plate oscillating parallel to itself (Sec. 2.4.1) well agrees with experimental data. This is unexpected because at first glance, the Stokes problem does not model intermittent flow. This example shows that even a relatively simple approach grounded on physical considerations may lead to significant results.

3.3.5 Baldwin–Lomax Model [18]

This model consists as well two layers with eddy-viscosity defined as

$$\mu_{\text{in}} = \rho l^2 |\omega|, \quad |\omega| = \left| \frac{\partial V}{\partial x} - \frac{\partial U}{\partial y} \right|, \quad \mu_{\text{ou}} = 0.027 F_{\text{wake}} F_K \left(y, y_{\text{max}}/0.3 \right)$$

$$F_{\text{wake}} = \min \left[y_{\text{max}} F_{\text{max}}, y_{\text{max}} U_{\text{dif}}^2 / F_{\text{max}} \right], \quad F_{\text{max}} = (1/\kappa) \left[\max \left(l|\omega| \right) \right], \tag{3.24}$$

where the mixing length l is defined by Van Driest's formula (3.22) with $A^+ = 26$, y_{max} is the value of ordinate at which the product $(l|\omega|)$ achieved its maximum value, $|\omega|$ is the magnitude of vorticity (Sec. 2.2.2.3), $F_K \left(y, y_{\text{max}} / 0.3 \right)$ is the equation (3.23) with $y_{\text{max}}/0.3$ replaced for δ, and U_{dif} is the maximum value of U.

This model was formulated special for the cases when the usual boundary-layer properties like thicknesses are difficult to estimate, for example, for separated flows. Due to that, δ in function F_K and some quantities in other relations are replaced by unusual characteristics. Nevertheless, the results obtained for separated flows by this as well as by other algebraic models are unsuccessful [3].

3.3.6 Application of the Algebraic Models

First as in the case of laminar flow, we consider three examples of free stream flows: the far wake behind a body, the jet, and the mixing layer between two streams. These problems for turbulent flow are described by similar differential equation but with Reynolds stresses and by the same boundary conditions. The ideas of solutions are also close consisting of the similarity variables to reduce the partial differential equation to an ordinary equation. The basic difference in this case is that the solution in fact is semi-empirical since one constant should be found by meeting the experimental data. The problems considered here are solved applying a mixing length [3]. Other solutions of these problems based on the second Prandtl's hypothesis may be found in [1].

3.3.6.1 The Far Wake

The governing equation, boundary conditions, and the integral defining the drag are the same equations (2.104) except the right-hand side of differential equation, where v is replaced by $v_{tb} = l^2 |\partial U/\partial y|$. Recalling from Chapter 2 that this problem is usually simplified by linearization using small velocity difference $u_1 = U_\infty - U$ (here U is a mean component of turbulent flow velocity, 3.2.2) and similarity variable $\eta = y/\delta$, where δ is the half of the wake width, we arrive at an ordinary differential equation

$$U_\infty \frac{\partial u_1}{\partial x} = \frac{\partial}{\partial y}\left(l^2 \left|\frac{\partial u_1}{\partial y}\right|\frac{\partial u_1}{\partial y}\right) \quad u_1 = u_0(x)f(\eta),$$

$$\frac{U_\infty \delta u_0'}{u_0^2} f(\eta) - \frac{U_\infty \delta'}{u_0}\eta\frac{df}{d\eta} = c^2 \frac{d}{d\eta}\left(\left|\frac{df}{d\eta}\right|\frac{df}{d\eta}\right). \tag{3.25}$$

The last equation is obtained from the first one by (i) assuming that in this case, solution for u_1 has a similar form as for laminar flow (Sec. 2.6.5.1) but with unknown function $u_0(x)$ instead of $(x/L)^{-1/2}$, (ii) transforming the first equation into variables $\xi = x$, $\eta = y/\delta(x)$ using relation for u_1, chain rule, and primes for denoting the derivatives with respect to x (Comment 3.2), (iii) assuming that the mixing length is proportional to the wake width $l = c\delta$, where c is constant determined by meeting the data (Exercise 3.27).

The solution of equation (3.25) should satisfy the boundary conditions (2.104) in the form $df/d\eta = 0$ at $\eta = 0$, $f = 0$ at $\eta \to \infty$ and a relation for drag per unit mass and unit width $D = 2u_0\delta U_\infty I$, where I is the integral of function (3.25) $f(\eta)$ (Exercise 3.28). Since the total drag should be independent of x, it is clear that the product $u_0\delta$ in the last equation is constant. Proceeding from this fact, we determine the coefficients in the last equation (3.25) and find that their absolute values are equal leading to equalities (3.26). Since we are looking for similarity solution, these quantities should be constant. Denoting such a constant as c_1, we get a equation (3.26) for δ (Exercise 3.29) and transform equation (3.25) into equation (3.27) with constant coefficients:

$$u_0\delta = \frac{D}{2U_\infty I}, \quad \frac{U_\infty \delta'}{u_0} = -\frac{U_\infty \delta u_0'}{u_0^2} = \frac{2U_\infty^2 I\delta\delta'}{D} = c_1, \quad \delta = \sqrt{\frac{c_1 Dx}{U_\infty^2 I}} \tag{3.26}$$

$$c^2 \frac{\partial[f'(\eta)]^2}{\partial\eta} + c_1[f(\eta) + \eta f'(\eta)] = 0, \quad \frac{df}{f^{1/2}} = -\frac{\sqrt{c_1}}{c}\eta^{1/2}d\eta,$$

$$f = \left(\frac{c_2}{2} - \frac{\sqrt{c_1}}{3c}\eta^{3/2}\right)^2 = \frac{c_1}{9c^2}(1 - \eta^{3/2})^2. \tag{3.27}$$

Equation (3.27) is twice integrated. The first integration results in the second equation (3.27), whereas the second equation gives the solution for function f containing constants c, c_1, and c_2 that is defined via c and c_1 (Exercises 3.30 and 3.31). To find the relation between c and c_1, the integral defining drag $D = 2u_0\delta U_\infty I$ is calculated. Using last equation (3.27) and knowing that this integral equals unity, we obtain $c_1 = 20c^2$ (Exercise 3.32). Then, the constants are found by comparing equation (3.26) for δ with known data $\delta \approx 0.805\sqrt{Dx/U_\infty^2}$ [1], resulting first in $c_1 = 0.648$ and then in $c = 0.18$. The mean velocity distribution in a wake is found from second equation (3.25) (Exercise 3.33):

$$U(x,y) = U_\infty - 1.38\sqrt{\frac{D}{x}}\left[1 - \left(\frac{y}{\delta}\right)^{3/2}\right]^2, \quad \delta = 0.805\sqrt{\frac{Dx}{U_\infty^2}} \tag{3.28}$$

This result well agrees with measurements (Fig. 3.3 in [3]). It shows also that solution (3.28) is not analytic function (Sec. 2.2.2.5) since derivatives of U with respect to y higher than second are discontinuous at $y/\delta = 1$ (Exercise 3.33) which means that there is a sharp interface at the wake edge observed also in experiments. This is unexpected remarkable result of mixing-length model.

3.3.6.2 The Two-Dimensional Jet

Solution of this problem is obtained in a similar way, but for the jet as well as in the case of laminar jet (Sec. 2.6.5.2), it is convenient to work with stream function rather than with velocity. Using formulae (2.17) for velocity components and recalling that in jet pressure is constant, one gets the steady-state boundary-layer equation (2.69) in terms of stream function (Exercise 3.34). The similarity solution of this equation is sought in the form similar to (3.25) $\psi = \psi_0(x) f(\eta)$ with $\psi_0 = \sqrt{(I/2)c_1 x}, \eta = y/\delta, \delta = c_1 x$. Using these variables reduces the equation for stream function to an ordinary differential equation

$$\frac{\partial \psi}{\partial y} \frac{\partial^2 \psi}{\partial x \partial y} - \frac{\partial \psi}{\partial x} \frac{\partial^2 \psi}{\partial y^2} = \frac{\partial}{\partial y}\left(l^2 \left|\frac{\partial^2 \psi}{\partial y^2}\right| \frac{\partial^2 \psi}{\partial y^2} \right), \quad \frac{I}{2} = \int_0^\infty U^2 dy, \quad c^2 f''^2 - \frac{1}{2} c_1 ff' = 0. \quad (3.29)$$

Here, as well as in equation (3.27), primes denote the derivatives with respect to η, and c is the coefficient of proportionality in the mixing length. Note that the sign of the absolute value in equation for ψ is replaced with minus since the velocity decreases across a jet. A solution of ordinary equation should satisfy the boundary conditions, $\eta = 0, f = f'' = 0, \eta \to \infty, \ f' \to 0$, and integral similar to that in the previous problem (Exercise 3.32) should be equal to unity providing as it follows from physics a constant energy in a jet. Ordinary equation (3.29) was solved numerically. To perform this, the equation divided by c_1 was integrated for different values of c^2/c_1, and the results were compared with experimental data. Such a procedure gave the constants $c = 0.098$ and $c_1 = 0.246$, providing basically an agreement with measurements but with some discrepancies (Fig. 3.5 in [3]).

Comment 3.11 Differential equation (3.29) is of the second order. Although there are four boundary conditions, including the requirement for integral, note that equation (3.29) satisfies two of them at $\eta = 0$, while two other are used to determine the constants.

3.3.6.3 Mixing Layer of Two Parallel Streams

This problem of two parallel streams is solved as well analogous to previous ones. In particular, governing equation (3.29) and the form of solution are the same as in problem of jet but with $\psi_0 = c_1 U_1 x$, while the boundary conditions are as those for laminar mixing layer $\eta \to \infty, U \to U_1; \ \eta \to -\infty, U = U_2$, where $U_1 > U_2$ and the slower stream is below the other. Similar to (3.29), ordinary differential equation (3.30) was obtained and simplified by differentiating the first term to get second equation (3.30). The last equation was solved numerically for the case of stagnation lower stream $U_2 = 0$ and suitable boundary conditions (3.30), giving the constants and solution, which well agrees with measured data (Fig. 3.4 in [3]):

$$c^2(f''^2)' + c_1 ff'' = 0, \quad 2c^2 f''' + c_1 f = 0, \quad \eta = 0, f = f' = 0, \eta \to \infty, f' \to 1,$$
$$c = 0.071, c_1 = 0.247. \tag{3.30}$$

Comment 3.12 Görtler [1] proposed solution of same second equation (3.30) in the form of series and using first two terms obtained an approximate analytical relation with other constants, also showing good agreement with another experimental result for the case $U_2 = 0$. This is a typical situation of the semi-empirical means when it is possible to obtain some solutions of the same problem approximately well-fitted different data.

3.3.6.4   Flows in Channel and Pipe

The fully developed flow in a channel or in a pipe is studied. Since fully developed velocity is independent of x, from the continuity equation follows that $\partial V/\partial y = 0$, and hence $V = 0$ because it is zero at the surface. In such a case, the inertia terms vanish (Exercise 3.35), and the boundary layer equation simplifies to first equation (3.31)

$$\frac{dp}{dx} = \frac{1}{r^j} \frac{\partial}{\partial r}\left[r^j \left(\mu \frac{\partial U}{\partial r} + \tau_{tb} \right) \right], \quad \mu \frac{\partial U}{\partial r} + \tau_{tb} = -\rho u_\tau^2 \frac{r}{R}, \quad (\mu + \mu_{tb}) \frac{\partial U}{\partial y} = \rho u_\tau^2 \left(1 - \frac{y}{R} \right). \tag{3.31}$$

Here, r is transverse coordinate (y for channel or r for pipe), R is half of height or radius and $j = 0$ or $j = 1$ for channel or pipe. To get the second and third equations from the first: (i) multiply both sides of this equation by r^j and integrate it knowing that in developed flow $dp/dx = const.$; (ii) write the first equation for the surface to get relation $dp/dx = -\tau_w (j+1)/R$; (iii) apply this relation and use y instead of $r = R - y$ to modify the second equation (Exercise 3.36).

Last equation (3.31) was solved numerically using Cebeci–Smith and Baldwin–Lomax models for eddy-viscosity estimation. Since U_e and δ_1 in the first as well as U_{dif} and y_{max} in the second models are unknown in advance, the relaxation iterative method was employed. The agreement with DNS (Chapter 9) and experimental results is gained with difference (1–8)% for velocity profiles and skin friction (Figs. 3.11 and 3.12 in [3]).

Comment 3.13 This problem in fact is the same as the Hagen–Poiseuille laminar flow in a tube (Sec. 2.4.3). Comparison shows how much complicated is the turbulent case even for such a relatively simple problem. While the Hagen–Poiseuille problem has an exact solution, the analogous turbulent task is solved numerically using at least one empirical constant. Due to that, along with semi-empirical methods in practical calculations, empirical correlations are widely employed, for example, simple relation for velocity profiles in a tube $u/U = (y/R)^{1/n}$, where U is the maximum velocity in the cross section and n is a function of $Re = \bar{u}D/\nu$ or Prandtl's formula $c_f^{-1/2} = 4\log(2\,Re\,\sqrt{c_f}) - 1.6$ for friction coefficient. Such correlations for different cases are usually given in engineering fluid flow courses; some examples with references may also be found in [1].

3.3.6.5 The Boundary-Layer Flows

We consider examples of prediction of boundary-layer characteristics by Cebeci–Smith and Baldwin–Lomax models. Comparisons of computed and measured results (Figs. 3.13–3.16 in [3]), including data from AFOSR Stanford Conference (Comment 3.14) shows that both models predictions are accurate as follows: (i) satisfactory for flows with zero and favorable (negative) pressure gradients; (ii) reasonable for flows with mild adverse (positive) pressure gradients; (iii) unsatisfactory for flows with strongly adverse pressure gradients. While the differences from data for velocity profiles are small in the first two cases, these for integral characteristics, such as friction coefficient and shape factor $H = \delta_2 / \delta_1$, are usually of 8–10%. At the same time in the third case, the results for c_f are higher than that measured by 22% and 36% for Cebeci–Smith and Baldwin–Lomax models, respectively.

Comment 3.14 The AFOSR 1968 (Air Force Office of Scientific Research) Conference on Computational of Turbulent Boundary Layer was held in Stanford University in 1968. There were 75 invited famous scholars in this area, and presented results became standard examples for checking the accuracy of theoretical and experimental researches.

A special case is the separated flows. When a separation occurs, the flow is no more attached to the surface, and a bubble forms between the flow and the surface. Since there no longer the equilibrium situation exists, the algebraic models could not describe properly the separated flows. Some attempts to modify the existing models to improve the prediction of separated flows were unsuccessful because studying such a complicated pattern requires a solution of full Reynolds-averaged Navier–Stokes equation (RANS).

3.3.6.6 Heat Transfer From an Isothermal Surface

The investigation results of heat transfer in the turbulent boundary layer obtained in 1970s of the last century using the Mellor–Gibson model just after it was published are presented in [12]. Here, we discuss a part of these results obtained for isothermal surfaces. The results for arbitrary nonisothermal surfaces are outlined in Section 4.3. The governing equation and boundary conditions are used in the Prandtl–Mises–Görtler form (2.77). The additional turbulent thermal diffusivity α_{tb} is computed using formulae (3.17) for eddy viscosity and assuming that the turbulent Prandtl number $\Pr_{tb} = \alpha_{tb} / \nu_{tb}$ is unity, whereas velocity profiles are defined by equations (3.18) and (3.19). The resulting partial differential equation (4.38) is reduced to a system of ordinary differential equations (4.39) that was solved numerically by Runge–Kutta method (details in [12], Sec. 4.2).

Calculations were performed for $\beta = -0.3$ (flow at stagnation point), $\beta = 0$ (zero gradient flow), $\beta = 1$ and $\beta = 10$ (flows with weakly and strong adverse pressure gradients), for $\Pr = 0.01, 0.1, 1, 10, 100, 1000$ and $\mathrm{Re}_{\delta_1} = 10^3, 10^5, 10^9$. The results are given in Figs. 4.1–4.4 in [12] as the Reynolds analogy coefficient $2St_* / c_f$, depending on the Prandtl number at different Re_{δ_1}, where St_* is the Stanton number for an isothermal surface. These data show that the coefficient of analogy: (i) equals unity for zero pressure gradient and $\Pr = 1$ for all Re_{δ_1} as it should be according to the Reynolds analogy; (ii) at zero pressure gradient increases for $\Pr < 1$ and decreases for $\Pr > 1$ at fixed Re_{δ_1}; (iii) decreases at $\Pr < 1$ and increases at $\Pr > 1$ as the Reynolds number grows; (iv) changes much intensively with Re at $\Pr < 1$ than at $\Pr > 1$;

(v) these features of behavior hold in both cases for favorable and adverse pressure gradients, but (vi) in contrast to a case of zero pressure gradient, the Reynolds analogy coefficient changes its behavior of increasing/decreasing with Pr variation not at Pr = 1, rather at less Prandtl numbers ($\approx 0.5 - 0.7$) at Pr < 1 and at larger Prandtl numbers (≈ 3 for $\beta = 1$ and ≈ 250 for $\beta = 10$) at Pr > 1.

Thus, the Reynolds analogy holds only at zero gradient flows and Pr = 1 while significantly fails in all other cases. The discussing results show that depending on pressure gradient, and Prandtl and Reynolds numbers, the Stanton number is greater or lower than half-friction coefficient being considerably greater than the skin friction at small Prandtl and Reynolds numbers. Much less these quantities differ at large Prandtl numbers and adverse pressure gradients.

Analysis of Reynolds analogy data yields the correlations for isothermal heat transfer rate

$$St_* = \mathrm{Pr}^{-1.35}(c_f/2)^{1-0.3\log\mathrm{Pr}}, \ St_* = 0.113\,\mathrm{Pr}^{-3/4}(c_f/2)^{1/2},$$

$$Nu_{x*}^{-0.023} = 1.04 - 0.0355\log\mathrm{Pe}_x. \tag{3.32}$$

These relations are valid for $1 < \mathrm{Pr} < 50$, $\mathrm{Pr} > 50$, $10^3 < \mathrm{Pe}_x < 10^{12}$, respectively, in flows with a zero pressure gradient and well agree with experimental results (Figs. 4.5, 4.6, and 4.7 in [12]). Analogous formulae for isothermal heat transfer in flows with favorable and adverse pressure gradients are also obtained and presented in [12].

3.3.6.7 The Effect of the Turbulent Prandtl Number

As was indicated in the previous section, the turbulent Prandl number is used to determine the turbulent thermal diffusivity through eddy-viscosity $\alpha_{tb} = v_{tb}/\mathrm{Pr}_{tb}$. Therefore to calculate heat transfer characteristics, one needs the turbulent Prandtl number distribution across the boundary layer. However, reviews (for example in [1, p.706]) of results obtained by different authors show that it is impossible now to estimate even an approximate function $\mathrm{Pr}_{tb}(y)$ so much that the existing data are inconsistent. As a result, in the majority studies, the turbulent Prandtl number is assumed to be constant and equal or close to unity (Exercise 3.37).

The above-described approach of investigating the Reynolds analogy was used to study the effect of turbulent Prandtl number. The isothermal heat transfer characteristics are numerically computed, assuming two values of turbulent Prandtl numbers $\mathrm{Pr}_{tb} = 1.5$ and $\mathrm{Pr}_{tb} = 0.5$, four Prandtl numbers $\mathrm{Pr} = 10^{-1}, 1, 10^2, 10^3$, and three Reynolds numbers $\mathrm{Re}_{\delta_1} = 10^3 (\mathrm{Re}_x = 2.95 \cdot 10^5)$, $10^5 (7.93 \cdot 10^7)$, $10^9 (2.56 \cdot 10^{12})$. Analysis of dependences $St_* / (St_*)_{\mathrm{Pr}_{tb}=1} = f(\mathrm{Pr}, \mathrm{Re}_{\delta_1})$ for both values (0.5 and 1.5) of turbulent Prandtl numbers plotted in Fig. 4.10 in [12] leads to following conclusions: (i) increasing or decreasing the turbulent Prandtl number yields decreasing or increasing in the rate of heat transfer compared to that at $\mathrm{Pr}_{tb} = 1$; (ii) the effect of the turbulent Prandtl number is independent of the Reynolds and Prandtl numbers at large Prandtl numbers ($\mathrm{Pr}_{tb} > 10^2$) so that the ratio $St_* / (St_*)_{\mathrm{Pr}_{tb}=1}$ is constant being ≈ 0.8 and ≈ 1.3 for $\mathrm{Pr}_{tb} = 1.5$ and $\mathrm{Pr}_{tb} = 0.5$, respectively; (iii) the effect of the turbulent Prandtl number grows as the Prandtl number decreases or/and the Reynolds number increases achieving the maximum at Prandtl numbers close to unity; (iv) the effect of Pr_{tb} decreasing (to $\mathrm{Pr}_{tb} = 0.5$) is much greater than that of Pr_{tb} increasing (to $\mathrm{Pr}_{tb} = 1.5$). The maximum value of the ratio $St_* / (St_*)_{\mathrm{Pr}_{tb}=1}$ achieved at $Pr = 1$ and $Re_{\delta_1} = 10^9$ is 67% in the first case, while only 25% in the second case. (v) The computed heat transfer coefficients for $Pr_{tb} = 1$ well agree with measured data, whereas the results for $Pr_{tb} = 1.5$ and to $Pr_{tb} = 0.5$ are lower and higher, respectively (Fig. 4.7 in [12]). We get an additional experimental confirmation on comparing the value of coefficient 0.115 in second formula (3.32) obtained from experimental data with computing results giving for this coefficient 0.096, 0.113, 0.136 for $Pr_{tb} = 0.5$, 1, 1.5, respectively. Thus, the computed value of this coefficient obtained assuming $Pr_{tb} = 1$ is close to empirical founded data, whereas the two other results are significant less for $Pr_{tb} = 0.5$ or larger for $Pr_{tb} = 1.5$ than 0.115 (Exercise 3.38).

Comment 3.15 Relatively small effect of the turbulent Prandtl number at large Prandtl and small Reynolds numbers is a result of the thin thermal boundary layer located basically in the laminar sublayer at a large Prandtl and significant role of the molecular heat transfer by conduction at low Reynolds numbers.

Comment 3.16 The computation results obtained using $Pr_{tb} = 1$ show well agreement with experimental data. This conclusion is in consistent with some other known similar findings. Nevertheless, the value $Pr_{tb} = 1$ should be considered only as suitable assumption for computations because there are experimental evidence that the turbulent Prandtl number is a function of transversal coordinate [1].

3.3.7 THE 1/2 EQUATION MODEL

This model proposed by Johonson and King takes an intermediate position between the algebraic and energy equation models. In fact, it is an algebraic model, like the Cebeci–Smith or the Baldwin–Lomax model, which applies in addition one ordinary differential equation. That is may be the reason way this model is called "1/2 equation" in contrast to one- or two-equations models based on the partial differential equations. The eddy viscosity is determined using the inner and outer layers similar to other algebraic models but with resulting relation for μ_{tb} containing both eddy-viscosities μ_{in} and μ_{ou}.

$$\mu_{tb} = \mu_{ou} \tanh\left(\frac{\mu_{in}}{\mu_{ou}}\right), \; \mu_{in} = \rho\left[1 - \exp\left(-\frac{u_D y}{v A^+}\right)\right]^2 \kappa u_s y,$$

$$\mu_{ou} = 0.0168 \rho U_e \delta_1 F_K(y, \delta) \sigma(x). \tag{3.33}$$

Here, $u_D = \max(u_m, u_\tau)$ and $u_m = \sqrt{\tau_m / \rho_m}$ are velocity scales, index m denotes the coordinate y_m at which the Reynolds stress achieves its maximum τ_m, the scale u_s is determined through several other quantities $u_\tau, u_m, y_m / \delta, \rho_w, \rho_m, L_m$ [3], where ρ_w is the density on the surface and $L_m = \kappa y_m$ if $y_m / \delta \leq 0.09 / \kappa$, $L_m = 0.09\delta$ if $y_m / \delta > 0.09 / \kappa$.

These additional scales are employed to improve the prediction of separated and reattachment flows.

Last equation (3.33) is equation (3.22) of the Cebeci–Smith model but with additional function $\sigma(x)$ that provides the departure from equilibrium results corresponding to $\sigma(x) = 1$. The maximum value of the Reynolds stress is defined by ordinary differential equation:

$$U_m \frac{d}{dx}\left(\frac{\tau_m}{\rho_m}\right) = 0.25 \frac{(u_m)_{eq} - u_m}{L_m}\left(\frac{\tau_m}{\rho_m}\right) - c\frac{(\tau_m/\rho_m)^{3/2}}{0.7\delta - y_m}\left[1 - \sqrt{\sigma(x)}\right],$$

$$\tau_m = (\mu_{tb})_m\left(\frac{\partial U}{\partial y} + \frac{\partial V}{\partial x}\right)_m. \tag{3.34}$$

Here, $c = 0.5$ if $\sigma(x) \geq 1$, otherwise $c = 0$ and $(u_m)_{eq}$ is the value corresponding to the case of equilibrium when $\sigma(x) = 1$. The function $\sigma(x)$ is found iteratively so that a maximum Reynolds stress obtained from first equation (3.34) is equal to the value given by the second relation showing that μ_{tb} distribution is adjusted to the maximum Reynolds stress.

Existing examples show that this model predicts better the attached and separated flows than corresponding algebraic models (Figs. 3.18 and 3.19 in [3]) (Exercise 3.39).

3.3.8 Applicability of the Algebraic Models

The algebraic models are the simple and easy in performing among modern turbulence models. Therefore, it is reasonable always to start with such a relatively simple approach. The other more complicated models are preferable only when alternatives are incomparable of accuracy or of other specific features. For the flows with zero, favorable and not very strong adverse pressure gradients not close to the point of separation, algebraic model predictions are usually satisfactory. At the same time, the user should keep in mind that algebraic models are incomplete since the mixing length is undefined. That is why in fact these models are semi-empirical requiring at least one constant to meeting experimental data. A range of applicability of such constants depends on the type of problem. For example, each of free stream flows considered above (Sec. 3.3.6) required a different constant defining the mixing length to agree the computed and measured results, whereas the data of isothermal heat transfer obtained in Section 3.3.6.6 with Prandtl assumption for mixing length well fit the experimental correlation in a wide range of Prandtl and Reynolds numbers. Moreover, as shown in [12], the same approach yields satisfactory results in more general case of heat transfer for the nonisothermal surface.

Physically, the reason of a greater range of model applicability in the case of the heat transfer lies in the effect of Prandtl number (Sec. 2.7.3) due to which the influence of velocity profiles on the thermal characteristics is relatively low (Exercise 3.40).

Despite limitation and shortages mentioned above, algebraic models are widely used in solving different turbulence problems, including complex flows without separation. In contrast to algebraic models, the 1/2 equation model is intended to separated flows computing and to provide much better agreement between calculated and measured results in this case. However, this improvement is achieved at the expense of simplicity by using several addition closure constants and iterative computational procedure. Besides, this model is also incomplete since there is no information of the length scale.

EXERCISES

3.23 Explain why it is necessary to consider the negative and positive pressure gradients separately using small and large values of parameter β, respectively.

3.24 Show that the quantities u_τ and u_p are measured in velocity units.

3.25 Derive formula for matching point coordinate y_m in the Cebeci–Smith model following direction from the text.

3.26 Show that according to Van Driest's formula, eddy viscosity near surface is proportional to y^4. Hint: use the Taylor series for function $\exp(-y^+)$.

3.27 Obtain ordinary differential equation (3.25) using the chain rule [see equations (2.36) and (2.37)] as directed in the text.

3.28 Show that drag in the wake is defined by integral similar to integral (2.104) in the form

$$D = 2u_0 \delta U_\infty \int_0^\infty f(\eta)\, d\eta$$

3.29 Find the expression for δ using the equation for u_0 that follows from the first equation (3.26).

3.30 Integrate first equation (3.27) to get the second equation (3.27). Hint: note that two last terms formed the full differential $d(\eta f)$ and use symmetry condition (2.104) to see that integration constant equals zero and then take negative sign at $\sqrt{c_1}$.

3.31 Integrate the result obtained in Exercise 3.30 to get the problem solution in the form of last equation (3.27). To determine the integration constant $c_2/2$ use the condition $f = 0$ at $\eta = 1$ (edge of the wake).

3.32 Compute the integral $\int_0^1 f(\eta)\, d\eta$ and find the relation between constants c_1 and c.

3.33 Obtain the solution (3.28). Hint: use equations (3.25) for u_1, (3.26) for $u_0\delta$ and δ and (3.27) for $f(\eta)$. Show that higher derivatives of this solution at $y/\delta = 1$ are discontinuous (infinite).

3.34 Transform the steady-state boundary-layer equation for the case of constant pressure using a stream function to get equation (3.29).

3.35 Show that for developed flows in the channel and pipe, the inertia terms become zero.

3.36 *Derive second and third equations (3.31) from the first equation following directions given in the text.

3.37 What is the turbulent Prandtl number? How is it used?

3.38 It is shown that calculations assuming $\mathrm{Pr}_{tb} = 1.5$ and $\mathrm{Pr}_{tb} = 0.5$ give, respectively, lower and higher results compared to measured data. At the same time, the results for corresponding coefficients for second formula (3.32) are opposite: it is higher in the first case and lower in the second than the experimental value. What is the reason of such a seeming contradiction?

3.39 What is a 1/2 equation model? Why it is called so? How this model differs from algebraic models?

3.40 Explain why the effect of the Prandtl number leads to slight dependence of heat transfer on the velocity distribution in the boundary layer. Hint: think about ratio of boundary-layer thicknesses.

3.4 One-Equation and Two-Equation Models

Since 1960s with computers advent, the one- and two-equations models become a basic power-ful tool of turbulence flows investigation. This type of models grounded on the kinetic energy equation and other differential equations simulate much closer to the real physical pattern of turbulence than algebraic models. While the one-equation models are still incomplete because they adapt the length scale from some typical flows, the two-equations models are complete, determining the turbulence length scale or some equivalent parameter by special differential equation. As always in modeling, the physically improved turbulence model becomes more mathematically complicated, and in the case of turbulence, the improved model also requires additional closure coefficients. However, using such models is worth because modern models enhance the computation results as well as take into account specific issues such as nonlocal and flow history effects, developing our insight into turbulence nature.

3.4.1 Turbulence Kinetic Energy Equation

Prandtl defined the kinetic energy per unit mass as a sum of square of fluctuation velocity com-ponents. Then, proceeding from the dimensional units, the eddy-viscosity and Reynolds stress are obtained in the terms of kinetic energy, density, and length scale

$$k = (1/2)\overline{u_i' u_i'} = (1/2)\left(\overline{u'^2} + \overline{v'^2} + \overline{w^2}\right), \quad \mu_{tb} = const.\rho k^{1/2} l, \quad \tau_{ii} = -\rho \overline{u_i' u_i'} = -2\rho k. \quad (3.35)$$

It is seen that Reynolds stress is proportional to the turbulence kinetic energy per unit volume. The differential equation defining the turbulence energy is called the transport equation since it describes that the changes arise as the turbulence moves with a flow. This equation is similar to average Reynolds equation (3.9) but consists of second and third specific terms in right-hand side instead of pressure gradient:

$$\rho \frac{\partial k}{\partial t} + \rho U_j \frac{\partial k}{\partial x_j} = \tau_{ij} \frac{\partial U_i}{\partial x_j} - \rho \varepsilon + \frac{\partial}{\partial x_j}\left[(\mu + \sigma_k \mu_{tb})\frac{\partial k}{\partial x_j}\right], \quad \varepsilon = C_D \frac{k^{3/2}}{l}. \quad (3.36)$$

Here, σ_k and C_D are closure coefficients. The terms in equation (3.36) physically signify the following. The two terms on the left-hand side define the substantial derivative (Example 2.2) of the turbulence energy. The first and the second terms on the right-hand side called produc-tion and dissipation give, respectively, the rate of energy that turbulence takes from mean flow and the rate of the loss of turbulence kinetic energy transforming into thermal energy. The last term in equation (3.36) consists of two parts. The first one as that in other analogous equations takes into account the molecular diffusion, whereas the second part of this term also looking similar to corresponding terms in other equations here represents two complex processes. The turbulent transport provided by turbulent fluctuations and the pressure diffusion occurring due to correlation of the pressure and velocity fluctuations (Exercise 3.41).

    The two terms on the left-hand side and the molecular diffusion term are exact, whereas the others in equation (3.36) are approximated using suggestions postulated by Prandtl in 1945. Accordingly, the Reynolds stress τ_{ij} in the production term and dissipation ε are determined via turbulence energy k by last equations (3.35) and (3.36), respectively. The second part of the last term representing the turbulent transport and the pressure diffusion are defined through Boussinesq approximation by eddy-viscosity μ_{tb} and gradient of dependent variable k. Thus, three approximate relations and two closure coefficients σ_k and C_D are needed to close the equation for turbulence energy.

3.4.2 One-Equation Models

Equation (3.36) for turbulence kinetic energy, two equations (3.35) for Reynolds stresses and equation (3.36) for dissipation form a basic set of equations for the one-equation model. To complete the model, the length scale should be specified. Prandtl used in his model the mixing length for this purpose. It can be seen applying equations (3.35) and (3.36) together with two last equations (3.10) that this assumption is valid if the production and dissipation are equal compensating each other and resulting in proportionality of both mixing length in equations (3.10) and (3.36) (Exercise 3.42). Due to that, the constant in formula (3.35) for eddy viscosity containing length scale may be taken unity giving $\mu_{tb} = \rho k^{1/2} l$. Other one-equation models using Prandtl's turbulence energy equation (3.36) differ from the above-considered Prandtl model basically by closure coefficients and length-scale functions [3].

Comment 3.17 Mentioned above assumption of lengths equality is acceptable only for equilibrium flows when parameter (3.11) β is constant. Otherwise, the eddy viscosity is not as simple; rather, it is a function of ratio of Reynolds stress and mean velocity derivative dU/dy.

A different type of one-equation model was introduced by Bradshaw, Ferriss, and Atwell. They employed the same equation (3.36), but instead of Boussinesq approximation in the last term used an experimental result according to which the Reynolds stress in the boundary layer, wakes, and mixing flows is approximately proportional to turbulence kinetic energy with constant factor $\tau_{tb} \approx 0.3k$. This modification changes the type of partial differential equation (3.36) from parabolic to hyperbolic (Sec. 1.2.4) remarkably simplifying the solution of this equation. This model shows well predictions, in particular, for the adverse pressure gradients and provides the best results among others models tested at AFOSR 1968 (Comment 3.14).

Comment 3.18 Hyperbolic equation has real characteristics: two lines crossing at each point of solution domain, which are used to build the method of solution (Exercise 3.43).

There are also one-equation models based on equations other than turbulence energy equation (3.36). Several models used equation similar to (3.36) but written for kinematics eddy viscosity $v_{tb} = \mu_{tb}/\rho$. Models employed other than (3.36) energy equations contain more closure coefficients (up to eight) and additional empirical functions, including definition for the length scale. Two models of this type, Baldwin–Barth and Spalart–Allmarass, are analyzed [3]. Comparing their computed result to measurements shows that the predictions of the Spalart–Allmarass model are satisfactory for special problems, such as airfoil and wings, for which the model was calibrated, whereas the results obtained by Baldwin–Barth model differ more from the measured data even than simpler algebraic models predictions. The general conclusion is that advantages and shortages of one-equation models are close to those of the mixing-length algebraic models. If the one-equation model is specified giving satisfactory results for some separated flows than it could not predict the other flows, like wake or mixing layers, with accuracy close to provided by mixing-length model.

3.4.3 Two-Equation Models

The modern two-equation models are currently the basic tools for solving complex engineering and scientific problems. The popularity of these models came due to the fact that two-equation models are the simplest complete models. This means that problem solution is found by the model without need of a constant for fitting the experimental data. Since the importance of the two-equation models, at the Conference of AFOSR 1980–81, basically these models are tested. Considering the one-equation models, we have seen the difficulties arising from the absence of clear length-scale definition. In contrast to that, the two-equation models provide along with turbulence energy equation the analogous equation for length scale or for some equivalent parameter. While any model of this type consists of the turbulence energy equation, there is no universal parameter for the second equation. Kolmogorov who first use for the turbulence model the second partial differential equation formulated it for specific dissipation rate ω, which is the dissipation energy per unit volume and time. Since this quantity has the dimension (1/s), it follows from dimensional analysis that the eddy viscosity, length scale, and the dissipation are defined as $\mu_{tb} \sim \rho k/\omega$, $l \sim k^{1/2}/\omega$, $\varepsilon \sim k\omega$. Other developers suggested the second equation for time $t \sim 1/\omega$, length scale, dissipation ε, or product kl. As was pointed at the Conference of AFOSR 1980–81, the uncertainty about the two-equation models is the vague choice of variable for the second equation. Although during the last time some clearness was gained, there still is no full answer what variable is most suitable [3].

3.4.3.1 The $k - \omega$ Model

The first two-equation model was suggested in 1942 by Kolmogorov. His second equation in terms of dissipation per unit volume and time ω was formulated in the form similar to the equation for turbulence kinetic energy (3.36):

$$\rho \frac{\partial \omega}{\partial t} + \rho U_j \frac{\partial \omega}{\partial x_j} = -\beta_\omega \rho \omega^2 + \frac{\partial}{\partial x_j}\left(\sigma \mu_{tb} \frac{\partial \omega}{\partial x_j} \right), \quad \omega = C \frac{k^{1/2}}{l}. \tag{3.37}$$

This equation as compared to equation (3.36) for k has no production and molecular diffusion terms. In the early days when it was difficult to use complex turbulence models, Kolmogorov did not develop the model complete; rather, he published an idea, including in equation for specific dissipation ω only the most important terms. The production term was not taken into account because the smallest eddies are responsible for dissipation (Sec. 3.2.1) for which production is not as important. Under the same reason, the molecular diffusion term was omitted which as well is not very important for the case of high Reynolds numbers considered by Kolmogorov.

Comment 3.18 Kolmogorov gave quite brief explanation to his model. Wilcox knowing an outstanding physical intuition of this great scholar has speculated the Kolmogorov's reasoning as follows [3]: (i) proceeding from the last equation (3.35), it is plausible that v_{tb} may be defined through k; (ii) the dimensions of v and k are m^2/s and m^2/s^2, so v/k has dimensions s; (iii) turbulence dissipation ε has dimensions m^2/s^3 (energy per unit time); (iv) consequently,

ε/k has dimensions $1/s$; (v) equations (3.35) for Reynolds stress and (3.36) for turbulence kinetic energy may be closed by introducing variable with dimensions s or $1/s$. This example shows how powerful is the simple method of dimensions if it is deeply physically grounded (Exercise 3.44).

Comment 3.19 History knows many prominent scholars who published their ideas and results without detailed explanation. One of the most famous examples is the theorem, which Pierre de Fermat published in 1637 saying that he proved it. After countless attempts, the first proof was published on 1995 by British mathematician Andrew Wiles. As another more recent example, we mention the Henry Poincare conjecture formulated in 1904, which was considered as one of the most important and difficult problems in topology. Recently, Russian mathematician Grigori Perelman solved this problem and was awarded for the first Clay Millennium Prize but declined to get it (Exercise 3.45).

In 1960s when an interest to turbulence models arises, the followers added to Kolmogorov's model missing terms and improved it. The production was included taking into account that the smallest eddies gained the kinetic energy from the larger ones for which production is important. The molecular diffusion was necessary to add because otherwise no-slip boundary condition could not be satisfied (Exercise 3.46). Further improving was achieved by testing models comparing the computed and measured results for estimating the closure coefficients. Currently, several versions of $k - \omega$ models exist, in particular, based on different variables for the $\omega -$ equation mentioned above. Here, we present the model developed by Wilcox as given in first edition (1994) of his book [3]:

$$\rho \frac{\partial k}{\partial t} + \rho U_j \frac{\partial k}{\partial x_j} = \tau_{ij} \frac{\partial U_i}{\partial x_j} - \beta_k \rho k \omega + \frac{\partial}{\partial x_j}\left[(\mu + \sigma_k \mu_{tb}) \frac{\partial k}{\partial x_j}\right], \quad \mu_{tb} = \frac{\rho k}{\omega}, \quad l = \frac{k^{1/2}}{\omega} \quad (3.38)$$

$$\rho \frac{\partial \omega}{\partial t} + \rho U_j \frac{\partial \omega}{\partial x_j} = \alpha \frac{\omega}{k} \tau_{ij} \frac{\partial U_i}{\partial x_j} - \beta_\omega \rho \omega^2 + \frac{\partial}{\partial x_j}\left[(\mu + \sigma_\omega \mu_{tb}) \frac{\partial \omega}{\partial x_j}\right], \quad \varepsilon = \beta_k \omega k, \quad (3.39)$$

where the closure coefficients are $\alpha = 5/9, \beta_k = 0.09, \beta_\omega = 3/40, \sigma_k = \sigma_\omega = 1/2$.

More complicated system of equations determines the $k - \omega$ model in the third edition (2006) of this book, which in addition consist of three closure coefficients: two new complex functions, two tensors, and equation for dissipation with one term more.

Comment 3.20 Comparing two versions of the $k - \omega$ model gives a typical example showing the way of improving existing models. To enhance the accuracy and/or extend the applicability, developers increase the number of closure coefficients and auxiliary functions. As a result the two-equations model (i) loses universality because new type of flow requires additional constants and functions like that for algebraic or one-equation models; (ii) loses initial physics clearness when it is possible to understand the role of model parts; (iii) becomes confused even for simple cases application; and (iv) becomes too complicated to use in other areas, for example, in heat or mass transfer. Perhaps some simpler models for specific cases of flow instead of one very complicated would be more convenient for using and understanding.

3.4.3.2 The $k - \varepsilon$ Model

The first version of this model was suggested as well in 1940s by Chou. The improved version known as the Standard $k - \varepsilon$ model was published in the middle of 1970s by Launder and co-authors and is widely used. The model is as follows:

$$\rho \frac{\partial k}{\partial t} + \rho U_j \frac{\partial k}{\partial x_j} = \tau_{ij} \frac{\partial U_i}{\partial x_j} - \rho \varepsilon + \frac{\partial}{\partial x_j} \left[\left(\mu + \frac{\mu_{tb}}{\sigma_k} \right) \frac{\partial k}{\partial x_j} \right], \quad \mu_{tb} = C_\mu \frac{\rho k^2}{\varepsilon}, \quad l = C_\mu \frac{k^{3/2}}{\varepsilon} \quad (3.40)$$

$$\rho \frac{\partial \varepsilon}{\partial t} + \rho U_j \frac{\partial \varepsilon}{\partial x_j} = C_{\varepsilon 1} \frac{\varepsilon}{k} \tau_{ij} \frac{\partial U_i}{\partial x_j} - C_{\varepsilon 2} \rho \frac{\varepsilon^2}{k} + \frac{\partial}{\partial x_j} \left[\left(\mu + \frac{\mu_{tb}}{\sigma_\varepsilon} \right) \frac{\partial \varepsilon}{\partial x_j} \right], \quad \omega = \varepsilon / C_\mu k, \quad (3.41)$$

with closure coefficients $C_{\varepsilon 1} = 1.44, C_{\varepsilon 2} = 1.92, C_\mu = 0.09, \sigma_k = 1, \sigma_\varepsilon = 1.3$ and additional correction function for low Reynolds numbers. Latter as well as for the $k - \omega$ model more complicated version was published [3, 2006] (Exercise 3.47).

3.4.3.3 The Other Turbulence Models

The other two-equation models applying different second differential equation, for example, $k - \tau$ or $k - kl$ models employing equation for time τ or for product kl are used and tested much less than $k - \varepsilon$ and $k - \omega$ models. Besides those, other special models and procedures are created to take into account additional effects, occurring in using turbulence models. In particular, special wall functions to ensure accurate satisfaction of the no-slip condition for both variables k and ε or k and ω at the surface as well as procedures for taking into account the surface roughness effect were developed. Many special models for correction the low Reynolds number effects are suggested. These arise close to the surface in the sublayer where the Reynolds number is small and gradients are high, resulting in difficulties to compute accurate parameters distribution. A detailed discussion of the problems mentioned here may be found, for instance, in [1. 2000, 3, 1994, 2006].

Comment 3.21 One may think that the closure coefficients estimation is a mystery. In fact, three things play a significant role in this process: physically grounded reasons, dimensions analysis, and solutions of simple problems. For example, the ratio β_k / β_ω in the $k - \omega$ model may be estimated, considering the isotropic turbulence. For this case, the equations (3.38) and (3.39) become $dk/dt = -\beta_k \omega k$ and $d\omega/dt = -\beta_\omega \omega^2$ because isotropic means independent of position. The asymptotic solution of this system $k \sim t^{-\beta_k / \beta_\varepsilon}$ after comparing to measurements data gives $\beta_k / \beta_\omega = 6/5$ [3].

3.4.4 Applicability of the One-Equation and Two-Equation Models

The one-equation models as well as algebraic models are uncompleted. Due to that, as was pointed in Section 3.4.2, using these models instead of the simpler algebraic gives no real advantages. In contrast, the completed two-equation models are superior to both one-equation and algebraic models because the whole information needed for solving the problem is inside the model.

Although the modern two-equation models are involved and requires the numerical solution, which leads to some difficulties (Comment 3.20), employing two-equation models is only one way to get solution of a complex problem. The turbulent flows with strong adverse pressure gradients, separated or/and reattachment flows, compressible flows with high Mach numbers, and other complications may be studied with reasonable accuracy only by two-equation models.

At the same time, the algebraic models are preferable for the prediction of behavior of the less complicated problems, such as turbulent free shear flows or zero and favorable pressure gradients flows. The significant advantage of algebraic models is the possibility to get a simple close analytic solution. Due to that in the cases when the accuracy of algebraic and other more complicated models are comparable, it is reasonable to use algebraic models, which yield not only simplification of the solution but also give a possibility for a better physical understanding of the studied phenomenon. Examples of such analytical solutions are the results obtained for the far wake and two-dimensional jet in Section 3.3.6. Another successful employing of algebraic model is the investigation of heat transfer presented in the same section (Exercise 3.48).

Comment 3.22 A special attention is given to analytic solutions because nowadays some researches lose the interest to analytic methods. Indeed, after numerical methods become widely used, the analytic solution not only retained their importance but gained a new role as one of the most reliable means of correction and checking the accuracy of software. We return to this topic in Chapter 8 when discussing the numerical approaches.

3.5   Integral Methods

The Karman–Pohlhausen integral method outlined and used in Section 2.7.1 for the laminar boundary layer is applicable to turbulent flows as well. Methods of this type widely used before computer era are still helpful due to their simplicity at least for preliminary estimation or for checking the order of magnitude of solution computed by more complicated methods. The name "integral methods" came from the fact that those methods use integral equations (2.112) and (2.113) obtained by integrating the conservation differential equations (2.111) across the boundary layer. As is explained in Section 2.7.1 owing to this procedure, the integral equations do not contain any information about parameter distribution across the boundary layer and required the knowing a priori type of velocity or/and temperature profiles or other equivalent information. Numerous different types of integral methods are suggested. Most of these are similar being intended to friction estimation mach less are pertained for heat transfer or for both friction and heat transfer evaluation. Surveys may be found in [1, 19–21].

Here instead of discussing in details one or two particular approaches, four different types of integral methods are considered briefly. The basic idea of each method is explained, and for an interested reader, the references for further reading are given. First, we consider some examples of the Karman-Pohlhausen original approach, when the velocity or temperature distribution across the boundary layer should be known a priori. The polynomial functions applied for solving laminar boundary-layer problems are not suitable for turbulent flows. For this case, several other functions describing parameter distributions were suggested. The one of the simplest methods applicable to pressure gradient flows and nonisothermal surfaces is based on the power-law function [21]. In this case, the velocity and temperature profiles are given as $u/U = (y/\delta)^{1/7}$ and $(T - T_\infty)/(T_w - T_\infty) = (y/\delta_t)^{1/7}$ (Comment 3.13). These expressions satisfactorily describe the parameter distributions across the major part of boundary layer, except the sublayer. Therefore, the friction and heat transfer coefficients could not be found by differentiating the profiles. Instead the empirical relations for basic coefficients

$$c_f = 0.045\,\mathrm{Re}_\delta^{-1/4},\ St = 0.0225\,\mathrm{Pr}^{-3/4}\,\mathrm{Re}_\delta^{-1/4}(\delta/\delta_t)^{1/7} \tag{3.42}$$

are used. Substituting these expressions in integral equations (2.114) along with ratios of boundary-layer thicknesses $\delta_1/\delta = 1/8,\ \delta_2/\delta = 7/72,\ \delta_{2t}/\delta = (\delta_t/\delta)^{8/7}$, obtained by second equation (2.114) using power-law profiles, yields two equations (Exercises 3.49 and 3.50):

$$\frac{d\delta^{5/4}}{dx} + \frac{115}{28}\frac{1}{U}\frac{dU}{dx}\delta^{5/4} = 0.289\left(\frac{v}{U}\right)^{1/4} \tag{3.43}$$

$$\frac{d}{dx}\left(\frac{\delta_t}{\delta}\right)^{9/7} + \frac{8}{9}\frac{1}{U\theta_w\delta}\frac{d}{dx}(U\theta_w\delta)\left(\frac{\delta_t}{\delta}\right)^{9/7} = \frac{0.260}{\mathrm{Pr}^{3/4}\delta^{5/4}}\left(\frac{v}{U}\right)^{1/4}. \tag{3.44}$$

These first-order linear ordinary differential equations may be solved analytically or numerically using one of the standard procedures (Exercise 3.51). Performing such a solution one gets the values of boundary-layer thicknesses δ and δ_t and then, after using equations (3.42), obtains the friction and heat transfer coefficients.

Comment 3.23 The empirical equations (3.42) may be derived from power-law profiles using well-known general expressions. Employing the general formula $c_f = c\,\mathrm{Re}_\delta^m$, law of the wall $u/u_\tau = f(yu_\tau/v)$ and transforming the velocity power-law profile to the wall law variables leads to the equation $u/u_\tau = (c/2)^{-4/7}\,\mathrm{Re}_\delta^{-(4m+1)/7}(yu_\tau/v)^{1/7}$. This equation becomes wall law only if the factor $\mathrm{Re}_\delta^{(4m+1)/7} = 1$ when $m = -1/4$ and hence the general formula for the friction coefficient becomes $c_f = c\,\mathrm{Re}_\delta^{-1/4}$, which with $c = 0.045$ coincides with expression (3.42) (Exercise 3.52). The second equation (3.42) may be derived from the temperature profile applying ratio $q/\tau = \lambda_{tb}(\partial T/\partial y)/\mu_{tb}(\partial u/\partial y)$ [12].

Comment 3.24 Not that to keep the form of well-known relations, like the wall law or boundary-layer equations, we return in what follows starting from equations (3.43) and (3.44) to usual notations u and U for the mean and external velocities instead of U and U_e that are used in discussing the turbulent flow characteristics.

Simple power-law profiles satisfactorily describe the real parameter distributions when they are regular, without singularities, as in the flows with zero or favorable pressure gradient. For the flows with adverse pressure gradients, better results give integral methods based on the wall law profile [22] or on the composite profile consisted of the wall layer for the inner and of the defect layer for the outer parts (Sec. 3.3.2) [20]. While the first method is applicable at low Reynolds numbers when the wall law is valid across almost whole boundary layer, the second one is used at high Reynolds numbers when the outer defect layer of the velocity profile is significant.

Another integral method suggested by Truckenbrodt and outlined in details in [1] is based as well on the known in advance type of velocity distribution. However, in this case, the particular profile is characterized by so-called shape factors rather than by using some function.

Three shape factors $H_{12} = \delta_1/\delta_2, H_{23} = \delta_2/\delta_3, H_{32} = \delta_3/\delta_2$, where δ_3 is energy thickness [see equation (3.45)], are employed, and it is assumed that some combination of shape factors H characterized a certain velocity profile. The other difference from the usual integral method is that the energy integral equation is used instead of the momentum integral equation. Such an energy integral equation and combination of shape factors are as follows:

$$\frac{d\delta_3}{dx} + 3\frac{\delta_3}{U}\frac{dU}{dx} = 2c_D, \quad H = \exp\left(-\int_{(H_{32})_\infty}^{H_{32}} \frac{dH_{23}}{(H_{12}-1)H_{23}}\right), \quad \delta_3 = -\int_0^\delta \frac{u}{U}\left[1-\left(\frac{u}{U}\right)^2\right]dy. \quad (3.45)$$

Here, c_D is the dissipation coefficient defined similar to the friction coefficient [see equation (3.46)]. Transforming differential energy equation (3.45) by employing some assumptions, experimental facts and algebra leads to the final formulae for Reynolds number and for shape factor

$$\mathrm{Re}_{\delta_3}(x) = \left(\frac{E(x)}{80\nu[U(x)]^{2(1+b)}}\right)^{1/1+b} \quad E(x) = E(x_1) + \int_{x_1}^{x}[U(\xi)]^n d\xi,$$

$$H = U(x)G(x)[N(x)]^{-1/4}, \quad (3.46)$$

determining friction and dissipation coefficients $c_f = \alpha/\mathrm{Re}_{\delta_3}H_{23}$ and $c_D = \beta/\mathrm{Re}_{\delta_3}$. Here b, α, β, H_{23} are functions of H [1, Figs. 22.6 and 22.7], $n = 3+2b$ and functions G and N are similar to E with $n = 2(1+b)$. Coordinate x_1 denotes a starting point, so that $E(x_1)$ depends on whether a laminar section precedes this point or turbulent flow starts here if, for example, the external flow velocity is close to its maximum value at this point. The analogous method based on the Reynolds number Re_{δ_1} and shape factor H_{21} is described in [1, 2000].

The vice versa integral method was developed by group of scientists from Central Aerugydrodynamics Institute (TsAGI). Although this method was used as well by other researchers [23,24], the most completed description was published in the book [25]. In this case, first the friction stress and heat flux profiles are approximated. It was shown that the ratios τ/τ_w and q/q_w distributions may be displayed by the polynomials satisfying the boundary conditions at the surface and at the edge of the boundary layer just as for the velocity and temperature in the case of laminar flow. After defining the polynomial coefficients, the velocity and temperature profiles are obtained by integrating expressions $\tau = (\mu + \mu_{tb})\partial u/\partial y$ and $q = -c_p\left[(\mu/\mathrm{Pr}) + (\mu_{tb}/\mathrm{Pr}_{tb})\right]\partial T/\partial y$ solved for $\partial u/\partial y$ and for $\partial T/\partial y$ along with Prandtl mixing-length formula for μ_{tb} and $\mathrm{Pr}_{tb} = 1$.

A simple integral method based on the conservative nature of the heat transfer laws is effective in studying and computing heat transfer from nonisothermal surfaces. Actually, we encountered with a conservative feature of heat transfer laws when we considered in Section 3.3.6 the application of algebraic models to heat transfer. There it was observed that the heat transfer results obtained by using the Mellor-Gibson model are valid for a wide range of Reynolds and Prandtl numbers without any corrections. This effect is achieved namely due to conservation property of heat transfer laws which means that heat transfer rate is slightly affected by the velocity field in boundary layer. There it was also indicated that physically the reason of conservation feature lies in the Prandtl number effect analyzed in Section 2.7.3.

Authors of the model in question employed the relation for the Stanton number as the first equation (3.47)

$$St = CRe_{\delta_{2t}}^{n} Pr^{m} \qquad Re_{\delta_{2t}} = \frac{1}{\theta_w}\left[C(1-n)Re_L Pr^m \int_0^{x/L} \theta_w^{1-n} d\left(\frac{x}{L}\right)\right]^{\frac{1}{1-n}}. \qquad (3.47)$$

This general dependence according to theoretical analysis and experimental data slightly depends on the pressure gradient and surface temperature distribution (e.g., conservation property) [26, 27]. Therefore, considering this relation as a universal dependence, one might apply the results obtained for the isothermal surface in a zero gradient flow to a general case of the nonisothermal surface in a gradient flow. Substituting equation (3.47) for the Stanton number into thermal integral equation (2.114) yields an ordinary differential equation for $Re_{\delta_{2t}}^{n-1}$ solution of which gives the second expression (3.47). Knowing the temperature head distribution $\theta(x/L)$ and using the constants $C = 0.0128, n = -1/4, m = -0.6$ obtained from a number of existing experimental data, one gets a value of $Re_{\delta_{2t}}$ from second equation (3.47), and then finds the Stanton number from the first equation (3.47). This simple approach was used for solving different problems with a satisfactory accuracy [23,27]. However, it was shown [24,26] that the following inequality [the first equation (3.48)] gives the estimation of a relative error in a Stanton value obtained in this approach

$$\frac{\Delta St}{St} \leq 0.1 \frac{Re_{\delta_{2t}}}{St\,Re_L} \frac{d\theta_w}{\theta_w d(x/L)}, \qquad \frac{h}{h_*} = 1 - g_1 x \frac{1}{\theta_w}\frac{d\theta_w}{dx}, \qquad \frac{c_f}{c_{f*}} = 1 - b_1 \frac{U_{av}}{U} x \frac{1}{U^2}\frac{dU^2}{dx}. \qquad (3.48)$$

It is seen from this expression that the value of error basically depends on the ratio θ_w'/θ_w of the temperature head gradient θ_w' in the flow direction to the temperature head itself θ_w. Due to that in the case of decreasing temperature head, the error significantly increases since the dominator (temperature head) in first equation (3.47) lessens.

Comment 3.25 In the next three chapters, it is shown that the ratio θ_w'/θ_w plays crucial role in heat transfer because the temperature head gradient determines the heat transfer characteristics, as well as pressure or square of velocity gradient determines the flow characteristics. This similarity is seen, in particular, from two last expressions (3.48) obtained by comparison of universal equations (1.23) and (2.128) (Exercise 3.56). Note that additional factor U_{av}/U in the last equation (3.48) arises because coefficients b_n in series (2.128) in fact depend on velocity since the dynamic boundary-layer equation in contrast to thermal equation is nonlinear.

EXERCISES

3.41 Repeat the physics meaning of terms in the turbulence kinetic energy. How are these terms presented via energy k?

3.42 Show applying equations indicated in the text that mixing length in equation (3.10) and length in equation (3.36) are proportional if the production and dissipation are equal to each other. Hint: denote the mixing length in equation (3.10) as l_m and find dU/dy from two last equations (3.10).

3.43 Study what are characteristics of the partial differential equations, how they are used to solve hyperbolic differential equations, and why the other type of equations could not be solve in the same way. Read the Chapter "Characteristics and the classification of partial differential equations" on page 695 of *Advanced Engineering Mathematics* by R. Wylie and L. Barrett, six edition or other similar source.

3.44 What is the dimensions method? How it works?

3.45 Read articles from Wikipedia about Farmat, Poinsare, Wiles, Perelman, and their outstanding achievements. Read also article to know what Clay Millennium Prize is.

3.46 Recall physical reasons due to which two terms in the Kolmogorov equation for dissipation were omitted, while latter of these were added.

3.47 Compare the $k-\omega$ and $k-\varepsilon$ models. What is the difference between ω and ε? What physically they mean?

3.48 Compare the range of application of algebraic and energy equation (one- and two-equation) models.

3.49 Derive formulae for $\delta_1, \delta_2, \delta_{2t}$ using expressions (2.114) and power-law profiles.

3.50 *Obtain equations (3.43) and (3.44) applying equations (3.42), integral boundary-layer equations (2.114), and expressions for boundary-layer thicknesses.

3.51 Recall or study analytical and numerical methods of solution of linear ordinary differential equations using for example *Advanced Engineering Mathematics*.

3.52 Show that first equation (3.42) may be derived from velocity power-law profile like it is described in the Comment 3.23. Hint: after transforming velocity profile to wall law variables use the general relation for friction coefficient.

3.53 How differ from each other the basic ideas of four types of integral methods considered here?

3.54 What is the shape factor? How many shape factors there are in the Truckenbrodt method? How are they used?

3.55 What are the conservation heat transfer laws like? How the benefits they offer may be employed?

3.56 Show that two last expressions (3.48) follow from universal equations (1.23) and (2.128) if one retains in these series, only two first terms taken into account that the first coefficients g_1 and b_1 are negative and significantly larger than others. Hint: $h = q_w/\theta_w$ and $\Phi/d\Phi = xU_{av}/Udx$.

REFERENCES

[1] Schlichting, H. (1979). *Boundary layer theory*. New York: McGraw-Hill.

[2] Gibbings, J. C. (2003). Diffusion of the intermittency across the boundary layer in transition. *Proceedings of the institution of Mechanical Engineering, Part C: Journal of Mechanical Engineering Science* 217/12/1339.

[3] Wilcox, D. C. (1994, 2006). *Turbulence modeling for CFD*. California: DCW Industries, Inc., La Canada.

[4] Clauser, F. H. (1956). The turbulent boundary layer. In *Advances in Applied Mechanics*, vol. IV (pp. 1–51). New York: Academic Press. doi: http://dx.doi.org/10.1016/S0065-2156(08)70370-3

[5] Rotta, J. C. (1962). Turbulent boundary layers in incompressible flow. *Progress in Aerospace Sciences* 2, 1. doi: http://dx.doi.org/10.1016/0376-0421(62)90014-3

[6] Coles, D. E., & Hirst, E. A. (1969). Computation of turbulent boundary layer, 1968 *AFOSR-IFP-Stanford Conference*, vol. II. CA: Stanford University.

[7] Townsend, A. A. (1976). *The Structure of turbulent shear flow* (2nd ed.). Cambridge: Cambridge University Press.

[8] Greenberg, M. D. (1998). *Advanced engineering mathematics*. NJ: Prentice Hall.

[9] Mellor, G. L. (1966). Effects of pressure gradients on turbulent flow near a smooth wall. *Fluid Mechanics 24*(2), 255–258. doi: http://dx.doi.org/10.1017/S0022112066000624

[10] Mellor, G. L., & Gibson, D. M. (1966). Equilibrium turbulent boundary layers. *Journal of Fluid Mechanics* 24(2), 225–253. doi: http://dx.doi.org/10.1017/S0022112066000612

[11] Laufer, J. (1954). The structure of turbulence in fully developed pipe flow. NACA, Rep. No. 1174.

[12] Dorfman, A. S. (2009) *Conjugate problems in convective heat transfer*. Boca Raton: CRC press.

[13] Cebeci, T., & Smith, A. M. O. (1974) Analysis of turbulent boundary layer, Ser. In *Applied Mathematics and Mechanics,* vol. XV. New York: Academic Press.

[14] Monin, A. S., & Yaglom, A. M. (1971). Statistical Fluid Mechanics, Vol.1, edited by John Lumley.

[15] Van Driest, E. R. (1956). On turbulent flow near a wall. *Journal of the Aeronautic Science* 23: 1007–1011.

[16] Corrsin, S., & Kistler, A. L. (1954). The free stream boundaries of turbulent flows, NACA TN 3133.

[17] Klebanoff, P. S. (1956). Characteristics of turbulence in a boundary layer with zero pressure gradient, NACA TN 3178.

[18] Baldwin, B. S., & Lomax, H. (1978). Thin-later approximation and algebraic model for separated turbulent flows, AIAA paper 78–257.

[19] Kestin, J., & Richardson, P. D. (1963). Heat transfer across turbulent incompressible boundary layers. *International Journal of Heat and Mass Transfer 6*(6): 147–189. doi: http://dx/doi.org/10.1016/0017-9310(63)90035-8

[20] Patankar, S. V., & Spalding, D. B. (1970). *Heat and mass transfer in boundary layers*. London: Intertext.

[21] Dorfman, A. S. (1983). Methods of estimation of coefficients of heat transfer from nonisothermal walls. *Heat Transfer-Soviet Research 15*(6), 35–57.

[22] Thomas, L. (1978). Simple integral approach to boundary layer flow. *Journal of Heat Transfer 100*(4), 744. doi: http://dx.doi.org/10.1115/1.3450894

[23] Kutateladze, S. S. (1970). *Fundamentals of heat transfer* (in Russian). Moscow, Leningrad: Mashgis Press.

[24] Kutateladze, S. S., & Leontiev, A. I. (1972, 1989). *Heat transfer, mass transfer and friction in turbulent boundary layers*, (in Russian). Moscow: Energiya Press; New York: Hemisphere Publishing Corporation.

[25] Fedyayevskiy, K. K., Ginevskii, A. S., & Kolesnikov, A.V. (1973). *Calculation of turbulent boundary layers in incompressible fluids* (in Russian). Leningrad: Sudostroenie Press.

[26] Iyevlev, V. M. (1952). Certain aspects of the hydrodynamic theory of heat transfer in incompressible flows. *Doklady Akademii Nauk SSSR 86*, 1077–1080.

[27] Kutateladze, S. S. (1973). *Wall Turbulence,* (in Russian). Novosibirsk: Naula Press.

PART II

Modern Conjugate Methods in Heat Transfer and Fluid Flow

Introduction

Concept of Conjugation

Let us begin with a simple problem. Consider a thin plane plate heated from one edge to answer two questions: (i) How the heat transfer coefficients differ depending on the flow direction in two cases: from heated to unheated edges (first case) and the opposite direction from unheated to heated edges (second case)? (ii) In what case, the total heat flux from the plate is greater?

This simple task is a typical conjugate problem because the answer of the questions and problem solution strongly depend on the temperature distribution along the plate and flow interface, which is unknown. If one ignores the interface temperature head distribution, there would not be any difference in both flow directions. However, that is, not the case. To see the real situation, consider the universal function (1.23) with the first two terms to get equation (3.48) $h/h_* = 1 + g_1 x (d\theta_w/dx)$. It is clear that in the first case, when the flow starts from a heated edge, the temperature head decreases in the flow direction. Hence, in this case, $d\theta_w/dx < 0$, and according to relation (3.48), the ratio h/h_* decreases along the plate. At the same time, it is known that the heat transfer coefficient on the isothermal surface h_* also lessens along the plate; therefore, the overall heat transfer coefficient h, determined as a product $h_*(h/h_*)$, strongly decreases along the plate in the first case. An opposite situation takes place in the second case. Since in this case, the flow starts from the coolest edge, the temperature head increases in the flow direction, resulting in $d\theta_w/dx > 0$ and subsequently, in increasing the ratio h/h_*, along the plate according to the same equation (3.48). Since in this case, the heat transfer coefficient h_* for an isothermal surface increases along the plate as well, the overall heat transfer coefficient $h = h_* \left(h/h_* \right)$ in the second case decreases or increases along the plate, depending on the ratio of two multiplying factors in the last equation.

Quantitative results as well as the answer to the second question might be obtained only by solving the conjugate problem. As it follows from such a solution, the total heat flux is greater in the first case despite the fact of sharply decreasing heat transfer coefficients. The reasons is that the high heat transfer coefficients and temperature heads at the heated edge in the first case and small temperature heads at the unheated edge in the other case (Example 6.5).

Why and When are Conjugate Methods Required?

Many engineering and natural systems are constructed of or include as a part two or more inter-acting bodies or/and substances. Properties and behavior of such systems largely depend on conditions developing on the interfaces of the interacting elements. However, these conditions are usually unknown a priori even if the properties of the elements are given. The conjugate procedure is an approach of determining the distribution of parameters arising on the interfaces as a result of elements interaction. This is achieved by solving the governing equation for each of interacting parts and following conjugation of these solutions at the interfaces.

There are areas where almost any problem is naturally conjugate. In particular, the conjugate nature is an inherent feature of any heat transfer problem because any heat transfer process is in essence an interaction of at least two mediums or/and subjects. For example, to get the temperature distribution on the interface between body and fluid flowing past it or to determine the temperature variation along a wall of a channel with flow, one should conjugate the temperature field in solid and in liquid obtained by solving the governing equations for each phase. The same procedure is needed in the case of thermal interaction of two fluids, for example, for studying heat transfer between the atmosphere and a sea or ocean and even in the simpler problem of conduction between two different solids or between subjects from the same substance but with different physical properties. The combustion system where cold mixture interacts with burnt gases may be used as an example of the last case. In such a combustion process, the characteristics of the final mixture flow may be found by conjugation of the velocity and temperature fields of gaseous flows up- and downstream of the flame.

There are countless other phenomena and engineering systems, accurate studying of which requires a conjugate procedure. As an example of biological systems, conjugate nature of which is an inherent feature of any problem, consider a peristaltic flow in channels of human body: blood vessels, urinary channels, or gastrointestinal tract. The Greek word peristaltikos means clasping and compressing. Corresponding English word peristalsis stands for fluid motion in channel with flexible walls when a progressive wave propagates along channel walls, result-ing in fluid motion in the wave direction. Mechanism of such a fluid movement consists of sequence transverse squeezing and following relaxing of walls portions, which arise due to the interaction between a fluid and the elastic channel walls. In physiology, this phenomenon is provided by intrinsic property of muscles of any tubular organ. The peristaltic pumping is also used in many engineering devices, such as a heart-lung machine, to transport blood and other biomedical fluids. Applying in such cases peristaltic pumping instead of common pumps pre-vents transporting substance from undesirable contact with machinery parts. Other engineering applications include mixing chemical reactions, transport clean, and sterile, as well as corrosive fluids and sewage to isolate conveying substance from environment, pharmaceutical production, food processing, and some other cases.

The term conjugate problem was coined in 1960s for specifying approaches considering the heat transfer with regard to a contacting-phase interaction. This term as well as two others, coupled and adjont problems, is now widely used denoting this modern effective avenue where the number of publications grows like an avalanche.

Indeed, the conjugate procedure was known long before 1960s when the intense interest in such a problem formulation arose with the advent of computers. It has been known long ago that exact solution of heat transfer problems requires satisfaction of the boundary conditions of the fourth kind which are, in essence, the same as conjugate conditions (Sec. 1.2.4). However, in practice before computer advent, availability of solutions of problems with such boundary conditions was restricted only to very simple cases. Due to calculation difficulties, instead of exact formulation, the boundary condition of the third kind defined by equation (1.24) has been used since the time of Newton. This relation was formulated assuming that the heat flux through a solid–fluid interface is proportional to difference between the body temperature and the fluid temperature far away from the solid. In such an approximate approach, accuracy of solution rests on successful definition of a proportionality factor-heat transfer coefficient, which is obtained experimentally. Since heat transfer coefficient depends on unknown a priori interface temperature distribution, experimental data for the simplest cases dealing with constant surface temperature or constant heat flux are applied in practical calculation. Such a simple method in which the effect of actual wall temperature distribution is neglected was acceptable when requested calculation accuracy was not as high as it was since the computer time. At the same time, there are cases when the effect of variable surface temperature is practically negligible so that using boundary conditions of the third kind is reasonable.

Contemporary conjugate methods of solutions of heat transfer problems substitute simplified approaches and give the temperature and heat flux distributions as calculation results. Then, there is no need in heat transfer coefficient, even though it can be calculated applying data obtained in the conjugate solution. Nevertheless, it is important to know when the simpler methods may be used with acceptable accuracy instead of a more complicated conjugate approach. The answer to this question and many others can be obtained applying conjugate methods, which are powerful tools for solving problems and modeling phenomena in many natural and engineering areas.

The next four chapters consist of the principles, general properties, and the basic methods of problem solution of conjugate heat transfer and peristaltic fluid motion. The discussion goes with a solution of typical relatively simple conjugate problems and physics analysis of results. A special chart for solving simple conjugate heat transfer problems with numerical examples is offered in Chapter 6. Applications of conjugate heat transfer in different areas from aerospace to food processing as well as of peristaltic fluid flows in biology, medicine, and engineering are presented in the fourth part of the book.

CHAPTER 4

Conjugate Heat Transfer Problem as a Conduction Problem

4.1 Formulation of Conjugate Heat Transfer Problem

A set of equations, initial and boundary conditions, governed the conjugate heat transfer problem consists of: (i) for fluid flow domain, the Navier–Stokes and energy equations (2.1–2.5) in general, or boundary-layer equations (2.68–2.70), or creeping flow equations (2.56) for high and low Reynolds numbers, respectively; for turbulent flows, the corresponding set contains Reynolds average equation (3.9) and analogous simplified boundary-layer equations; (ii) for body domain, one of the form of Laplace equation or some simplified conduction equation given in Sections 1.2.1 and 1.2.2; (iii) conjugate conditions (1.25) at the body-flow interface, some of boundary conditions (1.24), and initial conditions if required.

Any conjugate heat transfer problem is actually a question of heat transfer of some nonisothermal surface. To see that, consider the heat transfer phenomenon as a process of interaction between body and fluid flowing past or inside it. Such an interaction of heat results in some temperature distribution on the body-fluid interface that is generally a nonisothermal pattern. The characteristics of heat transfer specified by this temperature field are the same, no matter it is created in a heat transfer process or it is a giving temperature distribution of the same nonisothermal surface. Therefore, the investigated data, general properties, and theory of nonisothermal heat transfer are valid for conjugate heat transfer as well. The solution of conjugate heat transfer problem is also closely related to theory of arbitrary nonisothermal heat transfer. This becomes clear if we look at what we know about interface of interacting parts (for example, of body-fluid interface) at the beginning of solution. In fact, we know at the interface, only the conjugate conditions, that is, the temperatures and heat fluxes calculated from both sides of interface, are equal to each other.

In such a case, it is reasonable to seek a solution of conjugate problem by solving the set of equation for each domain separately to get the dependence of heat flux on the interface temperature for each domain following by conjugation of the finding results. This procedure gives two equations $q_w = f_1(T_w)$ and $q_w = f_2(T_w)$, which yield after their conjugation an equation $f_1(T_w) = f_2(T_w)$ that determines the interface temperature distribution. However, there is a question that was missing, namely, what boundary conditions should be used under solving the set of equations for each domain? It is obvious that it is necessary to use an arbitrary nonisothermal boundary condition since nothing more is known at the interface. Thus, trying to solve the conjugate problem, we arrived again to problems of nonisothermal heat transfer for a body and for a fluid.

The theory of nonisothermal heat transfer and related method of solution of boundary-layer conjugate heat transfer problems is grounded and outlined in details in [1]. The approach is based on universal functions of type (1.23) that give the expressions for heat flux in the form of series of successive derivatives of temperature head. These relations are obtained as solutions of boundary-layer equations and are applicable to arbitrary temperature head distribution as it is required for conjugate problems. Here, is presented a shortened description of this theory without proofs containing only basic ideas, properties analysis, practical methods, and examples of solutions. Some details are offered to a reader through exercises, indicating the relevant pages from the book [1].

However, the text is understandable without these details. More simple presentation as a first course of convective conjugate heat transfer is also available (http://abramdorfman.com/download/firstcours.pdf).

Examples of more complex conjugate problems based on solutions of Navier-Stokes and full-energy equations, such as two problems in Examples 2.8 where we analyzed specific problems of such solutions, one may find in Part IV of application. There are also application examples of conjugate heat transfer problem in creeping approximation.

4.2   Universal Function for Laminar Fluid Flow

As is mentioned in Section 1.2.3, universal is a function that is independent of particular boundary conditions being applicable to an arbitrary boundary condition. If such a function satisfies some differential equation, then it is a solution of any problem governed by this equation. In this section, we show how universal functions in form of series (1.23) are obtained in different cases and used to solve the conjugate heat transfer problems.

4.2.1   Universal Function for Heat Flux in Self-Similar Flows as an Exact Solution of a Thermal Boundary-Layer Equation

It is proved [p.56] (such numbers indicate the pages in the book [1]) that universal function (4.1) for laminar boundary layer

$$q_w = h_* \left(\theta_w + g_1 \Phi \frac{d\theta_w}{d\Phi} + g_2 \Phi^2 \frac{d^2\theta_w}{d\Phi^2} + g_3 \Phi^3 \frac{d^3\theta_w}{d\Phi^3} + \cdots \right) = h_* \left(\theta_w + \sum_{k=1}^{\infty} g_k \Phi^k \frac{d^k\theta_w}{d\Phi^k} \right) \quad (4.1)$$

follows from an exact solution of thermal boundary layer equation (2.77) in Prandtl–Mises–Görtler variables (2.75). This function as well as series (1.23), that is, a particular case of equation (4.1) is universal because it does not consist any specific relation of surface temperature or temperature head, rather it depends on arbitrary temperature head distribution $\theta_w(x)$ and its derivatives. Due to Görtler variable Φ, which depends on pressure gradient via integral of external flow velocity $U(x)$, equation (4.1) is more general and hence series (1.23) follows from equation (4.1) for the case of zero pressure gradient when $\Phi = \mathrm{Re}_x$ (Exercises 4.6).

In deriving function (4.1), we use equation (2.77) with additional term to take into account the mechanical energy dissipation [see the last term of equation (2.70)]. Equation (2.77) with such an additional term in Görtler variables (Exercise 4.7) and its exact solution for temperature excess $\theta = T - T_\infty$ distribution across boundary layer are

$$2\Phi \frac{\partial \theta}{\partial \Phi} - \varphi \frac{\partial \theta}{\partial \varphi} - \frac{1}{\mathrm{Pr}} \frac{\partial}{\partial \varphi}\left(\frac{u}{U}\frac{\partial \theta}{\partial \varphi}\right) = \frac{U^2}{c_p}\frac{u}{U}\left[\frac{\partial}{\partial \varphi}\left(\frac{u}{U}\right)\right]^2,$$

$$\theta = \sum_{k=0}^{\infty} G_k(\varphi)\Phi^k \frac{d^k \theta_w}{d\Phi^k} + G_d(\varphi)\frac{U^2}{c_p} \tag{4.2}$$

Substitution of solution (4.2) in the equation (4.2) results in two equations (Exercise 4.8)

$$\sum_{k=0}^{\infty} \Phi^k \frac{\partial^k \theta_w}{\partial \Phi^k}\left[2kG_k + 2G_{k-1} - \varphi G_k' - \frac{1}{\mathrm{Pr}}\frac{\partial}{\partial \varphi}\left(\frac{u}{U}G_k'\right)\right] = 0 \quad (k = 0, 1, 2 ...) \tag{4.3}$$

$$2\frac{\Phi}{U^2}\frac{dU^2}{d\Phi}G_d - \varphi G_d' - \frac{1}{\mathrm{Pr}}\frac{\partial}{\partial \varphi}\left(\frac{u}{U}G_d'\right) - \frac{u}{U}\left[\frac{\partial}{\partial \varphi}\left(\frac{u}{U}\right)\right]^2 = 0 \tag{4.4}$$

Here, prime denotes differentiation with respect to φ so that all terms in the brackets of equation (4.3) depend only on this variable. Thus, expressions in the brackets might be a set of ordinary differential equation, but the ratio u/U depends not only on φ rather on Φ also. The same situation one sees in equation (4.4) where besides ratio u/U, the first term also depends on φ and Φ. Therefore, we restrict further development to the flows with power-law external velocity $U = cx^{\beta/2-\beta}$ (Sec. 2.6.2). In that self-similar case, the velocity ratio depends on φ and β, and the factor of the first term in equation (4.4) equals 2β so that equations (4.3) and (4.4) become ordinary differential equations:

$$(1/\mathrm{Pr})\left[\omega(\varphi,\beta)G_k'\right]' + \varphi G_k' - 2kG_k = 2G_{k-1},$$

$$\varphi = 0, G_0 = 1, G_k = 0 \quad \varphi \to \infty, G_k = 0, k \geq 0 \tag{4.5}$$

$$(1/\mathrm{Pr})\left[\omega(\varphi,\beta)G_d'\right]' + \varphi G_d' - 2\beta G_d + \omega(\varphi,\beta)\left[\omega'(\varphi,\beta)\right]^2 = 0,$$

$$\varphi = 0, G_d = 0, \quad \varphi \to \infty, G_d = 0 \tag{4.6}$$

where $\omega(\varphi,\beta)$ denotes the ratio u/U in the self-similar flows. Boundary conditions (4.5) follow from solution (4.2): (i) at the surface ($\varphi = 0$) temperature excess $\theta = T - T_\infty$ should be equal to temperature head $\theta_w = T_w - T_\infty$, which corresponds to $k = 0$ (the derivative is equal to a function itself) and zero of all other terms; (ii) far away from the surface, the temperature should be equal to that of external flow T_∞ so that $\theta = 0$. The boundary conditions for equation (4.4) are zero because the part of solution (4.2) for energy dissipation is a particular solution of the inhomogeneous equation (4.2) (Exercise 4.9).

The heat flux on the surface is determined by Fourier law, which in the Görtler variables given below by the first expression (4.7) contains the limiting values of velocity $u_{\varphi=0}$ and of derivative $(\partial\theta/\partial\Phi)_{\varphi=0}$ at the surface. Defining these quantities, one gets equation for heat flux (Exercise 4.10):

$$q_w = -\lambda\left(\frac{\partial\theta}{\partial y}\right)_{y=0} = \frac{\lambda}{v\sqrt{2\Phi}}\left(-u\frac{\partial\theta}{\partial\varphi}\right)_{\varphi=0}, \quad q_w = h_*\left(\theta_w + \sum_{k=1}^{\infty}g_k\Phi^k\frac{d^k\theta_w}{d\Phi^k} - g_d\frac{U^2}{c_p}\right) \quad (4.7)$$

In the last relation (4.7), $g_k = \left(G_k'/G_0'\right)_{\varphi=0}$, $g_d = \left(G_d'/G_0'\right)_{\varphi=0}$, $G_0 = -2^{1/4}\left(\varphi^{1/2}G_0'\right)_{\varphi=0}$, h_* is the heat transfer coefficient of an isothermal surface in the case of negligible dissipation, and the first two terms are the same series (4.1), whereas the last one takes into account the dissipation effect.

The first five equations (4.5) were solved numerically. Since these equations are inhomogeneous and each function G_k depends on previous function G_{k-1}, the error of computation cumulates and as the number k increases becomes unacceptable. Therefore, equations (4.5) were transformed in a set of homogeneous equations and boundary conditions (4.8) by presenting function G_k as a sum of other function F_i (Exercise 4.11):

$$G_k = \sum_{i=0}^{i=k}\frac{(-1)^{k+i}}{(k-i)!}F_i, \quad (1/\text{Pr})\left[\omega(\varphi,\beta)F_i'\right]' + \varphi F_i' - 2iF_i = 0, \quad F_i(0) = 1/i!, \quad F_i(\infty) = 0 \quad (4.8)$$

Since problem (4.8) is a linear boundary value problem, it was reduced to initial value problem for two functions $V_i(z)$ and $W_i(z)$, satisfying equation (4.8) and special boundary conditions $V_i(0) = 0, V_i'(0) = 1$ and $W_i(0) = 1, W_i'(0) = 0$. Integrating numerically these initial value problems using, for example, Runge–Kutta method, one obtains functions $V(z)$ and $W(z)$ and then functions $F_i(z)$ [from equation (4.9)], $G_k(\varphi)$ [from sum of relations in equation (4.8)], and finally coefficients g_k using equation (4.9) (Exercises 4.12):

$$F_i(z) = \frac{1}{i!}\left[V_i(z) - \frac{V_i(\infty)}{W_i(\infty)}W_i(z)\right], \quad g_0 = \frac{V_0(\infty)}{W_0(\infty)},$$

$$g_k = \sum_{i=0}^{i=k}\frac{(-1)^{k+i}}{i!(k-i)!}\frac{V_i(\infty)/V_0(\infty)}{W_i(\infty)/W_0(\infty)} \quad (4.9)$$

Here, $z = \varphi^{1/2}$ is a variable that is used to remove the singularity, which equations (4.8) as well as the Prandtl–Mises equation (2.73) have at the surface (Exercise 4.13).

Coefficients g_k of series (4.1) obtained for different external flow velocities (different β) and Prandtl numbers are plotted in Figures 4.1 and 4.2.

In the limiting cases $\text{Pr} \to 0$ and $\text{Pr} \to \infty$, equations (4.5) simplify due to using plug velocity profile in the first case and the linear velocity profile $c\varphi^{1/2}$ in the second one (Sec. 2.7.3). Solving these equations analytically, one gets the following results [$\Gamma(x)$ is gamma function] (Exercises 4.14 and 4.15):

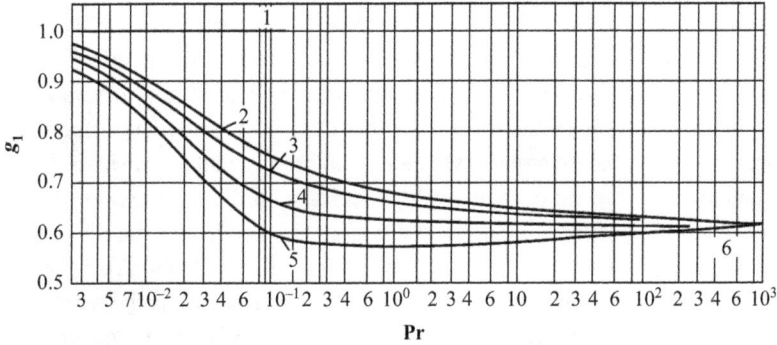

Figure 4.1. Coefficient $g_1(\text{Pr}, \beta)$ of universal function (4.1) for laminar boundary layer. Asymptotes $1 - \text{Pr} = 0$, $6 - \text{Pr} \to \infty$; β: 2-1 (stagnation point), 3-0.5 (favorable pressure gradient), 4-0 (zero pressure gradient), 5 (−0.16) (preseparation pressure gradient).

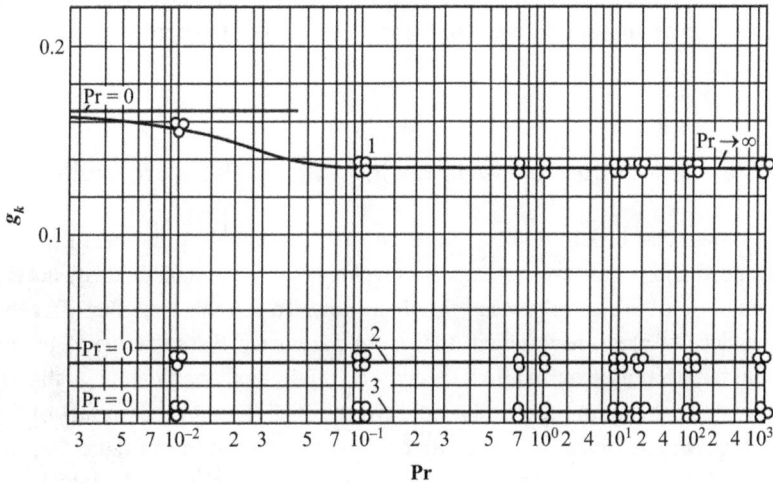

Figure 4.2. Coefficients $g_k(\text{Pr})$ of universal function (4.1) for laminar boundary layer: $1 - (-g_2)$, $2 - g_3$, $3 - (-g_4)$, ○ - numerical integration

$$g_k = \frac{(-1)^{k+1}}{k!(2k-1)} \quad \text{Pr} \to 0, \qquad g_k = \Gamma\left(\frac{2}{3}\right)\sum_{i=0}^{i=k} \frac{(-1)^{k+i}\,\Gamma(4i/3)+1}{(k-i)i!\Gamma\left[(4i/3)+2/3\right]} \quad \text{pr} \to \infty \quad (4.10)$$

$$g_1 = 1, g_2 = -1/6, g_3 = 1/30, g_4 = -1/168, g_5 = 1/1080, g_6 = -1/7920 \quad \text{Pr} = 0 \quad (4.11)$$

$$g_1 = 0.6123, g_2 = -0.1345, g_3 = 0.0298, g_4 = -0.0057 \qquad \text{Pr} \to \infty \quad (4.12)$$

Analysis of numerical results from Figures 4.1 and 4.2 and of limiting values shows that: (i) coefficient g_1 depends on the external flow velocity (via β) and on the Prandtl number (more significantly for $\text{Pr} < 0.5$) asymptotically tending for all β to the greatest value $g_1 = 1$ at $\text{Pr} \to 0$

and to the lowest one $g_1 = 0.6123$ at $Pr \to \infty$; (ii) coefficient g_2 is practically independent of the external flow velocity and depends slightly on the Prandtl number only for small Prandtl numbers asymptotically tending for all β as well to the greatest absolute value $|g_2| = 1/6$ at $Pr \to 0$ and to the lowest absolute value $|g_2| = 0.1345$ at $Pr \to \infty$; (iii) coefficients g_3 and g_4 are independent of both the external flow velocity and the Prandtl number so that their values obtained by numerical integration practically coincide with the limiting values $g_3 = 1/30$ and $g_4 = -1/168$; (iv) all subsequent coefficients are more independent of the external velocity and Prandtl number the greater is the number k. This follows from equation (4.5) because the first term in this equation, that only one depends on Pr and β, reduces as the number k increases; (v) it follows from the last two conclusions [(iii) and (iv)] that for $k \geq 3$ the coefficients g_k in general case might be calculated applying formula (4.10) for $Pr = 0$ because these are practically constant; (vi) coefficients g_k rapidly decrease as the number k increases.

Comment 4.1 There are different methods to estimate heat transfer coefficient for an isothermal surface h_* required for using universal function (1.23) or (4.1). These methods reviewed, for example, by Spalding and Pan [2], were created throughout many years during which this coefficient was applied for practical calculations, in particular, as a part of a boundary condition of the third kind.

4.2.2 Universal Function for Heat Flux in Arbitrary Pressure Gradient Flow

Below it is shown that exact universal function (4.1), determining the heat flux for arbitrary temperature head and power-law external flow velocity provides satisfactorily accurate results in the general case of arbitrary pressure gradient flow. To see this note that if coefficients g_k were independent of pressure gradient, that is, of parameter β, the series (4.1) would be an exact relation for arbitrary external flow velocity $U(x)$. In fact, coefficients g_1 slightly depends on β, whereas all others are practically independent of β. The greatest effect of the pressure gradient on the first coefficient is in vicinity of $Pr = 0.1$. As is seen in Figure 4.1, at this point, the deviation of the value of g_1 from average magnitude reaches $\pm 12\%$, which decreases as the Prandtl number grows or lessens getting to zero at $Pr = 0$ and $Pr \to \infty$. Consequently, using the average value of the first coefficient instead of real dependence $g_1(\beta)$ leads to error not greater than 12%, which is in a range of accuracy of other existing approximate methods.

In some cases, β and hence g_1 are known. For example, it is obvious that in a flow past a cylindrical body at the blunt nose, we have $\beta = 1$ (stagnation point), whereas for the rest of the body surface, the value of β is close to zero (zero presser gradient). We may also use some approximate relation for estimating β, for example, $\beta = 2(1 - \Phi/Re_x)$, which follows from equality of average velocities of given and self-similar flows (Exercise 2.106). However, there are no reasons to use different β because the final results obtained by various β from Figure 4.1 are very close as, for example, the data gained for flow past cylinder with $\beta = 1$ and $\beta = 0$ in considering below Example 4.4.

Thus, it turns out that for practical calculation, it is reasonable to employ: (i) the average values of g_1 for giving Prandtl number from Figure 4.1, and of $g_2 \approx 0.15$ [see above (ii)], and (iii) limiting values (4.11) for $Pr \to 0$ according to first formula (4.10) for other coefficients g_k [see above (v)].

Universal functions (4.1) and (1.23) are useful for studying nonisothermal and conjugate heat transfer. For this purpose, these relations are presented in the form of the nonisothermicity coefficient χ_t that one gets dividing both functions by $q_* = h_*\theta_w$:

$$\chi_t = \frac{q_w}{q_*} = \frac{h}{h_*} = 1 + \sum_{k=1}^{\infty} g_k \frac{\Phi^k}{\theta_w} \frac{d^k\theta_w}{d\Phi^k}, \qquad \chi_t = \frac{q_w}{q_*} = \frac{h}{h_*} = 1 + \sum_{k=1}^{\infty} g_k \frac{x^k}{\theta_w} \frac{d^k\theta_w}{dx^k} \qquad (4.13)$$

Here, q_* is the heat flux on an isothermal surface, so the nonisothermicity coefficient specifies how much the heat transfer intensity from a particular nonisothermal surface is more or less than that from an isothermal surface, and hence, for example, $\mathrm{St} = \mathrm{St}_* \chi_t$.

The second equation (4.13) as well as equation (1.23) shows that in the case of zero pressure gradient both the nonisothermicity coefficient and the heat flux are governed by the derivatives of temperature head with respect to the longitudinal coordinate x. At the same time, from the first equation (4.13) and equation (4.1), one sees that in the general case with pressure gradient $dp/dx \neq 0$, the same role plays the Görtler variable $\Phi(x)$, so the derivatives of temperature head with respect to this variable determine both the nonisothermicity coefficient and the heat flux. From these facts, it follows that the Görtler variable takes into account basic features, which distinguish the effect of nonisothermicity in pressure gradient flow from that in the case of zero pressure gradient. This result is achieved because the Görtler variable (2.75) is defined as $\Phi(x) = \mathrm{Re}_{(av)x} = U_{av}x/\nu$ (Exercise 4.16), where $U_{av}(x)$ is the average external flow velocity for the interval from leading edge $(x = 0)$ to a point of interest with coordinate x. Physically, this signifies that the Görtler variable takes into account the prehistory of the flow, which means that the flow characteristics at a given point are governed not only by the local quantities but as well by others in the considering interval of longitudinal variable $(0, x)$.

Comment 4.2 Depending on the problem, the characteristics at some point of interest are determined by data of parameter values: (i) at this point only (local data), or (ii) at this and behind points (local and historical data), or (iii) at this, behind, and advanced points (local, historical, and future data). These variants are in line with equations and boundary conditions, describing the problem. For example, a self-similar heat transfer with a boundary condition of the first kind in flow past a wage is described by the boundary-layer equations, and the friction and heat transfer coefficients, which depend only on local Reynolds and Nusselt number at the point of interest. That is so because only the wage angle β defining the velocity and the temperature head are required as the boundary conditions in this case (Sec. 2.6.2). More complicated problem of the nonisothermal surface in self-similar flow, described by the boundary-layer equations, required as the boundary conditions the temperature head distribution $\theta_w(x)$ and the angle β. Therefore in this case, the characteristics at some point depend on local (h_* or Nu_{x*}) and historical [derivatives of temperature head in the interval $(0, x)$] thermal data and only on local dynamic data (β) (Exercise 4.18). As an example of the case (iii), consider any Dirichlet problem for elliptic Laplace or Navier–Stokes equation, which requires the boundary condition of the first kind specified on the boundaries of the whole domain (Secs. 1.5.4 and 2.3.1). In such a problem, the characteristics at some point are determined by data from this point, behind and advanced points, as well as by data from the points above and under the point of interest giving the local, historical, and future information of whole boundary domain.

4.2.3 Integral Universal Function for Heat Flux in Pressure Gradient Flow

Although the coefficients of the universal functions in the differential form as series rapidly decrease with term number, and usually some first terms are enough to get satisfactorily accurate results, there are cases when these series converge slowly or diverge. This happened because the rate of convergence depends not only on the coefficients decreasing, but also on the behavior of the derivatives with increasing numbers. The simplest example when series of this type diverges is the power function with fractional exponent that we encountered computing the friction coefficient for self-similar flows in Section 2.7.2.2. In such a case, we apply the integral relation (2.129) saying that both forms differential as a series and integral form are equivalent.

To understand that these two relations indeed are equivalent, note that we have already seen that substitution of the variable Φ for x transforms equations (1.23) and second equation (4.13) for a zero pressure gradient into equations (4.1) and the first equation (4.13) that are applicable for the arbitrary pressure gradient. Issuing from this fact, one may expect that the same substitution in integral formula (2.110) for the zero pressure gradient gives the integral relation (4.14) valid for arbitrary pressure gradient flows:

$$q_w = h_* \left[\theta_w(0) + \int_0^\Phi f\left(\frac{\xi}{\Phi}\right)\frac{d\theta_w}{d\xi}d\xi \right], \quad \chi_t = \frac{1}{\theta_w}\left[\theta_w(0) + \int_0^\Phi f\left(\frac{\xi}{\Phi}\right)\frac{d\theta_w}{d\xi}d\xi \right] \qquad (4.14)$$

It is shown [p.65] that actually this expectation is truth because the first expression (4.14) is a sum of series (4.1). This assertion becomes evident if one modifies the integral (4.14) using consecutive integration by parts to see that each such procedure results in a proper term of series (4.1) (Exercise 4.19).

Two forms of universal function for the heat flux allow one performing accurate calculations, employing the differential form with several first terms in the case of the converged series and using integral form otherwise. The integral form (4.14) consists of the same influence function (2.129), which we applied in the integral universal function for the friction coefficient. Recall what is an influence function like (Sec. 2.6.8) and of three known approximate relation of this function: for the plate with a zero pressure gradient (Example 2.21) and two expressions (2.136) for limiting cases $Pr = 0$ and $Pr \rightarrow \infty$. Comparing these relations, one observes that function (2.129) is a general form for all three functions with different exponents C_1 and C_2. From equation (2.136), it follows that exponents depend on the Prandtl number decreasing from 1 and 1/2 at $Pr = 0$ to 3/4 and 1/3 at $Pr \rightarrow \infty$. To the same conclusion, we arrive knowing that (i) both form of universal functions are equivalent and (ii) the coefficients of the series depend basically on the Prandtl number. Moreover, in the process of consecutive integration by parts, we obtained the relation between coefficients g_k and exponents C_1 and C_2 via beta function (4.15) [p.66] (combination of gamma functions) $B(i,j) = \Gamma(i)\Gamma(j)/\Gamma(i+j)$ (Exercises 4.20–4.22):

$$g_k = \frac{(-1)^{k+1}}{k!}\left[\frac{k}{C_1}\sum_{m=0}^{k-1}(-1)^m \frac{(k-1)!}{m!(k-m-1)!}B\left(\frac{m+1}{C_1},1-C_2\right)-1 \right],$$

$$B(i.j) = \int_0^1 r^{i-1}(1-r)^{j-1}\,dr \qquad (4.15)$$

Equation (4.15) makes it possible to compare coefficients of series g_k with exponents in equivalent integral relation. Calculations show that coefficients g_k obtained by equation (4.15) using exponents C_1 and C_2 from both function (2.136) well agree with data (4.11) and (4.12) found as a result of exact solution (4.2) (Exercise 4.23). This tells us that approximate relations (2.136) are enough accurate as well as influence function (2.135) for a plate, which follows from (2.136) for $m = 0$.

The exponents C_1 and C_2 may be estimated according to known coefficients g_k by solving the inverse problem using equation (4.15). In this case, substitution of pair values of g_1 and g_2 gives two algebraic equations determining exponents C_1 and C_2. Solution of such a system of equations yields the dependence of exponents on the Prandtl number and parameter β in conformity with functions $g_1(\mathrm{Pr}, \beta)$ and $g_2(\mathrm{Pr}, \beta)$ plotted in Figures 4.1 and 4.2. The computation results are presented in Figure 4.3.

Comment 4.3 Two reasons are responsible for some difficulties in solving the inverse problem: (i) equation (4.15) is transcendental (is not linear), and it could not be solved for C_1 or C_2, and (ii) only two exponents C are needed, while the infinite number of coefficients g_k corresponds to those two quantities. Whereas the first difficulty is a technical question that can be overcome using for the solution graphic method or software, the other problem is an essentially complication because the system of equations is overdetermined. Since there are only two unknown exponents, each pair of equations (pair values of g_k) will give different values of C. Such a situation exists in general. However, in this particular case, there are only two coefficients g_1 and g_2 that are significant, whereas all others are negligible small. Therefore, a solution of a

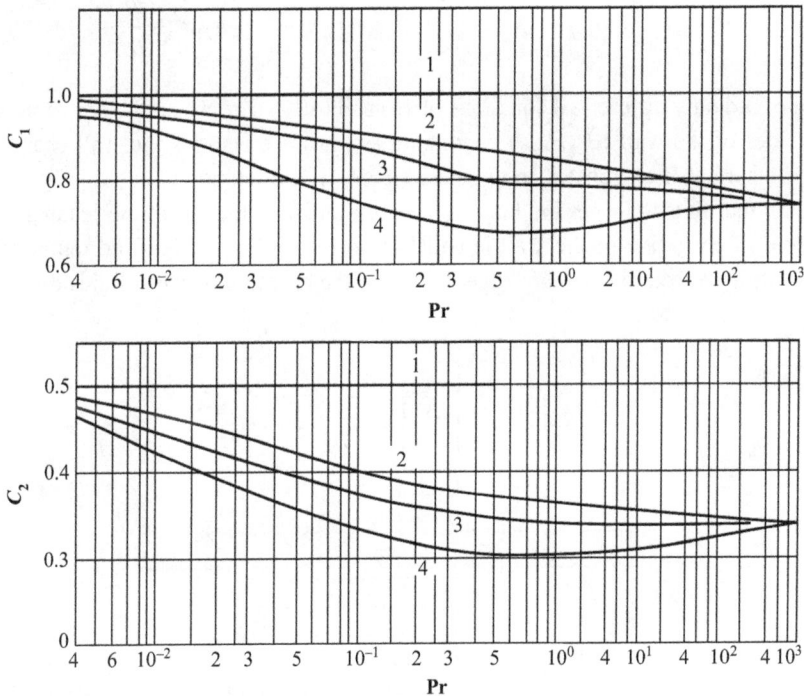

Figure 4.3. Dependence of exponents C_1 and C_2 on the Prandtl number and β on the laminar boundary layer: 1- Pr = 0, 2- β = 1, 3- 0, 4- (−0.16).

system of two equation (4.15) with coefficients g_1 and g_2 gives the proper approximate values of the exponents C_1 and C_2.

It is seen in Figure 4.3 that exponents required for integral universal function as well as coefficients of another its form as series depend on the Prandtl number and slightly on β. So the integral form is as well slightly affected by β and in it using the average or estimated by formula $\beta = 2(1 - \Phi / \mathrm{Re}_x)$ values of β may be applied.

4.2.4 Examples of Applications of Universal Functions for Heart Flux

Now, consider some applications of universal function (4.1) or (1.23) for conjugate problem solutions to see the difference from traditionally obtained results.

Example 4.1 Heat exchange process between two fluids.

Consider a conjugate solution of a classical problem of heat exchange between two fluids through a plane wall (Fig. 4.4). If the semi-infinite wall is thin ($\Delta / L \ll 1$), the longitudinal conductivity may be neglected. In such a case, the temperature distribution across the plane wall is practically linear, and due to that, the heat fluxes across the wall are constant (Sec. 1.2.1). As a result, two equations $-q_{w1} = q_{w2}$, $-q_{w1} = (\lambda_w / \Delta)(T_{w1} - T_{w2})$ are valid, which after using universal function (1.23) with two derivatives take the form

$$-h_{*1}\left(\theta_{w1} + g_{11}x\frac{d\theta_{w1}}{dx} + g_{21}x^2\frac{d^2\theta_{w1}}{dx^2}\right) = h_{*2}\left(\theta_{w2} + g_{12}x\frac{d\theta_{w2}}{dx} + g_{22}x^2\frac{d^2\theta_{w2}}{dx^2}\right) \quad (4.16)$$

$$\theta_{w2} - \theta_{w1} + (T_{\infty2} - T_{\infty1}) = \frac{h_{*1}\Delta}{\lambda_w}\left(\theta_{w1} + g_{11}x\frac{d\theta_{w1}}{dx} + g_{21}x^2\frac{d^2\theta_{w1}}{dx^2}\right) \quad (4.17)$$

A system of ordinary differential equations (4.16) and (4.17) determines the temperature heads on both sides of the wall (θ_{w1}, θ_{w2}) containing as known quantities the wall characteristics (Δ, λ_w), isothermal heat transfer coefficients (h_{*1}, h_{*2}) and temperatures ($T_{\infty1}, T_{\infty2}$) far from the wall of both fluids. To get the boundary conditions note that (i) at the leading edge, the temperature of each side of the wall is equal to corresponding initial fluid temperature and (ii) as the distance from the leading edge increases, the heat flux between fluids decreases due

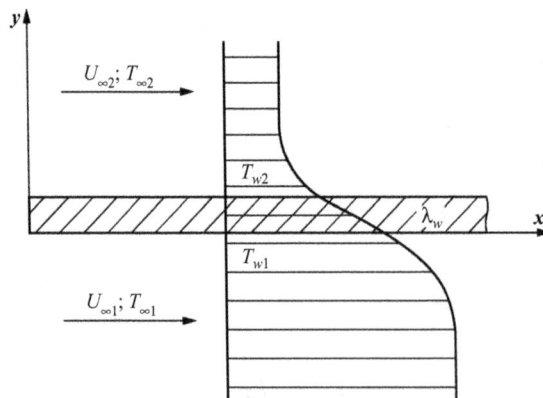

Figure 4.4. Scheme of heat exchange between two fluids through a thin wall.

to the increasing boundary layer, so that as the distance tends to infinity, the heat flux reaches asymptotically zero. This leads to asymptotically decreasing temperature head derivatives, which finally also becomes zero resulting in some constant temperature head of both fluids and corresponding wall temperature, which we denoted as $T_{w\infty}$. This temperature is defined from equation (4.16) by setting temperature head derivatives to zero and is used to form dimensionless variables (Exercise 4.24):

$$T_{w\infty} = \frac{Bi_{*1}T_{\infty1} + Bi_{*2}T_{\infty2}}{Bi_{*1} + Bi_{*2}}, \qquad Bi_* = \frac{h_*\Delta}{\lambda_w}, \qquad \Theta = \frac{T_w - T_\infty}{T_{w\infty} - T_\infty} \qquad (4.18)$$

It is reasonable to use function Θ and Biot number as dependent and independent new variables. While Θ changes along the wall from $\Theta = 0$ at $x = 0$ to $\Theta = 1$ at $x \to \infty$, the Biot number varies from infinity at $x = 0$ to zero at $x \to \infty$ since it is proportional to h_* and, hence, to $x^{-1/2}$ (Example 2.13). Equations (4.16), (4.17), and boundary conditions in new variables become (Exercises 4.25 and 4.26)

$$\Theta_1 + \hat{g}_{11}Bi_* \frac{d\Theta_1}{dBi_*} + \hat{g}_{12}Bi_*^2 \frac{d^2\Theta_1}{dBi_*^2} = \Theta_2 + \hat{g}_{21}Bi_* \frac{d\Theta_2}{dBi_*} + \hat{g}_{22}Bi_*^2 \frac{d^2\Theta_2}{dBi_*^2} \qquad (4.19)$$

$$Bi_{*2}(1-\Theta_1) + Bi_{*1}(1-\Theta_2) = Bi_{*1}Bi_{*2}\left(\Theta_1 + \hat{g}_{11}Bi_* \frac{d\Theta_1}{dBi_*} + \hat{g}_{12}Bi_*^2 \frac{d^2\Theta_1}{dBi_*^2}\right) \qquad (4.20)$$

$$Bi_* = \infty \quad \Theta_1 = \Theta_2 = 0, Bi_* \to 0 \quad \Theta_1 = \Theta_2 \to 1, \ \Theta_1' = \Theta_2' = \Theta_1'' = \Theta_2'' \to 0, \quad (4.21)$$

where the new coefficients $\hat{g}_k$ depend on the initial g_k (Fig. 4.5) (Exercise 4.25).

Equation (4.19) is simpler than the previous one consisting of only one Biot number, no matter Bi_{1*} or Bi_{*2}. Further simplification is possible in the case of the equal coefficients $\hat{g}_k$ for both fluids. Actually, this is truth for the first coefficients $\hat{g}_1$ for $Pr > 0.1$ (Fig. 4.5) and for the constant second coefficient $\hat{g}_2 = -0.11$. Therefore, employing for small Prandtl numbers ($Pr < 0.1$) an average value of coefficients $\hat{g}_1$ for both fluids, one arrives in a case when both sides of equation (4.19) are equal. Then, an equality of the dimensionless temperature heads $\Theta_1 = \Theta_2 = \Theta$ follows. In this case, equation (4.20) simplifies as well taking the form of the first equation (4.22) where dimensionless temperature heads Θ depends only on one variable Bi_Σ defined as a ratio [the second equation (4.22)] of plate thermal resistance and the resistances

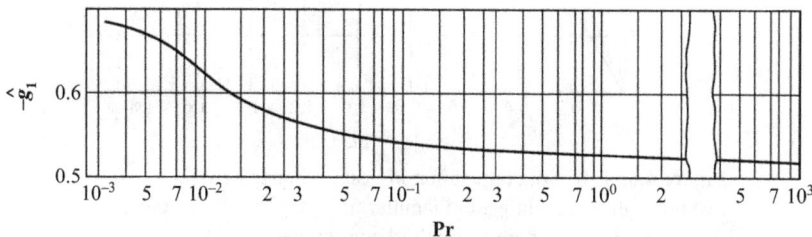

Figure 4.5. Coefficient $\hat{g}_1$ as a function of the Prandtl number.

sum of fluids. It can be shown (Exercise 4.27) that in this case, the dimensionless heat flux Bi_K through the wall depends only on Θ and hence depends only on the same variable Bi_Σ

$$\Theta\left(1+Bi_\Sigma\right)+\hat{g}_1 Bi_\Sigma^2 \frac{d\Theta}{dBi_\Sigma}+\hat{g}_2 Bi_\Sigma^3 \frac{d^2\Theta}{dBi_\Sigma^2}=1, \quad Bi_\Sigma=\frac{1}{1/Bi_{*1}+1/Bi_{*2}},$$

$$Bi_K=\frac{q_w \Delta}{\lambda_w\left(T_{\infty 2}-T_{\infty 1}\right)}=1-\Theta, \quad (4.22)$$

where Bi_* is defined by equation (4.18).

The first equation (4.22) was solved considering three approximations: the first one with only the first term in the left-hand side, the second including in addition the term with the first derivative, and the third without restriction using both derivatives. The third approximation shows that the second approximation is satisfactorily accurate, and the first two results are as follows (Exercise 4.28):

$$\Theta=\frac{1}{1+Bi_\Sigma}, \quad \Theta=\left(\frac{b}{Bi_\Sigma}\right)^{-b}\exp\left(-\frac{b}{Bi_\Sigma}\right)\int_0^{b/Bi_\Sigma}\xi^b \exp(\xi)d\xi, \quad b=-1/\hat{g}_1 \quad (4.23)$$

The computing results are plotted in Figure 4.6 as function $\Delta q_w/q_{w*}=f\left(Bi_\Sigma\right)$, where Δq_w is the difference between heat fluxes q_w and q_{w*} calculated with and without (as usually) effect of conjugation (Exercise 4.29). It is seen that the maximum effect of conjugation is 15–25% for large and small Prandtl numbers, respectively, and Biot numbers $Bi_\Sigma=0.5-0.7$.

The effect of conjugation is moderate in this problem because the temperature head increases in the flow direction on both sides of the wall. In what follows, we will see that, in general, the heat transfer coefficient is always relatively slightly greater than the isothermal one if the temperature head increases along the surface in the flow direction, whereas in the opposite case when the temperature head decreases in the flow direction, the heat transfer coefficient is highly less than the isothermal coefficient. In particular, two next examples clearly illustrate this general rule.

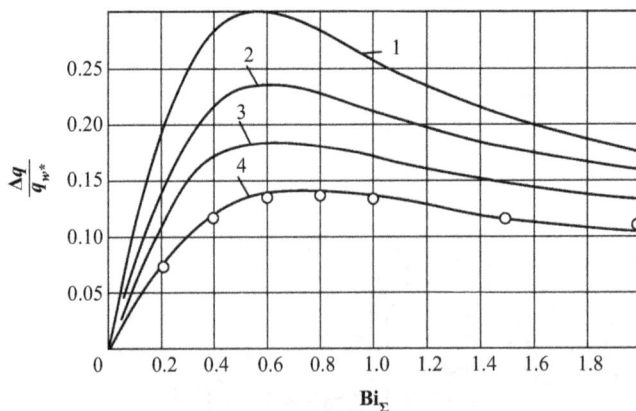

Figure 4.6. Effect of conjugation in heat exchange between two fluids through thin wall in laminar flow: $1-Pr-0, 2-0.01,$ $3-0.1, 4-Pr\to\infty$, points- the third approximation.

Example 4.2 Linear temperature head on the plate with a zero pressure gradient.

In this case, only first two terms remain in series (1.23) or (4.13). The calculation results in the form of nonisothermicity coefficient distribution along the plate for fluids with $Pr > 0.5$ ($g_1 \approx 0.63$) and zero pressure gradient are presented in Figure 4.7. This case is chosen because for laminar flows of such fluids, the value of coefficient g_1 is the lowest and is practically independent of the Prandtl number (Fig. 4.1). At the same time, these results are qualitatively valid in general for any laminar flow for which according to Figure 4.1 coefficient g_1 is not greater than unity. These data are qualitatively valid for turbulent flows as well, except turbulent flows with high Reynolds and Prandtl numbers when coefficients g_k are small (Fig.4.18) (negligible effect of nonisothermicity).

Each curve in Figure 4.7 is related to a fixed variation of temperature head determined by ratio θ_{we}/θ_{wi} of its ending θ_{we} and initial θ_{wi} values.

It is seen how strongly the effect of nonisothermicity differs for rising ($\theta_{we}/\theta_{wi} > 1$) and falling ($\theta_{we}/\theta_{wi} < 1$) temperature heads. For example, an increase of the temperature head in 1.5–2 times leads to the increase of heat transfer coefficient of about 20–30%, whereas similarly the decrease of temperature head results in lessening of heat transfer intensity in 1.5–2.5 times as compared to that for the isothermal surface. If the decreasing of temperature head becomes more than three times, the heat flux at the end of the plate reaches zero. For small Prandtl numbers, the nonisothermicity effect is even greater because coefficient g_1 grows as the Prandtl number decreases (Fig. 4.1) (Exercise 4.30).

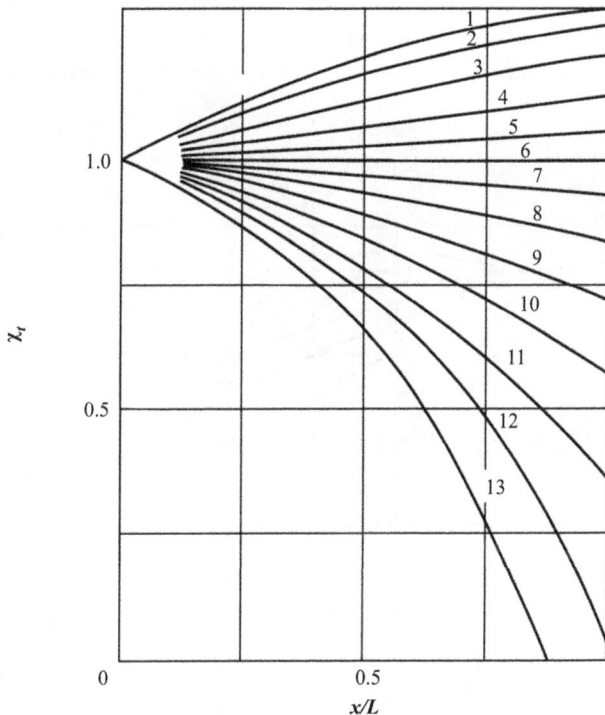

Figure 4.7. Effect of nonisothermicity for linear temperature head distribution on a plate:
$1 - \theta_{we}/\theta_{wi} = -1 - 2, 2 - 1.75, 3 - 1.5, 4 - 1.25, 5 - 1.1, 6 - 1.0,$
$7 - 0.9, 8 - 0.8, 9 - 0.7, 10 - 0.6, 11 - 0.5, 12 - 0.4, 13 - 0.3$

Example 4.3 Cylinder with linear temperature head in transverse flow.

Although the temperature head is also linear, this case is more complicated due to nonzero pressure gradient when the series (4.1) or (4.13) in Görtler variable Φ should be employed. Using formula (2.75) for variable Φ and approximate experimental velocity distribution around cylinder in the form of first polynomial (4.24) [2], one gets function $\Phi(x/D) = \Phi(\bar{x})$. Then, the inverse function $\bar{x}(\Phi)$ [the second equation (4.24)] required for applying series in the Görtler variable is obtained (Exercise 4.31):

$$U/U_\infty = 3.631\bar{x} - 4.275\bar{x}^3 - 0.168\bar{x}^5, \qquad \bar{x} = 0.74(\Phi/\mathrm{Re})^{1/2} + 0.1(\Phi/\mathrm{Re}) \quad (4.24)$$

Linear distribution $\theta_w/\theta_{wi} = 1 - K\bar{x}$ with $K = 1 - \theta_{we}/\theta_{wi}$ and the second equation (4.24) give the relation allowing to find derivatives with respect to Φ and get the result (Exercise 4.32):

$$\frac{Nu}{\sqrt{\mathrm{Re}}} = \frac{Nu_*}{\sqrt{\mathrm{Re}}}\chi_t = \frac{Nu_*}{\sqrt{\mathrm{Re}}}\left\{1 - \frac{K(\Phi/\mathrm{Re})^{1/2}\left[0.37(g_1 + 0.14) + 0.1g_1(\Phi/\mathrm{Re})^{1/2}\right]}{1 - K(\Phi/\mathrm{Re})^{1/2}\left[0.74 + 0.1(\Phi/\mathrm{Re})^{1/2}\right]}\right\} \quad (4.25)$$

The same significant effect of nonisothermicity, as in above example 4.2 can be seen in Figure 4.8. The curve of Nusselt distribution for an isothermal cylinder ($\theta_{we}/\theta_{wi} = 1$) is markedly deformed, especially, for the case of decreasing temperature head. Whereas increasing the temperature head slows the lessening of heat transfer intensity along the cylinder in comparison to that for the isothermal cylinder and even leads for $\theta_{we}/\theta_{wi} = 2$ to some increasing of the Nusselt number, the falling of heat transfer intensity in opposite case with $\theta_{we}/\theta_{wi} = 0.5$ is as strong that close to the point of separation the heat flux becomes almost zero.

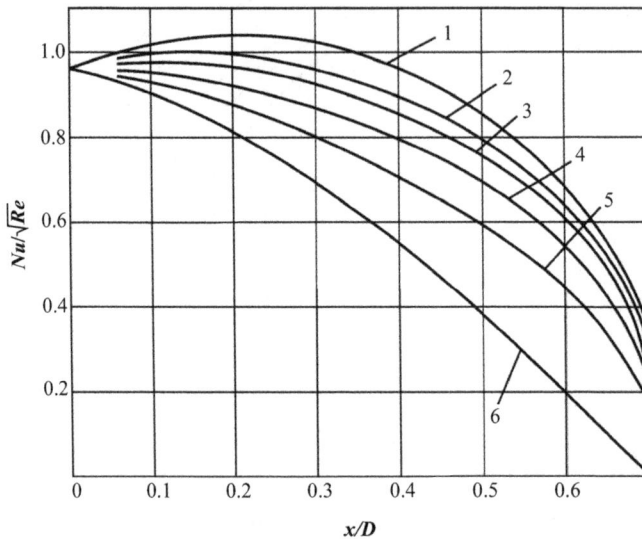

Figure 4.8. Heat transfer from a cylinder with linear temperature head in transverse flow of air (Pr = 0.7). $1 - \theta_{we}/\theta_{wi} = -1-2$, $2 - 1.5$, $3 - 1.25$, $4 - 1.0$, $5 - 0.75$, $6 - 0.5$

EXERCISES

4.1 What basic characteristics make the conjugate problem different from the others?

4.2 Compare the conjugate and the third-kind boundary conditions. What part of series (1.23) coincides with the boundary condition of the third kind?

4.3 It is known that the thermal interaction of atmosphere with a sea or ocean significantly affects the weather. May this process be investigated as a conjugate problem? Explain your answer.

4.4 Describe the set of equations and required conditions governing the conjugate problem. Why the creeping equation is not included in a set for turbulent flow equations?

4.5 Explain why the nonisothermal and conjugate heat transfers are closely related. Use the method of two equations for interface temperature estimation described in the text.

4.6 Show that function (4.1) becomes (1.23) in the case of a zero pressure gradient.

4.7 Derive equation (4.2) in Görtler variables (2.75) from the Prandtl–Mises equation (2.73) with the additional dissipation term using the chain rule (Exercise 2.58). Hint: first obtain equation (2.73) with a dissipation term from equation (2.70) (Exercise 2.56).

4.8 Show that substitution of solution (4.2) in equation (4.2) results in equations (4.3) and (4.4) [p.58]. Hint: Such a substitution results in two sums: one consists of derivatives with indices k, while the other contains the same derivatives with indices $k+1$; to sum up these sums, one should change the index $k+1$ for k (if it is difficult, try to do it for a sum with several first terms); after that, the terms with the same indices may be collected to get equations (4.3).

4.9 Recall what homogeneous and inhomogeneous equations are (Sec. 1.5) and study how the linear inhomogeneous ordinary differential equations are usually solved. Here and in what follows, the recommendation of studying means using at least some *Advanced Engineering Mathematics* without indicating this source.

4.10 *Repeat the derivation of expression (4.7) [p. 57].

4.11 Transform inhomogeneous equations (4.5) in homogeneous equations and boundary conditions (4.8) using sums of function F_i as is described on [p.59]. Explain why such transforming is reasonable.

4.12 Recall or study formulation of the initial value and two-point boundary value problems for ordinary differential equations and repeat the reducing the boundary value problem for linear second equations (4.8) to initial value problem for two other functions (V_i and W_i) by standard approach to integrate numerically these problem [p.59]. Hint: Present the solution as a sum $F_i = c_1 V_i + c_2 W_i$ and employ boundary conditions (4.8).

4.13 Equations (4.8) obtained from boundary-layer equations in the Prandtl–Mises–Görtler form (4.2) as well as the Prandtl–Mises equation (Sec. 2.5.2.2) have singularity at the surface. See how this singularity is removed and derive equations (4.8) without singularities (equation (3.19) in [1]) using a new variable $z = \varphi^{1/2}$ [p.59].

4.14 Obtain simplifying equations for limiting cases $Pr = 0$ and $Pr \to \infty$. See how these equations are solved to find the coefficients g_k of series (4.1) [pp. 59–61]. Think why the linear profile used in the case of $Pr \to \infty$ has the form $c\varphi^{1/2}$. Hint: analyze equations (2.74).

4.15 Calculate several coefficients g_k applying formulae (4.10).

4.16 Show that the Görtler variable (2.75) is defined as the average Reynolds number $\Phi(x) = Re_{(av)x}$ described in the text. Explain the benefits of using such a variable.

4.17 Explain to somebody what are local, historical, and future data. Where such data are used? Think about examples. Is any self-similar problem of type (i) (Comment 4.2)? Prove physically your answer.

4.18 Show that temperature head distribution $\theta_w(x)$ and the boundary condition containing the wage angle β are enough to calculate the heat flux distribution $q(x)$ on the wage using equation (4.1) despite that this relation required derivatives with respect to the Görtler variable and the heat transfer coefficient h_*.

4.19 *Learn how the expression (4.14) is modified using consecutive integration by parts [p.65] and perform yourself this procedure. Hint: do at first two or three integrations and note that there are two dummy variables ξ and ζ, while $\zeta = \xi/\Phi$ and $f(\zeta)$ is an influence function. Use for first integration $u_1 = \partial\theta_w/\partial\Phi$, $v_{k-1} = v_0 = \Phi f(\zeta)$ and be sure that the integrant depends on ζ, but not on Φ.

4.20 Transform equation (3.44) from [p.65] into two equations (3.46) Hint: perform this transformation for several times to guess how the general formula looks like.

4.21 Modify relation (3.45) from [p.65] between series coefficients and influence function using equations (3.46) to get equation (3.47) [p.66].

4.22 Obtain relation (4.15) between series coefficients and exponents C_1 and C_2 from equation (3.47) [p.66] using the second expression (2.129) for influence function and following the way outlined on [p.66].

4.23 Compute coefficients g_k applying relation (4.15) for $C_1 = 3/4$, $C_2 = 1/3$ and $C_1 = 1$, $C_2 = 1/2$. Compare the results to data (4.12) and (4.11) obtained by equations (4.10) for $\Pr \to \infty$ and $\Pr = 0$, respectively. To get values of beta and gamma functions, use tablets or, for example, Mathcad.

4.24 Explain why approaching the derivatives of temperature head to zero leads to constant wall temperature. Obtain relations (4.18).

4.25 *Derive equations (4.19) and (4.20). Hint: (i) transform pairs $(-\text{Bi}_1\theta_1 + \text{Bi}_2\theta_2)$, $(-\text{Bi}_1\theta_1' + \text{Bi}_2\theta_2')$ and $(-\text{Bi}_1\theta_1'' + \text{Bi}_2\theta_2'')$ to variable Θ, taking into account relation for $T_{w\infty}$; (ii) transform derivatives with respect to Bi_* (in fact to $x^{-1/2}$). Note that in this process, in general, each derivative with respect to Bi_* appears in all derivatives with respect to x beginning from the number of the derivative with respect to x. So that the first new derivative $d\Theta/d\text{Bi}_*$ appears in all initial derivatives with respect to x, and the second new derivative appears in all initial derivatives beginning from the second one, the third from third, and so an. In conformity with this, as it follows from formula (6.15) [p.168], the new coefficients $\hat{g}_k$ are defined as a sum of initial coefficients g_k with corresponding factors given by the second sum in the formula (6.15). However, in the case of two terms, that we consider, the first derivative with respect to Bi_* appears in the first and in the second derivatives with respect to x, and coefficient $\hat{g}_1$ is a sum of g_1 and g_2 with a proper factors, whereas the new second derivative corresponds to the second initial derivative only. The additional terms with higher derivatives for $k \geq 3$ in this case are neglected due to small coefficients g_k.

4.26 Formulate the boundary conditions (4.21).

4.27 Prove that in the case of equal coefficients $\hat{g}_1$ for both fluids the dimensionless temperature heads are equal, and obtain the first and the third relations (4.22) on the base of equation (4.20). Hint: use relations (4.18) and formula $\text{Bi}_\Sigma = C/\sqrt{x}$ (C is a constant), which follows from the second equation (4.22). Apply first equation (4.17) to get the heat flux through the wall.

4.28 Find the solutions of first equation (4.22) in the second approximation (without the last term of the left hand side) using standard formula for linear ordinary differential equation of the first order. Note that it is clear that the first approximation solution (4.23) satisfies both boundary conditions (4.21) as well as the second approximation solution satisfies the first of these conditions. It is shown in [3] that this solution satisfies the second boundary condition (4.21) as well. There are also given some other details including the solution of equation (4.22) in the third approximation.

4.29 Show that the first formula (4.23) together with second equation (4.22) gives the dimensionless heat flux calculated as usually without a conjugation effect. Compute some examples of $\Delta q_w / q_{w*}$ using relevant formulae (4.22), (4.23) and Mathcad.

4.30 Compute the curve $\chi_t(x/L)$ for $Pr = 0.01$ and $\theta_{we}/\theta_{wi} = 2$ and 0.5 (Example 4.2). Compare your results with corresponding curves for $Pr > 0.5$ (Fig. 4.7).

4.31 Obtain function $\Phi(x/D) = \Phi(\bar{x})$ using formula (2.75) and the first equation (4.24). Find the inverse function (4.24). Hint: if the first term of given function is $\Phi(\bar{x}) = a\bar{x}^2$, the first term on inverse function is $x = [(1/a)\Phi]^{1/2}$. Then for approximate solution it is assumed that the second term is $b\Phi$ (since $\Phi = (\Phi^{1/2})^2$). To estimate the value of b, plot the curve $\bar{x}(\Phi) - [(1/a)\Phi]^{1/2}$ and approximate this difference by $b\Phi$. To get the data for $x(\Phi)$, the calculation result of given function $\Phi(\bar{x})$ may be used.

4.32 *Derive relation (4.25) for $Nu/\sqrt{Re}$ following directions in the text and using two first terms of series (4.13) for χ_t in the Görtler variable. Note that your result will be slightly different from (4.25) by expression $(g_1 + g_2/2)$ instead of $(g_1 + 0.14)$. To understand the reason of that, consider the derivation of the same equation on the pages 126 and 127 in [1] where it was done in two steps: first, for the stagnation point using only the first term of inverse relation (4.24), and then for the rest cylinder part applying the second term of this relation. The amount of 0.14 came from equation (5.6) [p.126] in which all derivatives of series (4.13) are taking into account employing formula (4.10) for $Pr = 0$. Therefore, considering only $g_2/2$, one gets a little less amount as $1/12$ for $Pr = 0$ or as $0.1345/2$ for $Pr \to \infty$.

4.2.5 Universal Functions for a Temperature Head

Universal functions considered above determine the heat flux distribution, which corresponds to a given temperature head. Here, a universal function is obtained for the opposite case when the heat flux distribution is known, while corresponding temperature head should be found. To understand that such a universal function is given by second relation (4.26), consider the first one which is obtained dividing equation (4.1) by h_*

$$\theta_w + \sum_{k=1}^{\infty} g_k \Phi^k \frac{d^k \theta_w}{d\Phi^k} = \frac{q_w}{h_*} = \theta_{w*}(\Phi), \qquad \theta_w = \theta_{w*} + \sum_{n=1}^{\infty} h_n \Phi^n \frac{d^n \theta_{w*}}{d\Phi^n} \qquad (4.26)$$

Note that the first relation (4.26) is a differential equation defining the temperature head that we are looking for because the left part of this equation specifies θ_w via its derivatives, whereas it right side q_w / h_* is known since $q_w(x)$ and hence $q_w(\Phi)$ both are given. Note also that from physical view the ratio q_w/h_* determines the temperature head that would be established on the surface if the surface were an isothermal, and that is why we denote this ratio with an asterisk as $\theta_{w*}(\Phi)$.

Taking into account these facts, it is shown [p.70] that the solution of equation (4.26) is given by second expression (4.26) as a series similar to (4.1). Substitution of this series in the first equation (4.26) yields relations (3.62) [p.70] specifying the coefficients h_k in terms of coefficients g_k. The values of h_k founded by these relations corresponding to g_k from Figures 4.1 and 4.2 are presented in Figure 4.9 and the limiting values in conformity with data (4.11) and (4.12) are as follows (Exercise 4.33):

$$h_1 = -1/2, \quad h_2 = 3/16, \quad h_3 = -5/96, \quad h_4 = 35/1968, \quad h_5 = 0.00795 \quad \text{Pr} = 0 \quad (4.27)$$

$$h_1 = -0.38, \quad h_2 = 0.135, \quad h_3 = -0.037, \quad h_4 = 0.00795 \quad \text{Pr} \to \infty$$

Data from Figure 4.9 and values (4.27) indicate that coefficients h_k as well as g_k slightly depend on the Prandtl number and pressure gradient (β). Due to that, the universal function (4.26) for temperature head as well as function (4.1) for heat flux may be used in the general case with satisfactory accuracy. In this case, the universal function consist of derivatives of a given temperature head $\theta_{w*} = q_w / h_*$ with respect to Φ, like function (4.1) consists of the similar derivatives of known temperature head θ_w.

The integral form of universal function for the temperature head that corresponds to the differential relation (4.26) was also derived [p.71]:

$$\theta_w = \frac{C_1}{\Gamma(C_2)\Gamma(1-C_2)} \int_0^\Phi \left[1 - \left(\frac{\xi}{\Phi}\right)^{C_1}\right]^{C_2-1} \left(\frac{\xi}{\Phi}\right)^{C_1(1-C_2)} \frac{q_w(\xi)}{h_*(\xi)\xi} d\xi, \quad (4.28)$$

where the exponents are the same as shown in Figure 4.3. The case of $q_w = const.$ is one of the important cases when functions (4.26) and (4.28) are required. Dividing these relations by

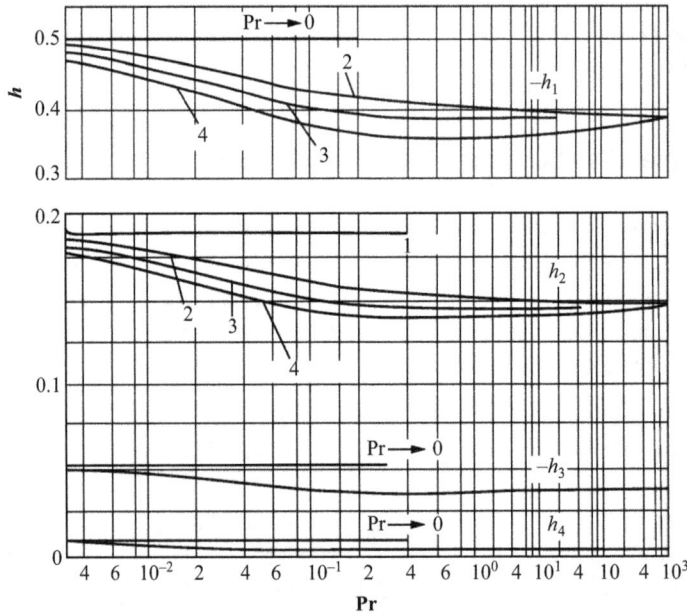

Figure 4.9. Coefficients $h_k(\text{Pr}, \beta)$ of universal function (4.26) for the laminar boundary layer. Asymptotes $1 - \text{Pr} = 0$, $\beta : 2 - 1$, $3 - 0$, $4 - (-0.16)$

q_w, one gets the connection between heat transfer coefficients h_q and h_* for the surface with $q_w = const.$ and $T_w = const.$ for general case of laminar flows with arbitrary pressure gradient and Prandtl numbers:

$$\frac{1}{h_q} = \frac{1}{h_*} + \sum_{k=1}^{\infty} h_k \Phi^k \frac{d^k(1/h_*)}{d\Phi^k} = \frac{C_1}{\Gamma(C_2)\Gamma(1-C_2)} \int_0^{\Phi} \left[1 - \left(\frac{\xi}{\Phi}\right)^{C_1}\right]^{C_2-1} \frac{\xi^{C_1(1-C_2)}}{\Phi} \frac{d\xi}{h_*(\xi)\xi} \quad (4.29)$$

For the simplest case of the zero pressure gradient and $Pr = 1$, equations (4.29) become well-known formulae (Exercise 4.35). The other important case of using formula (4.28) is the jump of heat flux described by relevant influence function. To get such a function, relation (4.28) is modified using consecutive integration by parts [p.72], resulting in the equation

$$\frac{1}{f_q(\xi,\Phi)} = \frac{h_*}{(h_q)_\xi} = \frac{C_1 h_*}{\Gamma(C_2)\Gamma(1-C_2)} \int_{\xi/\Phi}^{1} \left(1-\zeta^{C_1}\right)^{C_2-1} \zeta^{C_1(1-C_2)-1} \frac{d\zeta}{h_*(\Phi\zeta)} \quad (4.30)$$

where $f_q(\xi,\Phi)$ and $(h_q)_\xi$ are influence function of heat flux jump and heat transfer coefficient after jump, respectively, while $\zeta = \xi/\Phi$ is a dummy variable.

Comment 4.4 As is seen from (4.30), the influence function $f_q(\xi,\Phi)$ depends not only on ratio ξ/Φ but also on h_*. Therefore, unlike the influence function $f(\xi/\Phi)$ for temperature jump, the influence function for heat flux jump $f_q(\xi,\Phi)$ depends on both variables ξ and Φ separately.

Example 4.4 Heat transfer from cylinder at $q_w = const.$ and $T_w = const.$ in transverse flow.
Solution of this problem is governed by equations (4.29) and depends on the isothermal heat transfer coefficient distribution $h_*(\Phi)$. Due to that, the computation of integral in (4.29) in general should be determined numerically. If function $h_*(\Phi)$ is given analytically or may be approximated by some simple formula, this integral may be evaluated in a closer form. For example, in the case of $h_*(\Phi)$ given as a polynomial with coefficients a_i, the evaluation of integral (4.29) results in polynomial with coefficients b_i defined in terms of beta function through below-presented first relation (4.31) (Exercise 4.36). Thus, a polynomial [the second relation (4.31)] approximating the curve $\sqrt{Re}/Nu_*$ for the isothermal cylinder, as shown in Figure 4.8, produces a third polynomial (4.31) with coefficients b_i for heat transfer from cylinder at $q_w = const.$ with two coefficients related to exponents $C_1 = 0.92$ and $C_2 = 0.4$ for the area close to stagnation point $(\beta = 1)$ and $C_1 = 0.9$ and $C_2 = 0.38$ for the rest part of the cylinder $(\beta = 0)$ (Exercises 4.36 and 4.37):

$$b_i = \frac{a_i B\{[1-C_2+i/C_1],C_2\}}{B(C_2,1-C_2)}, \quad \sqrt{Re}/Nu_* = 1.04 + 0.75(\Phi/Re) - 0.83(\Phi/Re)^2 + 3.4(\Phi/Re)^3$$

$$\sqrt{Re}/Nu_q = 1.04 + (0.44 \div 0.46)(\Phi/Re) - (0.39 \div 0.42)(\Phi/Re)^2 + (1.4 \div 1.5)(\Phi/Re)^3 \quad (4.31)$$

The calculation using both values of exponents C (resulting via sign $\div$) yields the same data plotted in Figure 4.10, showing that in the case of employing variable Φ (that takes into account the flow prehistory) for flows with pressure gradient, final results slightly depend on β.

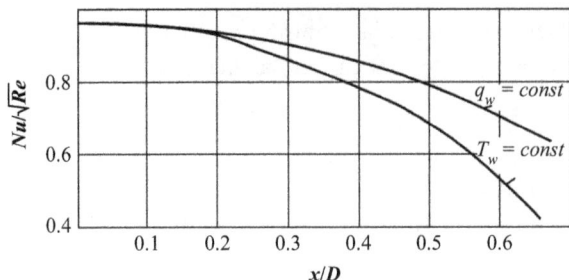

Figure 4.10. Two cases of heat transfer from the cylinder in transverse flow.

Example 4.5 Heat transfer after heat flux jump on a plate and cylinder in transverse flow.

For this case, the solution is given by function (4.30) containing isothermal heat transfer coefficient $h_*(\Phi\zeta)$ as well so that the integral evaluation like in the previous case depends on the form that this coefficient is given. For the polynomial presentation of the inverse heat transfer coefficient with terms $a_i\Phi^i$, equation (4.30) for $1/f_q$ yields polynomial solution with terms $b_i\Phi^i$ where b_i are coefficients defined through incomplete beta functions B_σ according to first equation (4.32) (Exercise 4.38):

$$b_i = \frac{a_i B_\sigma\{[(1-C_2)+i/C_1],C_2\}}{B(C_2,1-C_2)}, \quad \frac{1}{(h_q)_\xi} = \frac{B_\sigma\{[(1-C_2)+1/2/C_1],C_2\}}{h_* B(C_2,1-C_2)},$$

$$B_\sigma(i.j) = \int_0^\sigma r^{i-1}(1-r)^{j-1}\,dr \tag{4.32}$$

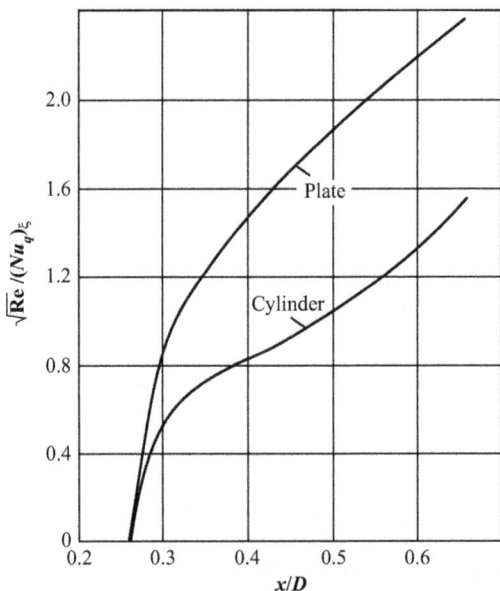

Figure 4.11. Heat transfer on a plate and a cylinder in transverse air flow after heat flux jump.

where $\sigma = [1-(\xi/\Phi)^{C_1}]$. A similar formula is obtained for a plate at zero pressure gradient. Taking into account that in this case $\Phi = \mathrm{Re}_x$, $h_* \sim \mathrm{Re}_x^{-1/2}$, and hence, $h_*(\Phi\zeta) = h_*(x)\zeta^{-1/2}$, one modifies the first formula (4.32) for the cylinder by substituting $1/2$ for i to get the second relation (4.32) (Exercise 4.39). The calculation results for both cases and $\mathrm{Pr} = 0.7$ when heat flux jump occurs at $30°(x/D = \pi/12)$ are presented in Figure 4.11 showing that the pressure gradient significantly increases the heat transfer intensity after heat flux jump.

Comment 4.5 The examples just considered show two types of heat transfer behavior at the leading edge. A body with blunt entrance reveals a finite heat transfer coefficient at the stagnation point (Figs. 4.8 and 4.10), while at the leading edge of a plate or at the starting point in the cases specifying by influence function, the heat transfer coefficient is infinite (Fig. 4.11). The cause of the last result is the parabolic type of boundary-layer equations, which could not satisfy boundary conditions at the leading edge (Sec. 2.5.2.1).

4.2.6 Universal Function for Unsteady Heat Flux in Self-Similar Flow

An exact solution of the unsteady thermal boundary-layer equation with an arbitrary unsteady temperature head $\theta_w(t, x)$ in self-similar flow shows [p.75] that the universal function in this case has the form similar to series (4.1) containing two types of derivatives with respect to variable Φ as in a steady case and with respect to dimensionless time $z = Ut/x$. Such a universal function, which is an exact result for self-similar flows, may be used for satisfactorily accurate calculation in the general case as well as analogous universal function (4.1) is used in the case of steady-state heat transfer. For self-similar flows, the derivatives with respect to Φ and z in the general form may be substituted by coordinate x and time t (Exercise 4.40), and an unsteady universal function takes the following form:

$$q_w = h_*\left(\theta_w + g_{10}x\frac{\partial\theta_w}{\partial x} + g_{01}\frac{x}{U}\frac{\partial\theta_w}{\partial t} + g_{20}x^2\frac{\partial^2\theta_w}{\partial x^2} + g_{02}\frac{x^2}{U^2}\frac{\partial^2\theta_w}{\partial t^2} + g_{11}\frac{x^2}{U}\frac{\partial^2\theta_w}{\partial x\partial t} + \cdots\right) \quad (4.33)$$

Series (4.33) contains three types of terms: depending on coordinate only, time only, and on both coordinate and time with coefficients g_{k0}, g_{0i}, and g_{ki}, respectively (Exercise 4.41). Coefficients g_{k0} are the same as g_k applied for the steady-state case. The two others depend on time, on the Prandtl number, and on β as well as g_k. For the simplest case of the zero pressure gradient and $\mathrm{Pr} = 1$, the first four coefficients g_{ki} are computed numerically (Fig. 4.12). These coefficients, similar to coefficients g_k, decrease rapidly with increasing numbers. It follows from Figure 4.12 that coefficients g_{ki} grow with time and, at $z > 2.4$, become practically independent of time being much greater for large times than coefficients g_k for the zero pressure gradient and $\mathrm{Pr} = 1$. In particular, the first coefficient $g_{01} = 2.4$ is four times larger than the corresponding value for steady state $g_1 = 0.6$, which means that the nonisotermicity effect caused by time temperature gradient is four times greater than that induced by coordinate temperature gradient.

Example 4.6 Unsteady heat transfer on a plate with time-linear temperature head.

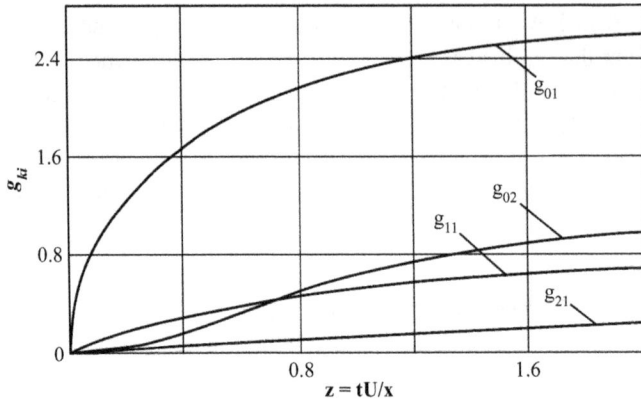

Figure 4.12. Coefficients g_{ki} $(i \neq 0)$ as functions of z for the zero pressure gradient and $\text{Pr} = 1$.

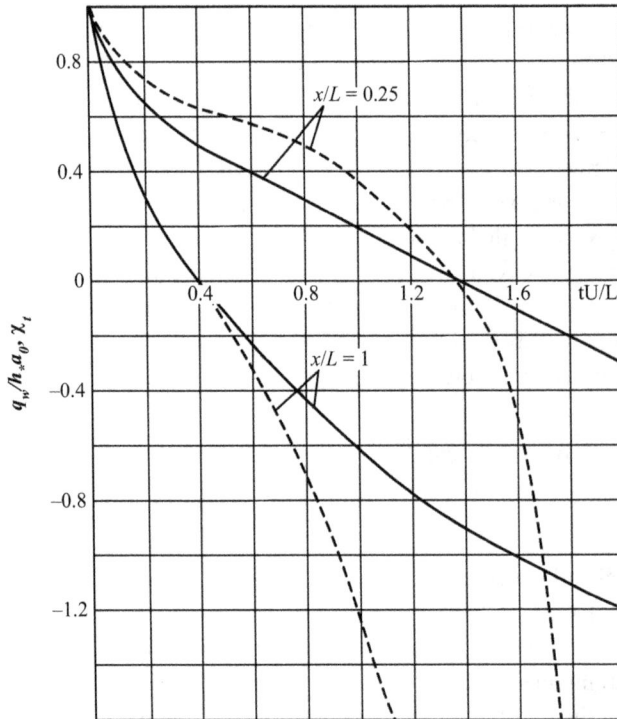

Figure 4.13. Unsteady heat transfer on a plate with time linearly variation temperature head: $\underline{\quad} q_w / h_* a_0, \text{---} \chi_t$.

Figure 4.13 presents time variation of unsteady heat flux and nonisothermicity coefficient in the case of linearly decreasing in time temperature head $\theta_w = a_0 - a_1 t$. Both results are gained by series (4.33) that for linear time dependence takes the forms

$$q_w / h_* a_0 = 1 - C\left[z + g_{01}(z)(x/L)\right], \quad \chi_t = 1 - \left[Cg_{01}(z)(x/L)/(1-Cz)\right], \quad C = a_1 L / a_0 U, \quad (4.34)$$

where the second formula is obtained dividing both sides of equation (4.33) by $h_* \theta_w$ (Exercises 4.42 and 4.43).

Some specific features of unsteady nonisothermal heat transfer follow from equations (4.34) and Figure 4.13: (i) heat flux and heat transfer coefficient (or nonisothermicity coefficient) strongly decrease with time and shortly reaches zero; (ii) despite that the given temperature head depends only on time, the both other characteristics $(q_w$ and $\chi_t)$ depend on time and coordinate; (iii) due to that, the physical pattern is complicated: at $t = 0$, the temperature head and the whole structure is steady, and then the temperature head varies with time remaining at each moment the same along the plate since it is independent of x; nevertheless, the heat flux and heat transfer coefficient depend on time and coordinate; (iv) thermal characteristics $(q_w$ and $\chi_t)$ become zero the sooner, the farther from leading edge the plate cross-section is; in particular, as shown in Figure 4.13, it occurs at $z = 0.4$ and 1.4 for $x/L = 1$ and 0.25, respectively.

The features just indicated analyzing an example are inherent in unsteady nonisothermal heat transfer, which one may see from general relation (4.33). In particular, the characteristics in Figure 4.13 depend on coordinate even when temperature head is a function only of time because the terms of series (4.33) with time derivatives depend on coordinate and velocity as well [for example, the term with the first time derivative is $(x/U)(\partial\theta_w/\partial t)$].

Comment 4.6 One more specific property of heat transfer at temperature head with variable coordinate or/and time is an inversion occurring when a heat flux becomes zero. We discuss that phenomenon in next Chapter 5; here we pay attention only to the fact that the temperature head becomes zero at time different from that for heat flux so that temperature head is finite when the heat flux is zero and vice versa. Note that this result follows from equations presenting the real relation between heat flux and temperature head, for example, universal functions (1.23), (4.1) or (4.33), but the usual relation of proportionality between these quantities (as in boundary conditions of the third kind) is in conflict with such a conclusion.

4.2.7 Universal Function for Heat Flux in Compressible Fluid Flow

The Dorodnitsyn's or Illingworth–Stewartson's variables (in Russian or English literature) transform boundary-layer equations for compressible fluid in equations of their incompressible form. Here, these variables are used to show that the universal function (1.23) or (4.1) for plate at a zero pressure gradient is valid for compressible flow as well if the temperature head is substituted by dimensionless adiabatic enthalpy i_{ad} to get

$$q_w = \frac{q_{w*}C_x}{\sqrt{C}}\left(i_{ad} + \sum_{k=1}^{\infty}g_k x^k \frac{d^k i_{ad}}{dx^k}\right), \quad i_{ad} = \frac{J_w - J_{ad}}{J_{\infty}} = i_w - \frac{r}{2}(k-1)\,\mathrm{M}^2,$$

$$C = \sqrt{\frac{T_{w.av}}{T_{\infty}} \cdot \frac{T_{\infty} + S}{T_{w.av} + S}}, \tag{4.35}$$

where J_{ad} is an adiabatic wall enthalpy (Example 2.12), and r is the recovery factor (Example 4.8).

q_{w*} is the isothermal heat flux of incompressible flow, C is the coefficient in relation $\mu/\mu_{\infty} = C(T/T_{\infty})$ known as Chapman–Rubesin viscosity law determined by formula (4.35), and C_x is the value of C defined for local temperatures. First relation (4.35) for heat flux is an exact solution of the thermal boundary-layer equation for the plate with an arbitrary temperature head distribution. The well-known Chapman-Rubesin's solution for a polynomial temperature follows from this relation as a particular case. Examples of using equations (4.35) are given in Chapter 6.

Comment 4.7 Special variables play a significant role in boundary-layer theory simplifying the initial form of equations. Thus, Blasius variable transforms the boundary-layer equation to the ordinary differential equation; Prandtl–Mises variables result in the form of boundary-layer equation close to the conduction equation; Görtler variable gives the form analogous to the self-similar equation; the Dorodnitsyn–Illingworth–Stewartson variables as it was just pointed transform the equations for compressible fluid in an incompressible form; and the Stepanov–Mangler variable simplifies the axisymmetric equations in two-dimensional forms (Example 4.9). There are other well-known variables, for example, Crocco variables that are useful in heat transfer of compressible flows studying.

4.2.8 Universal Function for Heat Flux for a Moving Continuous Sheet

Consider in Figure 4.14 a schematic presentation of production goods, such as synthetic films and fibers, metal sheets, glass, and so forth. In such a production system, a continuous sheet of material extruded from a die is pulled through a surrounding coolant with constant velocity U_w. At the middle of last century, it was shown [5] that the boundary layer forming on a moving continuous sheet differs from the well-known boundary layer existing on streamlined bodies. As is seen in Figure 4.14, in this case, the boundary layer grows in the moving direction, whereas on the streamlined or moving plate, the layer develops in the opposite direction. It may also be understood (Exercise 4.45) that in frame moving along with the sheet, the boundary-layer behavior is governed by unsteady equations with the same as for streamlined plate boundary conditions. At the same time, in a frame attached to a die, both governing systems of equations are identical, while the boundary conditions differ because the flow velocity in respect to unmoved frame is not zero.

The first calculations reveal that the friction coefficient and isothermal heat transfer coefficient for a continuous moving sheet are by 34% and 20%, respectively, greater. The universal

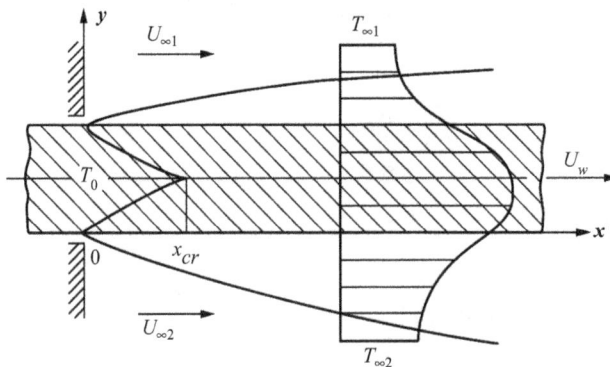

Figure 4.14. Schematic pattern of the boundary layer on a moving continuous sheet for symmetric ($U_1 = U_2$, $T_{\infty 1} = T_{\infty 2}$) and asymmetric flows.

functions for this case in the forms of series and integral are the same (4.1) and (4.14) but only in variable x since the flow pressure gradient is zero. The heat transfer coefficient for an isothermal surface is defined as $h_* = g_0\sqrt{U_w/vx}$. Coefficients g_k and exponents C_1 and C_2 were computed in [6] for different Prandtl numbers and stationary or blowing coolant with various

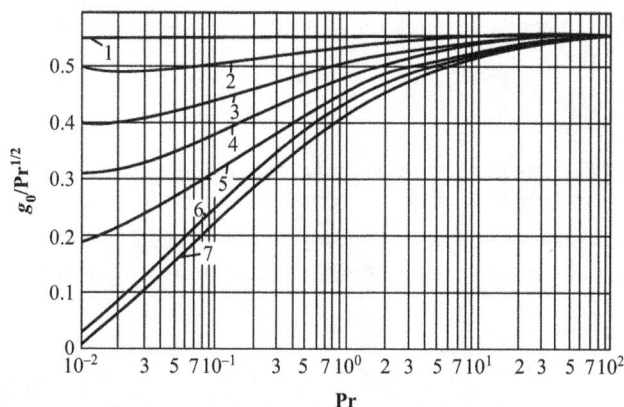

Figure 4.15. Coefficient $g_0/\mathrm{Pr}^{1/2}$ as a function of Prandtl number and a ratio $\varepsilon = U_\infty/U_w$ for a moving isothermal continuous sheet: 1-$\varepsilon = 1, 2 - 0.8, 3 - 0.5, 4 - 0.3, 5 - 0.1, 6 - 0, 7 - (-0.05)$.

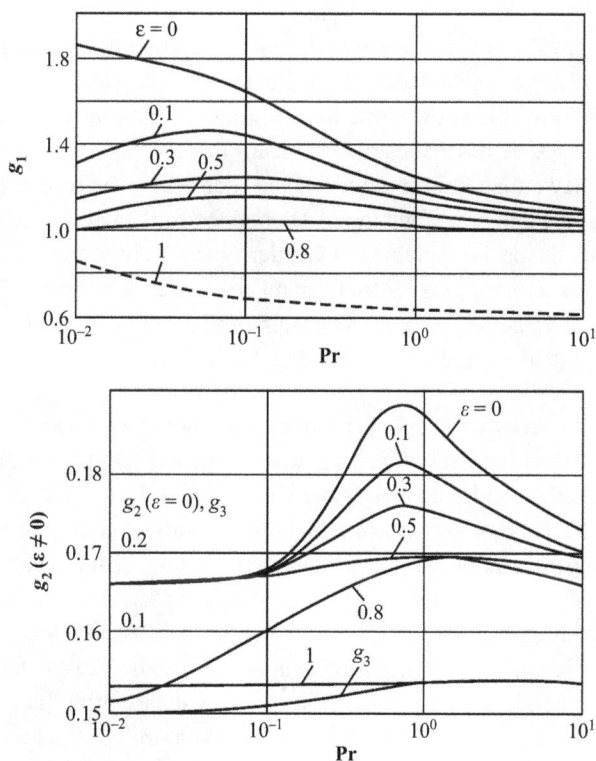

Figure 4.16. Coefficients g_k as a function of Prandtl number and a ratio $\varepsilon = U_\infty/U_w$ for a moving continuous sheet: 1- streamlined plate.

ratio $\varepsilon = U_\infty / U_w$ of the velocity of a coolant U_∞ and the moving sheet U_w. The coefficients g_k are given in Figures 4.15 and 4.16, while exponents C_1 and C_2 are not presented here may be found in [1, Fig. 3.10]. The coefficients g_k in this case are markedly greater than these for streamlined bodies, but decrease rapidly as well with increasing coefficient numbers. For instance, in the case of stationary surrounding, the first coefficient g_1 is twice greater, which means that the effect of nonisothermicity is significantly larger. Example 11.7 shows the application of these results.

4.2.9 Universal Function for Power-Law Non-Newtonian Fluids

Non-Newtonian fluids are those that do not obey Newton's and Fourier's laws.

Universal functions are obtained for fluids that behave according to power laws:

$$\tau = k_\varsigma \left(\frac{1}{2} I_2 \right)^{\frac{n-1}{2}} \mathbf{e}, \quad \mathbf{q} = k_q \left(\frac{1}{2} I_2 \right)^{\frac{s}{2}} \mathrm{grad}\, T, \quad I_2 = 4\left(\frac{\partial u}{\partial x} \right)^2 + 4\left(\frac{\partial v}{\partial y} \right)^2 + 2\left(\frac{\partial u}{\partial y} + \frac{\partial v}{\partial x} \right)^2 \quad (4.36)$$

Here, τ and $\mathbf{e}$ are the stress and rate of deformation tensors (Example 2.6), respectively, and $\mathbf{q}$ is the heat flux vector. Laws (4.36) satisfactorily describe the behavior of a group of substances, such as suspensions, polymer solutions and melts, starch pastes, clay mortars, and so on. Newton's and Fourier's laws follow from (4.36) at $n = 1$ and $s = n - 1 = 0$. The universal function for power-law fluids is constructed, as well as above for Newtonian fluids, on the base of exact solution of the thermal boundary-layer equation for an arbitrary temperature head on a body in self-similar flow. It is known, that for $n \neq 1$, the self-similar solutions exist only when the same condition $s = n - 1$ is satisfied, which means that the basic fluids properties (4.36) are (similar to that for Newtonian fluid) proportional to each other (Exercise 4.46).

It is shown that the universal function (4.1) obtained for Newtonian fluids is valid as well for power-law fluids in special variables of Görtler type (Exercise 4.47). The coefficients g_k depending on exponent n for large Prandtl numbers, which are typical for non-Newtonian fluids, and for several values of exponent m in self-similar flow velocity $U \sim x^m$ (instead of β, Exercise 4.48) are plotted in Figures (3.11) and (3.12) [p.85].

Example 4.7 Heat transfer from a cylinder in transverse non-Newtonian fluid flow.

Figure 4.17 presents the data gained for dilatant ($n > 1$) and pseudoplastic ($n < 1$) fluids using just outlined results and potential velocity distribution $U/U_\infty = 2 \sin 2(x/D)$ around the cylinder. It is seen that the rheology (n value) significantly affected the intensity of heat transfer and its variation. In particular, while at the stagnation point the heat transfer coefficient is finite for Newtonian fluids, it is zero for dilatant and infinite for pseudoplastic fluids. In this case as well as in the others discussed above, the distribution of the Nusselt number along the cylinder strongly differs for decreasing and increasing temperature heads. Thus, in the first case, the heat flux becomes zero at angle $\gamma = 80°$, whereas in the second case at the same location, the heat flux value is close to that in the case of a constant temperature head. As a result, the variation of the Nusselt number along the cylinder changes from the parabolic form for dilatant fluid in the case of constant or growing temperature head to a s-formed curve for a falling temperature head and pseudoplastic fluid.

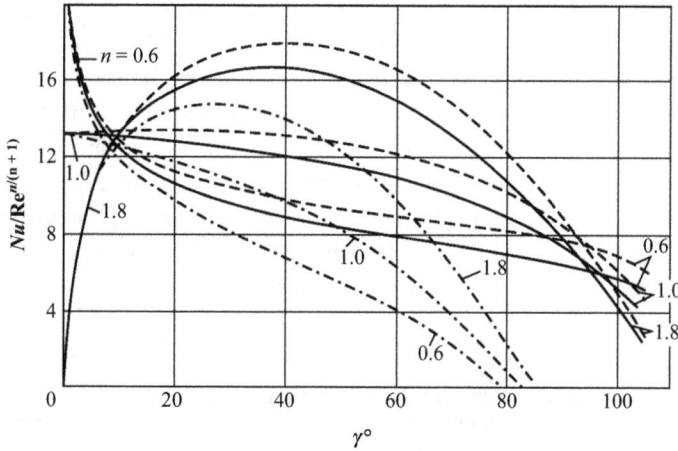

Figure 4.17. Heat transfer from nonisothermal cylinder in transverse flow of power-law non-Newtonian fluid: ———— $\theta_{we}/\theta_{wi} = 1$, – – –$\theta_{we}/\theta_{wi} = 1 + x/D$, – – – – –$\theta_{we}/\theta_{wi} = 1 - x/D$ (notations in Example 4.2) Pr = 1,000.

4.2.10 Universal Function for the Recovery Factor

According to the second relation (4.7), the effect of dissipation is proportional to square of the velocity. Due to that, for incompressible flows, for which the usual velocity is low, the dissipative term might be comparable with other terms of energy equation only for large Prandtl numbers. Therefore, the effect of dissipative heat was studied for non-Newtonian fluids having high Prandtl numbers. Analysis of these data shows (Exercise 4.49) that even in this case the dissipation heat should be taken into account only when the temperature head decreases, yielding rapidly lowering heat flux along the surface or in time. For the case of constant or increasing temperature head, the heat flux associated with dissipation is usually not important except some special cases (Example 4.8).

Example 4.8 Effect of mechanical energy dissipation.

Nevertheless, estimation of the adiabatic wall temperature and recovery factor requires the dissipation heat flux evaluation. This problem is in essence an estimation of a temperature head at given heat flux that was considered in Section 4.2.4. Thus, determining $\theta_{w*}(\Phi)$ in equation (4.26) with an additional dissipative term from equation (4.7) as the first sum (4.37) presented below and putting then $q_w = 0$ yields second equation (4.37), defining the universal function for recovery factor:

$$\theta_{w*}(\Phi) = \frac{q_w}{h_*} + g_d \frac{U^2}{c_p}, \quad r = \frac{T_{ad} - T_\infty}{U^2/2c_p} = 2g_d \left(1 + \sum_{k=1}^{\infty} h_k \frac{\Phi^k}{U^2} \frac{d^k U^2}{d\Phi^k}\right) \quad (4.37)$$

Here, h_k are coefficients given for Newtonian fluids in Figure 4.9. For non-Newtonian fluids, those may be calculated using the same approach and coefficients g_k (Sec. 4.2.4). Equation (4.37) presents the recovery factor for an arbitrary pressure gradient. The case of the zero pressure gradient follows for $U = const.$, when all derivatives are zero yielding for recovery factor $r = 2g_d$. The values of g_d are given in Figure 3.13 [p.87] (Exercise 4.50).

Comment 4.8 One merit of using the adiabatic temperature is that its substitution for wall temperature in common relations [for example, as in equation (4.35) for enthalpy] yields the expressions with regard to dissipation heat. The other advantage is that applied adiabatic quantities are important to understand the heat exchange process between the wall and fluid. As indicated in Section 2.4.6.2, the fluid cools the wall until it reaches the adiabatic temperature, while then the heat flux changes its direction so that the wall cools the fluid. The value of adiabatic temperature in this process is defined by recovery factor that even in the case of a zero pressure gradient depends on the Prandtl number (Example 2.12) being equal to unity for $Pr = 1$ and greater and lower than unity for $Pr > 1$ and $Pr < 1$, respectively (Exercise 4.50).

4.2.11   Universal Function for an Axisymmetric Body

According to the Stepanov–Mangler variables (comment 4.5), an universal function for a two-dimensional problem is valid for the same axisymmetric problem if the integrant in variable Φ, and the resulting values of heat flux are multiplied, respectively, by R^{n+1} and R^n, where $R(x)$ is the body radius and n is the exponent in power-law relations (4.36). Thus, for Newtonian fluid, an universal function for the axisymmetric case may be found using relation (4.1) and multiplying Φ and q_w by R^2 and R, respectively.

EXERCISES

4.33 Obtain relations defining coefficient h_1 and h_2 by substituting second equation (4.26) into first one with only two terms in both equations and following collection of similar terms in the resulting expression. Calculate these coefficients for limiting Prandtl numbers cases and compare the result with the data in the text.

4.34 Explain the difference between two temperature heads θ_w and θ_{w*} in expressions (4.26). When such temperature heads are used?

4.35 Show that in the case of a zero pressure gradient and $Pr \approx 1$, general formula (4.29) reduces to well-known relations given in the literature (for example in [4]). Calculate the ratio h_*/h_q for a zero pressure gradient and $Pr \approx 1$ and for $Pr = 0$. Check results looking in [4] or in [1].

4.36 Derive the first formula (4.31) for polynomial coefficients b_i by substituting polynomial $1/h_* = a_i \Phi^i$ into integral (4.29) and applying relation (4.15) for beta function. Hint: use a new variable $\zeta = (\xi/\Phi)^{C_1}$ and polynomial for $1/h_*$ expressed in this variable.

4.37 Calculate coefficients of the second polynomial (4.31) issuing from first polynomial (4.31) and formula (4.31) for coefficients b_i.

4.38 Show that first formula (4.32) defines coefficients of polynomial resulting from (4.30) in the case when the inverse isothermal heat transfer coefficient is used in the form of polynomial $a_i \Phi^i$. Compute the Nusselt number after heat flux jump on a cylinder at angle $30°$ if the inverse isothermal Nusselt number is given by the first polynomial (4.31). Compare your results with data given in Figure 4.11. Hints: use new variable $\sigma = 1 - (\xi/\Phi)^{C_1}$, corresponding dummy variable and note that integration is performed on dummy variable while variable Φ is independent on it. For determining beta functions use, tables or compute these using Mathcad.

4.39 Obtain second formula (4.32) following directions in the text and using hints given in the previous example. Calculate corresponding dependence presented in Figure 4.11.

4.40 Show that in the case of steady self-similar flow derivatives with respect to variable Φ and to dimensionless time z are proportional to those with respect to x and to t, respectively. Hint: use expressions for Φ and U in self-similar flows and note that they depend only on x.

4.41 Continue series (4.33) for $i = 0, k > 0, k = 0, i > 0, k > 0, i > 0$ knowing that the general term is $g_{ki}\left(x^{k+i}/U^i\right)\left(\partial^{k+i}\theta_w/\partial x^k \partial t^i\right)$. Hint: first check the given terms.

4.42 *Show that unsteady boundary-layer equation (2.70) in three variables, two Görtler variables Φ, φ (2.75) and dimensionless time $z = Ut/x$, takes the form (3.77) [p.75].

4.43 Obtain equations (4.34) from series (4.33) and compute curves plotted in Figure 4.13 for different values of x/L. Explain why characteristics depend on coordinate x despite that the given temperature head depends only on time.

4.44 Show that variables (3.88) transform the Prandtl–Mises equation for enthalpy i in equation (3.89) [p.78].

4.45 Compare two types of boundary-layer equations forming on a continuous sheet and on a streamlined plate.

4.46 What physically means the condition $s = n - 1$? How the proportionality of the basic properties (known as apparent viscosity and conductivity) may be seen from equations (4.36)?

4.47 *Consider how it is proved that series (4.1) in variables (3.96) is valid for power-law non-Newtonian fluids [p.84]: first, the variables (3.96) are used to transform the thermal boundary-layer equation (3.95) in form (3.97), and then it is shown that substitution of function (4.1) into equation (3.97) yields ordinary differential equations (3.98). Try to perform these steps yourself.

4.48 Explain why to determine the pressure gradient in the case of non-Newtonian fluids flows the exponent m is used instead of β. Hint: consider equation (3.100) [p.84].

4.49 Consider an analysis [1, Example 5.9, p.131] of dissipation energy data plotted on Figure 5.8 [p. 132] to understand the conclusions given in this text (Sec. 4.2.9).

4.50 Learn more about the recovery factor: (i) compare the last formula (2.53) with expression for adiabatic temperature given in example 2.12. Are quantities c_p and λ/μ the same? Check the units and order of magnitude for $Pr = 1$. What is the value of recovery factor in these equations? (ii) Examine equation (3.99) [p.84] to understand how many parameters define the recovery factor via formula $r = 2g_d$. What parameter is missed on the graph (3.13) [p.87]?

4.3 Universal Functions for Turbulent Flow

The universal function for heat flux for turbulent flow has the same form (4.1). The coefficients g_k are obtained in a similar way from the equation in Prandtl–Mises–Görtler form (4.2) with additional turbulent characteristics calculated using the Mellor–Gibson model as is described in Section 3.3.6.6 for the case of isothermal heat transfer:

$$\Phi \frac{\partial \theta}{\partial \Phi} - \varphi \frac{\partial \theta}{\partial \varphi} - \frac{\partial}{\partial \varphi}\left(\frac{u}{U}\varepsilon_a \frac{\partial \theta}{\partial \varphi}\right) = \frac{U^2}{c_p}\varepsilon_u \frac{u}{U}\left[\frac{\partial}{\partial \varphi}\left(\frac{u}{U}\right)\right]^2, \quad \varepsilon_a = \frac{1}{\Phi}\left(\frac{1}{Pr} + \frac{\varepsilon - 1}{Pr_{tb}}\right) \quad (4.38)$$

Here, $\varepsilon_a = \alpha_e/v\,\Phi, \varepsilon_u = v_e/v\,\Phi, \varepsilon = v_e/v$, the effective quantities (marked by index e) are found as sums of laminar and turbulent components, and the variable Φ is the same as (2.75), while another $\varphi = \psi/v\Phi$ is slightly different from variable (2.75) used for laminar flow in equation (4.2). Substitution of series (4.2), written in these variables, in the first equation (4.38) yields two differential equations similar to equations (4.3) and (4.4). These equations are transformed into a system of ordinary differential equations using the equilibrium boundary-layer solutions as well as it was done applying self-similar solutions for laminar flow (Sec. 4.2.1). As a result, the system of equations, which determined the coefficients g_k, similar to equation (4.5), is obtained:

$$\left(\omega\varepsilon_a G'_k\right)' + \varphi G'_k - kG_k = G_{k-1}, \quad \varphi = 0, G_0 = 1, G_k = 0 \quad \varphi \to \infty, G_k = 0, k \geq 0 \quad (4.39)$$

where $\omega\left(\varphi, \beta.\mathrm{Re}_{\delta_1}\right)$ and $\varepsilon_a\left(\varphi, \beta.\mathrm{Re}_{\delta_1}\right)$ are defined using results for equilibrium flows (Exercise 4.51). Further transformation of this equation is also the same that performed for the case of laminar flow, leading to relation similar to equation (4.9), which reduces the calculation of coefficients g_k to numerical solution of two initial value problems for two functions.

Calculation results for the same broad range of parameters as in Section 3.3.6.6 for isothermal heat transfer: $\beta = -0.3$ (stagnation point), $\beta = 0$ (zero pressure gradient), $\beta = 1$ and $\beta = 10$ (weakly and strong adverse pressure gradients), and $\mathrm{Re}_{\delta_1} = 10^3, 10^5, 10^9$, are plotted as functions of the Prandtl number in Figures 4.18 and 4.19. Analysis of these data indicates that coefficients g_k (i) rapidly decrease with increasing k as in the case of laminar flow; hence, one may use only a few of the first terms of series. (ii) They decrease with increasing Prandtl number as in the case of laminar flow; however, in contrast to the laminar flow, where at large Prandtl

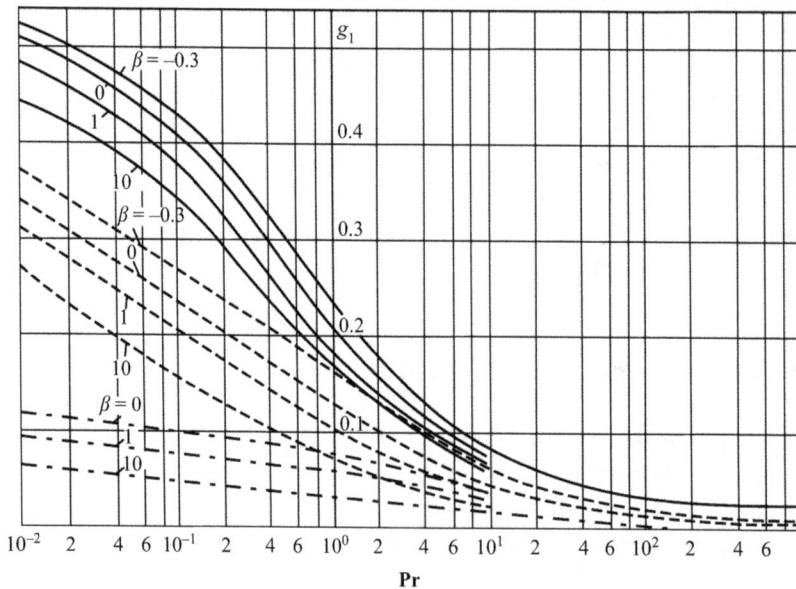

Figure 4.18. Coefficient g_1 for the turbulent boundary layer: ___ $\mathrm{Re}_{\delta_1} = 10^3$, ___ 10^5, ----- 10^9.

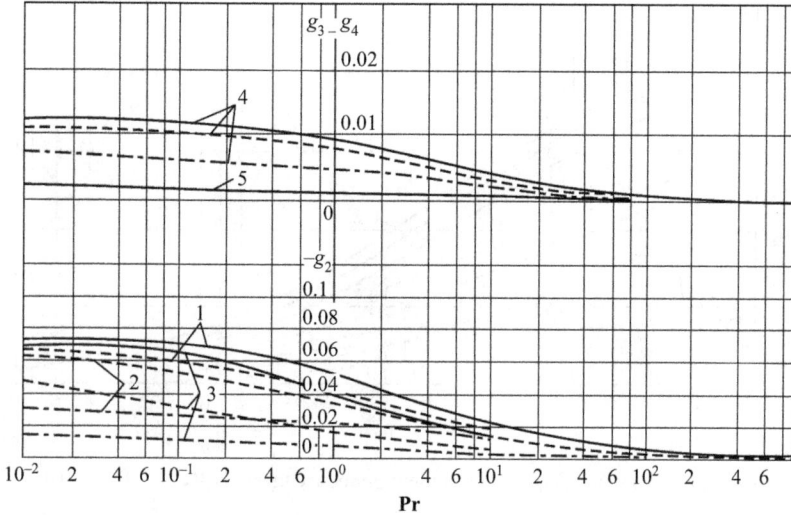

Figure 4.19. Coefficients g_k for the turbulent boundary layer: $g_2 : 1 - \beta = -0.3, 2 - \beta = 0, 3 - \beta = 1, 4 - g_3$ ___ $\mathrm{Re}_{\delta_1} = 10^3$, ___ 10^5, ----- 10^9.

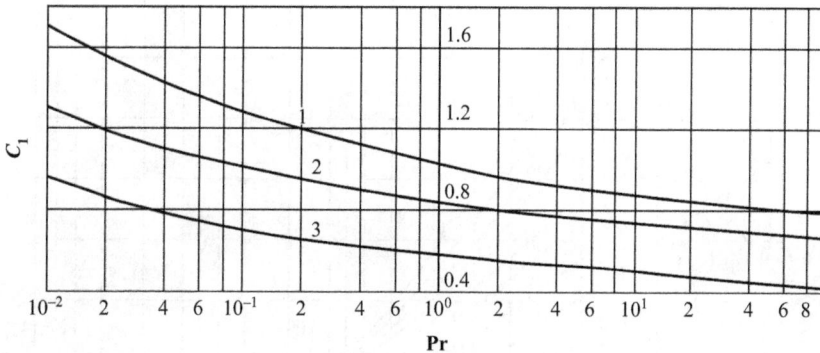

Figure 4.20. Exponent C_1 for the turbulent boundary layer: $1 - \mathrm{Re}_{\delta_1} = 10^3, 2 - 10^5, 3 - 10^9$.

numbers, the coefficients are independent of Pr, but finite, in this case, they tends to zero with increasing Pr so that beginning from some value of Prandtl number $\left(\approx 10^2 \right)$, the nonisothermicity effect becomes negligible. (iii) They are significantly smaller than those for laminar flow decreasing with increasing Reynolds number so that the nonisothermicity effects are lower than that in laminar flow; (iv) g_1 and g_2 depend weakly on β, while the others are practically independent of β. This allows one to use universal function (4.1) to calculate heat transfer in the case of arbitrary pressure gradient as well as for laminar flow (Sec. 4.2.2).

Since differential form (4.1) is valid for turbulent flow, the other forms are applicable as well with suitable coefficients and exponents. Those are found knowing g_k by the same way as for laminar flow. The exponents C_1 and C_2 for the integral forms (4.14) found by formula (4.15) and coefficients h_k for universal function for temperature head (4.26) are plotted in Figures 4.20, 4.21, 4.22, and 4.23. It is seen that exponent C_1 does not depend on the pressure gradient, while other characteristics does. Using these graphs, one might apply any of relations derived above for the laminar boundary layer. Relevant examples are presented in the next two chapters.

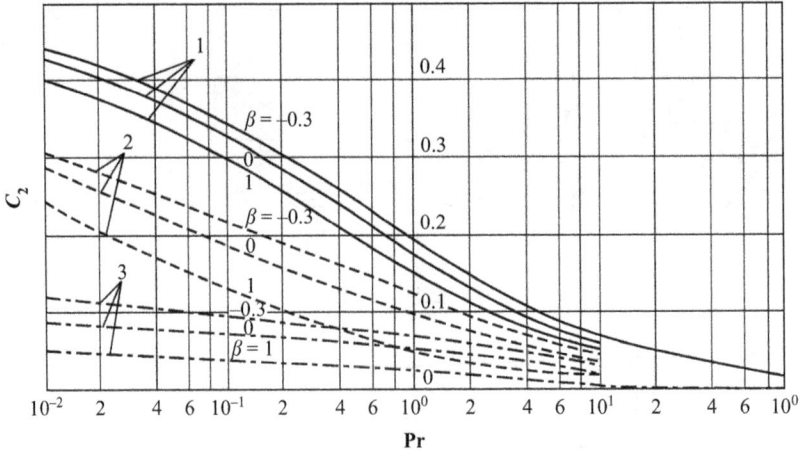

Figure 4.21. Exponent C_2 for the turbulent boundary layer: 1- $\text{Re}_{\delta_1} = 10^3, 2 - 10^5,$ $3 - 10^9.$

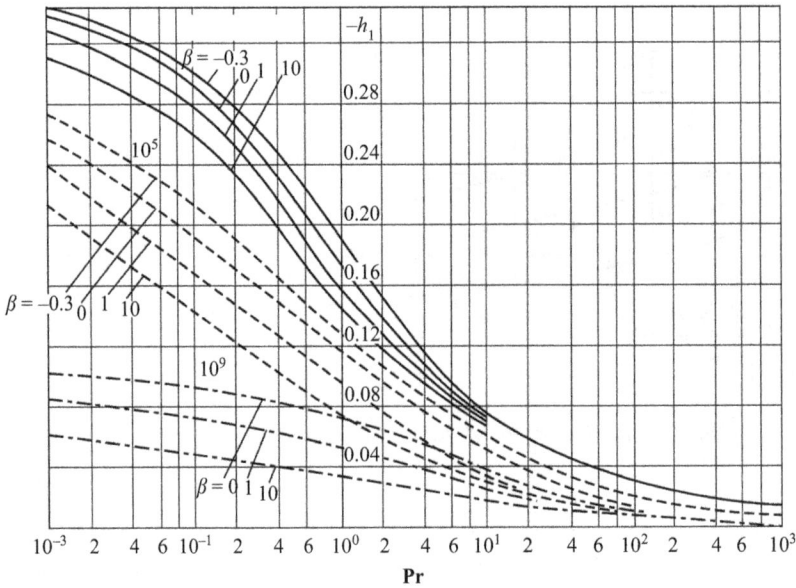

Figure 4.22. Coefficient h_1 and h_3 for the turbulent boundary layer: ___ $\text{Re}_{\delta_1} = 10^3,$ _ _ _ $10^5,$ _ _ _ _ _$10^9.$

Examples 4.9 Comparison computed and measured data for turbulent heat transfer.

1. It follows from Figures 4.20 and 4.21 that for $Pr = 1$ and zero pressure gradient $C_1 = 1,$ $C_2 = 0.18$ if $\text{Re}_{\delta_1} = 10^3,$ while for $\text{Re}_{\delta_1} = 10^5$ these exponents are $C_1 = 0.84, C_2 = 0.1.$ These data agree with well-known influence functions experimentally found at pertinent lower ($\text{Re}_x = 5 \cdot 10^5$) and higher ($10^8$) Reynolds numbers [details p.116].

2. Computational results shown in Figure 5.10 [p.134] are in reasonable consistent with corresponding experimental data obtained for increasing and decreasing linear and exponential temperature heads and exponential heat flux (Exercise 4.52).

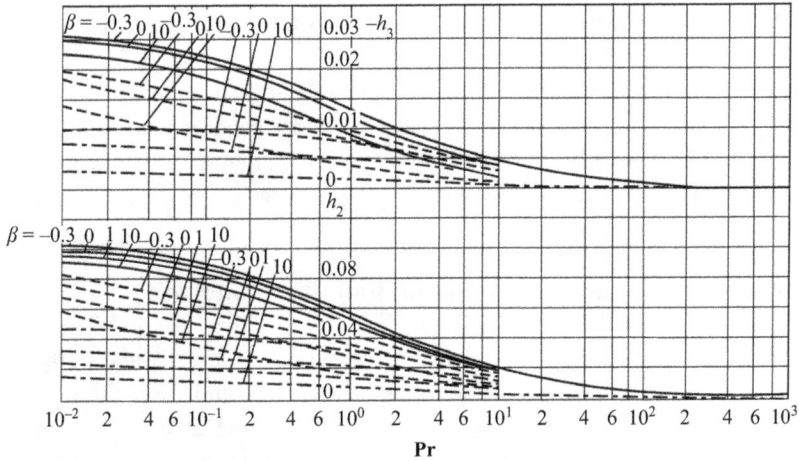

Figure 4.23. Coefficients h_k, $k = 2, 3, 4$ for the turbulent boundary layer ___ $Re_{\delta_1} = 10^3$, $---10^5$, $----10^9$.

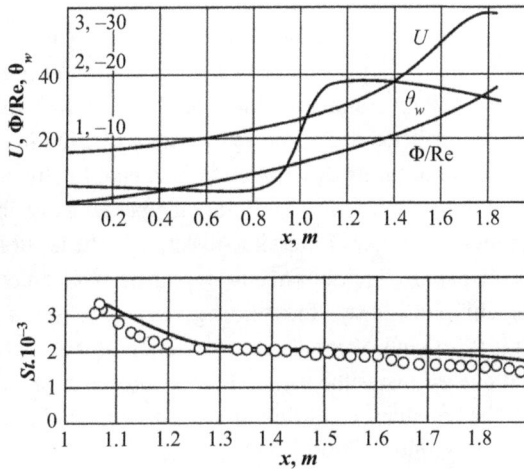

Figure 4.24. Comparison of calculated and experimental [7] data for stepwise temperature head.

3. Calculation results are also in agreement with measurements from the well-known investigation of heat transfer in the case of stepwise temperature heads in flows with different external velocity distribution [7]. One example of comparison of data obtained experimentally and by integral formula (4.14) is presented here in Figure 4.24, while four others and analysis may be seen in Figure 5.11 [pp. 135–137].

Comment 4.9 The methods of deriving basis equations, the procedure of calculation coefficients and exponents, as well as proofs of generalization of some relations in fact are the same for both laminar and turbulent flows. Therefore, these are not repeated here for turbulent flow, and due to that, this section presenting actually the same amount of final information as in the case of laminar flow is relatively short.

4.4 Reducing a Conjugate Problem to a Conduction Problem

In this section, we show that the universal function in the form of series, such as equation (4.1) or others of this type, may be used as a general boundary condition to (i) estimate the errors caused by boundary condition of the third kind and (ii) simplify the conjugate problem by reducing it to equivalent conduction problem.

4.4.1 Universal Function as a General Boundary Condition

The universal function for heat flux in the form of series of the derivatives might be considered as a sum of perturbations of a surface temperature distribution or of the surface boundary condition. In such an approach, the isothermal surface, for which all derivatives are zero and in series retains only the first term, is considered as an unperturbed boundary condition; the series containing only the first derivative is viewed as a linear boundary condition; the series with the first two derivatives is taken as a quadratic boundary condition and so on, finally resulting in the universal function with infinite derivatives presenting an arbitrary boundary condition.

Those reasoning led to a conclusion that universal function (4.1) or others of this type are actually general boundary conditions describing different types of boundary conditions. Such a function with only the first term is the boundary condition of the third kind, giving the first approximation of some surface temperature distribution. Retaining two first terms yields a linear boundary condition, describing better the same temperature distribution. The more terms are retained in the series, the more accurate description is obtained, finally resulting in the general boundary condition with infinite terms defining arbitrary temperature distribution (Exercise 4.53).

Such a calculation procedure makes it possible to estimate the accuracy of obtained results by comparing two consistent approximations at any step. In particular, comparing data obtained by boundary conditions with one and two first terms gives estimation of accuracy of using the boundary condition of the third kind. Some coarse estimation of that accuracy may be obtained in a much simpler way if one estimates the value of the second term of universal function using the solution gained with the boundary condition of the third kind and compares that value with result of this solution (see examples below).

4.4.2 Estimation of Errors Caused by Boundary Condition of the Third Kind

As was mentioned in the previous discussion, the boundary condition of the third kind is a basic relation in convective heat transfer since the time of Newton being widely used even now despite existing effective numerical methods and the modern conjugate approach.

Moreover, the heat transfer college textbooks even do not mention about modern methods remaining as many years ago with Newton's proportionality law only. There are two reasons of such a situation: (i) a simplicity of the boundary condition of the third kind and (ii) the fact that for some problems, this simple approach provides practically acceptable results. Therefore, it is necessary to have a tool for understanding whether for a particular problem the conjugate solution is required or the accuracy achieved by common method is satisfactory. The estimation of the second term of universal function in the form of series just described helps to answer this question. Some examples of such estimation are given next (Exercise 4.54).

Example 4.10 Heat exchange process between two fluids (Fig. 4.4).

The temperature head on the one side of a plate found by common approach using the boundary condition of the third kind is as follows:

$$\theta_{w1} = \frac{T_{w1} - T_{\infty 1}}{T_{\infty 2} - T_{\infty 1}} = \frac{q_w}{h_{*1}\left(T_{\infty 2} - T_{\infty 1}\right)} = \frac{1}{1 + h_{*1}/h_{*2} + h_{*1}\Delta/\lambda_w} = \frac{1}{D_1 + D_2 x^{-n}} \qquad (4.40)$$

The last expression where coefficients D_1 and D_1 are constant is obtained assuming that the flow regimes on both sides of the plate are the same and knowing that in both cases of laminar or turbulent flows the heat transfer coefficient decreases along the plate as x^{-n}.

The accuracy of result (4.40) is estimated comparing the second term $g_1 x(d\theta_{w1}/dx)$ of universal function (1.23) with result (4.40) θ_{w1}, which is in fact the first term of this universal function. Differentiation of equation (4.40) yields the corresponding relatively error and its maximum values for both cases of laminar and turbulent flows

$$\sigma = \frac{g_1 x}{\theta_{w1}}\frac{d\theta_{w1}}{dx} = \frac{g_1 n D_2}{D_1 x^n + D_2}, \quad \sigma_{\max} = g_1 n \text{ at } x = 0, \quad (\sigma_{\max})_{\text{lm}} = \frac{g}{2}, \quad (\sigma_{\max})_{\text{tb}} = \frac{g}{5} \qquad (4.41)$$

For laminar flow, at a zero pressure gradient and fluids with $\text{Pr} > 0.5$ (Fig. 4.5), the first coefficient is $g_1 \approx 0.62$, so for this range of Prandtl numbers, the maximum error is about 30%, while for turbulent regime, at a zero pressure gradient and $\text{Pr} = 0.5$ (Fig. 4.18), the maximum value of the first coefficient is $g_1 \approx 0.22$, which decreases as the Prandtl number grows. Hence, for this range of Pr, the maximum error is about 4%. At the same time, for turbulent flow with $\text{Pr} = 0.01$, we have $g_1 \approx 0.52$ that results in $\sigma_{\max} \approx 10\%$. Thus, for laminar flow, the error is moderate, while for turbulent flow, when $\text{Pr} > 0.5$, the use of the boundary condition of the third kind does not lead to significant errors. These estimations are in agreement with corresponding conjugate solutions which for laminar and turbulent flows give the maximum errors 25% (Fig. 4.6) and 7% [p.164], respectively.

The cause of relatively small errors is the increasing temperature head in the flow direction on both sides of the plate. As we saw above and will see later, the small effect on heat transfer characteristics of a growing temperature head and a strong effect on those of a falling temperature head are general features of nonisothermal heat transfer.

Example 4.11 Heat transfer from a thermally thin plate heated from one end.

A qualitative analysis of this problem was considered in the introduction to Part II. Error caused by common approach is estimated for the following particular problem.

A steel plate of thickness 0.01 m and of length 0.25 m is streamlined by a symmetric air flow of temperature 300 K and velocity 3 m/s The left-hand end is insulated, and the temperature of the right-hand end is maintained at T_w.

Equation (1.7) for a thermally thin plate for steady-state flow and boundary condition of the third kind with average heat transfer coefficient governing this problem lead to ordinary differential equation (4.42) of second order with constant coefficients. Solving this equation using the standard procedure for the case of flow from the right to left edges one obtains (we use notation Bi^2 instead of Bi to have Bi instead of $\sqrt{\text{Bi}}$ in the final result) (Exercise 4.56)

$$\frac{d^2\theta_w}{d\bar{x}^2} + \text{Bi}^2\theta_w = 0, \quad \theta_w = \frac{T_w - T_\infty}{T_w(L) - T_\infty} = \frac{1}{ch(\text{Bi})}ch\left(\text{Bi}\frac{x}{L}\right), \quad \text{Bi}^2 = \frac{Nu_{*av}L\lambda}{\lambda_w\Delta} \qquad (4.42)$$

An error and its maximum are found after differentiating the expression (4.42) for θ_w

$$\sigma = g_1 \frac{x}{\theta_w} \frac{d\theta_w}{dx}, \quad \sigma = g_1 \mathrm{Bi} \frac{x}{L} th\left(\mathrm{Bi} \frac{x}{L}\right) \qquad \sigma_{max} = g_1 \mathrm{Bi} \qquad (4.43)$$

For laminar flow at a zero pressure gradient and $\mathrm{Pr} > 0.5$, we have $g_1 = 0.62$. Calculating Nu_{*av} for $\mathrm{Re} = 5 \cdot 10^4$ yields $\mathrm{Bi}^2 = 0.65$, and $\sigma_{max} \approx 0.5$. Hence, the solution of this problem with the boundary condition of the third kind is unaccepted (Exercise 4.57). The reason for this is that in this case the temperature head decreases along the plate. The conjugate solution is given in Example 6.1.

Example 4.12 A thermally treated polymer continuous sheet at temperature T_0 is extruded from a die and passed at velocity U_w through a bath with water ($\mathrm{Pr} = 6.1$) at temperature T_∞ (see scheme in Fig. 4.14).

This problem with the boundary condition of the third kind is solved in Example 11.7. The result is presented as $\theta_w(\tilde{x})$, with $\tilde{x} = x\alpha_w / \Delta^2 U_w$ and $\theta_w = (T_w - T_\infty)/(T_0 - T_\infty)$.

The error is estimated for the case of the thermal resistance ratio $(c_p \rho \lambda)_w / c_p \rho \lambda = 8.51$ and quiescent coolant. The derivative $d\theta_w / d\tilde{x}$ is found by numerical differentiating the corresponding curve $\theta_w(\tilde{x})$ presented in Figure 11.1 (Exercise 4.58). Then, taking the value of coefficient $g_1 = 1.3$ from Figure 4.16 and using Equation (4.43) for error σ, one obtains the order of accuracy of the common approach. The error grows as the distance from the die increases and finally reaches $\sigma_{max} \approx 2.6$, which indicates that this problem must be solved as conjugate. The reason of this is the decreasing temperature head.

The examples show that in the case of decreasing temperature head, any problem should be considered as a conjugate. In another case, when the temperature head increases, the estimation of error may help to understand whether a particular problem requires the conjugate solution or the accuracy of common method is satisfactory.

4.4.3 Equivalent Conduction Problem with the Combined Boundary Condition

It was shown that the solution of conjugate heat transfer problem is a result of coupling two other solutions obtained for a body and fluid with arbitrary temperature or temperature head distribution on their interface (Sec. 4.1). On the other hand, the universal function (4.1) is a solution of thermal boundary-layer equations for fluid for arbitrary surface temperature distribution, which in fact is one of the just-mentioned two required solutions for coupling in the conjugate problem. Consequently, in the case of using universal function (4.1) or one of the others of this type, there is no need to solve the boundary-layer equation for fluid, and only the conduction equation for a body subjected to the same temperature distribution remains to be solved. In other words, using universal function yields the conjugate heat transfer problem reducing to an equivalent conduction problem with the boundary condition in the form of universal function (4.1) or others of this type.

As shown in the previous section, relation (4.1) may be treated as a general boundary condition. Hence, if the heat conduction equation is solved using this condition with the first term only, an approximate solution of the conjugate problem as with the boundary condition of the third kind is obtained. By retaining the first two terms in equation (4.1) and solving the heat conduction equation, a more accurate solution of the conjugate problem is obtained. This process of refining, in principle, can be continued by retaining more terms in equation (4.1). However, this entails difficulties posed by the calculation of higher order derivatives and series convergence. Therefore, for high-order approximations, the integral form of universal function (4.14) should be used.

In practical calculations, it is convenient to retain the first few terms of the series and to calculate the error using the results of previous approximation. When in this case, the first three terms of the series are retained, the conjugate problem is reduced to a heat conduction equation for solid with the following combined boundary condition:

$$q_w = h_* \left[\theta_w + g_1 \Phi \frac{d\theta_w}{d\Phi} + g_2 \Phi^2 \frac{d^2\theta_w}{d\Phi^2} + \varepsilon(\Phi) \right], \qquad \varepsilon(\Phi) = \frac{1}{h_*} \left(q_w^{\text{int}} - q_w^{\text{diff}} \right) \qquad (4.44)$$

In this boundary condition, combined of both forms of universal functions, heat fluxes q_w^{int} and q_w^{diff} are defined by integral relation (4.14) and by differential relation (4.44), respectively, where their difference $\varepsilon(\Phi)$, substituted the all terms of series neglected in the first expression (4.44). The first approximation is found by assuming that $\varepsilon(\Phi) = 0$. Calculating this term using the results of the first approximation makes it possible to introduce the error $\varepsilon(\Phi)$ into equation (4.44) and to find the second approximation. Repeating this procedure using the second approximation, one finds new values of q_w^{int} and q_w^{diff} and gets the third approximation. By continuing this process, the solution with a desired accuracy can be obtained (Exercise 4.59).

Retaining in equation (4.44) terms with derivatives not higher than second leads to differential equations of the second order in any approximation. As a result, in this case, the conjugate problem is reduced to the ordinary differential equation in the case of thin body and to Laplace or Poisson equation in the general case. This allows using the well-known effective methods of solving such equations.

The suggested method of reducing the conjugate problem to the equivalent conduction problem can be used to solve any linear conjugate convective heat transfer problem. Examples are given in the following chapters basically in applications.

4.4.4 Equivalent Conduction Problem for Unsteady Heat Transfer

The method of reduction of a steady conjugate heat transfer problem to an equivalent conduction problem outlined above applicable to similar unsteady problems if the thermal capacity of fluid is small in comparison with that for a body, as, for example, in the case of a pair nonmetallic fluid-metal wall. In such a case, the parameter $Lu = \rho c / \rho_w c_w$, which on [p. 200] was suggested to name the Luikov number, is small as well, and the so-called quasi-steady approach may be used. This means that unsteady effects in fluid might be neglected because time of the heating boundary layer is small in comparison with that for body so that at each moment, the boundary-layer parameters are considered as a steady value defined by universal function (4.1) or other of this type. Hence, in such a case, the only unsteady conduction equation remains to be solved using the same as for a steady conjugate problem combined boundary condition (4.44).

To find the condition when the quasi-steady approach is applicable, we employ the universal function (4.33) obtained for the unsteady thermal boundary layer. Considering the thermally thin plate (equation 1.7) and determining the unsteady effects using from series (4.33) the term with first time derivative only, we obtain the first equation (4.45) for unsteady surface temperature.

$$\frac{1}{\alpha_w}\frac{\partial T_{av}}{\partial t} - \frac{\partial^2 T_{av}}{\partial x^2} + \frac{h_*}{\lambda_w \Delta}\left[T_w - T_\infty + g_{01}\frac{x}{U}\frac{\partial T_w}{\partial t} + \cdots\right] = 0 \qquad g_{01}\frac{x}{\Delta}\frac{Nu_* Lu}{Pe} \ll 1 \qquad (4.45)$$

The quasi-steady approach is valid when the term with time derivative for fluid $\partial T_{av}/\partial t$ is negligible compared to that for a body $\partial T_w/\partial t$. This condition leads to an inequality [last equation (4.45)] which shows when the quasi-steady approach for unsteady boundary-layer problems is applicable (Exercise 4.61):

For boundary-layer heat transfer problems, the inequality (4.45) usually is satisfied due to the large Peclet number. However, in many situations, the satisfaction of this inequality depends on the Luikov number, especially when the Peclet number is not very high. In general, the ratio $Lu = \rho c/\rho_w c_w$ depends on properties of both subjects of a fluid and a body and, therefore, is more flexible than two other Nusselt and Peclet numbers in inequality (4.45). Due to that, the Luikov number $Lu = \rho c/\rho_w c_w$ often is considered as the basic characteristic of the unsteady conjugate heat transfer.

EXERCISES

4.51 *Variable $\varphi = \psi/v\Phi$ in the case of turbulent flow differs from $\varphi = \psi/v\sqrt{2\Phi}$ used for laminar flow. To understand the reason of that, look at the equation (4.3) for laminar flow. In this case, a variable φ was chosen in the form, which transforms initial partial differential equation in relation (4.3), where expressions in the brackets depend only on variable φ. These expressions for the self-similar flows when the velocity ratio u/U also depends only on φ one reduces to ordinary differential equations, finally determining coefficients g_k of series (4.1). To provide the same procedure in the case of turbulent flow when equilibrium flows are used instead of self-similar flows, it is necessary to employ different variable φ yielding the transformation of the initial partial differential equation in the system of ordinary differential equations, determining finally the coefficients g_k. Consider on [p.100] the proof that the velocity profile u/U and effective viscosity $\varepsilon = v_{tb}/v$ as well as other turbulent characteristics in equations (4.38) and (4.39) for equilibrium flows are functions only of φ, β and Re_{δ_1}, providing required transforming in the case using the Mellor–Gibson turbulence model (Sec. 3.3.3).

4.52 Consider an analysis of Figure 5.10 given on [pp. 133–134]. Derive formulas used in this analysis applying relations (4.13), (4.14), or others given above. Compute some examples of decreasing and increasing temperature heads from the caption to Figure 5.10 and compare those with data given in Figure 5.10 to see the heat transfer behavior in different cases.

4.53 What is a general boundary condition? What benefits offers such a condition?

4.54 Explain how universal function in the form of series with derivatives helps us to understand whether or not the conjugate solution is required.

4.55 *Estimate an error caused by the boundary condition of the third kind in the problem from Example 4.10 for the case when the flow is laminar on the one side of the plate and turbulent on another. Show that the result is the same as in Example 4.11 for laminar flow on both sides of the plate. Hint: consider the expression for relatively error σ taking

$h_1 = c_1 x^{-n_1}$ and $h_2 = c_2 x^{-n_2}$ to see that error becomes negative if the index 1 denotes side with turbulent flow, which does not fit the consider situation. Think and explain why.

4.56 Solve equation (4.42) and think why it is possible and convenient to take this equation in the form with Bi^2. Hint: since the plate is considered as semi-infinite, its length L containing in the Biot number may be chosen arbitrarily.

4.57 Estimate the error in Example 4.11 for turbulent flow at $Re = 10^6$ for flow with $Pr = 1$ and $Pr = 0.01$.

4.58 Estimate numerically the derivative $d\theta_w/d\tilde{x}$ of curve derivative $\theta_w(\tilde{x})$ in Figure 11.1 (curve 7) and calculate the error of using the boundary condition of the third kind in the case of treatment a polymer sheet in Example 4.12.

4.59 Expression (4.44) contains two derivatives with respect to variable Φ. Find connections between those and corresponding derivatives with respect to variable x. Compare your results with equation (6.10) [p.166].

4.60 Explain what combined boundary condition is like. How it works reducing a conjugate problem to a conduction problem for a body?

4.61 Derive the equation (4.45) using equation (1.7) and the term with first time derivative only from series (4.33) for heat flux. Obtain the inequality (4.45) comparing the terms with time derivatives for fluid and body as is described in the text. Explain the conclusions following from this inequality.

REFERENCES

[1] Dorfman, A. S. (2009). *Conjugate problems in convective heat transfer.* Boca Raton: CRC Press Taylor & Francis.

[2] Spalding, D. B., & Pun, W. M. (1962). A review of methods for predicting heat transfer coefficients for laminar uniform-property boundary layer flows. *International Journal of Heat Mass Transfer 5*, 239–244. doi: http://dx.doi.org/10.1016/0017-9310(62)90014-5

[3] Dorfman, A. S. (1972). Calculation of thermal fluxes and the temperatures of the surfaces of a plate with heat transfer between the fluids flowing around the platem. *High Temperature 10*(2), 293–298.

[4] Kays, W. M. (1969, 1980). *Convective heat and mass transfer.* New York: McGraw-Hill.

[5] Sakiadis, B. C. (1961). Boundary layer behavior on a continuous solid surface, *AIChE J 7 (Pt. 1),* 26–28; *(Pt. 2),* 221–225.

[6] Dorfman, A. S., & Novikiv, V. G. (1980). Heat transfer from continuously moving surface to surroundings. *High Temperature 18*, 898–901.

[7] Moretti, P. M., & Keys, W. M. (1965) Heat transfer to a turbulent boundary layer with varying free stream velocity and varying surface temperature—an experimental study. *International Journal of Heat and Mass Transfer 8*, 1187–1202. doi: http://dx.doi.org/10.1016/0017-9310(65)90062-1

CHAPTER 5

GENERAL PROPERTIES OF NONISOTHERMAL AND CONJUGATE HEAT TRANSFER

As mentioned at the beginning of Chapter 4, the behavior of conjugate heat transfer is basically determined by the temperature distribution, which is established on the body-fluid interface as a result of their interaction. Therefore, the general properties of conjugate heat transfer are actually the same as those of arbitrary nonisothermal temperature distribution, namely arbitrary because in conjugate problem a priori temperature distribution is unknown.

In this chapter we formulate general properties of conjugate and nonisothermal heat transfer using the universal functions obtained in Chapter 4. As we will see, these functions which present the dependences of heat flux on arbitrary temperature head distribution in different situations, consist of enough information required for such a procedure. The effect of different factors on the conjugate and nonisothermal heat transfer characteristics is considered, and general relations and conclusions are discussed.

5.1 Effect of Temperature Head Distribution: Temperature Head Decreasing-Basic Reason for Low Heat Transfer Rate

The structure of the universal functions gives the understanding of the effect of temperature head variation on the heat transfer intensity. Three factors determine this effect: (i) signs and values of temperature head derivatives; (ii) signs and values of coefficients g_k, which depend on Prandtl number for laminar flow and on Prandtl and Reynolds numbers for turbulent flow; and (iii) pressure gradient. The rate and character (favorable or adverse) of the temperature head effect are defined basically by the first two factors, while the pressure gradient specifies the independent variable (x or Φ) of derivatives. In the following sections we analyze the influence of these factors on heat transfer intensity and illustrate conclusions with relevant examples (Exercises 5.1 and 5.2).

5.1.1 Effect of the Temperature Head Gradient

As can be seen in Table 5.1, the first coefficient g_1 is significantly larger than the others, in all cases, considered in Chapter 4. Even in comparison with the second coefficient g_2, which is the greatest among others, the first coefficient is about 3 (unsteady laminar layer)–10 (turbulent layer) times larger. This means that, if all the derivatives are of the same order, the first derivative, which defines the temperature head gradient, basically specifies the effect of non-isothermicity (Exercise 5.3).

Because the first coefficient is positive, it means, according to universal function (4.1), that positive temperature head gradients lead to increasing of heat flux while negative gradients cause decreasing of heat flux. More precisely, if the temperature head increases in the flow direction or in time, the heat transfer coefficient is greater than the isothermal coefficient, whereas the decreasing of the temperature head along the flow direction or in time yields a decrease in the heat transfer coefficient when compared with the isothermal coefficient. Therewith, the temperature head gradient is determined by a derivative with respect to coordinate x in the case of a zero pressure gradient and with respect to the Görtler variable Φ in the case of a non zero pressure gradient.

Recall our discussion in Chapter 4 of the general features of temperature head gradient in the cases of the plate and cylinder in transverse flow with linear temperature head (Examples 4.2 and 4.3). There we saw that an identical change of increasing and decreasing temperature head leads to significantly different variations in the heat transfer coefficient. Formally this is easy to understand by examining relation (4.13) to see that the nonisothermicity coefficient χ_t (the ratio of nonisothermal and isothermal heat transfer coefficients) is basically defined by a sum of two first terms (other g_k are small), which is much less when the second term is negative (falling temperature

Table 5.1 Relation between coefficients g_1 and g_2

	g_1	g_2	$\dfrac{g_2}{g_1}$
1	1	1/6	1/6
2	0.6123	0.1345	0.22
3	0.380	0.135	0.36
4	1/2	3/16	3/8
5	2.4	0.8	1/3
6	≈ 0.5	≈ 0.05	≈ 0.1
7	≈ 0.1	≈ 0.01	≈ 0.1
8	≈ 0.2	≈ 0.04	≈ 0.4
9	≈ 0.8	≈ 0.2	≈ 0.25
10	≈ 0.4	≈ 0.06	≈ 0.15
11	1.25	0.15	0.12

Laminar layer: arbitrary $\theta_w - 1 - \mathrm{Pr} \to \infty$, $2 - \mathrm{Pr} \to 0$, arbitrary
$q_w - 3 - \mathrm{Pr} \to \infty$, $4 - \mathrm{Pr} \to 0$; unsteady laminar layer: $5 - \mathrm{Pr} = 1$; turbulent layer:
$6 - \mathrm{Pr} \to 0$, $\mathrm{Re}_{\delta_1} = 10^3$, $7 - \mathrm{Re}_{\delta_1} = 10^9$, $8 - \mathrm{Pr} = 1$, $\mathrm{Re}_{\delta_1} = 10^3$;
non-Newton fluid: $9 - n = 1.8$, $10 - n = 0.2$, $11 - \mathrm{Pr} \approx 1$, $\varepsilon = 0$ moving surface.

head) than in the opposite case. Physically the reason for such different results is that the same absolute change in falling and growing temperature heads yields much greater relative variation in the first case, when the temperature head itself is small, than in the second one. Moreover, as seen in mentioned examples from Chapter 4 as well as in relation (4.13), in the case of decreasing temperature head, the heat flux and the corresponding heat transfer coefficient become zero if the streamlined surface is sufficiently long. It is simple to understand that this occurs when the sum of all terms with derivatives in equation (4.13) reaches unity resulting in $\chi_t = 0$ (Exercise 5.4).

These general conclusions of heat transfer behavior clearly show, examples of nonisothermal heat transfer, seen in previous and next chapter, as well as these features are visible in conjugate problem solutions presented in Chapter 6 and applications.

5.1.2 Effect of Flow Regime

The first coefficient of series (as shown previously), which basically determined the nonisothermicity effect, significantly depends on flow regime. According to Table 5.1, its value varies from the largest $g_{01} = 2.4$, at the time derivative in equation (4.33) for the case of laminar unsteady flow, to negligible small magnitude for turbulent flow of fluids with large Prandtl numbers (Fig. 4.18). The other coefficients g_k for turbulent flow are less than those in the case of laminar regime. The higher the Reynolds and Prandtl numbers, the lesser the coefficients g_k, and correspondingly the lesser the nonisothermicity effect. Nevertheless, in all cases the qualitative effect of the temperature head gradient on the heat flux intensity is the same as in the case of laminar flow discussed in the previous section. The quantitative results can be seen in Example 5.1, a comparison of nonisothermicity effects in laminar and turbulent flows. The reliability of universal functions for turbulent flow, shown in Section 4.3, follows from summary of comparison of calculated and measured data (Example 4.9) (Exercise 5.5).

Example 5.1 Effects of nonisothermicity in turbulent and laminar flows.

Figure 5.1 shows the variation of nonisothermicity coefficient for $Pr = 1$ and two regimes in the case of linear temperature head $\theta_w/\theta_{wi} = 1 - K(x/L)$ with $K = 1 - \theta_{we}/\theta_{wi}$, where indices i and e denote initial and end values.

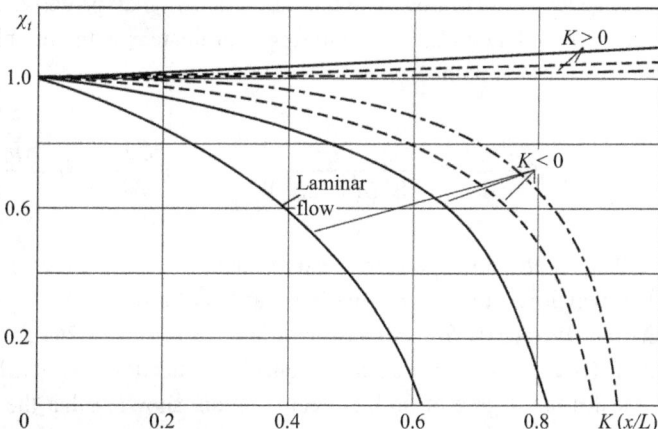

Figure 5.1 Nonisothermicity coefficient for different flow regimes.
——— $Re_{\delta_1} = 10^3$, – – – – – $Re_{\delta_1} = 10^5$, – · – · – $Re_{\delta_1} = 10^9$.

The pattern indicates that in spite of small coefficients of series (4.1) for turbulent flow, the general effect of nonisothermicity is the same as in the case of laminar flow. The increasing temperature head slightly increases the heat transfer coefficient, while the falling temperature head sharply decreases the heat transfer coefficient. These effects in turbulent flow are less intensive than in laminar one and become weaker as the Reynolds number grows. Nevertheless, if the surface is long enough, the heat flux finally reaches zero too, as seen on Figure 5.1. Nonisothermicity effects also decline with growing Prandtl number, but the whole situation remains the same until Prandtl numbers are large (say Pr >100), when the effects of nonisothermicity become negligible due to small coefficients g_k (Figs. 4.18 and 4.19).

5.1.3   Effect of Pressure Gradient

The effect of pressure gradient on nonisothermal heat transfer coefficient is determined basically by the second term of series (4.1), which specifies the temperature head gradient. To estimate the effect of pressure gradient, this term is modified using the formula (2.75) for Görtler variable Φ to get $\Phi(d\theta_w/d\Phi) = (U_{av}/U)\,x\,(d\theta_w/dx)$. From this expression one observes that the ratio U_{av}/U of an average external to local velocities modifies the temperature head gradient defining the effect of pressure gradient. It can be shown that this ratio is less or great than unity in accelerating (negative pressure gradient) and slowing (positive pressure gradient) external flows respectively (Exercise 5.6). Thus, external flow with decreasing pressure leads to lessening of the nonisothermicity effect, while on the contrary the nonisothermicity effect becomes more pronounced. The quantitative results depend on a particular function $U(x)$ specifying the external flow. For example, for self-similar flows $U_{av}/U = 1/(m+1)$. This means that by the other condition being equal, the second term of series for the flow near stagnation point ($m = 1$) is half as much as in the case of the zero pressure gradient ($m = 0$).

5.2   Biot Number-A Measure of Problem Conjugation

In Chapter 1, we analyzed different types of boundary conditions using relation (1.26) between Biot number (Bi) and temperature head gradient for the plate at zero pressure gradient. The same relation for the general case of flow with pressure gradient may be derived (Exercise 5.7) or formally obtained from relation (1.26) substituting coordinate x with variable Φ and using average body thickness for Biot number

$$\lambda_w \left.\frac{\partial T}{\partial y}\right|_{y=0} = h_* g_1 \Phi \frac{\partial \theta_w}{\partial \Phi} \quad \text{or} \quad \frac{1}{Bi}\left.\frac{\partial T}{\partial (y/\Delta_{av})}\right|_{y=0} = g_1 \Phi \frac{\partial \theta_w}{\partial \Phi} \quad Bi = \frac{h_* \Delta_{av}}{\lambda_w} \qquad (5.1)$$

This expression tells us that the temperature head gradient $\partial \theta_w/\partial \Phi$ is inversely proportional to the Biot number, which is a ratio of thermal resistances of a body (Δ_{av}/λ_w) and a fluid in the case of isothermal surface ($1/h_*$). Because relations (5.1) and (1.26) are obtained from conjugate conditions (Sec. 1.2.4), the existence of such a connection between Bi and $\partial \theta_w/\partial \Phi$ means that Bi indicates the degree of problem conjugation, showing that the larger the Biot number is the smaller the gradient $\partial \theta_w/\partial \Phi$ establishes in a corresponding conjugate problem (Exercise 5.7).

The analysis of relation (1.26), which holds for general cases as well [equations (1.26) and (5.1) follow one another], confirms this statement testifying the follows: (i) in both limiting cases $Bi_* \to \infty$ and $Bi_* \to 0$, the full conjugate problem degenerates because in these cases only one of the resistances is finite, while another is infinite as in the first case or becomes zero as in the second one; (ii) at $Bi_* \to \infty$, the conjugate problem transforms into a longwise one-dimensional problem with uniform temperature ($\partial\theta_w/\partial\Phi \to 0$) around the body, except a small area at $x = 0$, where $\partial\theta_w/\partial\Phi \to \infty$ (Exercise 1.9). We encountered this type of a problem considering heat transfer from semi-infinite solid or a thin laterally insulated rod (Sec. 1.3.2). To this type also belongs the problem of heat transfer in a hot poorly conducting slab abruptly immersed in high conducting fluid (quenching process) (Exercise 5.8); (iii) if the Biot number is large but finite, the conjugate problem also becomes one-dimensional, but in contrast to a previous case in a transverse direction (Sec. 1.2.4); in this case a thin body model is valid when one-dimensional equation (1.4) is applicable (Sec. 1.2.1), which we use in Example 4.1 considering heat transfer between two fluids; (iv) in the other limiting case, when $Bi_* \to 0$, the solid temperature is also uniform, but in this case due to the small thermal resistance of a body (Sec. 1.2.4). This situation is modeled by a thermally thin model when the averaged one-dimensional equation (1.7) might be used (Sec. 1.2.1). In practice we encountered such a problem when, for example, a high conductive body cools slowly transferring heat to a poorly conductive environment; (v) because in both limiting cases only one resistance component is finite, and the two-dimensional conjugate problem degenerates in one-dimensional, one usually expects that the greatest conjugate effect should be in the case when both resistances are of the same order, and the value of Bi is close to unity; (vi) although in limiting cases the problem becomes one-dimensional, it remains to be the conjugate, because in these cases, except the case (ii) when $Bi_* \to \infty$, the resistances are large or small but finite (Exercises 5.9 and 5.10).

In what follows we will see that the results of the outlined analysis are supported by a number of conjugate problems, which are taken into consideration in Chapter 6 and in applications

Comment 5.1 In deriving equations (1.26) and (5.1), we apply only the second term with the first derivative for heat flux instead of the full universal function (4.1). Such simplification is possible in this case because the effect of temperature head gradient basically is determined by the second term due to relatively large coefficient g_1 and a small contribution in nonisothermicity effect of the other terms: the first term ($h_*\theta_w$) defines the heat flux on an isothermal surface, and all other coefficients g_k (except g_1) are small.

Comment 5.2 Some authors use other criteria for characterizing the relation of body/ fluid thermal resistances. For example, Luikov in an early work suggested Brun number $Br = (\Delta/x)(\lambda/\lambda_w)Pr^{1/3} Re_x^{1/2}$ or later Cole proposed the criterion $(\lambda/\lambda_w)(Pe)^{1/3}$. In fact, both these criteria are Biot numbers (Exercise 5.11).

5.3 Gradient Analogy

The term *gradient analogy* stands for similar effects of external velocity gradient on friction coefficient and temperature head gradient on heat transfer coefficient. It is well known that positive external velocity gradient (negative pressure gradient) leads to increasing friction coefficient and opposite velocity gradient results in decreasing friction coefficient in comparison with that for zero gradient. As we have seen in previous discussions and relevant examples, the temperature head gradient affects analogously the heat transfer coefficient, increasing or lessening it when compared to an isothermal value when the temperature head grows (positive gradient) or decreases. Both dynamical and thermal effects significantly differ in the cases of positive and negative gradients. While in the first case both coefficients increase relatively slightly, in the second one, they intensively fall reaching zero if the gradient is strong enough or if the surface is sufficiently long.

These qualitative reasonings are confirmed by universal functions (2.128) and (4.13) in the forms of nonisotachicity $\chi_f = c_f / c_{f*}$ and nonisothermicity $\chi_t = h / h_*$ coefficients. Comparing equations (2.128) and (4.13), one concludes that both relations are similar, and for each dynamic term with derivative of pressure P_w (which is proportional to U^2), in the first relation, there is a corresponding and similar thermal term with derivative of temperature head θ_w in the second. Both first coefficients b_1 and g_1 are positive, and, therefore, positive or negative gradient results in growing or lessening of friction and heat transfer coefficients. The same effect holds for unsteady cases yielding growing or lessening heat transfer coefficient depending on positive or negative space and first time derivatives because both coefficients g_{10} and g_{01} in universal function (4.33) are positive.

It is known that the unlikely influence of the favored and adverse velocity gradients on the friction coefficient is a result of different velocity profile deformation in dynamic boundary layer in those cases. Similarly, the different effects of increases and decreases in temperature head on the heat transfer coefficient are a result of different temperature profile deformations in a thermal boundary layer.

Physically, the profile deformation occurs due to fluid inertness. Consider first the case when the surface temperature is higher than the fluid temperature. If the wall temperature increases in the flow direction (favored gradient), the descended layers of fluid of the adjoining wall come in contact with an increasingly hotter wall. Because of the fluid inertness, these layers warm up gradually. As a result, the cross-sectional temperature gradients near a wall turn out to be greater than in the case of constant wall temperature, which leads to higher heat transfer coefficients than corresponding isothermal coefficients. Analogously, in the case of decreasing surface temperature, the cross-sectional temperature gradients near a wall and the heat transfer coefficients as well become lesser than those for an isothermal surface. The same situation exists in the case of cooler than fluid surface temperature. The difference is only that in this case, the absolute values of the falling temperature head and lesser heat transfer coefficients correspond to increase in the flow direction wall temperature, and inversely, the growing absolute values of the temperature head and higher heat transfer coefficients correspond to decrease in the flow direction wall temperature.

Considering the second terms of the series for χ_t. and for χ_f, one concludes that since coefficients g_2.and b_2 are negative, the effect of the second terms, which depends on the curvature of the $\theta_w(\Phi)$ and $P_w(\Phi)$ curves, is opposite: a positive curvature leads to a reduction of the friction and heat transfer coefficients, and a negative curvature yields an increase in these coefficients under otherwise equal conditions.

Comment 5.3 This case is more complicated, because the change in curvature also yields changes in gradients, and the result of comparing depends on a concrete situation. For example, if we compare a nearly linear convex and concave $\theta_w(\Phi)$ and $P_w(\Phi)$ curves, we arrive at an opposite result: the friction and heat transfer coefficient turn out to be smaller in the first case (negative curvature) and larger in the second (positive curvature). The reason for this is that changes in gradients are more significant.

It is obvious from equations (2.128) and (4.13) that the effect of the third, fifth, and other odd derivatives is of the same nature as that of the first, while the effect of even derivatives is of the same nature as that of the second. This follows from the fact that all odd coefficients of both series are positive, while all even coefficients are negative.

In the case of unsteady heat transfer, the effect of the second derivatives is the same as in the case of steady heat transfer, because in this case as well the second coefficients g_{20} and g_{02} are negative. The other coefficients in equation (4.33) (g_{k0} and g_{0i}) are positive for odd and negative for even numbers, while coefficients g_{ki} are positive if $(k+i)$ is odd and negative if $(k+i)$ is even. According to the sign of coefficients g_{ki}, the derivatives of higher order influence the intensity of heat transfer similar to the first or the second derivatives.

Although equations (2.128) and (4.13) have similar form, there is significant difference between them: the coefficients g_k in equation (4.13) are either practically constant or depend weakly on Pr and β, but they do not depend on the temperature head; while the coefficients b_k of universal function (2.128) determining the effect of free stream velocity are functions on the same free stream velocity (on β). This dependence of series coefficients is a result of principle difference between nonlinear dynamic and linear thermal boundary layer problems, and this is the reason why the universal function (4.13) is the solution of linear thermal boundary layer equation, while the function (2.128) is the solution of artificial linear (linearized) dynamic boundary layer equation (Exercise 5.13).

5.4 Heat Flux Inversion

It is well-known that at negative free stream velocity gradient, the local friction coefficient intensively decreases and becomes zero, resulting in boundary layer separation. We analyzed some examples of flow with separation in Chapter 2. Here, we discuss the similar phenomenon existing in thermal boundary layer that is known as the heat flux inversion. It occurs in a similar case of negative temperature head when according to our previous analysis the heat transfer coefficient sharply goes down finally becoming zero.

Physically, both phenomena the flow separation and the heat flux inversion, are results of velocity or temperature field deformation. The schematic patterns of the deformation of the excess temperature $\theta = T_w - T$ in laminar and turbulent boundary layers for the linear temperature head decreasing are shown in Figure 5.2 (Exercise 5.14). These are analogous to well-known patterns of the velocity profiles deformation in dynamic boundary layer. It is seen how a usual initial temperature profile deforms into a profile with an inflection point and then converts into a profile with a vertical tangent at a wall. Although the temperature head is finite at this point, the local heat flux vanishes and changes its direction. The coordinate of this point, where $\chi_t = 0$ and the inversion of heat flux occurs, is obtained from equation (4.13), which for the case of linear temperature head $\theta_w / \theta_{wi} = 1 - x/L$ becomes $\chi_t = 1 - g_1 \bar{x}/(1 - \bar{x})$. Equating this expression to zero, one gets $(x/L)_{inv} = 1/(1 + g_1)$, which for laminar ($g_1 = 0.62$) and turbulent ($g_1 = 0.2$) boundary layers gives $(x/L)_{inv} = 0.62$ and $(x/L)_{inv} = 0.834$ respectively (Exercise 5.15).

Comment 5.4 An inflection point is a point on a curve at which a curvature sign changes from being convex to concave or vice versa. The vertical tangent at the wall indicates that the heat flux at the wall is zero.

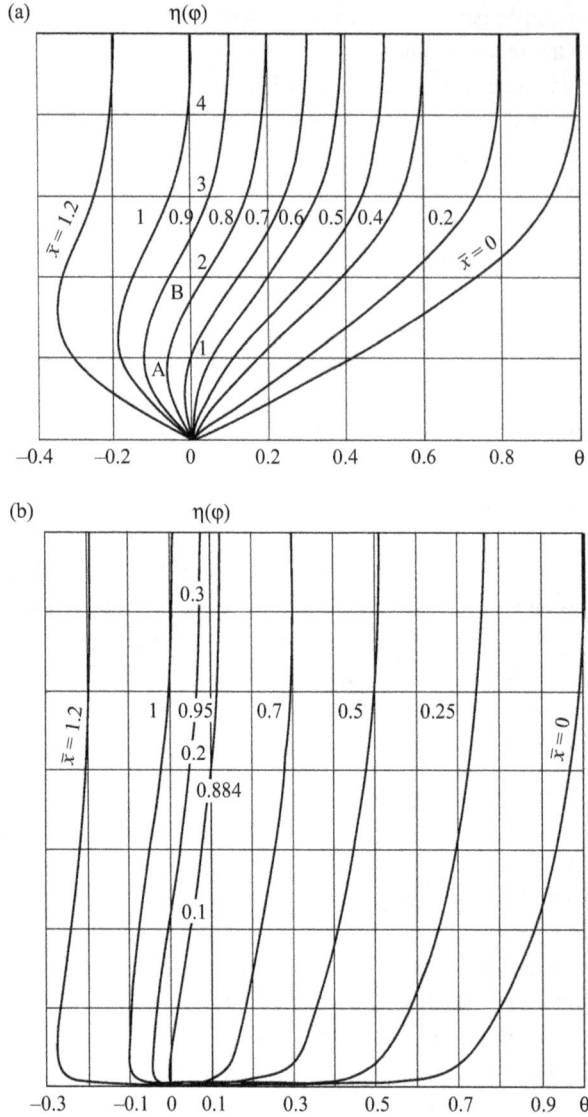

Figure 5.2 Deformation of the excess temperature profile for
the linear decreasing temperature head $\theta_w/\theta_{wi} = 1 - \bar{x}$, with
$\bar{x} = K(x/L)$ and Pr = 0.7, for laminar (a) and turbulent
$\text{Re}_{\delta_1} = 10^3$ (b) boundary layer.

The deformation-profile pattern for the temperature shown in Figure 5.2 is analogous to a well-known deformation-profile pattern of the velocity, which leads to a separation of the boundary layer. Nevertheless, these phenomena are radically different. Separation leads to restructuring of the flow, to the appearance of a reverse flow, and to the actual destruction of the boundary layer, so that boundary layer equations are no longer valid beyond the separation point. In contrast with this situation, the thermal boundary layer equations remain valid beyond the point of zero heat flux because only the direction of the heat flux changes at this point, but the hydrodynamics remains the same. Beyond the inverse point of the heat flux in the region before the point $\bar{x} = 1$, at which the temperature head vanishes, the heat flux is directed from the liquid to the wall, although the wall temperature is higher than that of the fluid outside of the boundary layer.

We present a physical analysis [1] of this deformation process for the same situation as discussed in the previous section when the surface temperature is higher than that of fluid, but for opposite negative temperature head gradient that specifies the case of heat flux inversion. Because the hot wall temperature decreases, the descended layers of fluid of the adjoining wall come in contact with a cooler wall. As a result, the temperature difference between the wall and the layers of fluid near the wall decreases and finally becomes zero at the inverse point with coordinate $\bar{x}_{inv} = 1 / (1 + g_1)$. After this point, the temperature of the fluid near the wall turns out to be above the wall temperature because the wall temperature continues to decrease. Thus, before the inverse point, the heat flux is directed from the wall to the fluid, while after this point, close to the surface the heat flux direction changes, so that the heat flux near the wall up to the point B in Figure 5.2 (a) is directed from the fluid to the wall. Because of this, the boundary layer near the wall is divided vertically by point A, at which the heat flux vanishes, into two regions. In the region adjacent to the wall, the heat flux is directed toward the wall, and in the other region it is directed away from the wall.

At the end of this region, at the point $\bar{x} = 1$, the flow temperature outside the boundary layer and the surface temperature become equal, and the temperature head vanishes. Nevertheless, the heat flux at this point does not vanish, so the concept of the heat transfer coefficient becomes meaningless, as it was first indicated by Chapman and Rubesin [2]. At point $x/L = 1$, the temperature head changes its direction, and the heat flux, which changed the direction before at point $(x/L)_{inv}$, continues to increase to infinity as $x \to \infty$ (Exercise 5.16).

Comment 5.5 The pattern of profile deformation described previously could not be explained in terms of common relation of heat flux and temperature head proportionality. In particular, at the point of inversion, where the heat flux is zero, the temperature head does not vanish, and vice versa; later at the point where the temperature head becomes zero the heat flux is finite. Because of that the concept of the proportionality of heat flux to temperature head and the definition of the heat transfer coefficient become meaningless.

5.5   Zero Heat Transfer Surfaces

In Chapter 2, we considered so-called preseparation flows providing relatively small energy losses in the case of increasing pressure due to small negative velocity gradients (Example 2.17). The shape of such devices is designed so that in each section the flow is close to separation. This provides small friction coefficients and, as a result, low total losses.

Analogous results of small total heat losses may be obtained employing the decreasing temperature head that provides small heat transfer coefficients becoming zero at the point of inversion. Moreover, in this case it is possible to specify a zero heat transfer coefficient in each section of flow because in contrast to separation the heat flux inversion is not destroyed by the flow structure. On the surface with such temperature distribution, the fluid layer adjoining the wall becomes nonconducting and serves as an insulating layer. The insulating effect is achieved because at the point of inversion the temperature across the fluid layer adjoining wall is constant. As a result, the heat flux becomes zero in the normal to the wall direction (Comment 5.4).

In the self-similar flows, the heat flux becomes zero in the case when the exponent of the power law temperature head is $m_1 = (1-m)/2$ (Example 2.14). Assuming that in a general case the zero heat flux is achieved also at power law temperature head with unknown exponent s, and using the integral form of universal function (4.14) yields

$$\theta_w = K\left(\frac{\Phi}{\text{Re}}\right)^s, \quad q_w = h_* \left[\frac{Ks(\Phi/\text{Re})^s}{C_1} \int_0^1 (1-\sigma)^{-C_2} \sigma^{(s/C_1)-1} d\sigma + \lim_{\Phi \to 0} K\left(\frac{\Phi}{\text{Re}}\right)^s\right] \quad (5.2)$$

where $\sigma = (\xi/\Phi)^{C_1}$. Since $C_1 > 0$ and $0 < C_2 < 1$, integral (5.2) in the case of $s > 0$ converges and can be expressed by beta function. For $s < 0$, integral (5.2) diverges. Considering this integral as an improper integral, one gets (Exercise 5.17)

$$q_w = h_* \left[\frac{Ks(\Phi/\text{Re})^s}{C_1} \lim_{\zeta \to 0} \int_\zeta^1 (1-\sigma)^{-C_2} \sigma^{(s/C_1)-1} d\sigma + \lim_{\Phi \to 0} K\left(\frac{\Phi}{\text{Re}}\right)^s\right] \quad (5.3)$$

As $\sigma \to 0$ the integrand tends to $\sigma^{(s/C_1)-1}$. Hence, since $s < 1$ (see below) this expression is also finite as $s < 0$ and after using equation (4.15) for the beta function becomes

$$q_w = \frac{q_{w*}s}{C_1} \frac{\Gamma(1-C_2)\Gamma(s/C_1)}{\Gamma(1-C_2+s/C_1)} \qquad \theta_w = K\left(\frac{\Phi}{\text{Re}}\right)^{-C_1(1-C_2)} \quad (5.4)$$

It can be seen from this equation that condition $q_w = 0$ may be achieved only if $\Gamma(1-C_2+s/C_1) = \pm\infty$ resulting in $1-C_2+s/C_1 = 0$ and finally in $s = -C_1(1-C_2)$ (Exercise 5.18). So, the zero heat transfer surface exists in general if the temperature head decreases according to the power law with this exponent according to equation (5.4).

This result is valid for laminar and turbulent flows as well as for other cases if the influence function has the form (2.135) with exponents C_1 and C_2. Using data from Figure 4.3, one estimates that for laminar flow the exponent $-C_1(1-C_2)$ in equation (5.4) is practically independent on β and Pr and equals $(-1/2)$ (Exercise 5.19).

The temperature distribution [equation (5.4)] is difficult to implement in reality because according to equation (5.4) the surface temperature is infinite at the starting point where $\Phi \to 0$ [3]. Therefore, only laws close to equation (5.4) can be realized. For instance, a law that gradually approximates the equation (5.4) as the distance from the starting point grows

$$\theta_w = K\left(K_1 + \frac{\Phi}{\text{Re}}\right)^s \quad q_w = q_{w*}\left\{(1-z)^s + s\int_0^z \left[1 - \left(\frac{\zeta}{z}\right)^{C_1}\right]^{-C_2} \frac{d\zeta}{(1-z+\zeta)^{1-s}}\right\} \quad (5.5)$$

Here $\zeta = \xi/(K_1 + \Phi/\text{Re})$ and z is the value of ζ when $\xi = \Phi$. As $\Phi \to \infty$, the last expression becomes equation (5.4) and heat flux becomes zero.

Comment 5.6 These results may be used to minimize heat losses in a car or other moving objects if the heat sources inside a wall are distributed so that heat releases according to equation (5.5). In such a case total heat remains inside the car because the outer side of the wall is practically isolated.

5.6 Examples of Optimizing Heat Transfer in Flow Over Bodies

According to the universal functions, the heat transfer rate depends not only on heat transfer coefficient, but also on the distribution of the temperature head. Therefore, by appropriate selection of the temperature head distribution, one can ensure that a given heat transfer system satisfies the desired conditions, for example, the maximum or minimum of heat transfer rate.

In the general case of gradient flow over a flat, arbitrary nonisothermal surface, the heat flux and temperature head are related by equations (4.14) and (4.28). Mathematically, the problems of optimum heat transfer modes reduce to finding a function $\theta_w(x)$ or $q_w(x)$ corresponding to the extreme of either integral (4.14) or (4.28) or quantities that are their functions, for example, the total heat flux. The solution to such problem in general is difficult and becomes more complicated if the conjugate problems should be considered. It is much simpler to use equation (4.14) or (4.28) for comparing different outcomes and selecting the best. Although this approach does not utilize all capabilities of optimization, the following examples show that the results are useful for practical applications.

Example 5.2 There are several heat sources or sinks, for instance, electronic components with linear varying strengths. How should these be arranged on a plate so that the maximum temperature of the plate would be minimal?

a) Zero pressure gradient flow.
For linear varying heat flux $q_w = K + K_1(x/L)$, in this case equation (4.28) becomes

$$\theta_w = \frac{1}{h_*}(KI_1 + K_1I_2), \quad I_i = \frac{C_1}{\Gamma(C_2)\Gamma(1-C_2)}\int_0^1(1-\zeta^{C_1})^{C_2-1}\zeta^{C_1(1-C_2)+n+i-2}d\zeta \quad (5.6)$$

where $\zeta = \xi/x$ and n is exponent in the expression $\mathrm{Nu}_{x*} = C\mathrm{Re}_x^{1-n}$. This integral can be expressed by using gamma functions like other similar integrals above (Exercise 5.20)

$$I_i = \frac{\Gamma[1-C_2+(n+i-1)/C_1]}{\Gamma(1-C_2)\Gamma[1+(n+i-1)/C_1]} \quad (5.7)$$

Coefficients K and K_1 in the relation for heat flux are expressed through the total heat flux Q_w and difference $\Delta q = q_{max} - q_{min}$, which should be the same for compared cases. Then, the relation for temperature head may be presented in the form (Exercise 5.21)

$$\theta_w = \frac{1}{h_*}\left\{I_1q_{av} \pm \Delta q\left[\frac{1}{2}I_1 - I_2\frac{x}{L}\right]\right\} \quad (5.8)$$

where $q_{av} = Q_w/BL$ is the average heat flux, B is the width of the plate, and plus sign is assigned to decreasing heat flux and the minus sign to the increasing heat flux along the plate.

For gradientless flow $h_* \to \infty$ as $x \to 0$. Hence, the temperature head at the beginning becomes zero, and therefore as it follows from equation (5.8) in the case of increasing heat fluxes the maximum temperature head is located at the end of the plate, at $x \to L$. This is not obvious for decreasing heat fluxes. Differentiating equation (5.8), taking into account that

$h_*(x) = h_*(L)(x/L)^{-n}$, and equating the result to zero, one finds the point coordinate at which the temperature head is maximum if heat flux decreases

$$x_m / L = \frac{I_1 n[1 + 0.5(\Delta q/q_{av})]}{I_2 (1+n)(\Delta q/q_{av})} \qquad (5.9)$$

Setting in this expression $x_m = L$, one finds the limiting value

$$\left(\frac{\Delta q}{q_{av}}\right)_{lim} = \frac{I_1 n}{I_2(1+n) - 0.5 I_1 n} \qquad (5.10)$$

For all ratios $\Delta q/q_{av}$ lower than the limiting, the maximum temperature head is located at the end of the plate, and for all those higher than the limiting, at $x < L$ (Exercise 5.22).

In the case of laminar flow at $\Pr \geq 1: C_1 = 3/4, C_2 = 1/3$ (Fig. 4.3), and $n = 1/2$; in the case of turbulent flow at $\Pr \approx 1: C_1 = 1, C_2 = 0.18$ at $Re_{\delta_1} = 10^3$ ($Re = 5 \cdot 10^5$), $C_1 = 0.84, C_2 = 0.1$ at $Re_{\delta_1} = 10^5$ ($Re = 10^8$) (Figs. 4.20 and 4.21) and $n = 1/5$; for laminar wall jet at $\Pr \approx 1: C_1 = 0.42$, $C_2 = 0.35$, and $n = 3/4$. At these values of exponents, equation (5.6) yields the following: for laminar flow $I_1 = 0.73$ and $I_2 = 5/9$; for turbulent flow $I_1 = 0.93$ and $I_2 = 0.8$ at $Re_{\delta_1} = 10^3$; $I_1 = 0.96$ and $I_2 = 0.88$ at $Re_{\delta_1} = 10^5$; for laminar wall jet $I_1 = 0.55$ and $I_2 = 0.43$ [4] (Exercise 5.23).

Calculation results obtained from equation (5.8) using these values of integrals are plotted in Figure 5.3. In Figure 5.3, the Percentage value of difference $\Delta \theta_{w.max} = |\theta_{w.max.in} - \theta_{w.max.de}|$ of the maximum temperature heads in the cases of increasing $\theta_{w.max.in}$ and decreasing $\theta_{w.max.de}$ heat fluxes, referred to the higher of these two is given, vs. $\Delta q/q_{av}$. If the strength of sources or sinks is

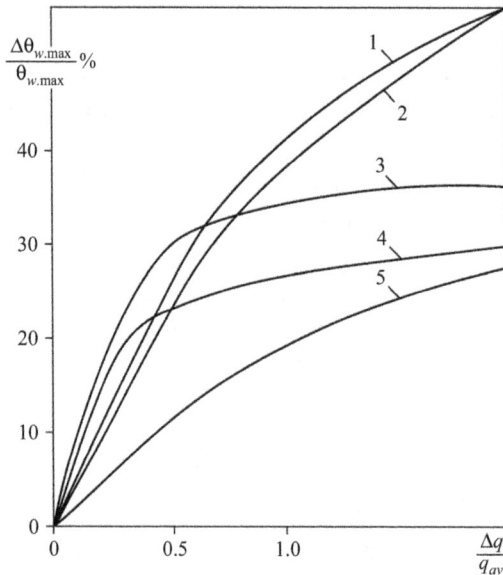

Figure 5.3 Difference between maximum temperature heads under linear decreasing and increasing heat fluxes: 1-laminar wall jet, 2-laminar flow, 3 and 4-turbulent flow $Re = 10^8$ and $Re = 5 \cdot 10^5$, 5-stagnation point.

known, the value of ratio $\Delta q/q_{av}$ may be calculated, and by using Figure 5.3, one can estimate the difference of the maximum temperature heads in the cases of increasing and decreasing heat fluxes.

According to these data the maximum of the temperature head in the first case is greater than in the second, and it can be seen from this fact that the strength of sources that encountered a cold coolant should decrease in the direction of flow, while the strength of the sinks encountered by a hot flow should increase in the flow direction. It can be seen from Figure 5.3 that by an appropriate positioning of the sources or sinks on the plate one can significantly decrease the maximum surface temperature without decreasing the heat flux. It is seen that, in this case, the profit is approximately proportional to ratio $\Delta q/q_{av}$ at small values of this parameter and reaches about 30% at $\Delta q/q_{av} \approx 0.5$ and continues increasing for laminar flow, remaining almost the same for turbulent flow as the $\Delta q/q_{av}$ increases.

b) The flow near stagnation point
The free stream velocity in this case is proportional to coordinate $U = Cx$. The heat transfer coefficient from isothermal surface is independent of x ($n = 0$), and Görtler variable is $\Phi = Cx^2/2$. Taking this into account, transforming the linear dependence for heat flux to variable Φ and substituting the result into equation (5.6) for I_i gives again equation (5.8). For integrals I_i the same formula as in equation (5.7) is valid if $(n+i-1)$ is replaced by $(n-1)/2$. Using this formula and neglecting the slight dependence of C_1 and C_2 on the pressure gradient, one finds $I_1 = 1$ and $I_2 = 0.73$. Equation (5.8) then becomes

$$\theta_w = \frac{1}{h_*}\left\{ q_{av} \pm \Delta q \left[\frac{1}{2} - 0.73(x/L) \right] \right\} \tag{5.11}$$

Since in this case h_* is independent of x, the temperature head under decreasing heat fluxes is maximum at $x = 0$, whereas in the case of increasing heat flux the temperature head is the highest at $x = L$. Unlike the previous case of a zero pressure gradient, the maximum temperature head at increasing heat flux is smaller than that at decreasing heat flux. The corresponding relationship for the relative absolute value of difference in maximum temperature heads is plotted in Figure 5.3, from which it is seen that the difference is smaller than in the case of a plate. It is obvious that in this case in order to reduce the maximum surface temperature the strength of the sources should increase in the direction of flow in the case of a cold coolant, while the strength of sinks should decrease in the direction of a hot flow (Exercise 5.24).

Example 5.3 Suppose the maximum allowable surface temperature is given. Find the mode of change of the temperature head at which the quantity of the heat removed (or supplied) from the surface is maximum.
   a) Zero pressure gradient.
       The sought distribution of the temperature head is approximated by a quadratic polynomial. To determine the heat flux, it is reasonable to use the differential form of universal function (4.1) because in this case only two terms are retained in a series

$$\theta_w = a_0 + a_1 (x/L) + a_2 (x/L)^2$$

$$q_w = h_*\left[a_0 + a_1 (1+g_1)(x/L) + a_2 (1+2g_1+2g_2)(x/L)^2 \right] \tag{5.12}$$

Integrating this equation after using $h_* = h_*(L)(x/L)^{-n}$, yields the total heat flux

$$Q_w = h_* LB \left[\frac{a_0}{1-n} + \frac{a_1(1+g_1)}{2-n} + \frac{a_2(1+2g_1+2g_2)}{3-n} \right] \qquad (5.13)$$

Two cases are considered

1. The maximum temperature head is located at the beginning or at the end of the plate. In this case the first equation (5.12) has one of two forms:

$$\theta_w = \theta_{w.max} + a_1(x/L) + a_2(x/L)^2$$

$$\theta_w = \theta_{w.max} + a_1[1-(x/L)] + a_2[1-(x/L)]^2 \qquad (5.14)$$

Because the first terms present the maximum value of the temperature head, the sum of the two other terms should be negative at any x in the range $0 \le x \le L$.

$$a_1(x/L) + a_2(x/L)^2 \le 0, \quad a_1[1-(x/L)] + a_2[1-(x/L)]^2 \le 0 \qquad (5.15)$$

If $a_2 \le 0$ then in order to satisfy the first of these inequalities at $x \to 0$ and the second at $x \to 1$, it is necessary that $a_1 \le 0$. It then can be seen from equation (5.13) that the maximum heat flux is attained at $a_1 = a_2 = 0$ and $a_0 = \theta_{w.max}$, that is, when heating the plate uniformly to the temperature equal to the specified maximum temperature. If $a_2 > 0$, then $a_1 < 0$. Satisfaction of the first inequality (5.15) at $x = L$ and the second at $x = 0$ requires $|a_1| > a_2$. It easy to check that under these conditions the sum of the two last terms in equation (5.13) for laminar (e.g., $Pr \approx 1$, $g_1 = 0.62, g_2 = -0.135$, and $n = 1/2$) or turbulent (e.g., $Pr \approx 1$, $Re_{\delta_1} = 10^3$, $g_1 = 0.2$, $g_2 = -0.05$ and $n = 1/5$) flow at zero pressure gradient is negative, and hence the total heat flux is again maximum at $a_1 = a_2 = 0$ and $a_0 = \theta_{w.max}$.

2. The maximum temperature head occurs at $0 < x < L$. In this case at $a_2 < 0$, the parabola [Equation (5.12)] is open down toward the negative ordinate. Therefore, after changing the sign to a minus at the last term in equation (5.12), only the case with positive a_2 should be studied. In such case, the coordinates of parabola vortex in the frame $\theta_w(x)$ are determined as $x_m/L = a_1/2a_2$ and $\theta_{wmax} = a_0 + a_1^2/4a_2$. From these equations, one finds a_0 and obtains instead of equation (5.13), the following expression

$$Q_w = h_* LB \left[\frac{\theta_{w.max}}{1-n} + a_2 F \left(\frac{a_1}{a_2} \right) \right],$$

$$F \left(\frac{a_1}{a_2} \right) = -\frac{1}{4(1-n)} \left(\frac{a_1}{a_2} \right)^2 + \frac{1+g_1}{2-n} \frac{a_1}{a_2} - \frac{1+2g_1+2g_2}{3-n} \qquad (5.16)$$

The discriminant of the last quadratic trinomial which is defined as

$$\Delta = \frac{1+2g_1+2g_2}{(1-n)(3-n)} - \left(\frac{1+g_1}{2-n} \right)^2 \qquad (5.17)$$

is positive for laminar, turbulent, and laminar wall jet over the plate, which means that trinomial [Equation (5.16)] has no roots, and because this trinomial is negative at $a_1/a_2 = 0$, it follows that $F(a_1/a_2)$ is negative at all ratios a_1/a_2. Then, according to the first equation (5.16), heat flux is maximum again at $a_1 = a_2 = 0$ (Exercise 5.25).

Thus, in the previous cases the largest heat flux is removed if the plate temperature is uniform and equal to the specified maximum temperature. At first sight this conclusion appears obvious. In fact this is not true because the result of the analysis depends on the relationship between coefficients g_1 and g_2, which govern the effect of nonisothermicity, and exponent n in the expression for the heat transfer coefficient on an isothermal wall. Two examples in which a nonuniform distribution of temperature head is optimal are given below (Exercise 5.26).

b) The flow near stagnation point
Because in this case $\Phi = Cx^2/2$, it is convenient using for heat flux equation (4.14) in the form of integral. Formula similar to equation (5.7) gives integrals $I_1 = 2.9$ and $I_2 = 1.62$. Knowing that for stagnation point $n = 0$, one gets, instead of relation (5.16), the following trinomial

$$F\left(\frac{a_1}{a_2}\right) = -\frac{1}{4}\left(\frac{a_1}{a_2}\right)^2 + 0.73\frac{a_1}{a_2} - 0.533 \tag{5.18}$$

The discriminant of this trinomial is $\Delta \approx 0$, and the root is 1.46. Therefore, the first equation (5.16) shows that $Q_{w.\max} = h_*\theta_{w.\max}$ is attained at any a_2, as long as $a_1/a_2 = 1.46$ giving $F(a_1/a_2) = 0$ and corresponding distributions of the temperature head and heat flux (Exercise 5.27).

$$\theta_w = \theta_{w.\max} - a_2\left[0.53 - 1.46(x/L) + (x/L)^2\right]$$
$$q_w = h_*\left\{\theta_{w.\max} - a_2\left[0.53 - 2.12(x/L) + 1.62(x/L)^2\right]\right\} \tag{5.19}$$

Thus, the same maximum heat flux that is transformed in the case of uniform heating of the plate to a maximally permissible temperature can be transferred in the case of non-uniform heating provided the temperature which is lower than maximally permitted in all the points of the plate except one. The physical explanation for this is that heat transfer coefficients in the case of increasing temperature head are higher than in the case of isothermal plate, and this compensates for the decrease in the temperature heads. Figure 5.4 shows distributions (Equation 5.19) for several values of $a_2/\theta_{w.\max}$. The minimum temperature head is zero at the beginning in the case of the highest value of this ratio.

c) The jet wall flow at low Prandtl numbers.
In this case, $C_1 = 4$, $C_2 = 0.03$, and $n = 3/4$ [4]. With these values, the sum of the two last terms in equation (5.13) becomes zero at any value of $a = -a_1 = a_2$. Hence, if $a_0 = \theta_{w.\max}$, then the transferred heat flux is maximum for the following distribution of temperature head and corresponding distribution of heat flux (Exercise 5.28).

$$\theta_w = \theta_{w.\max} - a(x/L)[1 - (x/L)] \qquad q_w = h_*\{\theta_{w.\max} - a(x/L)[5 - 9(x/L)]\} \tag{5.20}$$

Figure 5.4 (b) shows these distributions for several values of $a/\theta_{w.\max}$. For the case with the highest value of this ratio the minimum temperature head at $x = L/2$ is equal to zero.

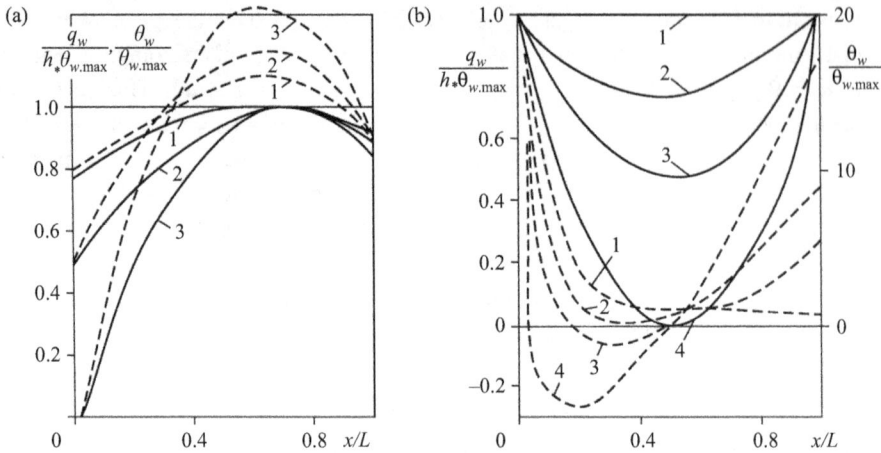

Figure 5.4. Different distributions of the temperature head (solid curves) and the corresponding heat fluxes distributions (dashed curves) providing the same total heat removed from surface a) stagnation point $a_2/\theta_{w.max}$: $1-0.4$, $2-1.0$, $3-1.88$ b) jet wall: $1-0$, $2-1.0$, $3-3$, $4-4$.

Example 5.4 It is required to remove (supply) from the plate a heat flux Q_w. Find a heat flux pattern that minimizes the maximum plate temperature.

The desired heat flux distribution is approximated by a quadratic polynomial. Integrating this expression gives the value of heat flux for uniform distribution

$$q_w = a_0 + a_1(x/L) + a_2(x/L)^2 \quad q_{w.un} = \frac{Q_w}{BL} = a_0 + \frac{a_1}{2} + \frac{a_2}{3} \tag{5.21}$$

The expression for the temperature head is found as before by equation (4.28)

$$\theta_w = \frac{1}{h_*}[a_0 I_1 + a_1 I_2(x/L) + a_2 I_3(x/L)^2] \tag{5.22}$$

Integrals I_i are given by equation (5.7). Equating the derivative of θ_w to zero and taking into account that $h_* = h_*(L)(x/L)^{-n}$, one obtains an equation for coordinates where the temperature is maximum (Exercise 5.29).

$$\left(\frac{x_m}{L}\right)^2 + \frac{I_2 a_1(n+1)}{I_3 a_2(n+2)}\left(\frac{x_m}{L}\right) + \frac{I_1 n}{I_3(n+2)}\left(\frac{q_{w.un}}{a_2} - \frac{a_1}{2a_2} - \frac{1}{3}\right) = 0 \tag{5.23}$$

Two cases are considered

1. Equation (5.23) has no roots. In this case the following inequality is valid

$$\left[\frac{I_2 a_1(n+1)}{2I_3 a_2(n+2)}\right]^2 - \frac{I_1 n}{I_3(n+2)}\left(\frac{q_{w.un}}{a_2} - \frac{a_1}{2a_2} - \frac{1}{3}\right) < 0 \tag{5.24}$$

The maximum temperature head is in this case attained at the end of the plate at $x = L$ and according to equations (5.21) and (5.22) can be represented as

$$\frac{\theta_{w.\max}}{\theta_{w.\max.un}} = 1 - \frac{a_2}{q_{w.un}}\left[\frac{a_1}{a_2}\left(\frac{I_2}{I_1} - \frac{1}{2}\right) + \frac{I_3}{I_1} - \frac{1}{3}\right] \tag{5.25}$$

Differentiating this equation with respect to a_1/a_2, substituting $q_{w.un}/a_2$ derived from condition (5.24), and equating the derivative to zero lead to the equation

$$\frac{n+2}{n}\left(\frac{I_2}{I_1} - \frac{1}{2}\right)\frac{I_3}{I_1}\left[\frac{I_2(n+1)}{2I_3(n+2)}\right]^2\left(\frac{a_1}{a_2}\right)^2 + 2\left(\frac{I_3}{I_1} - \frac{1}{3}\right)\frac{(n+2)}{n}\frac{I_3}{I_1}\left[\frac{I_2(n+1)}{2I_3(n+2)}\right]^2$$

$$\times \frac{a_1}{a_2} + \frac{1}{2}\frac{I_3}{I_1} - \frac{1}{3}\frac{I_2}{I_1} = 0 \tag{5.26}$$

The optimum a_1/a_2 is determined from equation (5.26). The results of calculations are listed in Table 5.2. It is seen that an appropriate selection of the distribution of the removed heat flux allows one to significantly reduce the maximum temperature, particularly in the case of a wall jet flow, in which this decrease may exceed 30%. The corresponding heat flux distribution in this case according to Table 5.2 is

$$q_w = q_{w.un}[1.95 - 2.85(x/L) + 1.45(x/L)^2] \tag{5.27}$$

2. The equation (5.23) has roots. It follows from equation (5.21) that

$$q_w = a_2\left[\frac{q_{w.un}}{a_2} - \frac{1}{2}\frac{a_1}{a_2} - \frac{1}{3} + \frac{a_1}{a_2}(x/L) + (x/L)^2\right] \tag{5.28}$$

To make sure that the heat flux does not change sign, the following expression should be satisfied

$$\frac{1}{2}\left(\frac{a_1}{a_2}\right)^2 - \left(\frac{q_{w.un}}{a_2} - \frac{1}{2}\frac{a_1}{a_2} - \frac{1}{3}\right) < 0 \tag{5.29}$$

Table 5.2 Comparison between maximum temperatures obtained by different distribution of removed heat flux

Kind of flow	I_1	I_2	I_3	n	a_1/a_2	$q_{w.un}/a_2$	$\theta_{w.\max}/\theta_{w.\max.un}$
Laminar flow	0.73	0.56	0.48	1/2	−2.14	1.18	0.79
Turbulent flow $Re_{\delta 1} = 10^3$	0.93	0.80	0.74	1/5	−2.31	3.23	0.88
Turbulent $Re_{\delta 1} = 10^5$	0.96	0.88	0.81	1/5	−2.38	3.33	0.86
Laminar wall jet	0.55	0.43	0.37	3/4	−1.97	0.69	0.67

Using equation (5.23), one obtains the inequality

$$\frac{1}{4}\left(\frac{a_1}{a_2}\right)^2 + \frac{I_3}{I_1}\frac{n+2}{n}(x_m/L)^2 + \frac{I_2}{I_1}\frac{a_1}{a_2}\frac{n+1}{n}(x_m/L) < 0 \qquad (5.30)$$

The above inequality should be satisfied if both conditions (5.24) and (5.29) are to be met. Obviously, at $a_1/a_2 > 0$ equation (5.30) is not satisfied at $x_m > 0$. This is not obvious in the case $a_1/a_2 < 0$. However, one can easily check by direct calculation that at $0 < x_m/L < 1$ equation (5.30) is not satisfied for the studying case also at $a_1/a_2 < 0$ (Exercise 5.30).

EXERCISES

5.1 What factors determine the effect of the temperature head on the heat transfer intensity according to the universal function?

5.2 How does universal function (4.1) tell us about different effects of decreasing and increasing temperature heads? Which signs determine whether the resulting nonisothermicity coefficient will be greater or lesser than unity?

5.3 Why does the temperature head gradient basically determine the intensity of nonisothermal heat transfer?

5.4 Explain why same temperature gradients result in different qualitative and quantitative changes in heat transfer intensity.

5.5 Describe how the flow regime and characteristic numbers affect the heat transfer intensity. What reasons caused those effects?

5.6 Derive an expression showing the influence of pressure gradient on the temperature head gradient. Prove that the ratio U_{av}/U is lesser or greater than unity depending on velocity gradient. Hint: present the ratio U_{av}/U using an integral and a product xU, and compare areas covered by each expression.

5.7 Obtain an equation (5.1) for flow with pressure gradient by applying the same approach used for deriving a similar equation (1.26).

5.8 Show that the quenching process is described by problem with $Bi_* \rightarrow \infty$ discussed in (ii) in analysis of equation (5.1).

5.9 Recall the term *lumped approach*, and find a case of this type among the cases considered in our discussion of equation (5.1).

5.10 There are two cases with uniform temperature head considered in analysis of equation (5.1): one at $Bi_* \rightarrow \infty$ and another at $Bi_* \rightarrow 0$. Compare the reasons of these results in both cases.

5.11 Show that the Brun number and criterion proposed by Cole are actually a Biot numbers. Hint: use the formula for Nusselt number. Note that the Cole relation did not consist of body thickness, and therefore it could not be a conjugate parameter. If the body thickness will be included in this criterion, it becomes Brun number since $Pe = PrRe$.

5.12 What does gradient analogy mean? How do the effects of positive and negative gradients differ? How do the other derivatives affect the values of the friction and heat transfer coefficients?

5.13 See details: compare equations (2.77) to understand the difference between nonlinear dynamic and linear thermal boundary layer equations. Then substitute in both equations self-similar velocity profile $\omega(\varphi, \beta)$ for real profile u/U to understand the difference between linearized dynamic and linear thermal equations. Note that in fact these modified equations are the same as equations (2.127) $(Z \sim P)$ and (4.2) (without right hand part), which were used to obtain the universal functions (2.128) and (4.1). Hint: think about connection between dependent variables in modified equations (2.77) and parameter β.

5.14 The patterns in Figure 5.2 are computed using equation (4.2) written for zero pressure gradient, linear decreasing temperature head, and parameters indicated in the caption. The result containing two first terms of series is as follows $\frac{T_w - T}{T_w - T_\infty} = \left[1 - G_0(\varphi) \right] \left(1 - K \frac{x}{L} \right) + G_1(\varphi) K \frac{x}{L}$ Functions G_0 and G_1 are found by the numerical integration of equation (4.5). Derive this equation.

5.15 Derive the expression for coordinate of heat flux inversion point using equation (4.13) for linear temperature head $\theta_w / \theta_{wi} = 1 - K(x/L)$ with $K = const$. And find the relation for coordinate of inversion point.

5.16 Using the drafts in Figure 5.2, explain the physical features of deformation temperature field outlined in text. In particular, estimate the heat flux direction in different regions between points A and B. Hint: moving along the curve $\bar{x} = const$. determine (recalling that on Figure 5.2 $\theta = T_w - T$) the direction in which the fluid temperature decreases.

5.17 *Recall or study, using Advanced Engineering Mathematics, the features of improper integrals to understand that integral (5.2) is improper. Determine the type of singularity of this integral and show when it converges or diverges. Derive the equation (5.2) from equation (4.14). Think why equation (5.2) is preferable.

5.18 Prove that both expressions in equation (5.4) follow from equation (5.2). Hint: take into account that $q_* = h_* K (\Phi / \mathrm{Re})^s$.

5.19 Calculate exponent $-C_1(1 - C_2)$ for several values of Pr and β to show that it slightly varies being about $(-1/2)$.

5.20 Obtain equations (5.6) and (5.7). Hint: in deriving equation (5.7), use a new variable ζ^{C_1} and property of gamma function

5.21 Derive equation (5.8) after expressing dependence $q_w = K + K_1(x/L)$ through total heat flux $Q_w = q_{av} BL$ and difference $\Delta q = q_{max} - q_{min}$.

5.22 Find equations (5.9) and (5.10), and explain the obtained conclusions.

5.23 Calculate integrals [equation (5.6)] for laminar flow at low Prandtl number and for turbulent flow at high Reynolds number. Use these results to obtain curves similar to those plotted in Figure 5.3, following directions given in the text.

5.24 Explain why in the cases of a plate and a stagnation point the strength of sources as well as strength of the sinks should be arranged in the opposite manners.

5.25 Recall the role of discriminant, and repeat the analysis of case 2 in Example 5.3.

5.26 Analysis in Example 5.3 shows that the largest heat flux is removed if the plate temperature is uniform and equal to the specified maximum temperature. Then, it is stated that this is not always true because in general the result depends on nonisothermicity effect and on the value of heat transfer coefficient. Explain physically why, from the last two facts, one may expect the existence of situations when other nonuniform distribution of temperature head is optimal.

5.27 Perform analysis of equation (5.18) leading to equation (5.19). Draw graphs 5.4 (a) with coordinates in which all curves coincide, and find the point coordinate corresponding to the function maximum.

5.28 Perform analysis of the case c) and obtain equation (5.20). Show that for the highest value of $a/\theta_{w.max}$, the minimum temperature head equals zero and is at $x = L/2$.

5.29 Obtain equation (5.23), and repeat an analysis of the case 1 when there are no roots in equation (5.23) to understand that the temperature maximum in this case is located at $x = L$.

5.30 What fact is proved in case 2 in analysis of equation (5.23)? Why is this proof necessary to perform?

REFERENCES

[1] Gröber, H., Erk, S., & Grigull, U. (1955). *Die Grengesetze der Warmeubetragung* (3rd ed.)

[2] Chapman, D., & Rubesin, M. (1949). "Temperature and velocity profiles in the compressible laminar boundary layer with arbitrary distribution of surface temperature. *Journal Aeronautical Science* 16, 547–65.

[3] Eckert, E. R. G., & Drake, R. M. (1959). *Heat and mass transfer*. McGraw-Hill

[4] Grechannyy, O. A., Nagolkina, Z. I., & Senatos, V. A., 1984. Heat transferring jet flow over an arbitrary nonisothermal wall. *Heat Transfer: Soviet Research* 16, 12–22.

CHAPTER 6

CONJUGATE HEAT TRANSFER IN
FLOW PAST PLATES

CHARTS FOR SOLVING CONJUGATE
HEAT TRANSFER PROBLEMS

In this chapter, we investigate conjugate heat transfer in flow past plates at different situations. The majority problems are considered for thermally thin plates when conduction equation (1.7) is valid. Such a simplification is justified because, as we will see, the results obtained in this way are qualitatively representative for exact data. Moreover, for the majority part of a plate, except a relatively small region at the leading edge, even quantitatively, the results are accurate. On the basis of such considerations, the charts and the simple method of solution of conjugate problems employing these charts are developed. Examples (1–3, 5–8) shown below illustrated the wide applicability of this method and comprehensible analysis of the results obtained.

6.1   Temperature Singularities on the Solid–Fluid Interface

In some early works ([13] and [14], in [3] p. 227), it was shown that in laminar flow, the wall temperature distribution on a thermally thin plate is not an analytic function near the leading edge; rather, it is presented at $x = 0$ as a series of variable $x^{1/2}$. In fact, as it follows from universal equation (4.1), the temperature distribution on the thermally thin plate at $x = 0$, in general, is not analytic of coordinate x but is presented in series in integer powers of variable $x^{1/s}$, where s is the denominator of exponent in relation for an isothermal heat transfer coefficient in the form $h_* = h_{*L} \left(x/L^{-r/s} \right)$.

Comment 6.1 We considered the analyticity of the function in Section 2.2.2.5 in connection with differentiability. An analytic function at some point x_0 may be presented in neighborhood of this point by Taylor series in integer powers of variable x. Otherwise, it is said that the function is singular at this point. We described this term in Section 2.5.2.2, considering the boundary layer equation in the Prandtl–Mises form. Most functions encountered in applications are analytic of all x or are singular at some specific points.

6.1.1 Basic Equations

Consider a thermally thin plate of finite length with given boundary conditions at the edges. For generality, it is assumed that the plate is streamlined past both sides by laminar or turbulent, gradient or gradientless flows of Newtonian or power-law non-Newtonian fluids. The flow may be symmetrical by one stream or asymmetrical by two different streams. Examine first the case when exponents r/s for both sides are the same (e.g. the same laminar or turbulent flows), but the stream temperatures and isothermal heat transfer coefficients h_{*L} are different. For such a case, substitution of equation (4.1) for a plate into equation (1.7) yields the following equation for the average plate temperature (Exercise 6.1):

$$\sum_{k=0}^{\infty} D_k \zeta^k \frac{d^k \theta}{d\zeta^k} - \zeta^{r/s} \frac{d^2 \theta}{d\zeta^2} - \mathrm{Bi}_{*L2} - \zeta^{r/s} \overline{q}_v = 0, \quad \theta = \frac{T_w - T_{\infty 1}}{T_{\infty 2} - T_{\infty 1}}, \quad \mathrm{Bi}_{*L} = \frac{h_{*L} L^2}{\lambda_w \Delta} \quad (6.1)$$

$$\overline{q}_v = \frac{q_{v.av} L^2}{\lambda_w (T_{\infty 2} - T_{\infty 1})} \quad D_0 = \mathrm{Bi}_{*L1} + \mathrm{Bi}_{*L2}, \quad D_k = g_{k1} \mathrm{Bi}_{*L1} + g_{k2} \mathrm{Bi}_{*L2}, \quad \zeta = x/L.$$

This equation remains valid for the case of equal fluid temperatures if the scale in definition of θ is changed to T_∞ and Bi_{*L2} in equation (6.1) is omitted. For the symmetric streamlined plate, one sets in addition $\mathrm{Bi}_{*L1} = \mathrm{Bi}_{*L2}$ and $g_{k1} = g_{k2}$.

The exponent r/s usually is not integer. Therefore, equation (6.1) has a singularity at point $x = 0$. If the source q_v near the origin is an analytic function of x, the solution of equation (6.1) at $x = 0$ can be presented as a series of the variable $x^{1/s}$:

$$\overline{q}_v = \sum_{i=0}^{\infty} b_i \zeta^i, \qquad \theta = \sum_{i=0}^{\infty} a_i \zeta^{i/s}. \qquad (6.2)$$

This follows from the fact that substituting series (6.2) into equation (6.1) gives equations for the series coefficients, a_i:

$$\sum_{i=0}^{\infty} [D_0 + D_1(i/s) + D_2(i/s)(i/s-1) + \cdots + D_k(i/s)(i/s-1) \cdots (i/s-k+1) + \cdots] a_i \zeta^{\frac{i}{s}}$$

$$-\sum_{i=0}^{\infty} (i/s)(i/s-1) a_i \zeta^{\frac{i+r}{s}-2} - \mathrm{Bi}_{*L2} - \sum_{i=0}^{\infty} b_i \zeta^{i+\frac{r}{s}} = 0. \qquad (6.3)$$

To collect the terms with equal powers of ζ, the indices in the first and third sums are changed so that they and those in the second sum become equal. This transforms indices i to $i - (2s - r)$ in the first and to $i/s - 2$ in the third sums. Then, because indices must be positive integers, it follows from new indices for the first sum that $i > 2s - r$. Hence, the first $2s - r$ coefficients a_i should be zero except coefficients a_0 and a_s, which are free because the terms in the second sum at $i = 0$ and $i = s$ vanish.

The new indices for the third sum show that there should be $i/s = k$, where k is an integer, which means that coefficients b_i should be taken into account only when i is divisible by

s (otherwise, the indices in the third sum would not be integers) and $i \geq 2s - r$ since the first $2s - r$ coefficients a_i are zero. The other coefficients are determined by the following equations (Exercise 6.2):

$$D_0 a_0 - (r/s - 1)(r/s - 2)a_{2s-r} - \text{Bi}_{*L2} = 0, \quad i = 2s - r$$

$$[D_0 + D_1(j-1) + D_2(j-1)(j-2) + \cdots + D_k(j-1)(j-2)\cdots(j-k) + \cdots] a_{s(j-1)}$$

$$-(i/s)(i/s - 1)a_i - b_{(i/s)-2} = 0, \quad j = (i+r-s)/s, \ i > 2s - r. \tag{6.4}$$

All coefficients a_i of series (6.2) can be calculated successively in terms of a_0 and a_s, whereas the last two are determined from the boundary conditions at the ends of the plate.

Thus, equation (6.2) gives actually the temperature distribution near the origin of the plate and shows that this temperature is an analytic function in integer powers of the variable $x^{1/s}$, where s is found from relation for an isothermal heat transfer coefficient.

In the case of asymmetric flow, the exponents r/s can be different on two sides. This happened, for instance, when the flow on the one side is laminar and on the other side turbulent, or when the pressure gradients or the exponents in the power law for non-Newtonian fluids are different on two sides. For this case, basic equation (6.1) becomes

$$\frac{d^2\theta}{d\zeta^2} - \text{Bi}_{*L1}\zeta^{-\frac{r_1}{s_1}} \sum_{k=0}^{\infty} g_{1k}\zeta^k \frac{d^k\theta}{d\zeta^k} - \text{Bi}_{*L2}\zeta^{-\frac{r_2}{s_2}} \left(\sum_{k=0}^{\infty} g_{2k}\zeta^k \frac{d^k\theta}{d\zeta^k} - 1 \right) + \bar{q}_v = 0. \tag{6.5}$$

In this case, the temperature head near $\zeta = 0$ is presented by the same series (6.2) in integer power of variable $\zeta^{1/s_1 s_2}$. If the liquids are numbered so that $r_1 s_2 < r_2 s_1$, then the first $(2s_1 s_2 - r_2 s_1)$ coefficients a_i are zero, except a_0 and a_{s1s2} determined from the boundary conditions. The rest of the coefficients are found analogously by equation such as (6.4) (Exercise 6.3):

$$(r_2/s_2 - 1)(r_2/s_2 - 2)a_{2s_1 s_2 - r_2 s_1} + (1 - a_0)\text{Bi}_{*L2} = 0, \quad i = 2s_1 s_2 - r_2 s_1$$

$$\text{Bi}_{*L1} \left[1 + g_{11}(j_1 - 1) + g_{21}(j_1 - 1)(j_1 - 2) + \cdots + g_{k_1}(j_1 - 1)\cdots(j_1 - k) + \cdots\right] a_{s_1 s_2 (j_1 - 1)}$$

$$+ \text{Bi}_{*L2} \left[1 + g_{12}(j_2 - 1) + g_{22}(j_2 - 1)(j_2 - 2) + \cdots + g_{k2}(j_2 - 1)\cdots(j_2 - k) + \cdots\right] a_{s_1 s_2 (j_2 - 1)}$$

$$-(i/s_1 s_2)[(i/s_1 s_2) - 1] a_i - b_{\left[(i/s_1 s_2) - 2\right]} = 0, \quad j_1 = (i + r_1 s_2 - s_1 s_2)/s_1 s_2, \ j_2 = (i + r_2 s_1 - s_1 s_2)/s_1 s_2,$$

$$i > 2s_1 s_2 - r_2 s_1. \tag{6.6}$$

6.1.2 Singularity Types

6.1.2.1 Laminar Flow at the Stagnation Point

It is obvious that the wall temperature at $x = 0$ is an analytic function of the coordinate x only when the exponent r/s in the relation for isothermal heat transfer coefficient is an integer. In particular, this is the case of the flow at a stagnation point when the free stream velocity is proportional to x, and the isothermal heat transfer coefficient is independent of coordinate ($r/s = 0$). Thus, in this case, there is no singularity, and the wall temperature is presented as a series in integer powers of x.

6.1.2.2 Laminar Flow at Zero-Pressure Gradient

In this case, $s = 2$, $r = 1$, and the wall temperature distribution is presented as a series in integer powers in variable $x^{1/2}$. The first three ($2s - r = 3$), except a_0 and a_s, should be zero. Therefore, $a_1 = 0$, a_0 and a_2 are determined from the boundary conditions, the coefficient a_3 is found from first equation (6.4), the rest of the coefficients are found from second equation (6.4), and the coefficients b are taken into account starting from index 4 and only those indices which are divisible by 2 and are greater than or equal to 4, that is, 4, 6, 8, and so on.

6.1.2.3 Turbulent Flow at Zero-Pressure Gradient

Since in this case $s/r = 1/5$, the temperature head is presented as a series in power of $\zeta^{1/5}$. The first nine coefficients, except a_0 and a_s, are zero. The coefficients b are taken into account starting from index 10 and only those indices which are divisible by 5 and are greater than or equal to 10, that is, 10, 15, 20, and so on.

6.1.2.4 Laminar Gradient Flow with Power-Law Free-Stream Velocity cx^m

The heat transfer coefficient for an isothermal surface is defined as $h_* = h_{*L}(x/L)^{(m-1)/2}$.

The values of s and r are determined after simplifying the fraction $(1-m)/2$. For example, for $m = 1/5$ or $m = 1/3$, one gets $s = 5, r = 2$ or $s = 3, r = 1$. In the latter case, the coefficients a_1, a_2, and a_4 are zero, whereas a_0 and a_3 are found from the boundary condition, b, are used when indices are divisible by 3 and are greater than or equal to 6.

6.1.2.5 Asymmetric Laminar-Turbulent Flow

In the case when the flow on the one side is laminar ($r_2/s_2 = 1/2$) and on the other side is turbulent ($r_1/s_1 = 1/5$), the temperature head distribution is described by series in integer power of $\zeta^{1/s_1 s_2} = \zeta^{1/10}$. Since $r_1 s_2 = 2 < r_2 s_1 = 5$, the first $20 - 5 = 15$ coefficients are zero, but a_0 and a_{10} are determined from the boundary conditions.

6.2 Charts for Solving Conjugate Heat Transfer

6.2.1 Charts Development

Using series (6.2), one calculates temperature head to some value of $\zeta > 0$. Then, the numerical integrating of equation (6.1) or (6.5) yields the solution for the rest part of a body. If these equations are used with only several first derivatives, the numerical integration can be performed by standard methods, for example, by Runge–Kutta method. In some cases, these equations can be reduced to well-investigated equations. In particular, equation (6.1) with the two first derivatives is reduced to a hypergeometric equation, solution of which is presented in the form

$$\theta = C_1 x F(\alpha, \beta, \gamma, D_2 x^{2-r/s}) + C_2 F(\alpha - \gamma + 1, \beta - \gamma + 1, 2 - \gamma, D_2 x^{2-r/s}) + \sigma_{\text{Bi}} + \vartheta_q \quad (6.7)$$

$$\alpha + \beta = \frac{D_1 + D_2}{D_2(2 - r/s)}, \alpha\beta = \frac{D_1 + D_0}{D_2(2 - r/s)}, \gamma = (3 - r/s)/(2 - r/s), \sigma_{\mathrm{Bi}} = \frac{\mathrm{Bi}_{*L2}}{\mathrm{Bi}_{*L1} + \mathrm{Bi}_{*L2}} \quad (6.8)$$

Here, two hypergeometric functions F are solutions of homogeneous equation (6.1) for the case when γ is not an integer, α and β are the roots of a quadratic equation, σ_{Bi} is the ratio of thermal resistances, and ϑ_q is a particular solution of inhomogeneous equation (6.1) with source (Exercise 6.7).

Two functions F in equation (6.7) are independent of boundary conditions, whereas constants C_1 and C_2 specified the particular problem. Therefore, these functions are universal in that respect and can be tabulated creating the charts. Using those charts, a simple method for solving the conjugate problems is developed. In the case when coefficients g_k for both sides of a plate are equal, equation (6.1) reduces to (Exercise 6.8)

$$\sum_{k=0}^{\infty} g_k z^k \frac{d^k \theta}{dz^k} - z^{r/s} \frac{d^2\theta}{dz^2} - \sigma_{Bi} - z^{r/s} \frac{\bar{q}_v}{z_L^2} = 0 \quad (6.9)$$

$$z = (\mathrm{Bi}_{*L1} + \mathrm{Bi}_{*L2})^{1/(2 - r/s)} (x/L) \; z_L = (\mathrm{Bi}_{*L1} + \mathrm{Bi}_{*L2})^{1/(2 - r/s)}. \quad (6.10)$$

According to this equation, function θ determining the temperature head in dimensionless form (6.1) on thermally thin plate depends only on single variable z. This function is used to create charts in the form of two hypergeometric equation (6.7)

$$\vartheta_1 = F(\alpha - \gamma + 1, \beta - \gamma + 1, 2 - \gamma, g_2 z^{2 - r/s}) \qquad \vartheta_2 = zF(\alpha, \beta, \gamma, g_2 z^{2 - r/s}). \quad (6.11)$$

For zero-pressure gradient laminar and turbulent flows, these functions and corresponding first two derivatives are calculated starting from their initial values:

$$\vartheta_1(0) = 1, \vartheta_1'(0) = 0, \vartheta_2(0) = 0, \vartheta_2'(0) = 1. \quad (6.12)$$

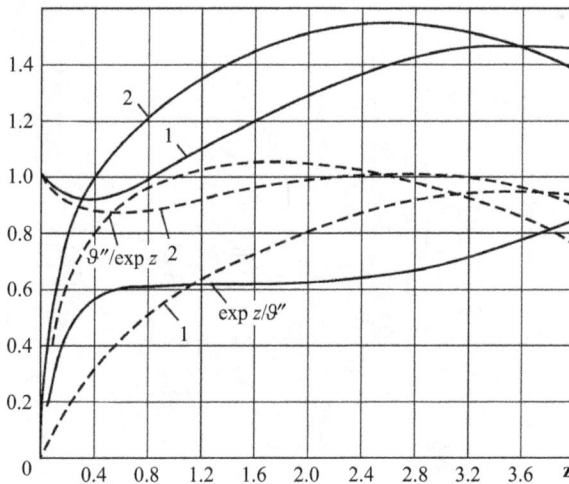

Figure 6.1. Universal functions ϑ_1 and ϑ_2 for laminar flow, Pr > 0.5, —— ϑ_1, - - - - ϑ_2, $1 - \vartheta/\exp z$, $2 - \vartheta'/\exp z$.

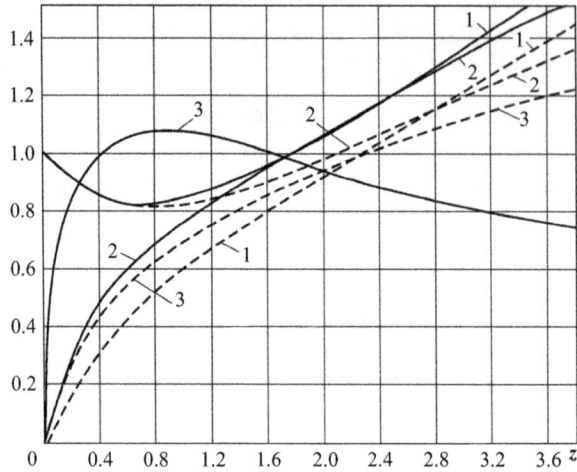

Figure 6.2. Universal functions ϑ_1 and ϑ_2 for turbulent flow: $Pr = 0.7$, $Re = 10^6 \ldots 10^7$, —— ϑ_1, — — —ϑ_2, $1 - \vartheta/\exp(3z/4)$, $2 - \vartheta'/\exp(3z/4)$, 3 —— $\exp(3z/4)/\vartheta_1''$, — — —$\vartheta_2''/\exp(3z/4)$.

The results are plotted in Figures 6.1 and 6.2 making the first part of charts (Exercise 6.9).

The second part of charts consists of functions ϑ_3 and ϑ_4, giving particular solutions of inhomogeneous equation (6.9) for linear source $\bar{q}_v = A + B(x/L)$ in the form

$$\vartheta_q = \frac{AL^2}{\lambda_w(T_{\infty 2} - T_{\infty 1})z_L^2}\vartheta_3 + \frac{BL^3}{\lambda_w(T_{\infty 2} - T_{\infty 1})z_L^3}\vartheta_4 = \bar{A}\vartheta_3 + \bar{B}\vartheta_4 \qquad (6.13)$$

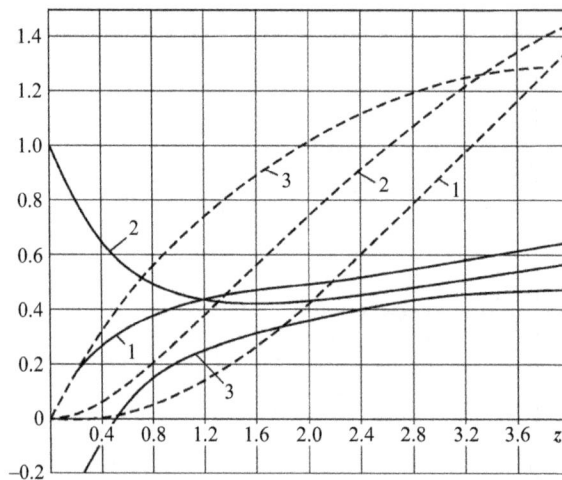

Figure 6.3. Universal functions ϑ_3 and ϑ_4 for laminar flow. $Pr > 0.5$, ——ϑ_3, - - - - ϑ_4, $1 - \vartheta/\exp(3z/4)$, $2 - \vartheta'/\exp(3z/4)$, $3 - \vartheta''/\exp(3z/4)$.

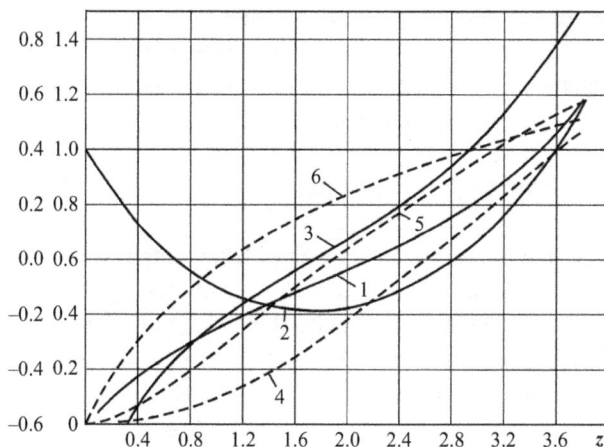

Figure 6.4. Universal functions ϑ_3 and ϑ_4 for turbulent flow. $\mathrm{Pr} = 0.7$, $\mathrm{Re} = 10^6...10^7$, —— ϑ_3, - - - ϑ_4, $1 - \vartheta_3 / 2$, $2 - \vartheta_3'$, $3 - \vartheta_3''$, $4 - [-\vartheta_4/\exp(3z / 4)]$, $5 - [-\vartheta'_4/\exp(3z/4)]$, $6 - [-\vartheta_4''/\exp(3z/4)]$.

As was noted above in the cases when the known functions, such as the hypergeometric, for instance, are not applicable, the data for chart may be computed using series (6.2) and numerical solution, applying boundary conditions (6.12) and equation (6.9) if the values r/s are the same for both sides of a plate or equation (6.5) if values of exponents r/s are different on two sides of a plate.

6.2.2 Using Charts

Problem solution begins from determining constants in equation (6.7), depending on given boundary conditions at the ends of a plate and taking into account initial conditions (6.12) for the universal functions. This procedure is simple and becomes clear from following examples and exercises. At known temperature head, the dimensionless local heat flux from a plate $\bar{q}_w$ is determined from equation (1.7) through the second derivative of the temperature head in form (6.14), which on integrating gives the total heat flux from the plate. The other equation (6.14) defines the dimensionless longitudinal heat flux along the plate $\bar{q}_x$ through the first temperature head derivative (Exercise 6.10):

$$\bar{q}_w = \frac{q_w L^2}{\lambda_w \Delta (T_{w0} - T_\infty) z_L^2} = \frac{\theta''}{2\theta_0}, \quad \bar{q}_x = \frac{q_x L}{\lambda_w \Delta (T_{w0} - T_\infty) z_L} = \frac{\theta'}{\theta_0}. \quad (6.14)$$

Here, index 0 denotes the leading edge, and derivatives are taken with respect to variable z (6.10).

Example 6.1 A steel plate of a length $L = 0.25\,\mathrm{m}$ and a thickness $\Delta = 0.01\,\mathrm{m}$ is in the air flow of velocity 3 m/s The left end of the plate is isolated; the temperature of the other end is at T_{wL}. The air temperature is $300\,\mathrm{K}$. Calculate the heat transfer characteristics.

It is convenient to use dimensionless temperature head based on the given temperature $\theta = (T_w - T_\infty)/(T_{wL} - T_\infty)$. Then, applying equation (6.7) with tabulated functions according to equation (6.11), boundary conditions at left $q_x(0) = 0$ and at right $\theta(z_L) = 1$ ends, and the initial conditions (6.12), we have two first equations (6.15):

$$q_x = C_1 \vartheta_1'(0) + C_2 \vartheta_2'(0) = 0, \quad \theta(z_L) = C_1 \vartheta_1(z_L) + C_2 \vartheta_2(z_L) = 1, \quad \theta = \vartheta_1(z)/\vartheta_1(z_L). \quad (6.15)$$

The problem solution (the last equation) is obtained taking into account that (i) from equation (6.12), $\vartheta_1'(0) = 0$; (ii) then from first equation (6.15), follows $C_2 = 0$; and (iii) from second equation (6.15), we get $C_1 = 1/\vartheta_1(z_L)$. Finally, $\theta = C_1 \vartheta_1(z) + C_2 \vartheta_2(z)$ gives solution (6.15). Since $\mathrm{Re} = 5 \times 10^4$, the flow is laminar, and the Nusselt number for the isothermal surface, corresponding Biot number, and variable (6.9) at the plate end are

$$\mathrm{Nu}_{*L} = 0.295\sqrt{5 \cdot 10^4}, \quad \mathrm{Bi}_{*L} = \frac{\mathrm{Nu}_{*L} \lambda L}{\lambda_w \Delta} = 0.656, \quad z_L = (2\mathrm{Bi}_{*L})^{1-(2-r/s)} = 1.2. \quad (6.16)$$

Heat fluxes are determined by equations (6.14): $\bar{q}_w = \vartheta_1''/2\vartheta_1(z_L)$, $\bar{q}_x = -\vartheta_1'(z)/\vartheta_1(z_L)$, taking into account that according to equation (6.15), $\theta_0 = 1/\vartheta_1(z_L)$. The numerical results are given in Table 6.1 (Exercise 6.11).

Comment 6.2 We considered this problem in Chapter 4 while estimating the errors caused by boundary condition of the third kind. There, we used the solution of this problem with boundary condition of the third kind. A comparison of the result of that solution with data from the conjugate solution shows that the maximum difference of both solution is 36%, which is in agreement with estimated error 0.4 (Example 4.12).

Example 6.2 A copper plate of length 0.5 m and 0.02 m in thickness is streamlined on the one side by air at temperature 313 K with velocity 30 m/s Another side of the plate is isolated. The temperatures of ends are maintained at $T_{w0} = 593$ K and $T_{wL} = 293$ K. Find local temperature and heat flux distributions.

Table 6.1. Heat transfer characterizes of a plate heated from one end (Example 6.1)

z	x/L	$\vartheta_1(z)$	$\vartheta_1'(z)$	$\vartheta_1''(z)$	θ	$-\bar{q}_x$	$\bar{q}_w$
0	0	1	0	∞	0.278	0	∞
0.2	0.167	1.12	0.949	2.75	0.311	0.264	0.382
0.4	0.334	1.37	1.48	2.66	0.388	0.411	0.369
0.6	0.501	1.72	2.04	3.02	0.478	0.567	0.429
0.8	0.668	2.19	2.70	3.62	0.608	0.750	0.503
1.0	0.835	2.81	3.50	4.42	0.780	0.969	0.614
1.2	1	3.60	4.48	5.12	1	1.244	0.753

If the dimensionless temperature is used in the form based on leading edge temperature $\theta = (T_w - T_\infty)/(T_{w0} - T_\infty)$, the boundary conditions are $\theta(0) = 1$ and $\theta(L) = \theta_L$. Then, according to condition (6.12), the equation for leading edge gives $C_1 = 1$, and after that, the similar condition for another end $\vartheta_1(z_L) + C_2 \vartheta_2(z_L) = \theta_L$ yields $C_2 = [\theta_L - \vartheta_1(z_L)]/\vartheta_2(z_L)$, resulting in solution $\theta = \vartheta_1(z) + [\theta_L - \vartheta_1(z_L)]\vartheta_2(z)/\vartheta_2(z_L)$.

Calculating $\mathrm{Re} = 0.88 \times 10^6$ shows that the flow is turbulent, $Nu_{*L} = 0.0255\,\mathrm{Re}^{4/5}$, $\mathrm{Bi}_{*L} = 2.53$, and $z_L = 2.53^{5/9} = 1.67$. Taken from charts the values of $\vartheta_1(z_L)$ and $\vartheta_2(z_L)$, one finds expression for temperature head as $\theta = \vartheta_1(z) - 1.22\vartheta_2(z)$ (Exercise 6.12). The results are plotted in Figure 6.5 (analysis in Example 6.6).

Example 6.3 Consider the same problem as in Example 6.2 for an aluminum plate of length $0.3\,\mathrm{m}$ and thickness of $0.002\,m$ streamlined by a flow of air of a velocity $250\,\mathrm{m/s}$ on an altitude $20\,\mathrm{km}$. Air temperature is $T_\infty = 223\,\mathrm{K}$, and kinematic viscosity is $v = 1.65 \times 10^4\,\mathrm{m}^2/\mathrm{s}$. The front end is at stagnation temperature $T_{\infty 0} = 254\,\mathrm{K}$, and the other is at $T_{wL} = 323\,\mathrm{K}$.

Since the Mach number is $\mathrm{M} = U/U_{sd} = 250/(20.1 \cdot \sqrt{223}) = 0.833$ ($U_{sd} -$ speed of sound), it is necessary to take into account the effect of compressibility. In such a case, instead of dimensionless temperature head, the enthalpy difference should be used (Sec. 4.2.6). Since the Mach number is not too high, the adiabatic temperature difference instead of enthalpy difference may be applied ignoring the dependence $c_p(T)$. Therefore, the following dimensionless temperature head is used: $\theta = (T_w - T_{wL})/(T_{ad} - T_{wL})$.

The boundary conditions $\theta(0) = 1$ and $\theta(L) = 0$ together with equations (6.12) give the constants $C_1 = 1$ and $C_2 = -\vartheta_1(z_L)/\vartheta_2(z_L)$, leading according to (6.7) to the temperature head $\theta = \vartheta_1(z) - \vartheta_2(z)\vartheta_1(z_L)/\vartheta_2(z_L)$. To find the value of z_L, the Biot number should be multiplied

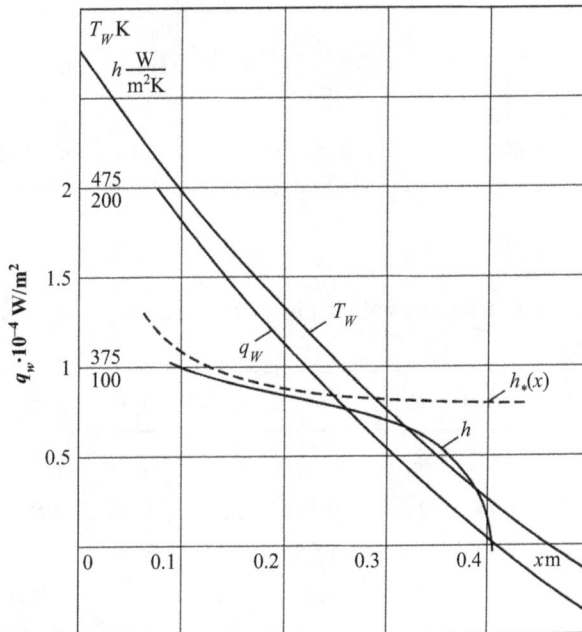

Figure 6.5. Heat transfer characteristics for a plate streamlined on the one side by turbulent flow.

by $C_x/\sqrt{C}$. This becomes clear if one writes equation (1.7) using equation (4.35) for heat flux to get an equation similar to equation (6.5) (Exercise 6.13):

$$\frac{d^2\theta}{d\zeta^2} - \frac{C_x}{\sqrt{C}} \mathrm{Bi}_{*L1} \zeta^{-\frac{r_1}{s_1}} \sum_{k=0}^{\infty} g_{1k} \zeta^k \frac{d^k\theta}{d\zeta^k} - \frac{C_x}{\sqrt{C}} \mathrm{Bi}_{*L2} \zeta^{-\frac{r_2}{s_2}} \left(\sum_{k=0}^{\infty} g_{2k} \zeta^k \frac{d^k\theta}{d\zeta^k} - 1 \right) + \overline{q}_v = 0 \quad (6.17)$$

Ratio $C_x/\sqrt{C}$ is obtained applying Chapman–Rubesin formula (4.35) that yields for left and right ends $C_x \approx 0.93$ and $C_x \approx 0.98$, respectively. Assuming that the average value of $C_x = 0.95$ is equal to C, we find that $C_x/\sqrt{C} = \sqrt{C} = 0.975$. Reynolds value $\mathrm{Re} = 4.55 \times 10^5$ tells us that the flow is laminar and corrected (multiplied by 0.975); the Biot number is $\mathrm{Bi}_{*L} = 2.91$. Then, $z_L = 2.04$, and temperature head is $\theta = \vartheta_1(z) - 1.69 \vartheta_2(z)$. The relations for heat fluxes can be obtained from equations (6.14) substituting $T_{ad} - T_{wL}$ for $T_w - T_\infty$. The results are presented in Table 6.2 (Exercise 6.14).

Example 6.4 Air flows ($\mathrm{Re} = 5 \times 10^4$) over the one side of a thin ($\Delta/L = 1/600$) radiating plate ($\lambda/\lambda_w = 0.135 \times 10^{-4}$) with uniform internal heat sources ($\overline{q}_v = 5.1$). Another side of the plate is isolated. The front end is at the free stream temperature T_∞. The radiation is taken into account by parameter $N = \sigma \varepsilon T_\infty^3 / \lambda_w \Delta = 0.07$, where σ and ε are the Stefan–Boltzmann constant and emissivity.

This problem is solved by Sohal and Howell using numerical integration of the integro-differential equation [1]. Here, we show how this problem is solved applying series (6.2) in integer powers of variable $\zeta^{1/2}$ ($\mathrm{Re} = 5 \times 10^4$–laminar flow). In the case of a flow past one side of radiation plate, equation (6.17) becomes

$$\frac{d^2\theta}{d\zeta^2} - \mathrm{Bi}_{*L} \zeta^{-r/s} \left(\theta - 1 + \sum_{k=1}^{\infty} g_k \zeta^k \frac{d^k\theta}{d\zeta^k} \right) - N(\theta^4 - 1) + \overline{q}_v = 0, \quad \theta = T_w / T_\infty. \quad (6.18)$$

The solution of this equation with derivatives not higher than second requires two boundary conditions. One condition is given at the left edge $\theta(0) = 1$. Another is obtained due to the fact

Table 6.2. Heat transfer characterizes of the plate streamlined by compressible flow (Example 6.3)

z	x/L	$\vartheta_1(z)$	$\vartheta_1'(z)$	$\vartheta_1''(z)$	$\vartheta_2(z)$	$\vartheta_2'(z)$	$\vartheta_2''(z)$	θ	$\overline{q}_x$	$\overline{q}_w$
0	0	1	0	∞	0	1	0	1	1.6	∞
0.4	0.196	1.37	1.48	2.66	0.446	1.29	1.18	0.656	0.584	0.772
0.8	0.392	2.19	2.70	3.62	1.08	1.94	2.10	0.462	0.404	0.260
1.2	0.588	3.60	4.48	5.42	2.06	2.02	3.37	0.304	0.352	0.028
1.6	0.783	5.80	7.16	8.14	3.58	4.71	5.17	0.152	0.378	−0.132
2.04	1	11.0	11.7	12.6	6.26	7.61	8.04	0	0.476	−0.264

that the heat transfer intensity diminishes as the distance from the leading edge increases and the boundary layer becomes thicker. This situation is quite the same as in the problem of heat exchange between two fluids (Example 4.1), resulting in some constant asymptotic temperature as the distance from the start end goes to infinity. To find this asymptotic temperature, one puts in equation (6.18) $\zeta = 1$ and equals the derivatives to zero, obtaining the following algebraic expression

$$\mathrm{Bi}_{*L}(1-\theta_{as})-N(\theta_{as}^4-1)+\overline{q}_v = 0. \tag{6.19}$$

Since in series (6.2), the first $(2s-r) = 3$ coefficients, except a_0 and $a_s = a_2$, should be zero, coefficients $a_1 = a_3 = 0$. The other coefficients are determined in terms of $a_0 = 1$ (according to $\theta(0) = 1$) and a_2 using equations (6.4) modified by additional term Nd_{i-2s}, where d_{i-2s} are coefficients in series in variable $\zeta^{1/s}$ presenting θ^4 in equation (6.18). Then, after substituting coefficients D by g_k (see (6.1)), equation (6.4) becomes

$$(r/s-1)(r/s-2)a_{2s-r}+(1-a_0)\mathrm{Bi}_{*L} = 0, \ i = 2s-r$$

$$\mathrm{Bi}_{*L}[1+g_1(j-1)+g_2(j-1)(j-2)+\cdots+g_k(j-1)(j-2)\cdots(j-k)+\cdots]a_{s(j-1)}$$

$$-(i/s)(i/s-1)a_i+Nd_{i-2s}+\overline{q}_v = 0, \ j = (i+r-s)/s, \ i > 2s-r. \tag{6.20}$$

Employing these expressions, one gets the coefficients for series (6.2) (Exercise 6.15):

$$a_1 = a_3 = 0, a_4 = -\overline{q}_v/2, \ a_5 = 0.432\mathrm{Bi}_{*L}a_2, \ a_6 = 0.667a_2N, \ a_7 = -0.115\mathrm{Bi}_{*L}\overline{q}_v,$$

$$a_8 = 0.079\mathrm{Bi}_{*L}^2 a_2 + 0.167N(3a_2^2 - \overline{q}_v) \tag{6.21}$$

The next step is to estimate $\mathrm{Bi}_{*L} = Nu_{*L}\lambda L/\lambda_w\Delta = 0.535$ and using coefficients (6.21) find the series determining the plate temperature:

$$\theta = 1+a_2\zeta - 2.55\zeta^2 +0.231a_2\zeta^{5/2}+0.0467a_2\zeta^3-0.314\zeta^{7/2}$$

$$+ (0.0226a_2+0.0351a_2^2-0.0596)\zeta^4 +\cdots \tag{6.22}$$

Assuming that this equation is valid for the whole plate up to the end where the second boundary condition $\theta = \theta_{as}$ at $\zeta = 1$ should be satisfied, we get an equation defining a_2 $\theta_{as} = 1+1.3a_2+0.0351a_2^2-1.924$, where θ_{as} is given by equation (6.19). Algebraic equation (6.19) is transcendental and may be solved graphically or by trial-and-error method. This solution gives the value of asymptotic temperature $\theta_{as} = 2.79$, and then a solution of quadratic equation for coefficient a_2 yields $a_2 = 3.33$. The numerical data obtained by equation (6.22) are given in Table 6.3. These are in good agreement with those calculated in [1, Figure 5] (Exercises 6.16 and 6.17).

We considered in details four examples of using charts and series for solving conjugate problems to help a reader employ this method, which is as simple as a common approach with boundary condition of the third kind. The below given examples of this procedure are discussed

Table 6.3. Dimensionless temperature $\theta = T_w / T_\infty$ along the radiating plate (Example 6.4)

$z = x/L$	0	0.1	0.2	0.3	0.4	0.5	0.6	07	0.8	0.9	1.0
θ		1.00	1.31	1.58	1.82	2.01	2.18	2.31	2.45	2.57	2.68	2.79

with much less details, focusing on physical interpretation of analyzing data. However, before that, we review applicability of charts considering accuracy and a range of applications of this technique.

EXERCISES

6.1 Obtain equation (6.1) using equation (1.7) for a thermally thin plate and universal equation (4.1) written for a plate at a zero-pressure gradient.

6.2 Recall or study the solution of differential equations using power series and method of undetermined coefficients. Obtain equation (6.3) by substitution series (6.2) in equation (6.1). Perform an analysis leading to series coefficients determination.

6.3 *Perform the same analysis as in previous exercise for the case of asymmetric flow to obtain equation (6.6), determining series coefficients in this case.

6.4 Determine the series structure for the case when the exponent in relation for an isothermal heat transfer coefficient h_* is (−0.4).

6.5 Determine the series structure for asymmetric flow with values of $r/s = 1/3$ and 1/2.

6.6 Consider the series structure for power-law non-Newtonian fluid at a zero-pressure gradient. According to equation (3.101) [1, p.84] in this case, the heat transfer coefficient for an isothermal surface is presented in the form $h_* = h_{*L}(x/L)^{-[n/(n+1)]}$, where n is exponent in the rheology power law (4.36). If n is an integer, then $s = n + 1$, $r = n$. If n is a fraction such that $n = n_1/n_2$, then $s = n_1 + n_2$, $r = n_1$. So for $n = 2$, we have $s = 3$, $r = 2$, and for $n = 3/5$, it will be $s = 8$, $r = 3$. Using these data determine as in examples in the text the type of series (6.2) for coefficients a_i in those cases. Estimate a series structure for non-Newtonian fluid with $n = 0.25$.

6.7 Recall or study the property of an hypergeometric function and series, for example, on Wikipedia, to understand the creation of charts. Derive equation (6.7) from equation (6.1).

6.8 Obtain equation (6.9) and then equation (6.10) of two functions using for the chart creating. Explain why charts of those two functions allow a simple solution of conjugate heat transfer problems for flows past thermally thin plates.

6.9 Obtain the universal functions ϑ_1 and ϑ_2 for the accelerated laminar flow with $U = cx^{0.2}$ using hypergeometric series for equations (6.11). Draw a chart as given in Figure 6.1. Hint: first determine r/s and then α, β and γ using formulae (6.8) and relation for γ from the text.

6.10 Derive equations (6.14) using definitions for heat fluxes, equation (1.7), and variable (6.10).

6.11 Solve the problem from Example 6.1 for the case of knowing leading edge temperature T_{w0} and the isolated right edge. Compare results obtained in both cases. What are physical reasons of such results?

6.12 Solve the problem from Example 6.2 for the case of the given leading edge temperature T_{w0} and known heat flux $\bar{q}_x$ at the plate end. Hint: use expression (6.14) and values of C_1 and z_L, which are known because temperature of leading edge and flow regime are given.

6.13 Show that in the case of compressible flow, the Biot number in the formula (6.9) for variable z should be multiplied by ratio $C_x/\sqrt{C}$. Hint: write equation using equations (1.7) for a thermally thin plate and (4.35) for heat flux in the case of compressible flow to obtain equation (6.17) and compare it with similar equation (6.5) for incompressible flow.

6.14 Solve the problem from Example 6.3 for the case of given right edge temperature T_{wL} and heat flux at the leading edge $\bar{q}_x$. Hint: use formula (6.13) $\bar{q}_x = \theta'/\theta_0$ and relations $\theta_0 = C_1\vartheta_1(0) + C_2\vartheta_2(0)$ and $\theta_0' = C_1\vartheta_1'(0) + C_2\vartheta_2'(0)$.

6.15 Compute coefficients (6.21) using equations (6.20), singularities of series (6.2) in variables $\zeta^{1/2}$ and first two coefficients g_k for a zero-pressure gradient and $\Pr \approx 1$.

6.16 Solve the problem from Example 6.4 using first six and seven coefficients. Compare results with that obtained, employing eight coefficients to estimate the accuracy of data gained in Example 6.4.

6.17 Solve the same problem for a radiating plate streamlined past one side by turbulent flow taking $\text{Re} = 2 \times 10^6$.

6. 3 Applicability of Charts and One-Dimensional Approach

6.3.1 Refining and Estimating Accuracy of the Charts Data

For developing charts, we use equation (6.1) based on the universal function with first two derivatives. Although it was shown in Chapter 5 that the other terms are small, it is necessary to estimate the error arising by such a simplification. This can be done calculating $\varepsilon(x) = (1/h_*)\left(q_w^{int} - q_w^{diff}\right)$, which according to first equation (4.44) substitute the neglected terms of universal function in the form of series. Knowing $\varepsilon(x)$ allows one to refine the results obtained using charts or/and to estimate accuracy of those data. In this case, the data of heat flux q_w^{diff} are given by charts, and the value of heat flux q_w^{int} is computed by equation (4.14). The first derivative $d\theta_w/dx$ required for equation (4.14) is estimated employing charts data that are considered as a first approximation. Such a procedure gives both the accuracy of the chart data in the form of $\varepsilon(x)$ and the second approximation of heat flux as q_w^{int}. The next approximation (if necessary) may be obtained using error $\varepsilon(x)$ and second equation (4.44). To perform calculations, it is reasonable to employ the simple software (Exercise 6.18).

6.3.2 Applicability of Thermally Thin Body Assumption

To estimate the error arising when the body is assumed to be thermally thin, we approximate the temperature distribution across the body, as in the Karman–Pohlhausen integral method, using the second-order polynomial $T/T_{av} = a_0 + a_1\eta + a_2\eta^2$, where $\eta = y/\Delta$ (Sec. 2.7.1). To find the coefficients a_n, one assumes that the heat fluxes on the body surfaces are known: q_{w1} at $y = 0$ and q_{w2} at $y = \Delta$. That gives two conditions. The third one is obtained knowing that the integral of T/T_{av} across a body thickness is equal to unity. Using these three conditions and the Fourier law for heat flux, one gets the coefficients a_n and then the temperature distribution:

$$T/T_{av} = 1 - (\omega_1/3) + (\omega_2/6) + \omega_1\eta - (\omega_1 + \omega_2)\eta^2/2, \quad \omega = \frac{q_w\Delta}{\lambda_w T_{av}} = \frac{\text{Bi}\theta_w}{T_{av}}. \tag{6.23}$$

Setting $\eta = 0$ and $\eta = 1$ gives the surface temperatures, and the condition $T/T_\infty = 1$ yields the inequalities determining thermally thin body (Exercise 6.19):

$$\frac{1}{T_{av}} \left| \frac{Bi_1 \theta_{w1}}{3} - \frac{Bi_2 \theta_{w2}}{6} \right| \ll 1, \quad \frac{1}{T_{av}} \left| \frac{Bi_2 \theta_{w2}}{3} - \frac{Bi_1 \theta_{w1}}{6} \right| \ll 1, \quad \frac{Bi \theta_w}{6T_{av}} \ll 1. \tag{6.24}$$

The last inequality is attributed to a symmetrical streamlined object. Calculating the left-hand side of these inequalities, one can estimate the error arising when the actual temperature distribution across the body thickness is substituted by an average constant temperature. For the thin metallic plates, inequalities (6.24) are usually satisfied.

Comment 6.3 In expressions (6.24), Biot numbers $Bi = h\Delta/\lambda_w$ are determined using the actual heat transfer coefficient rather than employing above $Bi_* = h_*\Delta/\lambda_w$ for an isothermal surface. Therefore, these inequalities are applicable only to cases where the heat transfer behavior is regular without special features, which are inherent, for example, for systems with the decrease in flow direction or in time temperature heads. Physically, this is associated with parameter distributions, which could not be described by simple polynomials applied in this analysis. We encountered such a physical situation in Section 2.7.1 while discussing the applicability of the Karman–Pohlhausen method. This reasoning shows that inequalities (6.24) are basically applicable to systems with isothermal and increasing temperature heads, characteristics of which are regular.

6.3.3 Applicability of the One-Dimensional Approach and Two-Dimensional Effects

In this section, we investigate where and how the one-dimensional results for a thin plate should be corrected, taking into account the two-dimensional effects. Such a correction is necessary, in particular, at the initial segment of any plate where, in fact, it is impossible to neglect the transverse thermal resistance of the plate or its longitudinal conductivity. One-dimensional approximation is not valid close to the leading edge because in that area on the one hand the thermal boundary layer is thin, and therefore, the transverse plate resistance is important compared to that of fluid. On the other hand, longitudinal temperature gradients are high due to vigorous variations of the heat transfer coefficient. Thus, the temperature field at the entrance edge is always two-dimensional.

To find the answer to the questions formulated above, we compare the two- and one-dimensional solutions, considering the heat exchange between two fluids separated by a plate. The result of comparison is presented in Figure 6.6. Curve 1 corresponds to the solution for a thin plate obtained in Example 4.1 for a different ratio of fluid resistances:

$$Bi_\Sigma = \frac{1}{1/Bi_{\Delta 1*} + 1/Bi_{\Delta 2*}}, \quad Bi_{\Delta *} = \frac{q_* \Delta}{\lambda_w (T_{\infty 1} - T_{\infty 2})}. \tag{6.25}$$

These results are considered as a first approximation. The second approximation is found using the solution of the two-dimensional Laplace equation for a plate in the form

$$\theta_1 + \theta_2 = \frac{1}{\pi} \int_0^\infty (Bi_{\Delta 1} + Bi_{\Delta 2}) \ln[4 sh|z - \zeta|\pi \cdot sh|z + \zeta|\pi] dz, \quad \theta = \frac{T_w - T_{w\infty}}{T_{\infty 1} - T_{\infty 2}}$$

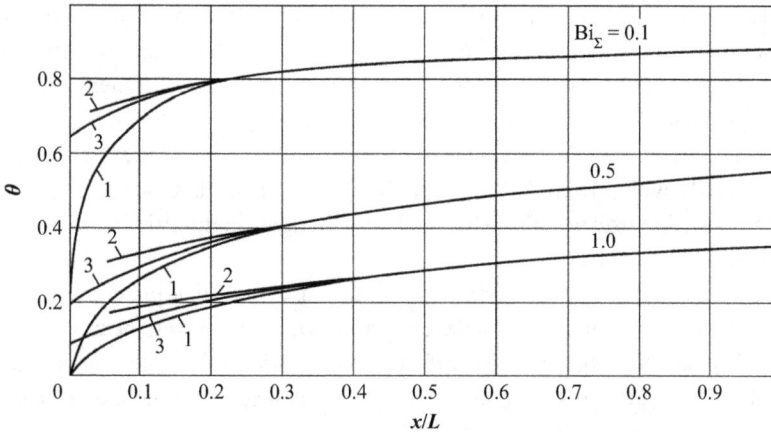

Figure 6.6. Comparison of one- and two-dimensional solutions for the heat exchange between two fluids separated by a plate: 1-thin plate, 2-second approximation for a thick plate according to equation (6.27), 3-final, two-dimensional result for a thick plate according to equation (6.29).

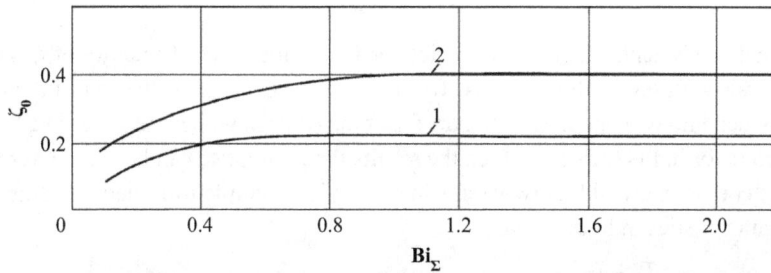

Figure 6.7. Initial length of a plate as a function of a ratio of thermal resistances of the plate and fluids: 1-turbulent flow and 2-laminar flow.

$$\theta_1 - \theta_2 = \frac{1}{\pi} \int_0^\infty (Bi_{\Delta 2} - Bi_{\Delta 1}) \ln[cth|z - \zeta|(\pi/2) \cdot cth|z + \zeta|(\pi/2)]dz, \qquad \zeta = x/\Delta. \quad (6.26)$$

The next refinement is achieved by substituting the result of first approximation into these equations. Considering, for simplicity, the case of symmetrical flow, we use equation (4.22) as the first approximation: $-Bi_{K1}^{(1)} = Bi_{K2}^{(1)} = 1 - \Theta^{(1)}$ and $\Theta_1^{(1)} = \Theta_2^{(1)}$, where Bi_K and Θ in Example 4.1 are related to Bi_Δ and θ in equations (6.26). Then, second equation (6.26) gives the second approximation (Exercise 6.20):

$$\theta_1^{(2)} = -\theta_2^{(2)} = \frac{1}{\pi} \int_0^\infty Bi^{(1)} \ln[cth|z - \zeta|(\pi/2) \cdot cth|z + \zeta|(\pi/2)]dz. \quad (6.27)$$

In Figure 6.6, the results obtained employing this equation are marked by 2.

It is seen that both approximations coincide beginning from relatively small initial distance $\zeta_0 = x_0/L$ so that for $\zeta > \zeta_0$, the semi-infinite thick plate may be considered as thin one using one-dimensional relations for computing heat transfer characteristics. Figure 6.7 presents the dependence of initial length ζ_0 on a ratio of plate and fluids thermal resistances Bi_Σ defined by equation (6.25).

In the case of a thick plate, the temperature distribution close to $x = 0$ as well as for a thin plate is described by series in integer power of variable $\zeta^{1/2}$. However, this series is different from that in the case of a thin plate, and the estimation of the series coefficients for a thick plate is much complicated. While the series for the thin plate contains all the powers of $\zeta^{1/2}$ and the series for the thermally thin plate involves only terms with $(\zeta^{1/2})^{3k}$, the series for a thick plate contains terms only with powers $(\zeta^{1/2})^{4k}$ and $(\zeta^{1/2})^{4k+1}$. Thus, both assumptions that a plate is thin or thermally thin lead us only to the qualitatively right singularities in vicinity of the start plate section (Exercise 6.21).

The complexity of coefficients estimation in the case of a thick plate arises because such a problem is governed by elliptical Laplace equation when the results near $x = 0$ depend not only on local parameters, but are also affected by situation of whole computation domain (Comment 4.2). Analysis shows that coefficients of such a series in the case of equal thermal resistances of both flows are determined as follows [2]:

$$a_0 = \frac{\pi - 4I_\infty}{\pi + 4d_{(-1)}\left[I_{(-1)} + 4d_0 I_0\right]}, \quad I_\infty = \int_\varepsilon^\infty \mathrm{Bi}^{(1)} \ln cth \frac{z\pi}{2} dz, \quad I_n = \int_0^\varepsilon z^{n/2} \ln cth \frac{z\pi}{2} dz$$

$$a_1 = 4a_0 d_{(-1)}, \quad a_2 = a_3 = 0, \quad d_n = \frac{\mathrm{Bi}_{\Delta*}}{2}\left[1 + \sum_{k=1}^\infty g_k \frac{(n+1)(n-1)...(n-2k+3)}{2^k}\right]. \tag{6.28}$$

One sees that coefficients a_n (6.28) of series for the temperature at vicinity of $\zeta = 0$ depend on the integrals with semi-infinite limits. To calculate these integrals, the semi-infinite interval $(0, \infty)$ is divided into two parts: the first part for limits $(0, \varepsilon)$ in which the series in $\zeta^{1/2}$ are valid and the second for limits (ε, ∞) in which the results for a thin plate can be used (Exercise 6.22). Using these coefficients and first two terms in series yields simple formulae for computing heat transfer characteristics in initial area:

$$\Theta = \frac{T_w - T_\infty}{T_{w\infty} - T_\infty} = a_0(1 + 4\mathrm{Bi}_\Sigma \zeta^{1/2}), \quad \mathrm{Bi} = a_0 \mathrm{Bi}_\Sigma (\zeta^{-1/2} + 4d_0). \tag{6.29}$$

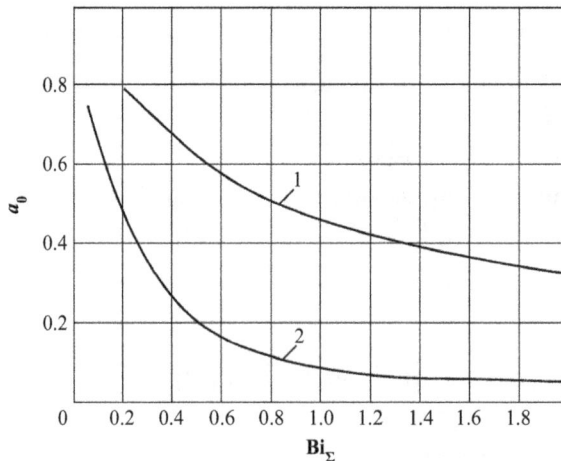

Figure 6.8. Dependence of the coefficient a_0 in equations (6.29) and (6.30) on the ratio of thermal resistances of the plate and fluids: 1-turbulent flow and 2-laminar flow.

The values of a_0 are given in Figure 6.8. The temperatures calculated by (6.29) are plotted in Figure 6.6 (curve 3). These differ in principle on the initial segment from the results obtained for a thin plate. The temperature head at the beginning is finite so that the plate temperature differs substantially from that of corresponding fluid, whereas under the assumption of a thin plate, the temperature head is zero, and, consequently, the temperature of the plate and corresponding fluid are equal at the beginning.

Due to the small length of the initial segment in a turbulent flow, it is possible to use only the first terms of the formulae (6.29) in this case

$$\Theta = \frac{T_w - T_\infty}{T_{w\infty} - T_\infty} = a_0 \qquad \mathrm{Bi} = a_0 \mathrm{Bi}_\Sigma \zeta^{-1/5}. \tag{6.30}$$

Comment 6.4 Strictly speaking, in the immediate vicinity of $\zeta = 0$, due to the smallness of the Reynolds and Peclet numbers, the boundary layer theory assumptions are not valid, and in this region, also with the full heat conduction equation, it is necessary to consider the full conservation equations for fluids. However, estimates show that at large values of the Reynolds number for the whole plate, the region in which the boundary layer equations are not valid is many times smaller than the initial segment over which the plate should be considered as a thick one.

According to Van-Dyke (Sec. 2.5), the error in the determination of the friction coefficient by the boundary layer theory does not exceed $1.7 / \sqrt{\mathrm{Re}_x}$. Assuming the maximum admissible error to be 10%, the ultimately admissible Reynolds number is about 300. The segment length over which the boundary layer theory is not valid according to this estimation is $x_{\lim}/L \approx 300/\mathrm{Re}$. On the other hand, as calculation shows, the initial segment length in laminar flow is $x_0/L \approx 0.4(\Delta/L)$ (Fig. 6.7). Thus, for example, at $\mathrm{Re} = 10^5$ and $\Delta/L = 1/100$, $x_{\lim}/L \approx 1/300$, $x_0/L \approx 1/250$. However, at $\Delta/L = 1/10$ and the same Re, the initial segment length turns out to be 12 times greater than $x_{\lim}$. With the increase in the plate thickness and Reynolds number, the difference between x_0 and $x_{\lim}$ increases rapidly. Thus, only when a plate is very thin, the both dimensions are close. In such a case, the initial segment is very small and only the value of the temperature at $x = 0$ is important.

6.4 Conjugate Heat Transfer in Flow Past Plates

According to the last results, the specific properties of conjugate heat transfer considered in this section in examples for thin plates are valid as well for the thick plates starting from short distance ζ_0 from a leading edge (Fig. 6.7). For the symmetrical flows, the heat transfer parameters at the initial segment of the length ζ_0 might be calculated using equations (6.29) or (6.30). Due to that, the results for a thin plate presented here give the understanding of conjugate heat transfer behavior in general for the case of flows past plates.

Example 6.5 Plate heated from one end in a symmetrical flow.

Recall that with qualitative analysis of this problem we start the second part of this book. If the plate is passed from heated end, the temperature head decreases in the flow direction. Otherwise, when the flow runs in the opposite direction, the temperature head grows in the flow direction. Since the Biot number for an isothermal surface is the same in both cases, this example clearly demonstrate the role of the temperature head variation.

Let the temperature head of heated end is θ_h and the other end is isolated. Determining the constants in equation (6.7) (without last two terms) from given conditions, one finds the temperature head in the first and the second cases:

$$\frac{\theta}{\theta_h} = \vartheta_1 - \frac{\vartheta_1'(z_L)}{\vartheta_2'(z_L)}\vartheta_2, \qquad \frac{\theta}{\theta_h} = \frac{\vartheta_1}{\vartheta_1(z_L)}. \tag{6.31}$$

Then, using equations (6.14), the local heat fluxes from the plate and along the plate are obtained. The total heat flux from the plate is found by integration equation (6.14) for local heat flux q_w. Using this result and knowing that in the first case $\theta'(z_L) = 0$ and in second one $\theta'(0) = 0$, the ratio of total fluxes is determined (Exercise 6.23):

$$Q_w = \frac{2L}{\lambda_w (T_h - T_\infty) z_L \Delta}\int_0^L q_w dx = \frac{\theta'(z_L) - \theta'(0)}{\theta_h} \qquad \frac{Q_{w1}}{Q_{w2}} = \frac{\vartheta_1(z_L)}{\vartheta_2'(z_L)}. \tag{6.32}$$

In Figure 6.9(a) are plotted the results for laminar flow. It is seen that heat transfer characteristics for both cases differ substantially. In the first case when the temperature head decreases, the heat transfer coefficients are significantly less than the isothermal coefficients and the heat flux sharply decreases along the plate so that the situation is close to inversion at the end. Here, the heat transfer coefficient is 4.5 time less than an isothermal one. In the second case, the temperature head increases, and according to this, the heat transfer coefficients are greater than the isothermal coefficients, but not more than 1.8 times. Nevertheless, the total heat flux in this case is less than that in the other case. This happened because in the first case, there are large temperature heads and heat transfer coefficients at the start of flowing, whereas in the second case at the beginning when the heat transfer coefficients are large, the temperature heads are small, and vice versa. As a result, the heat flux distribution curve has the minimum in this case. The value of the ratio of total heat fluxes Q_{w1}/Q_{w2} depends on the Biot number and in the case

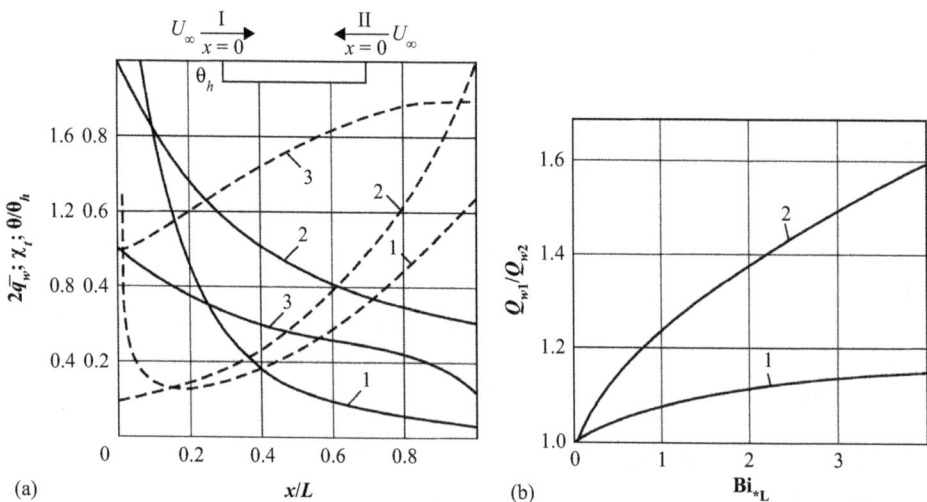

Figure 6.9. Heat transfer characteristics for the plate heated from one end: (a) local characteristics $\mathrm{Bi}_{*L} = 1.4$, I _____ first case, II- - - - second case, $1 - 2\bar{q}_w, 2 - \theta/\theta_h, 3 - \chi_t$; (b) ratio of total heat fluxes removed from plate 1-turbulent flow and 2-laminar flow.

of laminar flow reaches significant values [Fig. 6.9(b)]. For instance, for a steel plate with $\Delta/L = 1/10$ past air ($Bi_{*_L}= 0.8$) or water ($Bi_{*_L}= 4.5$), this ratio is 1.2 and 1.65, respectively. In the case of turbulent flow, the difference between total fluxes is smaller [Fig. 6.9(b)], but the distributions of the local characteristics along the plate in two cases of opposite direction of the flow differ in essence as well.

Comment 6.5 A traditional approach of using the boundary condition of third kind and average heat transfer coefficient results in the same solution for both cases independent of the flow direction.

Example 6.6 A plate streamlined on the one side and isolated on another.

Let the temperature θ_0 and heat flux $\bar{q}_x = \bar{q}_0$ are given at the starting end of a plate. Using these boundary conditions and last equations (6.14), one finds first equation (6.33), which gives the solution of the problem in question:

$$\frac{\theta'}{\theta_0} = \vartheta_1 + \bar{q}_0 \vartheta_2, \qquad \bar{q}_0 = \frac{\theta_L/\theta_0 - \vartheta_1(z_L)}{\vartheta_2(z_L)}, \qquad \bar{q}_0 = \frac{\bar{q}_L - \vartheta_1'(z_L)}{\vartheta_2'(z_L)}. \tag{6.33}$$

The same relation for θ'/θ_0 gives a solution for some other boundary conditions if the corresponding value of $\bar{q}_0$ at the starting end is first specified. For example, if the temperature θ_L or heat flux $\bar{q}_x = \bar{q}_L$ is given at the trailing edge, the proper value of $\bar{q}_0$ is determined by second or third expression (6.33), respectively (Exercise 6.25).

Figure 6.10 shows the variation of nonisothermicity coefficient and the temperature head for laminar (a) and turbulent (b) flows in the three cases $\bar{q}_0 = 10$, 0, and (-2). In the first two cases, the temperature head increases along the plate, whereas in the third one, the temperature head first decreases, and after reaching zero, its absolute value starts to increase. The same character of heat transfer variation as in other examples examined before is seen. For the increasing temperature head, the heat transfer coefficients are greater than those for an isothermal

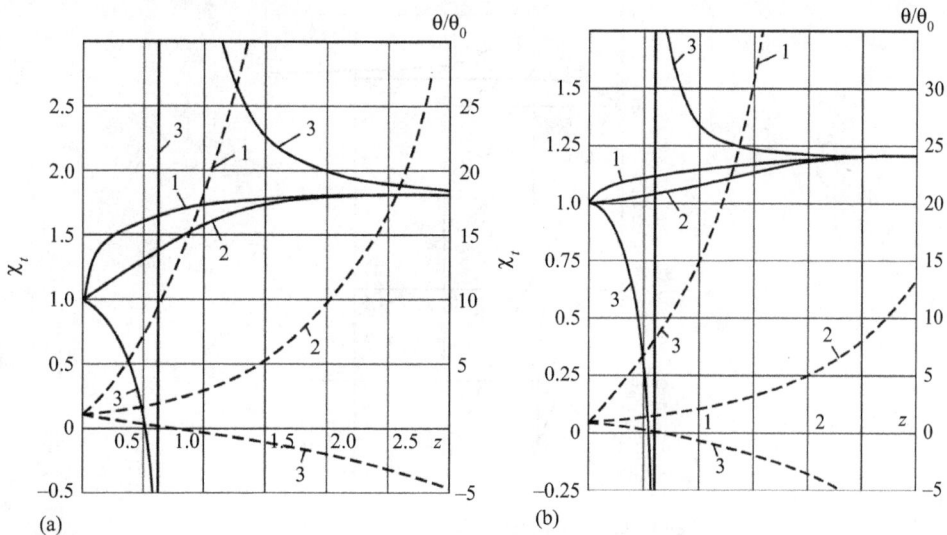

Figure 6.10. Heat transfer characteristics for the plate streamlined on the one side by laminar (a) and turbulent (b) flows _____ χ_t, - - - - θ/θ_0, $1 - \bar{q}_0 = 10$, $2 - \bar{q}_0 = 0$, $3 - \bar{q}_0 = -2$.

surface but not more than 75–80% in the case of laminar flow and not more than 20–25% for turbulent flow. In the third case in which temperature head decreases, these coefficients are so much smaller that in some points where the temperature head turns to zero, the heat transfer coefficient becomes meaningless, and the corresponding curve $\chi_t(z)$ undergoes discontinuity.

The same effect of the temperature head distribution on heat transfer coefficients variation along the plate shown in Figure 6.5 presented data obtained in Example 6.2 for a copper plate streamlined by air on the one side.

Example 6.7 A plate streamlined on both sides by turbulent flow and heated from leading edge.

In this case, the scales in equations (6.14) should be substituted by $T_{\infty 2} - T_{\infty 1}$. Considering the same case as in Example 6.6 with given temperature θ_0 and heat flux $\overline{q}_0$, we have $\theta_0 = 1$, and then from equation (6.14), first equation (6.34) is obtained. Two others for local heat fluxes on both sides of a plate, where σ_{Bi} is defined by (6.8), follow from equation (1.7) for a thermally thin plate and universal equation (4.1) for plate

$$\overline{q}_x = \theta', \qquad \overline{q}_{w1} = \theta'' + \sigma_{Bi} z^{-r/s}, \qquad \overline{q}_{w2} = q_{w1} - z^{-r/s}. \qquad (6.34)$$

In Figure 6.11 are plotted the computed results for the case when $\overline{q}_0 = -2$, and the thermal resistances of both flows are equal. Therefore, according to equation (6.8), $\sigma_{Bi} = 1/2$. The same pattern is observed. On the side on which the temperature head increases, χ_t is a little greater than unity, whereas on the other side where there is a section with decreasing temperature head, the value of χ_t sharply falls, becomes zero, and then reaches $\pm\infty$, resulting in discontinuous curve $\chi_t(z)$ for nonisothermicity coefficient (Exercise 6.26).

Example 6.8 A plate with inner heat sources

The problem with sources is an inhomogeneous problem solution of which is presented as a sum of general and particular solutions. The former is found similar to other solutions

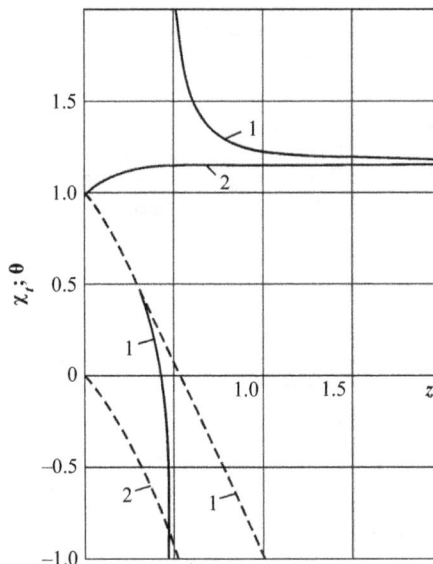

Figure 6.11. Variation of temperature head and nonisothermicity coefficient along the plate streamlined on both sides by turbulent flow
$\sigma_{Bi} = 0.5$, $\theta_0 = 1$, $\overline{q}_0 = -2$, _____ χ_t, - - - - θ,
1-2 different sides of a plate.

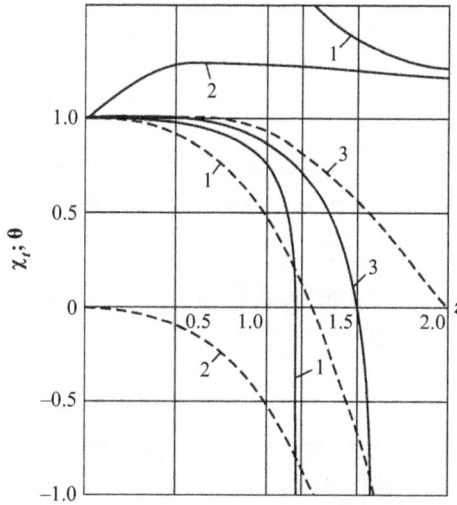

Figure 6.12. Heat transfer characteristics for the plate with inner heat sources streamlined by turbulent flow. _____ χ_t, - - - - θ, 1-2 different sides of a plate, $\sigma_{Bi} = 0.5$, 3-one side streamlined plate ($\sigma_{Bi} = 0$).

considered before using functions ϑ_1 and ϑ_2. For the case of uniform or linear distributed sources, the particular solution in form (6.13) can be obtained using functions ϑ_3 and ϑ_4 (Figs. 6.3 and 6.4). Figure 6.12 shows the results for turbulent flow with conditions at the leading edge $\theta_0 = 1$, $q_0 = 0$ and inner linear heat source defined by equation (6.13) and dimensionless coefficients $\overline{A} = 1$, $\overline{B} = 2$.

Again we see that the values and variation of heat transfer coefficients strongly depend on temperature head distribution. For the plate streamlined from the one side and for the side with temperature θ_1 of the plate past two sides, for which curves $\theta(z)$ have decreasing sections, the heat transfer coefficients distributions $h(z)$ essentially differ from that $h_*(z)$ for isothermal coefficients. At the same time, for another side of two sides of streamlined plate with increasing temperature head, the ratio h/h_* do not differ much from unity.

Example 6.9 Heat exchange process between two countercurrent fluids.

This problem is considered assuming that the thermal resistance of a plate is negligible in comparison with that of the fluids. This assumption considerably simplifies the problem that is more complicated than the similar concurrent conjugate problem considered in Example 4.1. At the same time, such a simplified problem retains the basic qualities of the same problem for the plate with finite resistance.

In the case of laminar flow after using universal equation (4.1) with the first three terms, the equality of heat fluxes from both sides of the plate leads us to an equation which defines the plate temperature:

$$[g_{22}\zeta^{3/2} + g_{12}\sigma(1-\zeta)^{3/2}]\theta'' + [g_{12}\zeta^{1/2} - g_{11}\sigma(1-\zeta)^{1/2}]\theta' + [\zeta^{-1/2} + \sigma(1-\zeta)^{-1/2}]\theta - \zeta^{-1/2} = 0.$$

$$(6.35)$$

Here, θ is defined by equation (6.1) and $\sigma = h_{*1}/h_{*2}$. Boundary conditions $\theta(0) = 1$ and $\theta(1) = 0$ follow from the fact that at each end, the temperature of the plate is equal to the temperature of the fluid for which this end serves as an initial (Fig. 6.13). Equation (6.35) was integrating numerically for different values of σ by solving two problems, satisfying [similar to conditions (6.12)] the following boundary conditions: $\vartheta_1(0) = \vartheta_1'(0) = \vartheta_2(0) = 1$ and $\vartheta_2'(0) = 0$. Initial part of the solution at $\zeta = 0$ were found using series (6.2). The coefficients of series and final solution in terms of functions ϑ_1 and ϑ_2 are

$$a_0 = 1, \ a_1 = a_3 = 0, \ a_2 = \vartheta'(0), \ a_4 = \frac{a_2 g_{11} - 1}{2 g_{21}}, \quad \theta = \frac{\vartheta_2(1)\vartheta_1(\zeta) - \vartheta_1(1)\vartheta_2(\zeta)}{\vartheta_2(1) - \vartheta_1(1)}. \quad (6.36)$$

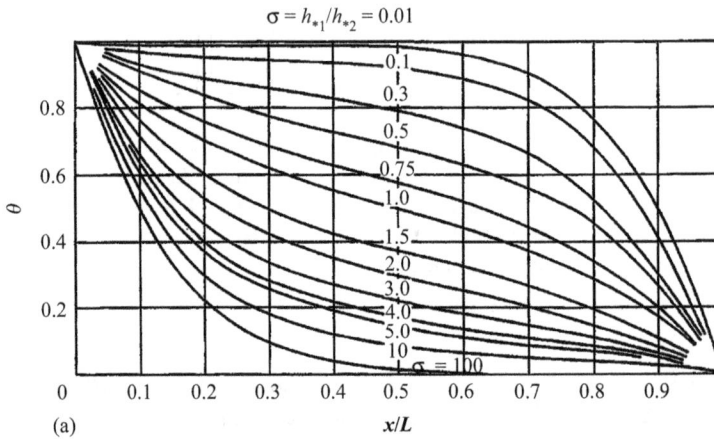

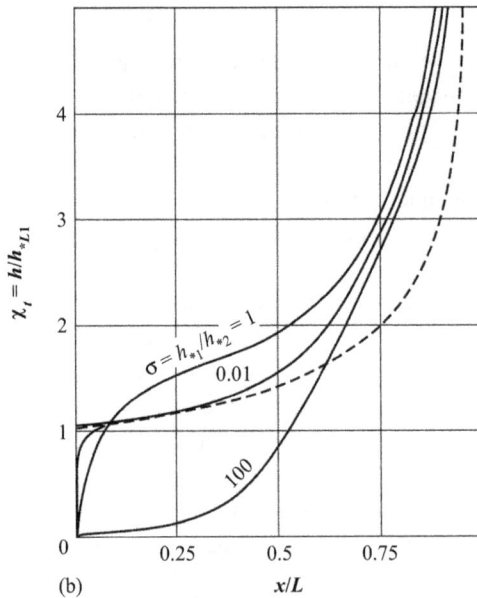

Figure 6.13. Variation along the plate of the heat transfer characteristics for two countercurrent flows (a) temperature head, (b) heat transfer coefficient, - - - - for the case of an isothermal plate.

Results for $Pr > 0.5$ for which coefficients g_k are independent of the Prandtl number are plotted in Fig. 6.13(a), showing that in the case of countercurrent flows the temperature along the interface changes significantly. In the case of equal thermal resistances ($\sigma = 1$) when the variation is maximal, the heat flux is about 30% bigger than that calculated with an isothermal heat transfer coefficient. However, because the temperature head grows in the flow direction along the interface, the distribution of heat transfer coefficient does not have singularities and not differs much from that for an isothermal plate [Fig. 6.13(b)].

At the same time, the results at the ends obtained by conjugate and traditional approaches differ in essence. The heat transfer coefficient at each of the ends obtained in the latter problem is equal to that of one of fluids because the heat transfer coefficient of another fluid $h_* \to \infty$. Therefore, the corresponding heat fluxes at the ends are finite. In contrast to that, the heat fluxes at the ends obtained in the conjugate problem are zero. It follows from the fact that according to first coefficient (6.36), the temperature head at the end where $x = 0$ tends to zero as x, whereas the heat transfer coefficient tends to the infinite value as $x^{1/2}$. Thus, the heat flux tends to zero as $x^{1/2}$. Due to the conjugate conditions, the heat fluxes on the other side tends to zero as well (Exercise 6.29).

As a result, the temperature profile on this side deforms close to end in the same way as in the case when a flow impinges on an adiabatic wall (Exercise 6.30). The scheme of such a profile deformation is plotted in Figure 6.14. This occurs because the fluids after the end of contacts arrive on the isolated surface. In such a case, the heat flux drops abruptly to zero, while the temperature head decreases gradually, becoming practically zero only at a certain distance from the entrance point.

Comment 6.6 In reality, the heat fluxes at the ends are not zero due to finite boundary layer thickness. In consequence, the described deformation process starts not at edges rather on some small distance from those.

The results presented here show that the effect of conjugation in the case of a countercurrent flow is more significant than that for concurrent fluids. In particular, in the latter case (Example 4.1), the surface at zero resistance is isothermal (Exercise 6.31), while for the countercurrent flows even in this simplest case, the interface is considerably nonisothermal so that the effect of conjugation should be taken into account.

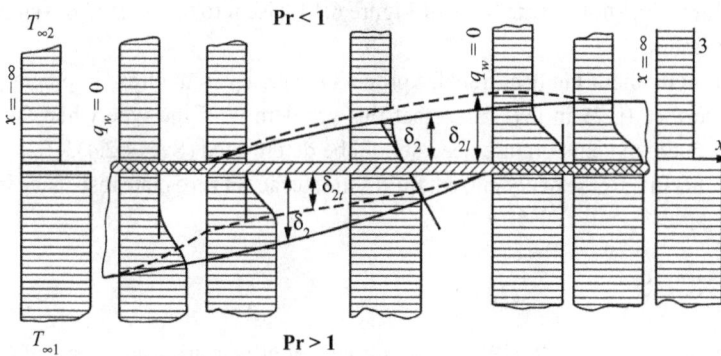

Figure 6.14. Scheme of temperature profile deformation in the heat transfer process between countercurrent fluids.

EXERCISES

6.18 Create software for solving conjugate heat transfer problems (i) following directions from the text to compute error $\varepsilon(x)$, (ii) using equations (6.10), (6.11), and (6.12) instead of charts and equations (4.14) and (4.44) for farther approximations. Calculate one of the problems considered in Examples 6.1–6.3. Estimate accuracy of data obtained by chart in the problem you chose.

6.19 Obtain equation (6.23) and inequalities (6.24) following directions from the text.

6.20 Derive equation (6.27) using second equation (6.26) and parameters values of first approximation. Hint: first find connections between notations in Example 4.1 and in equations (6.26).

6.21 Show that the series describes the temperature (or temperature head) at $x = 0$ for a thin plate containing all powers of variable $\zeta^{1/2}$. Hint: substitute the series in the power $\zeta^{1/2}$ in equation (4.16) for θ_1 or θ_2 and collect terms with the same power of variable $\zeta^{1/2}$.

6.22 Learn how coefficients (6.28) for temperature series in vicinity of $x = 0$ are derived (p. 173 in [3]).

6.23 Derive equation (6.32) for total heat flux by integrating equation (6.14) and then obtain second equation (6.32), determining ratio of total fluxes taking from a plate in two cases. Hint: note that derivatives in expressions (6.14) are taken with respect to variable (6.10) z.

6.24 Calculate and plot temperature distribution in the second case from Example 6.5. Determine the coordinate of minimum and explain physically why minimal value occurs.

6.25 Show that two last relations (6.33) give proper values for $\bar{q}_0$ if the temperature or heat flux is given at the trailing edge. Hint: find the solution of corresponding problem using a usual approach by defining constants in relation in equation (6.7) and compare the result with first expression (6.33).

6.26 Derive expressions (6.34) following directions from the text. Explain physically why the curve $\chi_t(z)$ for nonisothermicity coefficient becomes discontinuous. Hint: see Section 5.4.

6.27 Solve the problem from Example 6.7 for laminar flow and parameters indicated on Figure 6.11. Compare both results.

6.28 *Solve the problem from Example 6.8 for laminar flow and parameters indicated in the text. Compare both results. Hint: see Section 1.5.1.

6.29 Draw scheme of heat transfer between countercurrent flows similar to that for concurrent flows (Fig. 4.4) on the basis of Figure 6.14. Use it to understand physical analysis of flows features.

6.30 Learn more about the heat transfer process occurring in the flow impinges on the adiabatic surface (p.73 in [3]). Note that this problem is of the type when the heat flux is given, while the temperature head should be determined (Sec. 4.2.4).

6.31 Show that in the case of concurrent fluids, the surface at zero plate resistance is isothermal.

REFERENCES

[1] Sohal, M. S., & Howell, I. R. (1973). Determination of plate temperature in case of combined conduction, convection and radiation heat exchange. *International Journal of Heat Mass Transfer 16*, 2055–2066. See also Example 6.25 (p.223 in [3]). doi: http://dx.doi.org/10.1016/0017-9310(73)90108-7

[2] Dorfman, A. S. (1985). Combined heat transfer over the initial segment of a plate in a flow. *Heat Transfer-Soviet Research 18*, 52–74.

[3] Dorfman, A. S. (2009). *Conjugate Problems in Convective Heat Transfer*, Boca Raton: CRC Press, Taylor & Francis Boca Raton.

CONCLUSION OF HEAT TRANSFER INVESTIGATION (CHAPTERS 4–6)

SHOULD ANY HEAT TRANSFER PROBLEM BE CONSIDERED AS A CONJUGATE?

- At the beginning of Chapter 4, it was stated that nonisothermal and conjugate heat transfer processes are closely related to each other so that both phenomena are actually characterized by the same general properties. Their physical analysis as well as a number of examples reviewed in following chapters proved this statement showing that:

  (i) Temperature head distribution mainly determines heat transfer character and intensity and
  (ii) Decreasing temperature head is the basic reason of low heat transfer rate.

- Detailed analysis in Section 5.2 reveals that the Biot number is a primary measure of degree (or level) of a problem conjugation, indicating that the greatest conjugate effect exist when both resistances are of the same order, and the Biot number is close to unity.
- At a given value of the Biot number, other factors affect more or less the degree of conjugation. The table below listed main of those factors where they are arranged so that next to the right issue, each represent a subject with a lower degree of conjugation. For instance, because the turbulent flow is located to the right of laminar flow, this means that the conjugation effect in the problems with turbulent flows is less than that in corresponding problems with laminar flows.

Table C.1. Effect of different factors on the level of conjugation

Decreasing temperature head		Increasing temperature head
Comparable thermal resistance		Incomparable thermal resistance
Contourcurrent flows		Concurrent flows
Unfavorable pressure gradients	Gradientless flows	Favorable pressure gradients
Laminar flows	Transition flows	Turbulent flows
Small Reynolds numbers	Mean Reynolds numbers	High Reynolds numbers
Unsteady heat transfer		Steady heat transfer
Non-Newtonian fluids with $n > 1$	Newtonian fluids	Non-Newtonian fluids with $n < 1$
Small surface curvature		Large surface curvature
Porous surface with injection	Nonporous surface	Porous surface with suction
Continuously moving surface		Streamlined surface

- The data from the Table C.1 and other conclusions presented above are of critical importance for answer the question formulated in the heading: What method should be used for solving a particular problem, conjugate or common simple approach? Taking into account facts given here together with regard to the aim of a problem in question and the desired accuracy of results allow a researcher making a decision in a specific case.

- More reliable way to answer the key question is described in Section 4.4.2. In this method, one starts with the common solution using boundary conditions of the third kind. Then, the error arising by using this traditional approach may be approximately estimated by computing the second term of universal function applying the knowing data of traditional solution. Examples show that such found estimations are in reasonable agreement with corresponding conjugate solutions, giving the understanding of necessity of the conjugate solution.

- Studies indicate that for two types of problems the general rules may be stated:

  (i) Convective heat transfer problems, containing temperature head decreasing in flow the direction or in time, should be considered as a conjugate because in this case, the effect of conjugation is usually significant (Sec. 5.1).

  (ii) For the turbulent flow of fluids with high (say higher than 100) Prandtl numbers, the convective heat transfer problems may be solved using a traditional approach with boundary condition of the third kind because for such fluids effect of conjugation is negligible (Figs. 4.18 and 4.19).

CHAPTER 7

PERISTALTIC MOTION AS A CONJUGATE PROBLEM

MOTION IN CHANNELS WITH FLEXIBLE WALLS

7.1 What is the Peristaltic Motion Like?

The word peristalsis came from the Greek peristaltikos and means clasping and compressing. Peristalsis or peristaltic motion arises when progressive wave moving along the flexible tube results in contraction and expansion of the channel wall, inducing the motion of fluid inside the channel in the wave direction. Although there are a lot of engineering applications of peristaltic motion, the original idea of this phenomenon was adapted from creation. Peristalsis is an inherent property of human or primates animal organs transporting physiological fluids, such as blood or urine. Muscular walls of such organs in the form of a tube provide consecutive narrowing and relaxing of a wall portion, which travels in the longitudinal direction, resulting due to conjugation in movement of fluid inside a tube. Some examples are: (i) the gastrointestinal tract, in which muscles provide the swallowing and movement of the chyme from the mouth, through esophagus, stomach, small and large intestines to rectum and anus, resulting in the digestive process; (ii) the ureter transporting due to its muscles the urine from the kidney to the bladder against the pressure gradient; (iii) the urethra, which discharges the urine outside from the body; (iv) male reproductive tract supplying the spermatic flows ejection; (v) the ovum motion inside the fallopian tube and the transport of embryo in uterus; (vi) small blood vessels like venuels and arterioles as well as lymphatic channels (Exercise 7.2).

Peristaltic motion is also widely used in engineering devices and systems, for examples: (i) an artificial heart-lung device that maintains the circulation of the blood and the oxygen content of the body repeatedly drawing off the blood from the veins, re-oxygenates it, and pumps it into the arterial system—(ii) an artificial kidney machine (hemodialysis) providing waste products, such as urea and creatinine, diffuse through the membrane into the dialysis fluid and are discarded—(iii) devices for pumping biomedical fluids, such as blood, clean and sterile stuff, pharmaceutical production, and food to prevent the transported substance from contact with the parts of the mechanical pump as well as to isolate environment from conveying materials, such as corrosive fluid, slurries, and sewage, (iv) microdevices for improving the mixing process of chemical reagents and facilitate preparing biological and other mixtures and (v) microdevices for enhancing a mass transfer rate through the porous channels.

7.2 Formulation of the Conjugate Problem

We consider the peristaltic phenomenon as a conjugate problem because any peristaltic system is inherently a conjugate. This follows from the fact that the peristaltic flow occurs due to the close interaction between a flexible wall and fluid containing inside the channel. Similar to the approach for conjugate heat transfer described above, we use in this case a structure of conjugate problem consisting of two domains and conjugate conditions on the interface: (i) the fluid domain governed in general case by Navier–Stokes equation or by simplified versions of this equation, that is, the creeping (Sec. 2.5.1) or boundary layer (Sec. 2.5.2) equation for low and high Reynolds number, respectively; (ii) the wall domain governed for the rigid wall by given form of propagation wave (often sinusoidal) or by dynamic equation for a thin flexible wall; and (iii) the conjugate relations containing the no-slip condition and the balance of forces on the interface instead of equalities of temperatures and heat fluxes in the case of the heat transfer conjugate problem.

   Although the formulations of both conjugate problems, heat transfer and peristaltic flow, are similar, those problems differ significantly because in contrast to the former, the latter is nonlinear and due to that is much complicated to be solved. In particular, the efficient superposition method widely used in the case of conjugate heat transfer is not applicable to the peristaltic problems, so in this case, the known nonlinear results are approximate solutions of perturbation series (usually as one or two first terms) (Sec. 2.5.2.4) or numerical solutions. Due to complexity, the majority solutions of peristalsis problems are also not full conjugate, rather those are semi-conjugate solutions. We use this term to specify the approach when for simplicity instead of flexible wall one considers a rigid boundary with assigned wave propagating along the wall. In such a semi-conjugate problem, the only effect of the propagation wave on fluid flow is studied, whereas the backward effect of fluid flow on the wall motion is neglected (Exercise 7.4).

   A typical semi-conjugate model presents a peristalsis as a flow in a plane two-dimensional channel with walls flexible in transverse direction on which the longitudinal progressive

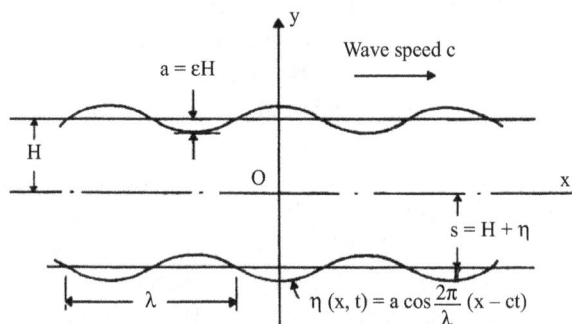

Figure 7.1. Scheme of the two-dimensional channel of a semi-conjugate model.

sinusoidal waves are imposed (Fig. 7.1). Such a model specifies a particular problem by four dimensionless parameters: (i) the amplitude ratio $\varepsilon = a/H$, where a is the amplitude of the wave and H is a half of channel cross-section; (ii) the wave number $\alpha = 2\pi H/\lambda$; (iii) the Reynolds number $R = cH/v$, where c is a wave velocity; and (iv) the dimensionless pressure gradient or the dimensionless time-mean flow rate.

The flow is unsteady in a fixed coordinate system (the laboratory frame) because of moving boundary. However, in a coordinate system moving with the wave (the wave frame), the flow is steady. The longitudinal velocities and coordinates in both systems are connected as $u = \hat{u} + c$ and $x = \hat{x} + ct$, $y = \hat{y}$, where overscore denotes the values in the wave frame. There is no analytical solution for arbitrary values of all four parameters; rather, the known results are found using assumption that at least one of parameters is zero or small. Most studies, especially in early years, are performed for the small or zero Reynolds number and long wavelength. Typical values of parameters are [1] $R \approx 1$, $\alpha = 0.02$, $\varepsilon < 1$ and $R \approx 10$, $\alpha = 2$, $\varepsilon \approx 0.3$ for ureter and gastrointestinal tract, respectively, $R \approx 1$, $\alpha = 0.5$, $\varepsilon \approx 1$ for the roller pumps.

There are countless articles of different peristaltic pumping applications. Here, we consider examples presenting the basic methods of solution of the problems. Examples of applications employing those methods are given in Chapter 12.

7.3   Early Works

The review of the early results published up to the 1970 was given by Jaffrin and Shapiro [1]. These results are obtained basically in studying the peristaltic flow in the ureter using simple linear models in which the peristalsis is considered as a flow induced by propagating sinusoidal wave in an infinite, two-dimensional channel. The main objectives of the early studies (e.g., [1–3]) were the understanding of the peristalsis mechanism to get some insight into physical processes in the ureter. A special interest has the flow reflux, in particular, the conditions when this phenomenon occurs because it might be one of the reasons why bacteria sometimes go opposite to the main flow from bladder to kidney. Early researches also contributed to engineering applications investigating, for example, the peristaltic flow in the blood roller pumps and experimentally confirming the early theories of peristaltic pumping.

Example 7.1 Peristaltic motion in the two-dimensional channel at low Reynolds number and long wavelengths (linear model) [1,2].

In the frame moving with wave, the peristaltic flow is steady, and at low Reynolds number, the problem is similar to that for steady flow in the two-dimensional channel or in a tube, seen in Section 2.4.3. However, there is a difference which becomes clear if one looks at the relation between the flow direction and pressure gradient. In both flows in the fixed frame, the pressure decreases in the flow direction because of energy losses. At the same time, in moving frame, peristaltic flow is directed opposite to that in the fixed frame, while the pressure gradient remains unchanged. Thus, for the problem in question, the Navier–Stokes equation simplifies, as in creeping flow, taking the same form (2.47) as in the case of the exact solution of the Navier–Stokes equation for the flow in channel but with a different sign at the pressure gradient:

$$\frac{\partial p}{\partial \hat{x}} + \frac{\partial^2 \hat{u}}{\partial \hat{y}^2} = 0, \qquad u = \frac{1}{2\mu}\left(\frac{\partial p}{\partial x}\right)\left(y^2 - s^2\right), \qquad s = H + \eta, \qquad \eta = a\cos\frac{2\pi}{\lambda}(x - ct). \qquad (7.1)$$

Comment 7.1 The differential equations (7.1) and (2.47), except signs, look identical. Nevertheless, from the viewpoint of physics, these equations differ in essence because the latter is an exact Navier–Stokes equation for the case of flow inside the two-dimensional channel (see beginning of Sec. 2.5), whereas the former is obtained using some assumption to simplify the Navier–Stokes equation (Exercise 7.5).

Double integration of differential equation (7.1) subjected to boundary conditions $\hat{u} = -c$ (wave velocity) and $(\partial\hat{u}/\partial\hat{y})_{\hat{y}=0} = 0$ (symmetry) results after returning to stationary frame in second equation (7.1), defining the distribution of flow velocity (Exercise 7.6). It is seen that this distribution is of Poiseuille type (Sec. 2.4.3); however, profile (7.1) unlike Poiseuille one [for example, (2.48) or (2.49)] depends not only on transverse coordinate but through variable s is also a function on the longitudinal coordinate $\xi = 2\pi x/\lambda$ and time $\tau = 2\pi ct/\lambda$. The instantaneous dimensionless flow rate in the fixed frame is obtained by integration of profile (7.1). Then, the corresponding flow rate in the wave frame is found using relation $u = \hat{u} + c$ and expression for dimensionless pressure $\bar{p}$:

$$Q(\xi,\tau) = \frac{1}{Hc}\int_0^s u\,dy = -\frac{2}{3}\frac{d\bar{p}}{d\xi}s^3, \ \bar{p} = \frac{2\pi H^2}{\mu\lambda c}p, \hat{Q}(\tau) = Q(\xi,\tau) - s, \frac{d\bar{p}}{d\xi} = -\frac{3(\hat{Q}+s)}{s^3}. \quad (7.2)$$

It follows from these equations that the flow rate in the moving frame depends only on time, and the pressure gradient $dp/d\xi$ strongly depends on wave characteristics so that high-pressure gradients occurs at $s \ll 1$ when the wave is intensive and contractions are severe.

Analysis of Lagrangian trajectories of flow particles shows that under certain conditions reflux occurs near the walls. At the same time, the negative Eulerian time-mean velocity could not be interpreted as reflux because calculation and experimental data presented examples when despite the negative Eulerian average velocity, the particle Lagrangian trajectories undergo positive displacement (Exercise 7.7).

Comment 7.2 The Lagrangian specification of the flow field is the way when an observer moves with flow, looking at a particular particle (moving frame). The trace of such a particle gives a trajectory. The Eulerian specification of the flow field is the way when an unmoving observer is focused on a specific location in a space, recording the parameters of passing flow in this location (laboratory frame).

7.4 Semi-Conjugate Solutions

Example 7.2 Peristaltic motion in the two-dimensional channel at a finite Reynolds number and moderate amplitude (nonlinear model) [3].

The nonlinear approach applying perturbation series developed in this early publication is widely used up to now for applications (Examples 7.5 and 7.6), including studying peristaltic motion of Newtonian and non-Newtonian fluids (Examples 7.4–7.7).

In this approach, the problem is governed by Navier–Stokes equation (2.21) in the form of stream function, and the perturbation series in power of $\varepsilon = a/H$ are used for the solution. Both relations in dimensionless variables are

$$\frac{\partial}{\partial t}\nabla^2\psi + \psi_y\nabla^2\psi_x - \psi_x\nabla^2\psi_y = (1/R)\nabla^2\nabla^2\psi, \ \ \psi = \psi_0 + \varepsilon\psi_1 + \varepsilon^2\psi_2 + \varepsilon^3\psi_3 + \cdots \quad (7.3)$$

Here, all linear and velocity parameters are scaled by half wide of channel H and wave speed c, respectively, and for others, the corresponding scales are used: cH for ψ, ρc^2 for p, H/c for t, and Reynolds number is defined as $R = cH/v$. In the same scales, the sinusoidal wave has the form $\eta = \varepsilon \cos \alpha (x - t)$. The solution of equation (7.3) should satisfy boundary conditions on the upper and lower waves (upper signs pertain to upper wave):

$$\psi_y = 0, \quad \psi_x = \mp \alpha \varepsilon \sin \alpha (x - t) \text{ at } y = \pm (1 + \eta). \tag{7.4}$$

Substituting series (7.3) into equation (7.3) and collecting terms with the same powers of ε yield equations determining coefficients ψ_n (Exercise 7.8):

$$\frac{1}{R} \nabla^2 \nabla^2 \psi_0 = \frac{\partial}{\partial t} \nabla^2 \psi_0 + \psi_{0y} \nabla^2 \psi_{0x} - \psi_{0x} \nabla^2 \psi_{0y}, \tag{7.5}$$

$$\frac{1}{R} \nabla^2 \nabla^2 \psi_1 = \frac{\partial}{\partial t} \nabla^2 \psi_1 + \psi_{0y} \nabla^2 \psi_{1x} + \psi_{1y} \nabla^2 \psi_{0x} - \psi_{0x} \nabla^2 \psi_{1y} - \psi_{1x} \nabla^2 \psi_{0y}, \text{ and so on.}$$

To get boundary conditions for these equation (7.4) are presented in Taylor series in power of η. For the first relation, one gets $\psi_y + \eta \psi_{yy} + (\eta^2 / 2) \psi_{yyy} + \cdots = 0$. Substituting expressions for η and for ψ [i.e., (7.3)], in this and similar equation obtained from (7.4), yields boundary conditions for equations (7.5) (Exercise 7.9):

$$\psi_{0y} = 0, \quad \psi_{1y} + \psi_{0yy} \cos \alpha (x - t) = 0, \tag{7.6}$$

$$\psi_{0x} = 0, \quad \psi_{1x} + \psi_{0xy} \cos \alpha (x - t) = -\sin \alpha (x - t), \text{ and so on.}$$

For the case of constant pressure gradient $\partial p / \partial x$, first equation (7.5) with two first boundary conditions (7.6) and symmetry condition give the Poiseuille profile for ψ_0, whereas second equations (7.5) and (7.6) lead us to the Orr–Sommerfeld differential equation:

$$\psi_0 = -R \left(\frac{\partial p}{\partial x} \right) \left(y - \frac{y^3}{3} \right), \left\{ \frac{d^2}{dy^2} - \alpha^2 + i\alpha R \left[1 + R \frac{dp}{dx} (1 - y^2) \right] \right\} \left(\frac{d^2}{dy^2} - \alpha^2 \right)$$

$$\times \Phi + 2i \frac{dp}{dx} R^2 \Phi = 0 \tag{7.7}$$

defining $2\psi_1 = \Phi \exp[i\alpha(x - t)] + \tilde{\Phi} \exp[-i\alpha(x - t)]$, where the equation determining function $\tilde{\Phi}(y)$ is conjugate to equation (7.7) for $\Phi(y)$.

Comment 7.3 Here, the term conjugate is used in another common meaning, indicating that the complex function $z = x - iy$ is conjugate to function $z = x + iy$ (Exercise 7.10).

In addition to equation (7.7) that determines the two first functions in solution (7.3), in [3] is found the third function ψ_2 which has the same structure. Thus, these functions giving the solution of the problem up to second order of ε require integration of the Orr–Sommerfeld equation. Recall that the Orr–Sommerfeld equation is the basic equation in the theory of laminar flow stability, which we briefly considered in Section 3.1.2. Since this equation is a fourth-order nonlinear ordinary differential equation with variable coefficients, it may be integrated only numerically (Exercise 7.11). However, the numerical integration of Orr–Sommerfeld equations was not performed in the reviewed paper; rather, the free pumping regime at zero initial pressure gradient ($dp/dx = 0$) was considered showing that in this case equation (7.7) may be integrated analytically. Using the solution obtained after integration, authors derived an expression for the mean (averaged over time) axial velocity distribution and show that it significantly depends on the average pressure gradient $(dp/dx)_2$ of order ε^2 induced by peristaltic flow. Analysis indicates that this distribution: (i) at negative pressure gradient $(dp/dx)_2 < 0$ is close to the Poiseuille profile; (ii) as the pressure gradient grows, it deforms becoming concave at $(dp/dx)_2 \approx 0$; (iii) as the pressure gradient continue increasing and reaches some positive value, which authors identify as the critical, the velocity of a caved-in profile becomes zero on the central line of a channel and then turns to be negative when the pressure gradient $(dp/dx)_2$ exceeds a critical value.

This reversal moving flow in the neighborhood of the channel central line was interpreted in [3] as a reflux, which may be considered as a possible reason of bacteria pass from the bladder to the kidney. However, later on, it was shown [2] that such a backward flow could not be seen as a reason of bacteria moving; rather, considering the Lagrangian trajectories of fluid particles is a more reliable way to estimate the direction of bacteria transmission (see Example 7.2).

Example 7.3 Peristaltic motion in the two-dimensional channel at a moderate Reynolds number (numerical solution) [4].

The problems for the moderate Reynolds number are usually solved numerically because in this case both inertia and viscose terms are of the same order, and the full Navier–Stokes equation should be considered. In this study, the Navier–Stokes equation is used in the forticity-stream function ($\omega - \psi$) variables (Sec. 2.2.2.3). In the wave frame, the problem in steady, and these equation and relevant boundary conditions are as follows:

$$\frac{\partial \psi}{\partial y}\frac{\partial \omega}{\partial x} - \frac{\partial \psi}{\partial x}\frac{\partial \omega}{\partial y} = \frac{1}{\text{Re}}\left(\alpha^2 \frac{\partial^2 \omega}{\partial x^2} + \frac{\partial^2 \omega}{\partial y^2}\right), \quad \alpha^2 \frac{\partial^2 \psi}{\partial x^2} + \frac{\partial^2 \psi}{\partial y^2} = -\omega. \tag{7.8}$$

Central axis: $\psi = 0$, $\dfrac{\partial \psi}{\partial y} = -1$, $\dfrac{\partial^2 \psi}{\partial y^2} = 0$, wave: $\psi = q$, $\dfrac{\partial \psi}{\partial y} = -1$, input–output $\dfrac{\partial \psi}{\partial x} = 0$.

Here, dimensionless variables are scaled using λ for x, H for y, c for velocities, cH for ψ, c/H for ω, cH for flow rate q in wave frame, and the Reynolds number is defined as $\text{Re} = (cH/v)\alpha$, where $\alpha = H/\lambda$. Equations (7.8) specified conditions on all boundaries of considering domain as is required for the Navier–Stokes equation (Sec. 2.3): (i) the no-slip condition on the lower (centre axis $\psi = 0$) and the upper (wave $\psi = q$) boundaries, (ii) the symmetry condition on the lower boundary ($\psi = 0$), (iii) assuming that there is no-flow at the leading and trailing edges of domain as if it were infinite long (Exercise 7.12).

Comment 7.4 To carry out the last condition (7.8) practically, one obtains numerical results using increasing numbers of longitudinal points. Comparing the desired accuracy with the difference of these data allows estimation of the number of points at which both values are of the same order, showing that this condition is approximately realized.

The problem was solved numerically applying the finite-difference approach (Chapter 8). The details of numerical scheme may be found in [4]. The longitudinal velocity profiles in both ending sections were compared with experimental data from [5] showing qualitatively agreement.

The basic results are as follows: (i) In the major central part with forward flow, the longitudinal velocity profiles are almost parabolic close to that obtained in the simple linear model (Example 7.1); at the leading and trailing edges exist relatively small regions with retrograde flow. (ii) At $Re < 1$, the velocity profiles are independent of Re, but as the Reynolds number grows, they decrease reaching asymptotically the potential flow profiles; (iii) the Lagrangian particles trajectories indicate that the reflux occurs near the wall at $Re < 1$, but it exists near the axis for $Re > 1$. (iv) The pressure rise per wavelength is independent of Re at $Re < 1$ and monotonically decreases as Re groves at $Re > 1$. The relation between the dimensionless pressure rise per wavelength and the time-mean flow rate is close to linear, which is important in using for engineering applications. (v) Shearing stresses vary slightly along the wall for small values of $\alpha = H / \lambda$ and are steep distributed having remarkable large maximum for large values of α, which should be taken into account in designing the peristaltic pumps. (vi) The results for the pressure rise and shearing stresses plotted via product $\alpha \varepsilon = (H / \lambda)(a/H) = a/\lambda$ show that curves for different ε closely coincides forming one general relation, indicating that this product is a significant factor in controlling the peristaltic flow behavior.

Example 7.4 Peristaltic motion in closed rectangular container (perturbation solution) [6].

Practical interest in the efficient microelectromechnical system (MEMS) with small scales results in variety of design suggestions for such devices. This study is also aimed to understand the possibility of using the peristaltic flow for mixing processes in small channels to design MEMS without mechanical elements. In conformity with the prototype of MEMS, the applied model consists of a closed, long rectangular container with typical for MEMS high oscillating peristaltic flow. To simulate the peristaltic flow, it is assumed that one of the walls undergoes sinusoidal oscillation in the form of traveling wave $y = 2\varepsilon H \cos(kx - \omega t)$, where k is the wave number and ω is the frequency.

The problem is governed by Navier–Stokes equation in vector form (2.13) and by boundary conditions on the wave $\mathbf{V}(x, y = 2\varepsilon H \cos(kx - \omega t), t) = 0, y = 2\varepsilon H \omega \sin(kx - \omega t)$ and on the other wall $\mathbf{V}(x, y = 2H, t) = 0$. In dimensionless variables, the governed system of equation and boundary conditions become (Exercise 7.13)

$$\alpha^2 \frac{\partial \mathbf{V}}{\partial t} + \varepsilon \alpha^2 \mathbf{V} \cdot \nabla \mathbf{V} = -\nabla p + \nabla^2 \mathbf{V}, \quad \mathbf{V}(x, \varepsilon \cos(\kappa x - t), t) = 0, \sin(\kappa x - t), \mathbf{V}(x, 1, t) = 0. \quad (7.9)$$

Here, lengths are scaled by $2H$, velocities by $2H\varepsilon\omega$, time by $1/\omega$, pressure by $\varepsilon\mu\omega$, and $\kappa = 2Hk$ and $\alpha = 4H^2\rho\omega/\mu$ are the dimensionless wave number and frequency, respectively.

Comment 7.5 Boundary condition is written in vector form $\mathbf{V}(x, y, t) = u, \mathbf{v}$ which means vector $\mathbf{V}$ depending on variables x, y, t has components $u, \mathbf{v}$.

The perturbation series are employed for the solution, taking into account that $\varepsilon \ll 1$:

$$\mathbf{V} = \mathbf{V}_0 + \varepsilon \mathbf{V}_1 + \varepsilon^2 \mathbf{V}_2 + \cdots, p = p_0 + \varepsilon p_1 + \varepsilon^2 p_2 + \cdots, \quad \int_{\varepsilon \cos(\kappa x - t)}^{1} u dy = \frac{\cos t - \cos(\kappa x - t)}{\kappa}. \quad (7.10)$$

The high frequency also required that $\alpha \gg 1$. The last equation (7.10) defined that the flow rate should be considered in addition because of the closed domain. This relation is obtained knowing that velocities on the walls are zero, including sidewalls at $x = \pm L/2$ (assuming that in a long container, the effect of flow at these positions is negligible).

It is shown that the dominant part of solution $\mathbf{V}_0$ satisfied the unsteady Stokes equation (Sec. 2.5.1). Although this problem is similar to Stokes' second problem (Exercise 7.14), analysis indicates that in this case an asymptotic solution consisting of inner and outer perturbation expansions should be used. That is relevant because the problem consists of two small parameters ε and $1/\alpha$. The situation is similar to that we encountered in delivering the second-order boundary layer equations where we employed the technique developed by Van Dyke (Sec. 2.5.2.4). Using the same approach and long mathematical manipulations, the authors obtain the inner and outer expansions for close to and away from walls areas and matching those get the leading-order series term for velocity.

Analysis of final expressions allows formulating several basic features specific for peristaltic flows with high-frequency oscillations: (i) the time-averaged velocity consist of two components—one independent of position x and another which is periodic in x. (ii) The x-independent component is dominant for a wave with a short dimensionless wavelength ($\kappa \gg 1$), and the x-periodic component is dominant for a wave with a long wavelength ($\kappa \ll 1$); for moderate wavelengths ($\kappa \approx O(1)$), both components are comparable. (iii) The time-average velocity along the channel in the boundary layer of thickness $O(2H/\alpha)$ near the oscillating boundary is in the direction of wave and is of order $O(2H\alpha\varepsilon^2\omega)$. (iv) The opposite to wave direction flow in a closed channel for $\alpha \gg 1$ is independent of α and is of order $O(2H\varepsilon^2\omega)$. (v) For high frequencies, Eulerian and Lagrangian descriptions are in consistent, whereas at low frequencies, the results confirm the conclusions (Example 7.2) obtained by Jaffrin and Shapiro (Exercise 7.15).

Example 7.5 Peristaltic flow in closed cavity (perturbation and numerical solutions) [7].

This investigation was as well motivated by application of peristaltic flows to enhance the mixing process in microdevices. The model is similar to that discussed in Example 7.4, but (i) the peristaltic motion is produced by both upper and lower vibrating walls moving with the same amplitude and frequency but with the opposite phase and (ii) cavity is shorter, which requires accounting for sidewalls effect. The problem is solved analytically for small amplitudes using the perturbation approach developed by Fung and Yin (Example 7.2) and numerically for the arbitrary amplitude, employing the finite-element approximation (Chapter 8).

The problem is governed by the Navier–Stokes equation in the same form using the same dimensionless variables for length, time, and velocities, and a different variable for the pressure scaled by $\varepsilon R^2\mu\omega$, where $R = 2H\sqrt{\omega/\nu}$ is the Strouhal number applied in this study instead of the Reynolds number. Thus, the equation and boundary conditions are similar to system (7.9) but with regard to literal (second equation) walls as well as to both horizontal upper (third equation) and lower (fourth one) moving walls (Exercise 7.16):

$$\frac{\partial \mathbf{V}}{\partial t} + \varepsilon \mathbf{V} \cdot \nabla \mathbf{V} = -\nabla p + \frac{1}{R^2}\nabla^2 \mathbf{V}, \quad \mathbf{V}(0, y, t) = \mathbf{V}(L, y, t) = 0, \tag{7.11}$$

$$\mathbf{V}(x, \varepsilon\cos(\kappa x - t), t) = 0, \sin(\kappa x - t), \quad \mathbf{V}(x, 1 - \varepsilon\cos(\kappa x - t), t) = 0, -\sin(\kappa x - t).$$

Using series and relevant integral condition (7.10) [with upper limit $1 - \varepsilon\cos(\kappa x - t)$ instead of 1] and long mathematical performance following the procedure from [3] (Example 7.2), the authors calculated the first two terms and presented the solution with order up to $O(\varepsilon^2)$. A detailed description of analytic solution as well as of the scheme and the technique of numerical procedure are given in [7].

Several conclusions specific for closed, short cavities are derived: (i) the first-order solution consists of two terms—one the same as in open channel (Example 7.2) and the other induced by the pressure oscillating occurring due to sidewalls; this conclusion is in contrast to previous results based on an assumption that the sidewalls effects are of second order. (ii) The displacement of passive traces over one-period oscillation predicted by the second-order approximation is of the same order of magnitude as that given by the first-order approximation so that both data are necessary to be taken into account; however, only the time-independent term of the second approximation is of order ε^2, whereas the time-dependent term of this solution is of order ε^3 and hence it may be neglected. (iii) In contrast to the case of infinity long channel, the peristaltic flow in closed conduit is unsteady even in the moving frame as well. (iv) Comparison with numerical results shows that the analytical solution is valid in the middle cavity zone which is one wavelength distance away from both sidewalls. (v) The Eulerian flow next to the vibration walls is in the wave direction, whereas the flow in the central area is in the opposite direction. (vi) For finite amplitudes, the particles trajectories indicate the global circulation in the entire cavity as a fluid moving in the direction of traveling waves next to vibrating walls, turning around next to the sidewalls, and moving in the opposite direction in the central region. (vii) The flow forms also stationary circulation bubbles, which next to vibrating walls rotate in a clockwise direction, whereas far away from the vibrating walls, they rotate in the counterclockwise direction (Exercise 7.17).

7.5 Conjugate Solutions

There are relatively a few conjugate solutions investigating the interaction between flexible walls and fluid inside the channel. Here, we analyze two results of that type: one is the earlier publication (Example 7.6) and another presents an example of contemporary investigations (Example 7.7). While the former research is performed assuming that the peristaltic flow is produced by sinusoidal wave traveling along the walls, the latter is based on a more realistic model assuming that the walls are oscillating and studying the interaction between oscillating walls and fluid flow. Both solutions are obtained applying perturbation series. Some other examples of conjugate solution are discussed in applications (Chapter 12).

Example 7.6 Two-dimensional peristaltic flow induced by sinusoidal waves [8].

This early research is an extension of the Fung and Yih study (Example 7.2), including effect of walls in purpose to understand the inherent dynamic solid–fluid interaction. This is achieved by considering equations of motion of the fluid and flexible walls. It is assumed that the peristaltic flow in an infinite long channel is produced by traveling sinusoidal waves of moderate amplitude $\eta = a\cos(2\pi/\lambda)(x - ct)$ imposed on the walls, which are taken as either thin elastic plates or membranes. This relatively simple conjugate approach is applied by other authors up to now (for example, [9]).

The problem is governed by the same, as in Example 7.2, Navier–Stokes equation (7.3) in the form of stream function and boundary conditions (7.4) written in the same dimensionless variables. The additional boundary condition specifies the equality of stresses on the solid–fluid interface. This condition is present applying the derivative of fluid pressure $\partial p/\partial x$ defined from the Navier–Stokes equation in form (2.2) and the operator method for determining stresses acting from the walls. Employing equation (2.17) for velocity components and operator signified as $\tilde{L}(\eta)$, one gets for interfaces

$$\frac{\partial}{\partial x}\tilde{L}(\eta) = \frac{\partial p}{\partial x} = \mu\nabla^2\psi_y - \rho(\psi_{yt} + \psi_y\psi_{yx} - \psi_x\psi_{yy}) \text{ at } y = \pm H \pm \eta. \tag{7.12}$$

The operator method is a means of indicating what procedures should be done using the variable shown in the parentheses (Exercise 7.18). In this study, such an operator is used to define the equations of motion of the walls to obtain for membrane or for a thin plate

$$\tilde{L}(\eta) = -T\frac{\partial^2}{\partial x^2} + m\frac{\partial^2}{\partial t^2} + C\frac{\partial}{\partial t} \qquad \tilde{L}(\eta) = D\frac{\partial^4}{\partial x^4} + m\frac{\partial^2}{\partial t^2} + C\frac{\partial}{\partial t}, \qquad (7.13)$$

where T is the tension in the membrane, D is the flexural rigidity of the plate, m is the mass per unit area, and C is the coefficient of viscous damping. Employing these equation and dimensionless variables from Example 7.2 yields instead of equation (7.12) the boundary condition on the interface at $y = \pm(1+\eta)$ (Exercises 7.19)

$$D\frac{\partial^5\eta}{\partial x^5} - T\frac{\partial^3\eta}{\partial x^3} + m\frac{\partial^3\eta}{\partial x\partial t^2} + C\frac{\partial^2\eta}{\partial x\partial t} = \frac{1}{R}\nabla^2\psi_y - (\psi_{yt} + \psi_y\psi_{yx} - \psi_x\psi_{yy}). \qquad (7.14)$$

Here coefficients are scaled: D by $\rho c^2 H^3$, T by $\rho c^2 H$, m by ρH, and C by $\rho Hc^2/v$. Comparing equations (7.14) and (7.13) makes it clear that equation (7.14) is valid for a membrane and for a plate if one takes in the first case $D = 0$ and in the second $T = 0$.

In the case of pure peristaltic flow occurs at a zero initial pressure gradient dp/dx, the solution of this problem is given by the same simplified Orr–Sommerfeld equation as in the Fung and Yih study (Example 7.2). However, the part of the boundary conditions depending on walls' properties is different. Using the solution for the time-average axial velocity, the authors investigated the effect of a thin elastic plate in the cases of negligible dissipation and with dissipation occurring due to viscous damping. The interaction between the Poiseuille flow, presenting in a channel when the initial pressure gradient is not zero, and peristaltic motion was also studied.

The same approach was used to investigate the peristaltic flow of different fluids and conditions. Here, we consider, as an example, a study of peristaltic flow of non-Newtonian couple-stress fluid [9]. Such a model of fluid describes the behavior of rheologically complex liquids such as blood, infected urine, or liquid crystals, taking into account the size of liquid particles (Comment 12.1).

Below we summarize the basic results obtained in both studies: (i) the mean velocity is maximum at the centre and remains constant over some range, which increases as the Reynolds number grows; at the boundaries, the mean velocity decreases with the increase in damping and couple-stress parameter but increases with increasing wall tension and elastance. (ii) The mean velocity perturbation increases with increasing wall damping, wall tension, and wall elastance but decreases with increasing couple-stress parameter. (iii) At higher wavelength, a reversal in the flow direction may occur for very high rigidity of the walls; the damping may cause the mean flow reversal at the walls, which is not possible for the elastic walls when the reversal at the centre occurs in the case of pure peristalsis. (iv) Higher level of tension would enhance the efficiency of peristaltic pumping and at lower tension increases the possibility of flow reversal. (v) Pumping a couple-stress fluid is tougher than a Newtonian fluid. (vi) In the limiting case, when the couple-stress parameter vanishes, the results pertaining to couple-stress fluid agree with those of Newtonian fluid.

Example 7.7 Steaming flows in a channel with elastic vibrating walls [10].

As mentioned above, this study is based on a more realistic conjugate model analyzing the interaction between vibrating walls and flowing fluid without assumption of the waves imposed on the walls. The problem is governed by two-dimensional Navier–Stokes equations (2.1)–(2.3) in Cartesian coordinates and dynamic equation for the thin elastic plate. For the thin plate, the elongate deformations are usually small in comparison with those in bending. Therefore, in thin approximation, a thin plate is modeled by surface $y = \eta(t, x)$, which is supposed to vibrate in the transverse direction.

The equation describing such oscillating surface is taken in [10] as it is suggested in [11].

$$\sigma_+ - \sigma_- = 2\rho_w \Delta \frac{\partial^2 \eta}{\partial t^2} + \frac{2\Delta^3 E}{3(1 - v_w^2)} \frac{\partial^4 \eta}{\partial x^4}, \frac{\partial \eta}{\partial t} = v, u = 0, -\tau_{xy}\frac{\partial \eta}{\partial x} + \tau_{yy} - p = \sigma_-. \tag{7.15}$$

In the first relation, Δ is the wall thickness, E is the Young modulus, v_w is the Poisson coefficient, σ_+ and σ_- are normal stresses at the external and internal wall surfaces, respectively, so difference $(\sigma_+ - \sigma_-)$ is the specific surface loading (Exercise 7.20). The second and third equations (7.15) are the kinematic (equality of velocities) and dynamic (equality of forces) conjugate conditions on the solid–fluid interface $y = \eta(t, x)$, where τ_{ij} are the viscous stresses in the fluid. Another condition [the first condition (7.16)] that should be satisfied follows from symmetry of channel. To complete the problem formulation, two more equations (7.16) are specified: one determines harmonic oscillations imposed on the left edges of walls and another defines the time-average pressure drop in the channel:

$$\partial u/\partial y = v = 0 \text{ at } y = 0, \quad \eta(0, t) = H + a\cos\omega t, \quad p(x = L) - p(x = 0) = \Delta p. \tag{7.16}$$

The following dimensionless variables are introduced: $x/L, y/H, u/\omega L, v/\omega H, \eta/H, p/\rho\omega^2 L^2$ Re and time ωt, where Re $= H^2\omega/v$ is the vibrational Reynolds number, which is the main dimensionless parameter in this problem. Using the same notations for dimensionless variables, one gets unchanged continuity equation (2.1) and Navier–Stokes equations together with equation (7.15) for walls as follows (Exercises 7.21 and 7.22):

$$\text{Re}\left(\frac{\partial u}{\partial t} + u\frac{\partial u}{\partial x} + v\frac{\partial u}{\partial y}\right) = -\frac{\partial p}{\partial x} + \frac{\partial^2 u}{\partial y^2} + \left(\frac{H}{L}\right)^2\frac{\partial^2 u}{\partial x^2}$$

$$\text{Re}\left(\frac{H}{L}\right)^2\left(\frac{\partial v}{\partial t} + u\frac{\partial v}{\partial x} + v\frac{\partial v}{\partial y}\right) = -\frac{\partial p}{\partial y} + \left(\frac{H}{L}\right)^2\frac{\partial^2 v}{\partial y^2} + \left(\frac{H}{L}\right)^4\frac{\partial^2 v}{\partial x^2} \tag{7.17}$$

$$-\tilde{\gamma}\frac{\partial^2 \eta}{\partial t^2} = -p + \left(\frac{H}{L}\right)^2\frac{\partial v}{\partial y} - \left(\frac{H}{L}\right)^3\left(\frac{\partial u}{\partial y} + \frac{H}{L}\frac{\partial v}{\partial x}\right)\frac{\partial \eta}{\partial x} + \tilde{\beta}\frac{\partial^4 \eta}{\partial x^4},$$

$$\tilde{\gamma} = 2\,\text{Re}\frac{\rho_w H\Delta}{\rho L^2}, \quad \tilde{\beta} = \frac{2E\Delta^3 H^3}{3(1 - v_w^2)\omega\mu L^6}, \eta(0, t) = 1 + \varepsilon\cos t, \quad \varepsilon = \frac{a}{H}.$$

If the channel is relatively long, as in this problem, such that the height–length ratio is small $(H/L \ll 1)$, equations in form (7.17) may be significantly simplified by neglecting terms with ratio H/L in power two or higher to obtain

$$\text{Re}\left(\frac{\partial u}{\partial t} + u\frac{\partial u}{\partial x} + v\frac{\partial u}{\partial y}\right) = -\frac{\partial p}{\partial x} + \frac{\partial^2 u}{\partial y^2}, \quad \frac{\partial p}{\partial y} = 0, \quad -\tilde{\gamma}\frac{\partial^2 \eta}{\partial t^2} = -p + \tilde{\beta}\frac{\partial^4 \eta}{\partial x^4}. \tag{7.18}$$

Comment 7.6 The approach using to simplify equations (7.17) is similar to that in the boundary layer theory (Sec. 2.5.2). Both approaches are based on applying scales proportional to main sizes of domain: small lateral scale (boundary thickness or channel height) and large longitudinal one (length) so that the former procedure is applicable to external (flow past bodies), and the latter is suitable to internal (channel flows) problems.

A simplified problem of the boundary layer type with nonlinear equation and complicated boundary condition (7.18) is linearized to obtain the analytical solution. This is achieved by neglecting in first equation (7.18) the nonlinear terms. The solution of such a linear problem is built using perturbation series, as (7.3), but starting with term of order ε due to reducing of initial wall oscillations by wall–fluid interaction:

$$u = \varepsilon u_1(x,y,t) + \varepsilon^2 u_2(x,y,t) + \cdots, \quad v = \varepsilon v_1(x,y,t) + \varepsilon^2 v_2(x,y,t) + \cdots,$$

$$\eta = 1 + \varepsilon \eta_1(x,t) + \varepsilon^2 \eta_2(x,t) + \cdots \tag{7.19}$$

Differentiating the linearized first equation (7.18) with respect to y, the authors get the partial differential equation (7.20) and show that its solution for the first-order term (7.19) u_1, satisfying the second (7.15) and the first (7.16) conditions, has the following form where function $F(\tilde{x})$ depends on variable $\tilde{x} = xL/H$:

$$\mathrm{Re} \frac{\partial^2 u_1}{\partial t \partial y} = \frac{\partial^3 u_1}{\partial y^3}, u_1 = e^{it} F(\tilde{x})(e^{\varphi y} + e^{-\varphi y} - e^{\varphi} - e^{-\varphi}), F'' = -\gamma \frac{\varphi}{\tanh \varphi - \varphi} F + \beta F''' \tag{7.20}$$

$$\varphi = \sqrt{\frac{\mathrm{Re}}{2}}(1+i), \quad \gamma = \frac{\rho H}{2\rho_w \Delta}, \quad \beta = \frac{\Delta^2 E}{3H^4 \rho_w (1 - v_w^2) \omega^2}, \quad \beta k^6 - k^2 - \gamma \frac{\varphi}{\tanh \varphi - \varphi} = 0.$$

Here, parameters γ and β characterize the ratio of masses of the fluid and the wall and the elasticity of the walls, respectively.

An ordinary linear differential equation with constant coefficients (7.20) for function $F(\tilde{x})$ is analyzed using its characteristic equation given by last equation (7.20) (Exercises 7.23 and 7.24). It is shown that the roots of characteristic equation together with vibrational Reynolds number $\mathrm{Re} = H^2 \omega / v$ determine variety oscillation regimes of the fluid–walls peristaltic system, which principally differ from oscillations in usually considered problems of this type. Analysis reveals that in the first-order approximation of order $\mathrm{Re}^{-1/2}$, three roots of cubic characteristic equation (7.20) for k^2 correspond to three feasible regimes: one root pertains to pair of traveling waves and two others describe the standing waves. Similar analysis for low Reynolds numbers ($\mathrm{Re} \ll 1$) indicates the three possible regimes as well. One is a relatively slow damping behavior of the system, another corresponds to the standing wave, and all three weak modes depend on problem parameters, such as initial frequency, the walls rigidity, fluid and walls densities, and so on. The drift velocities and mass transfer are also investigated. A detailed analysis is given in the paper [10].

The following basic conclusions are formulated: (i) the two-phase (conjugate) model of peristaltic motion is basically different from the usual models (semi-conjugate) assuming unchangeable channel walls; (ii) the walls–fluid interaction leads to waves damping and to essential decreasing of the bending walls oscillations; (iii) the fluid motion is not periodic in space as it is in usual models, and the cross-section velocity profiles are not constant changing from boundary layer type at the entrance to Poiseuille profiles at the end of the channel; (iv) three different regimes are possible—one is the traveling wave and two others are standing waves; (v) at high Reynolds numbers, the standing waves damp rapidly along the channel, surpassing the mass flux induced by traveling waves—the drift velocity profile exposes in this case a counter-flow stream near the walls at the entrance and the Poiseuille flow at the outlet; (vi) at low Reynolds number, the traveling waves with rather low damping give the maximal fluid mass flux, and the drift velocity profile exposes an intense stream along the walls and counter flow at the axis; (vii) in general, the muss flux intensity is much higher at high Reynolds numbers; (viii) average pressure gradient consists of two terms: one is the constant value of external pressure drop providing the Poiseuille flow, and another is a variable pressure gradient part existing due to vibrating walls with zero average pressure difference at the ends of channel; (ix) the study leads to the conclusion of efficiency of the mass flux at standing wave regimes rapidly damping along the channel, which is contrary to existing opinion that the traveling wave regime is the basic mechanism for fluid pumping.

EXERCISES

7.1 Explain how tubular human organs works moving physiological fluids.

7.2 Learn more about artificial human organs using Wikipedia.

7.3 Compare heat transfer and peristaltic conjugate problems. Explain why the latter is much complicated than the former.

7.4 What is the difference between peristaltic conjugate and semi-conjugate problems? Why the second is much simpler?

7.5 Explain physically the difference between differential equations (2.47) and (7.1).

7.6 Derive the velocity distribution (7.1) using integration of differential equation (7.1). Compare the result with the usual Poiseuille velocity profile.

7.7 Analyze equations (7.2) and following text to understand the conclusions made there.

7.8 Derive equations (7.5) using equation (7.3) and collecting terms with the same powers of ε.

7.9 Obtain two first conditions (7.6) by the way described in the text.

7.10 Recall or study the basic features of functions of complex variable using Advanced Engineering Mathematics.

7.11 At the beginning of Example 7.2, it is said that the model is nonlinear. Show that this is true. Recall from previous chapters what properties specified nonlinear model.

7.12 Explain physical reasons of boundary conditions (7.8). Recall what was said about such boundary conditions in previous chapters and analyze the last condition, taking into account, in particular, the information from comment 7.4.

7.13 Analyze the boundary conditions on the wave and another on the wall of container written in the vector form. Hint: take into account the comment 7.5.

7.14 Learn about the second Stokes' problem using, for example "Fluid Mechanics, SG2214,HT2010" on internet.

7.15 Compare the basic features of two peristaltic flows: at low frequency in channel and at high frequency in closed container according to results of Example 7.4.

7.16 Derive the system of equations (7.11) similar to system (7.9) using dimensionless variables from Example 7.5.

7.17 Compare results obtained for close containers in two studies reviewed in Examples 7.4 and 7.5.

7.18 Learn more about the operator method using, for instance, the article on the internet: http://qedinsight.wordpress.com/2011/02/25/an-operator-method-for-solving-second-order-differential-equations/. Derive equation (7.12) as it is directed in the text.

7.19 Show that equation (7.14) is valid for the membrane and for a plate and determine the scales for coefficients at derivatives, knowing that all variables in equation (7.14) are dimensionless. Hint: take into account that in equations (7.13) written in the dimension variables all terms are in the same units.

7.20   Study or recall basic terms of strength materials from Wikipedia or more fundamental using, for example, book [11].

7.21   Obtain first two Navier–Stokes equations (7.17) in dimensionless variables using equations (2.2) and (2.3).

7.22   * Derive third equation (7.17) which is dimensionless version of equation

$$(-2\rho_w)\Delta\frac{\partial^2\eta}{\partial t^2} = -p-\mu\left(\frac{\partial u}{\partial y}+\frac{\partial v}{\partial x}\right)\frac{\partial\eta}{\partial x}+2\mu\frac{\partial v}{\partial y}+\frac{2\Delta^3 E}{3(1-v_w^2)}\frac{\partial^4\eta}{\partial x^4}$$ written in the dimensional

variables. To get the last equation, (i) substitute the third equation (7.15) instead of σ_- in the first relation neglecting σ_+ and (ii) specify the stresses in the fluid using for-

mulae $\tau_{xy} = \mu\left(\frac{\partial u}{\partial y}+\frac{\partial v}{\partial x}\right), \tau_{yy} = 2\mu\frac{\partial v}{\partial y}$. Finally, apply dimensionless variables to obtain

equation (7.17). Read about stresses in laminar flow on the internet: http://www.engr .sjsu.edu/glyoung/CHE190/handouts/Shear%20Stress.pdf or more fundamental using, for example, book *Schlichting Boundary layer theory* (Reference [1] in Chapter 2).

7.23   Show that solution (7.20) satisfied the second (7.15) and the first (7.16) boundary conditions for u. Hint: note that on the wall dimensionless variable $\eta = 1$ (see Fig. 7.1).

7.24   Recall or study in *Advanced Engineering Mathematics* the integration of linear ordinary differential equations with constant coefficients, applying characteristic equations.

REFERENCES

[1]   Jaffrin, M. Y., & Shapiro A. H. (1971). Peristaltic pumping. *Annual Review of Fluid Mechanics 3*, 13–37. doi: http://dx.doi.org/10.1146/annurev.fl.03.010171.000305

[2]   Shapiro, A. H., Jaffrin, M. Y., & Weinberg, S. L. (1969). Peristaltic pumping with long wavelengths at low Reynolds number. *Journal of Fluid Mechanics 37*, 799–825. doi: http://dx.doi.org/10.1017 /S0022112069000899

[3]   Fung, Y. C., & Yih, C. S., 1968, Peristaltic transport. *Journal of Applied Mechanics-Transactions of the ASME 35*, 669–675. doi: http://dx.doi.org/10.1115/1.3601290

[4]   Takabatake, S., & Ayukawa, K. (1982). Numerical study of two-dimensional peristaltic flows. *Journal of Fluid Mechanics 122*, 439–465. doi: http://dx.doi.org/10.1017/S0022112082002304

[5]   Ayukawa, K., Kawai, T., & Kimura, M. (1981). Streamlines and path lines in peristaltic flows at high Reynolds numbers. *Bulletin of the Japan Society of Mechanical Engineers 24*, 948–955. doi: http://dx.doi.org/10.1299/jsme1958.24.948

[6]   Selverov, K. P., & Stone, H. A. (2001). Peristaltically driven channel flows with applications toward micromixing. *Physics Fluids 13*, 1837–1860. doi: http://dx.doi.org/10.1063/1.1377616

[7]   Yi, M., Bau, H. H., & Hu, H. (2002). Peristaltically induced motion in a closed cavity with two vibrating walls. *Physics Fluids 14*, 184–197. doi: http://dx.doi.org/10.1063/1.1425841

[8]   Mittra, T. K., & Prasad, S. N. (1973). On the influence of wall properties and Poiseuille flow in peristalsis. *Biomechanics 6*, 681–693. doi: http://dx.doi.org/10.1016/0021-9290(73)90024-9

[9]   Pandey, S. K., & Chaube, M. K. (2011). Study of wall properties on peristaltic transport of a couple stress fluid. *Meccanica 46*, 1319–1330. http://dx.doi.org/10.1007/s11012-1010-9387-8

[10] Shugan, I. V., Smirnov, N. N., & Legros, J. C. (2002). Streaming flows in a channel with elastic walls. *Physics Fluids 14*, 3502–3511. doi: http://dx.doi.org/10.1063/1.1504081

[11] Timoshenko, S. (1974). Vibration problems in engineering (4th ed.). New York: Wiley.

PART III

NUMERICAL METHODS IN FLUID FLOW AND HEAT TRANSFER

CHAPTER 8

CLASSICAL NUMERICAL METHODS IN FLUID FLOW AND HEAT TRANSFER

8.1 WAY ANALYTICAL OR NUMERICAL METHODS?

Finite-difference methods (FDM) were developed long before the numerical methods became a powerful tool for solving differential equations due to computers [1]. Since the understanding and the technique of the FDM seems to be simpler than that of analytical methods, it was believed that the time of analytical methods was over.

Although the techniques of analytical and numerical approaches indeed are different, both methods are based on the same fundamental principles. The only distinction between both approaches is that these basic principles are applied in the former to infinite-small differences, whereas in the latter, they are used for small but finite size values. For example, both analytical and numerical derivatives are determined by the same principle. However, to calculate an analytical derivative, one needs to have some knowledge, and nevertheless, sometimes that might be not easy. At the same time, to obtain the finite-difference derivative using the difference between function values at two grid points is not at all a problem. This feature of numerical methods gives an impression that the numerical approach is much simpler than analytical methods.

In the early 1960s, this seeming simplicity led to many unsuccessful attempts to solve numerically complex contemporary problems, showing that only deeply physics understanding together with careful software testing may yield the proper solution. It became clear that only an investigator who adopted a corresponding part of current knowledge can possess the complex technique of the numerical solution, which just seems to be simple, and then interpret the obtained results. Otherwise, an insufficiently considered and prepared program can give any unrealistic outcome.

After the applications of computers were expanded, analytical methods not only retained their importance, but gained new functions as well. In particular, despite there are many recommendations and rules for preparing and qualitatively checking numerical programs, the best way to test and control the accuracy is to compare the result obtained by some software with results of a proper, simple analytical solution [2].

There are some cases when analytical solutions are especially useful in preparing and test-ing the software:

- The formulae for the finite-difference derivatives are usually obtained using a Taylor series. For a grid point i located midway between points $i-1$ and $i+1$, one obtains using the first three terms of the Taylor series the following two expressions:

$$f_{i-1} = f_i - (x_i - x_{i-1})\left(\frac{df}{dx}\right)_i + \frac{1}{2}(x_i - x_{i-1})^2\left(\frac{d^2 f}{dx^2}\right)_i + \cdots \qquad (8.1)$$

$$f_{i+1} = f_i + (x_i - x_{i-1})\left(\frac{df}{dx}\right)_i + \frac{1}{2}(x_i - x_{i-1})^2\left(\frac{d^2 f}{dx^2}\right)_i - \cdots \qquad (8.2)$$

By adding and subtracting these equations, one gets formulae for the first two derivatives:

$$\left(\frac{df}{dx}\right)_i = \frac{f_{i+1} - f_{i-1}}{x_i - x_{i-1}}, \qquad \left(\frac{d^2 f}{dx^2}\right)_i = \frac{f_{i-1} - 2f_i + f_{i+1}}{(x_i - x_{i-1})^2} \qquad (8.3)$$

These formulae may be used only if the function in question is analytic. If at some grid points the function is singular, such as, for example, at these points one or more derivatives become infinite, equations (8.3) cannot be applied (Exercise 8.1). Such a case is considered in Section 6.1, where it is shown that the wall temperature distribution on a thermally thin plate near $x = 0$ is not an analytic function of the coordinate x. It is rather presented as a series in integer powers of variable $x^{1/s}$, where S is the denomina-tor of the exponent in the relation for an isothermal heat transfer coefficient. For laminar and turbulent flows, this variable is $x^{1/2}$ and $x^{1/5}$, respectively. It is obvious that in this case the derivative with respect to x is proportional to $x^{-1/2}$, or to $x^{-4/5}$ for laminar, or turbulent flow, or to $x^{(1/s)-1}$ for any other value of s. Hence, this derivative becomes infinite at $x = 0$ for laminar, turbulent flows or for other cases in which an exponent r/s in the relation for an isothermal heat transfer coefficient is less than unity. However, if one introduces a new variable $z = x^{1/s}$, the temperature distribution turns into analytical function, the derivative with respect to z becomes finite, and equations (8.3) can be used.

Another similar situation takes place for the velocity and temperature boundary layer equations in Prandtl–Mises form (2.73). These equations have a singularity at the body surface where the stream function is zero, $\psi = 0$. This follows from the fact that the longitudinal velocity near the surface is proportional to the transverse variable: $u \sim cy$. This relation holds true because on the surface $u = 0$, but $\partial u/\partial y \neq 0$ due to the friction. Then, according to definition (2.17), the stream function near the surface $\psi \sim uy$ and because $u \sim cy$, one obtains for stream function $\psi \sim cy^2$. From this, it follows that $y \sim \psi^{1/2}$, and finally, $u \sim c\psi^{1/2}$. Similarly, according to continuity equation (2.1), we have $\partial u/\partial x \sim \partial v/\partial y \sim cy$, and hence, $v \sim cy^2$. Using these estimations and knowing that both convective terms in the thermal boundary layer equation (2.70) are of the same order, one gets that $u(\partial T/\partial x) \sim v(\partial T/\partial y)$ and hence $T \sim cy \sim c\psi^{1/2}$ (Exercise 8.2).

Therefore, near the surface, both derivatives of the velocity and of the temperature with respect to variable ψ become infinite. Due to this singularity, equations (8.3) can-not be applied. However, introducing a new variable $z = \psi^{1/2}$ solves this problem again.

Thus, analytical analysis leads us to understanding what variable should be used to overcome the singularities in general and particularly in the case of numerical solution.

- The other difficulty that usually arises in preparing a program for the numerical solution is attaining the proper distribution of the grid points inside a considering domain. This distribution should correspond to the studied function gradient distribution. In achievement of this conformity, one can get significant information analyzing a field of analytical solution of a similar problem. In particular, it is known from exact solutions of the boundary layer problems that the maximum values of the velocity and temperature gradients are at the wall. These maximum gradients gradually decrease as the distance from the wall grows until they become zero at $y \to \infty$.

    On the basis of this information, one creates a grid with corresponding distribution of points or applies a special variable, for instance, $z = \Delta / \left[1 - \left(y/\delta \right) \right]$, where Δ is a size of a mesh at the wall. Such a variable increases mesh size as the distance from the wall grows (Exercise 8.3).

- Analytical solutions are useful as well when one needs to choose a function for approximately describing the solution behavior between the grid points. It is clear that this function should be as close as possible to actual dependence. Since this actual function is unknown, one way to find a relation close to it is to examine a proper analytical solution. In particular, for the case of a numerical solution of the boundary layer equation, it is reasonable to apply the polynomial or some other approximate distribution, for example, self-similar profiles that are usually used in integral methods (Sec. 2.7.1).

- Apparently, the most important function of the analytical solutions is that they can be used as a reference in checking and testing the software. Comparing the computation results of the same problem with an exact analytical solution, one determines not only the usefulness of the program in general, but estimates as well the deviations of obtained values from exact data.

The examples presented here demonstrate the role of the analytical results in developing the software for numerical solutions. At the same time, there is no doubt of the significance of the numerical approaches in obtaining the digital data for analytical solutions. Therefore, there is no reason to oppose the analytical and numerical methods; rather, it is reasonable to consider both approaches as a united, combined method of investigation and solution of contemporary theoretical and practical problems. In fact, the numerical and analytical methods are two means supplementing each other. Whereas the former is a powerful technique for an approximate solution of almost any complex problem using a known mathematical model, the latter gives a possibility of finding some exact solutions of relatively simple problems, investigating general properties of studied phenomena, and developing models on the basis of these data. Since any contemporary problem is a challenge, only by using both methods in combination, one could expect to obtain the adequate results (Exercise 8.4).

8.2 Approximate Methods for Solving Differential Equations

Approximate methods for solving differential equations were developed and widely used many years before they became a basis of modern numerical methods being now a powerful contemporary tool for calculation owing to computers. However, before the computer era, these methods can be used only for entire computation domain of interest being actually the analytical approaches. The use of computers makes it possible to divide the computation domain on small subdomains and apply the same approximate methods for each one. It is obvious that this pattern vastly increased the possibilities and the accuracy of these simple analytical approaches and converted them, in essence, into the contemporary numerical methods.

Numerical methods differ from each other by methods of discretization of the computation domain and by the analytical approximate methods for solving the differential equations for each subdomain. Depending on the first procedure, the numerical methods can be classified in three basic groups: the FDM, finite-element method (FEM), and boundary element method (BEM). The old one, the FDM [1], usually uses for discretization the uniform grids and calculates in the points of these grids numerical derivatives by formulac (8.3), which are finite-difference version of the usual derivatives in calculus. The two other techniques compute the values of the studied function in each usually irregular distributed subdomains, employing different approximate methods, as before these were used analytically for entire domains.

The distinction between two last approaches results from various numbers of subdomains that are needed to obtain a solution. While the FEM requires employing the subdomains of the whole field of an investigating function, the boundary-element method uses only a part of the whole number of subdomains located just on the boundaries. There are also two of the most widely used modern modifications of the finite-difference method (CVFDM) and finite-element method (CVFEM), which differ from earlier versions by using the control-volume (CV) formulation for deriving the equations for grid points or elements (discretization equations [2]). In that approach, the dicretization equations are obtained as a result of integrating the differential equations over each control volume. Here, the basic idea is that such an equation expresses the conservation laws for finite volume just as the differential equation expresses these laws for an infinitesimal volume (Exercises 8.6).

The modified FDM (CVFDM), for simplicity, is called FDM, but the modified FEM is in essence with the finite-volume method (FVM) since the volume coincides as an element. Thus, there are three commonly used types of finite-difference approaches: FDM, FVM, and BEM (Exercise 8.7).

The distinction between different analytical approximate methods employed in numerical methods is convenient to describe using the weighted residual approach (see, for example, [2], [3], or [4]). The weighted residual method is a generalized well-known method of moments that was widely used before computers became common, in particular, in the boundary layer theory by the name integral methods (Secs. 2.7.1 and 3.5). The concept of the weighted residual method can be explained as follows.

Let there be a need to find an approximate solution of differential equation $F(u) = 0$ subjected to a given boundary condition. First, the given boundary condition should be converted to a homogeneous one (Sec. 1.5.1). Then, a function $\tilde{u} = f(x)$ is chosen that exactly satisfies the boundary condition but contains one or more unknown parameters, for instance, a polynomial with undefined coefficients or some other function. Substituting this function into a differential equation yields a residual $R = F(\tilde{u})$ because $\tilde{u}$ is an approximate solution and, hence, does not satisfy the equation under consideration. Multiplying this residual by some weighted function w and integrating over the considering domain S, one tries to minimize an average error:

$$\int_S wR\,dx = \int_S wF(\tilde{u})\,dx = 0 \qquad (8.4)$$

Employing this equation to a series of weighted functions, one obtains as many algebraic equations as are required to determine the unknown parameters. Substituting these parameters into function $\tilde{u} = f(x)$ completed the approximate solution. Various approximate methods applied in numerical solutions differ from each other by classes of these weighted functions. For example, the method of moments results from series of weighted function: $1, x, x^2, \ldots$. The above-mentioned integral method is a case of method of moments when only the first one is used, that is, $w = 1$ (Exercise 8.8).

Example 8.1 Consider a simple conduction problem for a plane wall governed by one-dimensional equation and ordinary boundary conditions:

$$\lambda \frac{d^2T}{dx^2} + q = 0 \qquad x = 0 \quad T = T_0 \quad x = L \quad T = T_L, \tag{8.5}$$

where T_0 and T_L are temperatures of surfaces and q is the uniform heat generation. To solve the problem using the method of moments, one introduces a new variable ϑ, which transforms boundary conditions to homogeneous ones to get the problem in the form

$$\vartheta = T - T_0(1-\xi) - T_L\xi, \quad \frac{d^2\vartheta}{d\xi^2} + \bar{q} = 0 \quad \xi = 0, \xi = 1, \vartheta = 0 \quad \xi = \frac{x}{L}, \quad \bar{q} = \frac{qL^2}{\lambda} \tag{8.6}$$

For using first two moments, one should choose function with two parameters that satisfies boundary conditions. For instance, $\vartheta = a_1\xi(\xi - 1) + a_2\xi^2(\xi^2 - 1)$. Substituting this approximate relation into second equation (8.6) leads us to residual $R = 2a_1 + (12\xi^2 - 2) + \bar{q}$. Then, using functions $w = 1$ and ξ for first two moments, one obtains from equation (8.4)

$$\int_0^1 \left[2a_1 + \left(12\xi^2 - 2\right)a_2 + \bar{q} \right] d\xi = 0 \int_0^1 \left[2a_1 + \left(12\xi^2 - 2\right)a_2 + \bar{q} \right]\xi d\xi = 0 \tag{8.7}$$

These two equations determine coefficients $a_1 = \bar{q}/2$ and $a_2 = 0$, resulting in solution

$$T = \frac{qL^2}{\lambda} \frac{x}{L}\left(\frac{x}{L} - 1\right) + T_0\left(1 - \frac{x}{L}\right) + T_L \frac{x}{L} \tag{8.8}$$

In this particular case, the approximate method yields an exact solution.

Similar solutions can be obtained using other approximate methods. The different results arise from various series of weighted functions used to obtain the required system of algebraic equations. Thus, for Galerkin's method, the weighted functions are the same as the functions chosen for satisfying the boundary conditions so that equations (8.7) according to expression for ϑ become

$$\int_0^1 \left[2a_1 + \left(12\xi^2 - 2\right)a_2 + \bar{q} \right]\xi(\xi - 1)d\xi = 0,$$

$$\int_0^1 \left[2a_1 + \left(12\xi^2 - 2\right)a_2 + \bar{q} \right]\xi^2\left(\xi^2 - 1\right)d\xi = 0 \tag{8.9}$$

Solving these equations yields the same solution (8.8). One gets analogous solution using the point collocation method with the Dirac delta weighted function or the subdomain collocation method. In the last case, a calculation domain is divided into several subdomains. For instance, for two domains instead of equation (8.7), one obtains a system

$$\int_0^{1/2} \left[2a_1 + \left(12\xi^2 - 2\right)a_2 + \overline{q} \right] d\xi \qquad \int_{1/2}^1 \left[2a_1 + \left(12\xi^2 - 2\right)a_2 + \overline{q} \right] d\xi, \qquad (8.10)$$

which leads to the same result (8.8).

Comment 8.1 Dirac delta function is zero everywhere except at $x = 0$, where it is infinite resulting in integral of the entire x-axis that equals unity [equation (1.122)]. The third formula presents one of the properties of delta function:

$$\delta(x) = \begin{cases} 0 & x \neq 0 \\ \infty & x = 0 \end{cases} \qquad \int_{-\infty}^{\infty} \delta(\xi) d\xi = 1 \qquad \int_{-\infty}^{\infty} \delta(\xi) f(\xi) d\xi = f(0) \qquad (8.11)$$

Comment 8.2 Collocation method consists of choosing function satisfying the boundary conditions with some unknown parameters, which are defined by fitting the problem solution in the corresponding number of grid points (Exercise 8.9).

Comment 8.3 The relation $\vartheta = a_1\xi(\xi - 1) + a_2\xi^2(\xi^2 - 1)$ that is used in the above-analyzed examples is not unique so that there are many others with two free coefficients satisfying the same boundary conditions and resulting in slightly different approximate solutions. We consider several approximate methods with the same relation for ϑ to show that the basic distinction between those methods lies in the different weighted functions (Exercises 8.10–812).

A special case is the subdomains method when the domain under consideration is divided in numbers of subdomains, and one assumes that the weighted function is $w = 1$ for one of subdomains at a time and $w = 0$ for all others. Physically, this approach implies that the average residual error is zero over the each small domain. In particular, the CV formulation pertains to this type of methods. In this case, the system of algebraic equations is obtained by integrating the governing differential equation over each of subdomains. Applying this approach to the same simple one-dimensional conduction equation (8.5), one gets after its integration (Exercise 8.13):

$$\lambda \left[\left(\frac{dT}{dx} \right)_b - \left(\frac{dT}{dx} \right)_a \right] + \int_a^b q\, dx = 0, \qquad \frac{T_{i+1} - T_i}{x_{i+1} - x_i} - \frac{T_i - T_{i-1}}{x_i - x_{i-1}} + \overline{q}(x_b - x_a) \qquad (8.12)$$

Here, a and b denote the midways points between x_{i-1}, x_i, and x_{i+1}. In deriving the last equation (8.12) from the first one, it is assumed that the temperature changes linearly between grid points (Exercise 8.13). This example shows the usual way of using the CV approach in finite-difference numerical methods.

Comment 8.4 The first part of the second expression (8.12) presents the way of defining the finite-difference derivatives using the CV approach which differs from formulae (8.3) determine these derivatives via the Taylor series. Note that this way is preferable being more grounded due to the CV approach that provides an exact satisfaction of integral conservation laws over each subdomain and, hence, for the whole studied domain.

Another specific case, which is called weak formulation, is employed in the finite-element and boundary-element numerical methods. Whereas in FDM, an approximate solution is obtained by satisfying the differential equation at the grid points, in FEM and BEM, the approximate solution is found by distribution of the error of this solution over the each of the subdomains.

To introduce the basic concepts of the finite-element and boundary-element approaches, consider again one-dimensional conduction equation (8.5) for domain $(0,1)$. Multiplying this equation by some weighted function w after integrating and transforming the result by double integration by parts, one obtains two relations (Exercise 8.14):

$$\int_0^1 \left(-\frac{dT}{dx}\frac{dw}{dx} + \frac{q}{\lambda}w \right) dx + \left[\frac{dT}{dx}w \right]_0^1 = 0, \quad \int_0^1 \left(T\frac{d^2w}{dx^2} + \frac{q}{\lambda}w \right) dx + \left[\frac{dT}{dx}w \right]_0^1 - \left[T\frac{dw}{dx} \right]_0^1 = 0 \quad (8.13)$$

The first expression is the starting statement for the FEM, and the second is the beginning statement for the BEM. These expressions are named weakening expressions because the process of integrating by parts reduces the requirement of continuity of an applied function (Sec. 2.2.2.5).

Comment 8.5 Weak solution of a differential equation means that despite such a solution may be not differentiable, one is allowed to use it due to reducing usual requirements of solution continuity (Exercise 8.15).

The basic idea of using second equation (8.13) is that such a form of relation makes it possible to replace the search of an approximate solution by choosing a proper weighted function to get the boundary problem in which only boundary conditions should be satisfied. This can be done by two following ways [3,4]:

- Selecting such a weighted function that satisfies the governing equation in its homogenous form. For equation (8.5) and its homogeneous version, one gets a differential equation and its solution for a weighted function. Then, second equation (8.13) gives

$$d^2T/dx^2 = 0, T_0 = T_L = 0, \qquad d^2w/dx^2 = 0 \qquad w = a_1x + a_2 \qquad (8.14)$$

$$\int_0^1 \frac{q}{\lambda} w dx + \left[\frac{dT}{dx}w \right]_0^1 - \left[T\frac{dw}{dx} \right]_0^1 = \int_0^1 \frac{q}{\lambda}(a_1x + a_2) dx + \left(\frac{dT}{dx} \right)_1 (a_1 + a_2) - \left(\frac{dT}{dx} \right)_0 a_2 = 0 \quad (8.15)$$

Since the last equation should be satisfied for arbitrary a_1 and a_2, collecting terms containing these coefficients leads us to two equations, determining the derivatives at $x = 0$ and $x = 1$, and then yields the solution of the problem (Exercise 8.16):

$$\int_0^1 \frac{q}{\lambda} x dx + \left(\frac{dT}{dx} \right)_1 = 0, \qquad \int_0^1 \frac{q}{\lambda} dx + \left(\frac{dT}{dx} \right)_1 - \left(\frac{dT}{dx} \right)_0 = 0, \qquad (8.16)$$

$$\left(\frac{dT}{dx}\right)_1 = -\frac{q}{2\lambda}, \quad \left(\frac{dT}{dx}\right)_0 = \frac{q}{2\lambda}, \quad T = \frac{q}{2\lambda}x(1-x) \tag{8.17}$$

- Using a function (usually Dirac delta function) satisfying governing equation.
  In this case, one assumes that delta function satisfies the governing equation regardless of boundary conditions, and then second equations (8.13) and (8.18) give (Exercise 8.17)

$$\frac{d^2w}{dx^2} = -\delta_i, \quad \delta_i = \begin{cases} 1, & x = x_i \\ 0, & x \neq x_i \end{cases} \quad w = \begin{cases} x \text{ at } x \leq x_i \\ x_i \text{ at } x > x_i \end{cases} \tag{8.18}$$

$$-\int_0^1 T\delta(x)dx + \int_0^1 \frac{q}{\lambda}wdx + \left(\frac{dT}{dx}\right)_1 w_1 - \left(\frac{dT}{dx}\right)_0 w_0 = 0 \tag{8.19}$$

Since the first integral according to third equation (8.11) equals T_i for $\delta = \delta_i$ and $w_0 = 0$ [see equation (8.14)], this equation determines function T_i as follows:

$$T_i = \frac{q}{\lambda}\left(\int_0^{x_i} xdx + \int_{x_i}^1 x_i dx\right) + \left(\frac{dT}{dx}\right)_1 w_1 = \frac{q}{\lambda}\left(\frac{x_i^2}{2} + x_i - x_i^2\right) + \left(\frac{dT}{dx}\right)_1 x_i \tag{8.20}$$

Using the boundary condition $T_i = 0$ at $x_i = 1$, one finds unknown derivative $(dT/dx)_1$ at $x_i = 1$ and then obtains the same solution (8.16) using x for (8.20) instead of x_i:

$$\left(\frac{dT}{dx}\right)_1 = -\frac{q}{2\lambda}, \quad T = \frac{q}{\lambda}\left(x - \frac{x^2}{2}\right) - \frac{q}{2\lambda}x = \frac{q}{2\lambda}x(1-x) \tag{8.21}$$

Transforming second equation (8.13) using delta function as it just was done is a common approach in BEM. Modifying the first term of this equation using delta equation (8.18) leads to the usual form of equation in boundary element approach:

$$T_i = \int_0^1 \frac{q}{\lambda} wdx + \left[\frac{dT}{dx}w\right]_0^1 - \left[T\frac{dw}{dx}\right]_0^1 \tag{8.22}$$

This relation shows that the unknown function in this case is defined only by boundary conditions. In contrast to this, the FEM is based on first equation (8.13) from which follows that the unknown function is defined by information of the whole domain (Exercise 8.18).

Comment 8.6 Presentation of different numerical methods is done here using simple one-dimensional equations. However, the described procedures and features in principle are valid in more complicated cases for two- and three-dimensional problems [3,4].

8.3   Some Features of Computing Flow and Heat Transfer Characteristics

There are some specific difficulties and procedures in applying numerical methods to convective heat transfer and flow problems, which are discussed below [2,5].

8.3.1 Control-Volume Finite-Difference Method

8.3.1.1 Computing Pressure and Velocity

The main difficulty in solving the Navier–Stokes equations is that the pressure is unknown. Although there is no a special equation for pressure, it is indirectly controlled by the continuity equation. This is achieved due to connection between the velocity field calculation and satisfying the continuity equation. As is clear from Navier–Stokes equations (2.2) and (2.3), the velocity field can be calculated only if the pressure is known. At the same time, the continuity equation can be satisfied only when the velocities are computed using a proper pressure. Thus, to calculate the velocities and satisfy the continuity equation, the pressure should be known.

One well-known method to overcome this difficulty is to use the Navier–Stokes equation in the stream function form that is obtained by cross-differentiating equations (2.2) and (2.3) to eliminate the pressure (Sec. 2.2.2.3). Unfortunately, such an approach is applicable only for a two-dimensional problem because in the three-dimensional case the stream function does not exist.

The direct pressure calculation is based on determining the pressure difference for a control volume created for grid points. However, as is shown in [2], such a usual procedure in standard three-point CV approach with midway located grid point fails. It turns out that such a standard approach yields only uniform zero pressure.

A real-pressure distribution can be obtained by using special so-called staggered control volumes [2]. In staggered control volume, the velocity components and pressure are calculated for main points located on the CV faces that are set midway between two adjacent points normal to velocity component direction. Thus, the x-directed component u is calculated at the y-directed faces, and, vice versa, the y-directed component v is calculated at the x-directed faces. Such an approach results in the u-components located on the left and on the right faces from the main point and in the v-components located above and below the main point. The location of the main point with respect to two adjacent points is not important. It is important that only the main point is on the CV face. For example, in the grid with control volumes having different lengths, it is immaterial that the main points lying on the faces are not in the midst of two neighboring points.

Using the staggered grid eliminates the difficulties of calculating the pressure field, but a corresponding computer program becomes more complicated because it must record all information about the location of the velocity components and must perform tiresome interpolations (Exercise 8.19). Taking into account just discussed peculiarities of the numerical defining the velocity components and pressure, the special software named the SIMPLE (Semi-Implicit Method for Pressure-Linked Equations) was developed. This iterative procedure starts from guessing of the pressure field [2]. Then, using the finite-difference technique and the guessed pressure field, the Navier–Stokes equations in form (2.2) and (2.3) for velocity components are solved. The second iteration consists of applying just-found velocity components to calculate the pressure difference between adjacent grid points. This gives the new pressure field that is used to get new velocity components. These iterations are carried on until the continuity equation is satisfied. To control the process of satisfaction, the special equation is derived. The continuity equation is integrated over the CV giving the following equation:

$$\left[u_{(i+1/2),j} - u_{(i-1/2),j} \right] \Delta y + \left[v_{i,(j+1/2)} - v_{i,(j-1/2)} \right] \Delta x = 0 \qquad (8.23)$$

This equation is written for main points (i, j) using four neighboring points: $(i+1), j$ and $(i-1), j$ points from the left and from the right $i, (j+1)$ and $i, (j-1)$ points above and below the main point, whereas $i+1/2$ and $i-1/2$ denote points located midway between points $i+1$ and i and between points i and $i-1$, respectively, and likewise for points $j+1/2$ and $j-1/2$. Substituting calculated results in equation (8.23), one estimates the residue and stops iterations when the residue becomes sufficiently small (Exercise 8.20).

Later, to improve the process of convergence, some revised versions were developed. The most used among others are: SIMPLER (SIMPLE Revised) [2], SIMPLEC (SIMPLE Consistent) [6], and SIMPLEM (SIMPLE Modified) [7]. A noniterative procedure PISO (Pressure-Implicit with Splitting of Operators) is also developed [8]. The comparison shows that the differences in the efficiency of these approaches are generally modest. The analysis of available studies reveals that family SIMPLE is a reliable, practical calculating tool for variety applications and solving current complicated technological and theoretical problems [8].

8.3.1.2 Computing convection-diffusion terms

Consider a steady one-dimensional equation with only the convection and diffusion (or conduction, Sec. 2.2.1) terms, which on integration over a CV yields

$$\frac{d}{dx}(\rho c u T) = \frac{d}{dx}\left(\lambda \frac{dT}{dx}\right), \quad (\rho c u T)_{i+1/2} - (\rho c u T)_{i-1/2} = \left(\lambda \frac{dT}{dx}\right)_{i+1/2} - \left(\lambda \frac{dT}{dx}\right)_{i-1/2} \quad (8.24)$$

Using the piecewise-lineal approximation between grid points and equation (8.3) for the first derivative, the corresponding finite-difference equation is obtained:

$$\frac{1}{2}(\rho u c)_{i+1/2}(T_{i+1} + T_i) - \frac{1}{2}(\rho u c)_{i-1/2}(T_i + T_{i-1}) = \frac{\lambda_{i+1/2}(T_{i+1} - T_i)}{x_{i+1} - x_i} - \frac{\lambda_{i-1/2}(T_i - T_{i-1})}{x_i - x_{i-1}} \quad (8.25)$$

It can be shown that this fine-looking central-difference scheme leads to unrealistic results [2]. To see this, rearrange the last expression by collecting terms with the same T_n:

$$T_i\left[\frac{(\rho c u)_{i+1/2}}{2} - \frac{(\rho c u)_{i-1/2}}{2} + \frac{\lambda_{i+1/2}}{x_{i+1} - x_i} + \frac{\lambda_{i-1/2}}{x_i - x_{i-1}}\right] = T_{i+1}\left[\frac{\lambda_{i+1/2}}{x_{i+1} - x_i} - \frac{(\rho c u)_{i+1/2}}{2}\right]$$

$$+ T_{i-1}\left[\frac{\lambda_{i-1/2}}{x_i - x_{i-1}} + \frac{(\rho c u)_{i-1/2}}{2}\right] \quad a_i = a_{i+1} + a_{i-1} + (\rho c u)_{i+1} - (\rho c u)_{i-1} \quad (8.26)$$

Here, a stands for expressions in the brackets, and two last terms do not change the last expression (8.26) because according to the continuity law, $(\rho u)_{i+1} = (\rho u)_{i-1}$. Consider now a simple example. Let, for instance, we have

$$\lambda_{i+1}/(x_{i+1} - x_i) = \lambda_{i-1}/(x_i - x_{i-1}) = 1/2, (\rho c u)_{i+1} = (\rho c u)_{i-1} = 3/2,$$

$$T_{i+1} = 200, T_{i-1} = 100 \quad (8.27)$$

Then, $T_i = 75$, but if $T_{i+1} = 100$, $T_{i-1} = 200$, then $T_i = 225$. These results are clearly unrealistic because T_i cannot fall outside its neighbors T_{i+1} and T_{i-1}. It is obvious that such unrealistic results are possible in any case if $|\rho c u|$ exceeds $2\lambda/\Delta x$ (Exercise 8.21).

There are some possibilities to overcome this difficulty. The simplest way is to apply the upwind scheme where the midway between i and $i \pm 1$ points are determined as follows: $T_{i \pm 1/2} = T_i$ if $(\rho c u)_{i \pm 1/2} > 0$ and $T_{i \pm 1/2} = T_{i \pm 1}$ if $(\rho c u)_{i \pm 1/2} < 0$ with unchanged diffusion terms. The exact solution of first equation (8.24) subjected to boundary conditions $x = 0$, $T = T_0$, $x = L$, $T = T_L$ presented in the form

$$\frac{T - T_0}{T_L - T_0} = \frac{\exp \mathrm{Pe}(x/L) - 1}{\exp \mathrm{Pe} - 1}, \qquad \mathrm{Pe} = \frac{\rho c u L}{\lambda}, \tag{8.28}$$

gives an understanding of the applicability of the upwind scheme (Exercise 8.22). Analysis shows that for large values of $|\mathrm{Pe}|$, the temperature at the middle is nearly equal to that at upwind boundary, and this is the assumption used in the upwind scheme. However, in this scheme, the boundary temperature remains the same for all values of the Peclet number. The other imperfection is that for large $|\mathrm{Pe}|$, at the middle, the derivative dT/dx is almost zero so that the diffusion is almost absent. At the same time, in the upwind scheme, the diffusion is calculated always applying a linear profile, which overestimates it for large Peclet numbers.

Two schemes with qualitative behavior close to solution (8.24) are usually used instead of the central-difference scheme. One is the hybrid scheme that reasonably approximated such a behavior working alike the central-difference scheme for the range $-2 \le \mathrm{Pe} \le 2$ and reducing the diffusion to zero similar to the upwind scheme outside of this range. That is the reason why this scheme is named the hybrid scheme. A better approximation of exact solution (8.24) provides the power-law scheme, which is constructed using four ranges of the Peclet number [2]: $\mathrm{Pe} < -10$, $-10 \le \mathrm{Pe} < 0$, $0 \le \mathrm{Pe} \le 10$, and $\mathrm{Pe} > 10$ (Exercise 8.23).

8.3.1.3 False Diffusion

There are two types of false diffusion. The first type is the common misunderstanding: when by comparison of central-difference and upwind schemes using the Taylor expansion, one concludes that the upwind scheme produces the false diffusion. Indeed, the central-difference scheme is better than the upwind one only in the case of small Peclet numbers when the Taylor expansion is applicable. For large Peclet numbers, the truncating Taylor series cannot be used for analyzing convention-diffusion dependence because in this case that dependence is of exponential type. At the same time, as is discussed above the central-difference scheme leads to unrealistic results in the case of large Peclet numbers.

The other type is the real false diffusion that arises in situation, when the calculation result shows the presence of diffusion, despite that the diffusion coefficient is zero. For example, if two parallel, two-dimensional streams of equal velocities and different temperatures come in contact, the diffusion process forms a mixed layer only in the case of a nonzero diffusion coefficient. However, when the diffusion coefficient is zero, such two streams remain separated with temperature discontinuity at the interface. Thus, if in such a situation, the computer program shows a smeared profile in the cross-section, it is obvious that a real false diffusion is taking place.

In general, false diffusion arises when the flow is oblique to the grid lines and there is a gradient in the direction normal to the flow. Since the false diffusion is the most severe when the flow direction makes an angle of 45° with the grid lines [2], the intensity of false diffusion can be reduced by adjusting the flow along the grid lines. The other way to reduce the false diffusion is to use a small Δx and Δy since it is known that this results in small Peclet numbers when the central-difference scheme works perfectly (Exercise 8.24).

8.3.2 Control-Volume Finite-Element Method

The CVFEM is basically very close to the CVFDM described above and should not be considered as a different method. The only advantage that distinguishes the former from the latter is the ability to use the irregular grids since in the finite-difference approach, mainly uniform grids are employed. Such irregular, for example, triangular grids are more flexible and allow providing local grid refinement.

The difficulties discussed in the last subsection are inherent in the FEM as well. These difficulties have been resolved, and a CVFEM similar to the control-volume finite-difference one has been developed [2,5,9]. This method is characterized by the following features:

- For the triangular grids, the values of dependent variables are calculated for the grid points that lie at the vertices of the triangles, which plays a role of main points. The lines joined the centroid of each triangular element with midpoints of its sides dividing each element in three equal areas, regardless of the form of triangle element. These areas collectively construct the nonoverlapping contiguous polygonal volume elements that are similar to these in the finite-difference approach.
- Many CFD (computational fluid dynamics) codes used the staggered grids that do not have the problems of central-difference schemes. However, the staggered grids cannot be used for nonorthogonal grids and unstructured meshes, which are typical for the finite-element approach. Therefore, in the FEM, instead of the staggered grids, two other approaches are used [5, 9]. One consists of unequal-order formulations for pressure and velocity components in which for the former are used a sparser grid and a lower-order interpolation than that used for the latter. In the second technique, the equivalent of the co-located momentum-interpolation scheme is applied. However, the first approach requires two sets of control volumes that make the calculation awkward and the excessively fine grids in the case of high Reynolds numbers or pressure gradients. As a result, the co-located momentum-interpolation scheme has been adopted in computational practice. In these schemes, both the velocities and the pressure are calculated at the same set of nodes located at the centre of the control volume in contrast to the staggered grids when, as discussed above, the pressure and the velocities are determined at different faces of the CV. The formulae for velocities and pressure using co-located methods are derived, employing the discretization of the momentum and continuity equations for the staggered grids. Since staggered grids do not have problems of the central-difference schemes, these formulae do not have such problems as well (Exercise 8.25).

The methods of solving flow and heat transfer governing equations discussed above are based on single-grid solutions. These approaches are flexible, reliable, potential, and convenient and are working well. They have been employed usefully in science and industry for many years. However, the convergence of solutions in the single-grid schemes was acceptable some decades ago when the computer memory was not enough to use fine grids. Nowadays, the computational technique requires very fine grids [up to $(1-10)\times10^6$ fine volumes] to ensure the accuracy and reduce the errors.

There are several ways to accelerate the convergence of the family SIMPLE methods. Since the entire set of equation is highly nonlinear and coupled, any one of the equations can slow down the convergence of the entire set, and this required considering the convergence of whole set of equation. For example, the multigrid technique is based on the idea that convergence of iterative schemes arises from the low-frequency components of the error because the high-frequency components converge rapidly during the initial iterations. Therefore, this approach consists of eliminating the poor converging part of error and increasing its convergence by putting it on the coarse grid. Then, the correct variant is interpolated beck to the finer grid to correct the previously obtained poor converge part of the error. A review of the studies of multigrid schemes is presented in [5].

The other way to improve the convergence and reduce the time of mesh generation of SIMPLE methods is to use the meshless methods. The meshless methods apply the compact or global interpolation on nonordered irregular spatial domains, employing special radial basis functions (RBFs) [10, 11]. The primary feature of a radial basis function is that it depends only upon the radial distance between knots or nodal points. There are several commonly used global radial basis functions. The interpolation with these functions has been successfully used and demonstrated striking convergence. In particular, the scheme with multiquadratic radial basis functions (MQ-RBFs) shows exponential convergence. The MQ-RBFs scheme was successfully applied for the solution of ordinary and partial differential equations [10]. The local collocation meshless method using MQ-RBFs scheme have been proposed as well [11] (Exercise 8.26).

8.4　Numerial Methods of Conjugation

The following governing equations are used in the numerical investigations of conjugate convective heat transfer:

- For steady and unsteady flows in the fluid domain:
  Two- or three-dimensional continuity equation
  Two- or three-dimensional Navier–Stokes and energy equations
  or
  Two- or three-dimensional dynamic and thermal boundary layer equations for high Reynolds number
  or
  Two- or three-dimensional dynamic and thermal creeping flow equations for low Reynolds number

  One-dimensional velocity and temperature equations

- For steady and unsteady conduction in the body domain:
  Two-or three-dimensional unsteady conduction equations

  Two-or three-dimensional steady Laplase's or Poisson's equations

  One-dimensional unsteady and steady equations for a thin or for a thermally thin body.

  Although many combinations of these equations are possible, the majority of the known investigations of the conjugate convection heat transfer are using, for a flow, two- or three-dimensional (rare) steady or unsteady elliptic Navier–Stokes and energy equations or two- or three-dimensional steady or unsteady parabolic boundary layer equations and, for a body, two-dimensional steady Laplace's or Poisson's equations or steady or unsteady (more often) one-dimensional conduction equations.

There are several methods to perform the conjugation between solution domains.

- A direct approach is that in which the governing equations for both the fluid and body are conjugated by simultaneous solving one large set of equations. For example, a method proposed by Patankar [2, 12] is of that type. In that approach, the one generalized conservation expression is considered instead of both separated equations for the fluid and solid. If both depended variable are defined by ϕ, this general differential equation is

$$\frac{\partial}{\partial t}(\rho\phi) + div(\rho\mathbf{u}\phi) = div(\Gamma grad\phi) + \tilde{S} \tag{8.29}$$

where $\mathbf{u}$ is the velocity vector, Γ is the diffusion coefficient, and $\tilde{S}$ is the source term. A large calculation domain includes the fluid and body domains, conjugate conditions for their interface, and outer boundary conditions. To ensure that the velocity is zero in the solid when the velocity field is calculated, one puts for the grids points in the body domain a very large value of diffusion coefficient Γ, whereas at the same time for the grid points in the fluid domain, the diffusion coefficient Γ is made equal to the real fluid viscosity. When the temperature field is computed, one specifies in general equation the real values of diffusion coefficient Γ for the fluid and for the body, which gives the entire temperature field as a result of matching the temperature distribution in the fluid and the body.

In some cases, applying the direct conjugation involves difficulties; in particular, these occur due to the mismatch in the structure of the coefficient matrices in BEM, FEM, and/or FEM solvers [13].

- The iterative conjugation method consists of a separated solution of each set of equations for the fluid and for the body, respectively. An idea of such a strategy is that each solution of the body or the fluid equation produces the boundary condition for the other along the body–fluid interface. It starts from guessing of one of the fields on the interface, which is used as a boundary condition for the other. For example, one begins assuming the temperature or heat flux distribution along the interface that is used as a boundary condition for solving the equation for the body. Then, the obtained variable distribution along the interface is applied as a boundary condition for solving the system of equations for the fluid, and so on. If the process would converge, the iterations would continue until the required accuracy criterion is achieved. This iterative approach is widely used in FDM, FEM, or FVM.

One of the disadvantages of these commonly applied numerical approaches is that they require entire mashing of both the fluid and body domains and using the numerical differentiating to get the heat fluxes. An alternative algorithm that does not require these procedures was developed in [13]. Although this approach as many others uses the FVM to mesh the whole fluid domain and applies the iterative procedure for conjugation, it employs for the body domain the boundary element method, which requires the discretization only of the boundaries (Sec. 8.2). Besides, in this case, the interfacial heat fluxes, which are needed for providing the continuity in the conjugate problem, are obtained when the body conduction is carried out, and hence, the numerical differentiating is avoided. Solutions obtained by this method may be found in [13].

- The superposition method gives an expression for heat flux on the gradientless stream-lined plate in the case of arbitrary continuous temperature distribution (Sec. 2.6.8):

$$q_w = h_* \left\{ T_w(0) - T_\infty + \int_0^x \left[1 - \left(\frac{\xi}{x} \right)^{C_1} \right]^{-C_2} \frac{dT_w}{d\xi} d\xi \right\}, \qquad (2.110)$$

- where $C_1 = 3/4, C_2 = 1/3$ for laminar and $C_1 = 9/10, C_2 = 1/9$ for turbulent flows. On the other hand, it was shown (Sec. 4.1) that using a relation for heat flux with an arbitrary temperature distribution reduces the conjugate problem to the solution of conduction equation. Consequently, in the case of using equation (2.110), the solution of conjugate heat transfer problem is obtained by substituting this relation in a solution of conduction equation for a plate. In the case of the thermally thin plate, such a procedure reduces the conjugate problem to integro-differential or nonlinear differential equations. These equations are solved numerically using standard or special simple approaches. An example of early solutions of this type may be found in [14].

A general approach of this type that reduces a conjugate convective heat transfer problem to an equivalent heat conduction problem is outlined in Section 4.4. This method is based on the exact solutions given for an arbitrary nonisothermal surface in the laminar incompressible and compressible flows. This technique is also applicable to unsteady laminar flow, continuously moving surface, flow of non-Newtonian power fluids, and for turbulent flow for which the analogous solutions are derived in Sections 4.2 and 4.3.

The general forms of the Duhamel's integral (2.110), such as equation (4.14) for laminar flow and similar formulae for turbulent flow and other cases, are applicable to flows with a nonzero pressure gradient due to using the Görtler's variable Φ, which takes into account the presence of the pressure gradient (Sec. 4.2.3). Another advantage is that this approach contains the heat flux expression in two forms: the above-mentioned integral form (4.14) and equivalent differential form in series, such as (4.1) or (4.13). These two forms allow performing accurate calculations, employing the differential form with several first terms in the case of the con-verged series and using integral form otherwise (Exercise 8.27).

Examples of solutions applying numerical methods described and discussed in this chapter are considered in *Applications* (Part IV).

EXERCISES

8.1 Obtain formulae (8.1)–(8.3) using Taylor series. Explain why these could not be used if there are singularities.

8.2 Show that temperature near the surface has the same order as velocity $T \sim cy$. Hint: use the relation $u(\partial T/\partial x) \sim v(\partial T/\partial y)$ and estimations of order of velocity components.

8.3 Calculate how the mesh size varies with y and along the plate according the equation $z = \Delta / \left[1 - (y/\delta) \right]$ knowing how the boundary thickness δ changes in laminar or turbulent flow.

8.4 Discuss with colleagues the relationship between analytical and numerical methods. Compare those describing the advantages of each method showing that they are supple-menting each other.

8.5 Explain why the finite-difference formulae for derivatives could not be used if there are singularities. What is the way to overcome this difficulty, why it works?

8.6 Describe how contemporary numerical methods are connected with approximate analytical methods using widely before computer advent.

8.7 How many finite-difference methods are there? How they differ from each other?

8.8 Explain in words what the idea of weighted residual is. How this concept is used to construct different approximate methods?

8.9 Read articles about the collocation method and Dirac function in Wikipedia or in *Advanced Engineering Mathematics*.

8.10 Find intuitively or using a polynomial with unknown coefficients another relation for ϑ with two free coefficients satisfying the boundary condition $\vartheta = 0$ at $x = 0$ and $x = 1$ in Example 8.1. Hint: find polynomial with minimum required terms.

8.11 Repeat the solution of problem 8.1 and solve similar problem using for ϑ expression obtained in previous exercise. Compare both solutions.

8.12 Solve problem 8.1 with the same boundary condition at $x = 0$ and isolated plate end $dT/dx = 0$ at $x = L$. Hint: use polynomials to find both relations first for transforming boundary conditions to the homogeneous and then an expression with two unknown coefficients.

8.13 Derive equations (8.12). Explain how the assumption that the temperature changes linearly between grid points is used.

8.14 Obtain two expressions (8.13) by double integrating by parts the result obtained after integration of equation (8.5) multiplied by some weighted function w. Hint: employ $u = w, dv = \left(d^2T/dx^2\right) dx$ and $u = dw/dx, dv = \left(dT/dx\right) dx$ for the first and the second integration, respectively.

8.15 Learn more about weak solution reading article "Weak solution" from Wikipedia.

8.16 Derive equations (8.17) starting from relations (8.14). Explain each step of process.

8.17 Explain the derivation of equation (8.21) starting from equation (8.18). Consider each step. Hint: use equations (8.11).

8.18 Compare considered here basic contemporary numerical methods for solving differential equations. Explain how these methods differ in discretization schemes and in using grid of elements.

8.19 What causes the difficulties in computing pressure and velocity components? Why special staggered grids are needed to overcome these problems? How staggered grids differ from usual ones?

8.20 Describe the SIMPLE procedure. Draw the scheme of four points using for writing equation (8.23). How this equation is employed?

8.21 Show that the central-difference scheme is not applicable for convection-diffusion computing. Obtain first equation (8.26) and check the second one. Consider examples and explain why these results are unrealistic. Show that this difficulty arises when $|\rho c u|$ exceeds $2\lambda/\Delta x$.

8.22 Integrate differential equation (8.24) subjected to the boundary condition indicated in the text. Hint: use the new variable $z = dT/dx$ and then variable separation (Sec. 1.5).

8.23 Describe the two schemes usually used to substitute the central-difference scheme. What behavior these schemes show?

8.24 What is the real and seeming false diffusion? What are the ways to reduce the possibility of the false diffusion occurring?

8.25 Compare finite-difference and finite-element control-volume formulation. What are the basic features distinguish both methods?

8.26 What are ideas of multigrid and meshless methods? What was the reason to create these methods?

8.27 Describe and compare numerical conjugation methods. Think what approach you would prefer if it were necessary.

REFERENCES

[1] Panov, D. (1951). *Handbook of numerical treatment of partial differential equations* (5th ed.) (in Russian). Isdatel'stro ANSSR, Moscow.

[2] Patancar, S. V. (1980). *Numerical heat transfer and fluid flow*. London: Taylor & Francis.

[3] Brebbia, C.A., & Walker, S. (1980). *Boundary element technique in engineering.* London-Boston: Butterworths.

[4] Brebbia, C. A., & Dominguez, J. (2001). *Boundary elements: An introductory course* (2nd ed.). Boston: WIT Press.

[5] Acharya, S., Baliga, B. R., Karki, K., Murthy, J. Y., Prakash, C., & Vanka, S. P. (2007). Pressure-based finite-volume methods in computational fluid dynamic. *The ASME Journal of Heat Transfer 129*, 407–424. doi: http://dx.doi.org/10.1115/1.2716419

[6] Van Doormaal, J. P., & Raithby, G. D. (1984). Enhancements of the SIMPLE method for predicting incompressible fluid flows. *Numerical Heat Transfer 7*, 147–163. doi: http://dx.doi.org /10.1080/01495728408961817

[7] Moukalled, F., & Acharya, S. (1989). Improvements to incompressible flow calculation on a non-staggered curvilinear grids. *Numerical Heat Transfer* Part B *15*, 131–152. doi: http://dx.doi .org/10.1080/10407798908944897

[8] Issa, R. I, Gosman, A. D., & Watkinc, A. P. (1986). Computation of compressible and incompressible recirculating flows by a non-iterative implicit scheme. *Journal of Computational Physics 62*, 66–82. doi: http://dx.doi.org/10.1016/0021-9991(86)90100-2

[9] Baliga, B. R., & Patankar, S. V. (1983). A control-volume finite-element method for two-dimensional incompressible fluid flow and heat transfer. *Numerical Heat Transfer 6*, 245–261. doi: http://dx .doi.org/10.1080/01495728308963086

[10] Kansa, E. J., & Hon, Y. C. (2000). Circumventing the ill-conditioning problem with multiquadratic radial basis functions: Applications to elliptic partial differential equations. *Computers & Mathematics with Applications 30*, 123–137. doi: http://dx.doi.org/10.1016/S0898-1221(00)00071-7

[11] Divo, E., & Kassab A. J. (2007). An efficient localized radial basis function meshless method for fluid flow and conjugate heat transfer. *ASME Journal of Heat Transfer 129*, 124–136. doi: http://dx .doi.org/10.1115/1.2402181

[12] Patankar, S. V. (1978). A numerical method for conduction in composite materials, Flow in irregular geometries and conjugate heat transfer. Proceedings of 6th International Heat Transfer Conference, Toronto *3*, 297.

[13] Divo, E., Steinthorsson, E., Kassab, A. J., & Bialecki, R. (2002). An iterative BEM/FVM protocol for steady-state multi-dimensional conjugate heat transfer in compressible flows. *Engineering Analysis with Boundary Elements 26*, 447–454. doi: http://dx.doi.org/10.1016/S0955-7997(01)00106-0

[14] Sohal, M. S., & Howell, I. R., 1973. Determination of plate temperature in case of combined conduction, convection and radiation heat exchange. *International Journal of Heat and Mass Transfer 16*, 2055–2066. doi: http://dx.doi.org/10.1016/0017-9310(73)90108-7

CHAPTER 9

MODERN NUMERICAL METHODS IN TURBULENCE

9.1 Introduction

The classical numerical methods considered in previous chapters are applicable to both laminar and turbulent flows. However, the level of accuracy of governing equations for both flows is different. Although the unsteady three-dimensional Navier–Stokes equation describes, in principle, the turbulent flow as well as laminar one, the range of scales of eddies sizes is so wide that before sixtieth of the last century only Reynolds-average models, discussed in Chapter 3, might be used for studying and applications. The advent of computers changed the situation, and in the last 50 years, the numerical methods of solving the exact Navier–Stokes equations without averaging have been developed. The basic value of such methods is that they provide insight into the physics of turbulence, increasing our understanding of its nature.

In this chapter, we consider three such methods: Direct Numerical Simulation (DNS), Large Eddy Simulation (LES), and Detached Eddy Simulation (DES). Each method uses a special restriction in order to be in line with possibilities of today's computers. In DNS, the exact Navier–Stokes equations are solved for all range of scales of eddies, but the possible solution is confined to moderate Reynolds numbers. Two other methods achieved such restriction by separating large and small eddies, computing exactly the first and Reynolds-average modeling the second. The difference between LES and DES lies in variation of the treatment of small-size eddies.

Three characteristic scales of turbulence are used in modern methods: integral length l, which is appropriate to the energy-bearing eddies; the Kolmogorov scale η, which characterizes the smallest eddies; and the Taylor microscale λ, which is relevant to median size eddies. These scales are related to each other as follows:

$$\eta/l \sim \mathrm{Re}_{tb}^{-3/4}, \ \lambda/l \sim \mathrm{Re}_{tb}^{-1/2}, \ \eta = (v^3/\varepsilon)^{1/4}, \ \lambda \sim (l\eta)^{1/3}, \ \mathrm{Re}_{tb} = k^{1/2}l/v \tag{9.1}$$

where ε is dissipation (Sec. 3.4.3.1). A comparison shows that $\eta \ll \lambda \ll l$, and, since $l \sim 0.1\delta$, one estimates that the Kolmogorov length scale outside the viscose wall region is $1/10,000$ times less than the thickness of the boundary layer [1].

9.2 Direct Numerical Simulation

A direct numerical simulation is a method to solve the Navier–Stokes equations in order to obtain the complete space- and time-dependent field of turbulent flow. The practical and fundamental importance of such simulation is eminent because these solutions give numerically exact results, which may be considered like experimental data. These accurate results on the microscale level may be used for different studies: for analyzing the turbulence structure and instantaneous characteristics such as fluctuations correlations, to test the other approximate approaches, to create new turbulence models for applications, and to study specific phenomena of turbulent flows that are usually difficult or impossible to obtain in laboratory (Exercise 9.1).

However, performing DNS required fine grid sizes. Estimation of the number of grid points and time steps needed for getting accurate results shows that such calculation is a complex computation problem. If the increment along the mesh direction is h, the number N of points should satisfy the inequality $Nh > l$, provided that the integral scale l is contained within the computational domain. On the other hand, to resolve the Kolmogorov scale, it is necessary to have $h \leq \eta$. Using these two inequalities and relation (9.1) for η, and knowing that $\varepsilon \approx (u')^3 / l$, the expression determining the number of points $N \geq (u'l/v)^{9/4} = \mathrm{Re}_{tb}^{9/4}$, which is required to perform the three-dimensional DNS, can be derived. In addition, to obtain accurate results applying explicit methods of integration, the time step Δt should satisfy inequality for the Courant number $C = u'\Delta t / h < 1$, which ensures that fluid particle path in each step is less than a mesh spacing h. Because the turbulence time scale is of order l/u' and $h \sim \eta$, a simple calculation yields the order of the number of time steps $N_t \sim l/u'\Delta t = l/Ch = l/\eta C = \mathrm{Re}_{tb}^{3/4}/C$, where the result is obtained applying relation (9.1) (Exercise 9.2).

More specific estimations for flow in channel of height $2H$ are given in [1]: $N \approx (3\,\mathrm{Re}_{tb})^{9/4}$, $N_t \approx 0.006H/u_\tau \sqrt{\mathrm{Re}_{tb}}$, where $\mathrm{Re}_{tb} = u_\tau H/v$. According to these formulae, the numbers of grid points and time steps for the channel of one of Laufer's experiments [11, Chap. 3] depending on Reynolds number $\mathrm{Re}_{tb} = 360$ ($\mathrm{Re}_H = 0.61 \cdot 10^4$) -14500 ($3.1 \cdot 10^4$) are $N = 6.7 \cdot 10^6 - 1.5 \cdot 10^8$ and $N_t = 32 \cdot 10^3 - 63 \cdot 10^3$.

From these estimations, one can conclude that in DNS the required number of grid points and time steps grows fast with Reynolds number, resulting in necessity of enhanced computer memory and very high computational cost. Because the Reynolds numbers in most engineering applications are higher than moderate Re value for which the DNS may be performed using currently available computers, until now, only relatively simple problems were investigated using DNS. Some of these results are listed in [1] show that they are basically simple geometric flows in channels or in boundary layers (e.g., channel flow at $\mathrm{Re}_H = 6875$ or two-dimensional boundary layer with pressure gradient and $\mathrm{Re}_{\delta_2} = 1410$). Nevertheless, the importance of contribution of direct Navier–Stokes equation solutions in turbulence research is difficult to overestimate: significantly improving the exactness of results, this approach has opened a new chapter in understanding turbulence, leading to fresh ideas and thoughts. As more powerful computers appear, the possibilities of DNS would grow, resulting in fundamental studies and engineering applications (Exercise 9.3).

9.3 Large Eddy Simulation

A large eddy simulation is a method of reducing the number of grid points and time steps in order to solve directly Navier–Stokes equation for higher Reynolds numbers. Estimations show that usually the number of grid points required in LES is about ten times less than that in DNS [1]. The procedure of LES first was proposed in 1963 by Smagorinsky for the atmospheric motion study. The main idea of such a procedure is to separate the treatment of large eddies from small eddies. Because the large eddies carry the majority of the energy and are mostly affected by boundary condition, they should be computed directly using the DNS approach. On the other hand, the small eddies turbulence is weaker, nearly isotropic having almost universal characteristics, and therefore is more tractable for modeling applying Reynolds-average models.

To realize such scale separation, a special method known as filtering was developed. A simple example helps to understand the notion of filtering and, in particular to see that the integration may be used to perform filtering. Consider equation (8.3) for numerically determining the first derivative of some function $f(x)$. Using the Leibniz rule of interchanging the integration and differentiation (Exercise 1.49), this relation may be transformed in the integral presentation (Exercise 9.4):

$$\frac{df}{dx} = \frac{f(x+\Delta x) - f(x-\Delta x)}{2\Delta x} = \frac{1}{2\Delta x} \int_{x-\Delta x}^{x+\Delta x} \frac{df(\xi)}{d\xi} d\xi = \frac{d}{dx} \left[\frac{1}{2\Delta x} \int_{x-\Delta x}^{x+\Delta x} f(\xi) d\xi \right] \quad (9.2)$$

This expression gives an average value of the derivative of function $f(x)$. The derivative is constant within one grid step Δx, but its value is variable changing from one step to another. Due to that, relation of type (9.2) may be considered as a filter of the derivative sizes. One of the first filters employed in 1970 for three-dimensional turbulence model, known as volume-average box filter, was of this type (Exercise 9.5)

$$\bar{u}_i(\mathbf{x},t) = \frac{1}{\Delta^3} \int_{x-(1/2)\Delta x}^{x+(1/2)\Delta x} \int_{y-(1/2)\Delta y}^{y+(1/2)\Delta y} \int_{z-(1/2)\Delta z}^{z+(1/2)\Delta z} u_i(\xi,t) d\xi \varsigma \zeta \,,$$

$$\bar{u}_i(\mathbf{x},t) = \iiint G(\mathbf{x}-\xi,\Delta) u_i(\xi,t) d^3\xi \quad (9.3)$$

Since then, many other types of filters are suggested. A generalized filter defined by second relation of equation (9.3) was created applying the convolution integral used in Fourier transform (Example 1.17). This expression may present different filter types by specifying the filter function $G(\mathbf{x}-\xi,\Delta)$. Two examples of such a function given below pertain to the volume-average box filter [equation (9.3)] and another based on Fourier transform [1]

$$G(\mathbf{x}-\xi,\,\Delta) = \begin{cases} 1/\Delta^3, & |x_i - \xi_i| < \Delta x_i/2 \\ 0, & \text{otherwise} \end{cases} \qquad G(\mathbf{x}-\xi,\,\Delta) = \frac{1}{\Delta^3} \prod_{i=1}^{3} \frac{\sin(x_i - \xi_i)/\Delta}{(x_i - \xi_i)/\Delta} \quad (9.4)$$

Here, Δ denotes the smallest of the turbulence scales rated by filter as large ones, which are computed, while the others, under scale Δ, termed as subgrid scales (SGS), are eliminated and modeled. Because the relations in equation (9.3) determine the filter velocity $\bar{u}_i$, the subgrid velocity is obtained as a difference $u_i' = u_i - \bar{u}_i$ (Exercise 9.6).

Applying filtering ($\bar{}$) to Navier–Stokes equation (2.15) in Einstein notation, we get

$$\frac{\overline{\partial u_i}}{\partial x_i} = 0, \quad \frac{\overline{\partial u_i}}{\partial t} + \frac{\overline{\partial u_i u_j}}{\partial x_j} = -\frac{1}{\rho}\frac{\overline{\partial p}}{\partial x_i} + v\frac{\overline{\partial^2 u_i}}{\partial x_j \partial x_j}, \quad \frac{\partial \overline{u}_i}{\partial t} + \frac{\overline{\partial u_i u_j}}{\partial x_j} = -\frac{1}{\rho}\frac{\partial \overline{p}}{\partial x_i} + v\frac{\partial^2 \overline{u}_i}{\partial x_j \partial x_j} \quad (9.5)$$

Equation (9.5) is obtained from the previous one assuming that for linear terms filtering and differentiating operations are commuted. For the second nonlinear term, this assumption does not hold. Since the quantities u_i are unknown, the nonlinear term is defined in terms of filtered quantities $\overline{u}_i$ using the relation $\overline{u_i u_j} = \overline{u}_i \overline{u}_j + \tau_{ij}$, where τ_{ij} is a tensor representing interactions among and between large and small scales. Applying this relation and the rate-of-strain tensor S_{ij}, one obtains equation (9.5) in the following form:

$$\frac{\partial \overline{u}_i}{\partial t} + \overline{u}_j\frac{\partial \overline{u}_i}{\partial x_j} = -\frac{1}{\rho}\frac{\partial \overline{p}}{\partial x_i} + 2v\frac{\partial}{\partial x_j}\overline{S}_{ij} - \frac{\partial \tau_{ij}}{\partial x_i} \qquad \overline{S}_{ij} = \frac{1}{2}\left(\frac{\partial \overline{u}_i}{\partial x_j} + \frac{\partial \overline{u}_i}{\partial x_i}\right) \qquad (9.6)$$

Comment 9.1 The rate-of-strain shows how fast the components of velocity change in the three directions. Because there are three components and three directions, the rate-of-strain tensor is tensor of the second order defined by nine components (Sec. 2.2.2.6) (Exercise 9.7).

Equation (9.6) presents the filtered large scales filed modified by the SGS stresses through tensor τ_{ij}. Because this tensor is unknown, it should be modeled. Such modeling involves developing a model for the SGS stresses, which is a fundamental task in LES. During the past half century, many models, from the first simple to the more complicated nonlinear models, have been suggested [1]. The simplest Smagorinsky model is created using analogy with mixing-length formula (Sec. 3.3.1). In this case, the tensor τ_{ij} is proportional to the strain-rate tensor S_{ij} and is determined as [1,2]

$$\tau_{ij} = 2v_{tb}\overline{S}_{ij}, v_{tb} = (C_s\Delta)^2 \left|\overline{S}\right| \left|\overline{S}\right| = \sqrt{2\overline{S}_{ij}\overline{S}_{ij}} \qquad (9.7)$$

where v_{tb} is Smagorinsky eddy viscosity, C_s is constant coefficient, and $\left|\overline{S}\right|$ is the norm of the filtered strain-rate tensor. The Smagorinsky model as well as mixing-length one is not universal; rather it is calibrated by adjusting the coefficient, which usually varies $C_s \approx 0.1 - 0.2$. Some other models used Smagorinsky formula with different eddy viscosity. For example, the Lilly model with turbulence viscosity depending on SGS kinetic energy is similar to Prandtl's one-equation model (Sec. 3.4.2). The dynamic SGS model applies variable coefficient C_s, defining it by additional filtering (Exercise 9.8).

9.4 Detached Eddy Simulation

Large eddy simulation significantly widened the application of the direct solutions of Navier–Stokes equation by increasing the possible Reynolds number. This was achieved by using the Reynolds-average modeling for small eddies instead of computing those in DNS. However, important engineering applications such as airfoil, ground, or marine vehicle require much higher Reynolds number and so demand great numbers of grid points and time steps that lie far beyond the resources of current computers. The basic reason for the growth of such grid number is related to near-wall region with the smallest eddies, whose role increases as Reynolds number grows. Estimation shows that increase of Reynolds number by factor 10 leads to increase in the compute work by factor about 30 [1] (Exercise 9.9).

These problems motivated the attempts to develop improved LES methods by changing the treatment of small eddies. Some of these methods are published, for example, the article [2] in which the modified Smagorinsky model with dynamically computed filter width was suggested. The detached eddy simulation (DES) introduced by Spalart et al. in 1997 [3] is the most promising method of improved LES; it gained acceptance due to showing successful application of computing separation flows past bluff bodies and vehicles with blunt fore body at a manageable cost.

DES is a hybrid approach combining the RANS equation (Reynolds-averaged Navier–Stokes equation, Sec. 3.2) and LES methods in a united technique by using the former procedure for near-wall region and the latter one for the domain with large eddies. Because DES is a single solution for the entire computation domain, the essential issue in this method is the means of distinguishing between RANS and LES for treatment in corresponding regions. This is achieved using a special function called blending function, $f = \min(d, C_{DES}\Delta)$, where d is distance to the closest surface, Δ is the largest grid cell, and C_{DES} is constant of proportionality. The blending function impels that the model behave as RANS in regions close to walls, where $d \ll \Delta$, and perform as a subgrid model, such as Smagorinsky or other such models, away from the walls at $\Delta \ll d$. That feature of changeable procedure in DES instead of a fixed one in LES basically furnishes better simulation and reduces computing cost (Exercise 9.10).

One of the most important accomplishments of DES is the successful simulation of massive flow separation at high Reynolds numbers. In this technique, the RANS model is applied to study the attached boundary layer, while the LES approach is used in the separation region. Some results in the form of patterns showing the contours and isosurfaces of instantaneous vorticity and graphs of pressure coefficient distribution may be seen, for example, in [4]. The results for the sphere in subcritical flow at $Re = 10^4$ and supercritical flow at $Re = 1.14 \cdot 10^6$ are presented as a typical example of a flow past bluff bodies. In the first case, the laminar flow exists and calculation show early separation at the central angle about $82°$, which is in agreement with experimental data. In the second case, the turbulent flow separates at $120°$, which is also in agreement with experimental data. The distribution of pressure coefficient around the sphere obtained by DES agrees with measurements as well. The position of the minimum value in the case of turbulent flow is at about $90°$, according to experimental data. The back pressure at $180°$ gained for both regimes is also reasonably accurate and yields relatively correct drag. The other examples show computation solutions for aircraft presenting flows around: (i) the forebody of modern fighter aircraft at $90°$ angle of attack, (ii) the fighter aircraft F-15E at $65°$ angle of attack, (iii) F-18E in the case of the abrupt wing stall, and (iv) F-18C at $30°$ showing the vortex burst (Exercise 9.11).

9.5 Chaos Theory

Chaos theory is a field in mathematics, that studies the behavior of dynamical systems highly sensitive to initial conditions. This feature known as the butterfly effect means that small differences in initial conditions yield widely different outcomes (a butterfly flapping its wings today may result in a storm tomorrow far away from here). Chaotic behavior can be observed in some natural and engineering systems, such as weather. One remarkable example of the success of chaos theory is a qualitative simulation of Rayleigh–Bénard flow by a simple system of three ordinary differential equations. This phenomenon occurs in fluid present between two horizontal plates in gravitational field if the lower plate is heated (Example 2.24).

Currently, the chaos theory is not a tool for turbulence modeling; however, some turbulence characteristics are of the conditions that specify the chaotic regime.

This indicates that there are hopes of using the chaos theory for attacking the turbulence problems. At the same time, some researchers think that spectrum of wavelengths in turbulence, ranging from Kolmogorov length scale to the dimension of flow, is so broad that describing turbulence by chaotic methods would require a system of several hundred ordinary differential equations and hence is unrealistic [1] (Exercise 9.12).

9.6 Concluding Remarks

Modern methods of solving unsteady Navier–Stokes equation without averaging signify a great progress in understanding and applying turbulence. We have seen that even now, despite the restricted computer resources, the computation results obtained by modern methods, especially due to detached eddy simulation, in principle differ form solutions obtained from turbulence models based on the Reynolds-average approach. As computer efficiency increases, newer possibilities owing to higher Reynolds numbers will be achieved, leading to original thoughts and ideas. However, estimations show that direct numerical simulation of flow past airborne or ground vehicle requires computer efficiency that may be possible only by, approximately, 2045 (Spalart, [5]) . When it proved true, the turbulence, "the most important unsolved problem of classical physics," as it was characterized by well-known physicist Nobel laureate Richard Feynman, would be finally solved. Then one can claim that the first complete solution of Navier–Stokes equation at real practical values of Reynolds numbers took two centuries to obtain after its first publication in 1840.

For further reading about modern methods, the references of original articles may be found in the book [1]. More simple presentations are given in articles [2,4].

EXERCISES

9.1 Explain why classical Reynolds-average models and modern computation methods in turbulence differ in essence.

9.2 Obtain estimations of numbers of grid point and time steps required for three-dimensional DNS.

9.3 Explain why DNS is critically important despite the fact that it may be used basically for small Reynolds numbers.

9.4 What is the concept of filtering? Why is the filtering process used in LES?

9.5 Derive equation (9.2) and show that it may be used as a filter.

9.6 Explain what Δ means in equation (9.4) and for what the abbreviation SGS stands for.

9.7 Repeat the derivation of equation (9.6) to understand all notations, terms, assumptions, and operations. Read the article "Velocity vector and strain-rate tensor" online.

9.8 Think to understand why modeling the SGS is very important in LES. Hint: recall the relationship between eddies considered in Chapter 3. Describe the Smagorinsky model. Read at least two first paragraphs of the article [2].

9.9 What are the problems in LES that limit the increase of Reynolds number, and mention some attempts to improve this technique?

9.10 Explain the basic features of DES. What are the properties that differ the DES approach from pure LES in essence? What is blending function? How does it work?

9.11 Learn more about remarkable application results of DES by reading article [4]. Think: Is it always true that turbulent flow separates later than the laminar flow? Recall how Prandtl used the later separation of turbulent flow around bluff bodies to reduce the resistance.

9.12 Learn more about chaos theory by reading about it in Wikipedia.

REFERENCES

[1] Wilcox, D. C. (1994, 2006). *Turbulence modeling for CFD*. DCW Industries, Inc: La Canada, California.

[2] Tejeda-Martinez, A. E., & Jansen. K. E., (2004). A dynamic Smagorinsky model with dynamic determination of the filter width ratio, https://www.scorec.rpi.edu/REPORTS/2004-4.pdf

[3] Spalart, P. R., Jou, W. H., Strelets, M., & Allmaras, S. R. (1997). Comments on the feasibility of LES for wings, and on a hybrid RANS/LES approach. *First AFOSR International Conference on DNS/LES*, Ruston, LA, *Advanced in DNS/LES* Columbus, OH: Greyden Press.

[4] Squires, K. D. (2004). Detached-eddy simulation: Current status and perspectives. In *Direct and Large-Eddy Simulation* (5th ed.). Rainer Friedrich, Bernard Geurts, Oliver Metais, pp. 465–481.

[5] Spalart, P. R. (2009). Detached-eddy simulation. *Annual Review Fluid Mechanics* 41, 181–202. doi: http://dx.doi.org/10.1146/annurev.fluid.010908.165130

PART IV

APPLICATIONS IN ENGINEERING, BIOLOGY, AND MEDICINE

Examples here considered are a small part of huge number of application, which in fluid flow and heat transfer grows like an avalanche. Only examples of modern applications in two areas are discussed: conjugate heat transfer in engineering and peristaltic and turbulent flows in biology and medicine. Although the chosen examples present different methods of solution and subjects of applications, it is clear that (i) the most relevant solutions are outside of survey introduced here and (ii) the choice of examples is random and depends at least on preferences and background of a concrete person. Nevertheless, author hopes that the considered results give a reader primary understanding of situation in modern applications in fluid flow and heat transfer.

Unlike the previous text, examples of application are considered briefly without detailed comments presenting only statement of the problem in its original form, basic conclusions, and notes about the method of solution. This shortness allows reviewing more examples, at the same time giving a reader enough information to find the article of interest and get it using indicated references. Besides, examples are marketed by letters **a** (analytical) or **n** (numerical) showing at once what type of solution is employed.

CHAPTER 10

HEAT TRANSFER IN THERMAL AND COOLING SYSTEMS

10.1 Heat Exchangers and Pipes

10.1.1 Pipes and Channels

Example 10.1a Pipe with fully developed laminar flow heated symmetrically at the outer surface by uniform heat flux [1].

The governing equations and boundary conditions for the fluid (first line), for pipe (second line) and conjugate condition (third line) in cylindrical dimensionless coordinates, are as follows:

$$u\frac{\partial \theta}{\partial x} = 4\left(\frac{\partial^2 \theta}{\partial r^2} + \frac{1}{r}\frac{\partial \theta}{\partial r}\right), \qquad \theta(0,r) = \frac{\partial \theta}{\partial r}(x,0) = 0$$

$$4\left(\frac{\partial^2 \theta_s}{\partial r^2} + \frac{1}{r}\frac{\partial^2 \theta_s}{\partial r^2}\right) + \frac{1}{Pe^2}\frac{\partial^2 \theta_s}{\partial x^2} \qquad \frac{\partial \theta_s}{\partial x}(0,r) = \frac{\partial \theta_s}{\partial x}(L,r) = 0 \qquad (10.1)$$

$$\frac{\partial \theta_s}{\partial r}(x+2\Delta) = \frac{\lambda}{\lambda_w(1+2\Delta)}, \qquad \theta_s(x,1) = \theta(x,1), \qquad q_w = \frac{\lambda_w}{\lambda}\frac{\partial \theta}{\partial r}(x,1)$$

Variables are scaled by the pipe radius R, velocity U, Peclet number $Pe = 2RU\rho c/\lambda$, initial temperature T_i, outer heat flux q_0, and the temperatures for fluid and for pipe are used in the form $\theta = (T - T_i)\lambda/q_0 R$ and $\theta_s = (T_s - T_i)\lambda_w/q_0 R$. The problem is solved using the method of superposition for fluid equations, the finite-element method for the pipe equations. The conjugation procedure is performed by iterations (Sec. 8.4).

The following basic results are formulated:

- The temperature distribution on the interface shows that there is an isothermal region in the nearness of the inlet. In this case, the wall-to-fluid temperature difference is such that the heat is transferred from the fluid to the wall. This is similar to what is observed in other studies in the case of uniform $q_w = const.$ heating. An almost isothermal temperature distribution exists along the heated section for the higher conductivities and wall thicknesses and for the lower lengths of the heated section and Peclet numbers. The plots for the same heated section length have a common point of intersection in the second part of this section, which is shifted to the end of the pipe with the increasing pipe length.
- In the initial part of the heated section, the heat flux decreases from very high values to nominal magnitude $q_w = 1$. Then, it varies in two ways. The first is when it reaches the minimum and then goes up back to the nominal value but does not reach it. In this case, the temperature distribution inside the wall is close to one dimension, and the wall may be considered as a thermally thin object.
  The other type is when the heat flux decreases monotonically along the heated section so that the axial component of the temperature gradient decreases sharply from the outer to the inner wall surface. In that case, the wall should be seen as thermally thick, and disregarding wall heat conduction in this case leads to large errors.
- The dependence $Nu_x - Nu_{x*} = f(x)$ starts from zero at some axial position and close to the end decreases sharply reaching the curve $Nu_{x,q} = f(x)$ in such a way that Nu_x always remains lower than $Nu_{x,q}$ (index q means $q = const.$).
- The increasing Peclet number reduces the effect of axial conduction much more than the corresponding decreasing wall-to-fluid conductivity ratio or the wall thickness. The other way to considerably reduce the effect of axial wall conductivity is to increase the pipe length.

Example 10.2a *A parallel plate duct with outer wall subjected to convection with environment and turbulent flow at varying periodically inlet temperature [2].

Considering quasi-steady approximation (Sec. 4.4.4) and thermally thin radial lumping (spatially uniform) walls, the complex original system of equations and boundary conditions are reduced using the Laplace transform (Sec. 1.6. 2) to relatively simpler periodic problem for $\theta(\bar{r}, \bar{x})$ instead of initially unknown $\Theta(\bar{r}, \bar{x}, \bar{t})$:

$$\bar{u}(\bar{r})\frac{\partial \theta(\bar{r}, \bar{x})}{\partial \bar{x}} = \frac{\partial}{\partial \bar{r}}\left[\varepsilon(\bar{r})\frac{\partial \theta(\bar{r}, \bar{x})}{\partial \bar{r}}\right] - i\bar{\omega}\theta(\bar{r}, \bar{x})$$

$$\frac{\partial \theta}{\partial \bar{r}}(1, \bar{x}) + \left(\text{Bi} + \frac{i\bar{\omega}}{Lu}\right)\theta(1, \bar{x}) = b\frac{\partial^2 \theta(1.\bar{x})}{\partial \bar{x}^2}$$

$$\Theta(\bar{r}, \bar{x}, \tau) = \theta(\bar{r}, \bar{x})\exp(i\bar{\omega}\bar{t}), \qquad 0 < \bar{r} < 1, \ \bar{x} > 0 \qquad \bar{\omega} = \frac{\omega R^2}{\alpha} \tag{10.2}$$

$$\theta(\bar{r}, 0) = 1, \quad \frac{\partial \theta}{\partial \bar{r}}(0, \bar{x}) = 0, \quad \Theta = \frac{T(\bar{r}, \bar{x}, \bar{t}) - T_\infty}{\vartheta_0}, \bar{x} = \frac{x}{r_1 Pe}, \ \bar{r} = \frac{r}{R}, \ \bar{t} = \frac{\alpha t}{R^2}, \ \bar{\Delta} = \frac{r_2 - r_1}{r_1}$$

$$\varepsilon = 1 + \frac{v_{tb}}{\alpha}(\text{Pr} = \text{Pr}_{tb} = 1), \ Lu = \rho c_p / \rho_w c_{pw}, \ b = \frac{\bar{\Delta}}{\text{Pe}^2}\left(\frac{\lambda_w}{\lambda}\right), \ \text{Bi} = \frac{h_\infty R}{\lambda}, \ \text{Pe} = \frac{u_{max} R}{\alpha}$$

Here, ε is the eddy diffusivity, Lu is the Luikov number (Sec. 4.4.4), θ is the quasi-steady temperature, and $\bar{u} = u/u_{\max}$; r_1, r_2 are distances of inner and outer plates from the centerline, R is the hydraulic radius, ω, ϑ_0 are the inlet flow frequency and temperature amplitude at centerline, respectively.

A solution of this even simplified problem strongly depends on the evaluation of the complex eigenvalues and eigenfunctions of corresponding nonclassical Sturm–Liouville problem (Sec. 1.5.3) in the complex domain. For such a problem, no solution is available. To find approximate solution, the generalized integral transform is used. This approach consists of employing subsidiary problem close to the considered one. In this case, the following problem with boundary condition at $\bar{r} = 0$ and at $\bar{r} = 1$, which relates to classical steady Graetz problem (Example 10.5), is used:

$$\frac{d}{d\bar{r}}\left[\varepsilon\left(\bar{r}\right)\frac{d\psi\left(\mu_i,\bar{r}\right)}{d\bar{r}}\right] + \mu_i^2\bar{u}\left(\bar{r}\right)\psi\left(\mu_i,\bar{r}\right) = 0, \quad \frac{d\psi\left(\mu_i,\bar{r}\right)}{d\bar{r}} = 0,$$

$$\frac{d\psi\left(\mu_i,\bar{r}\right)}{d\bar{r}} + \mathrm{Bi}\,\psi\left(\mu_i,\bar{r}\right) = 0 \tag{10.3}$$

The eigenfunctions set of this system yields the pair of integral transform θ and corresponding inversion $\theta_{i,\,av}$:

$$\theta\left(\bar{r},\bar{x}\right) = \sum_{i=1}^{\infty} \frac{\psi\left(\mu_i,\bar{r}\right)}{N_i^{1/2}}\theta\left(\bar{r}\right), \quad \theta_{i,\,av}\left(\bar{x}\right) = \int_0^1 \bar{u}\left(\bar{r}\right)\frac{\psi\left(\mu_i,\bar{r}\right)}{N_i^{1/2}}\theta\left(\bar{r},\bar{x}\right)d\bar{r},$$

$$N_i = \int_0^1 \bar{u}\left(\bar{r}\right)\psi^2\left(\mu_i,\bar{r}\right)d\bar{r} \tag{10.4}$$

Applying these relations to transform equations (10.2) with some algebraic operations given in [2] leads to the relations for the wall and fluid bulk temperatures and wall heat flux in polar form in terms of amplitudes A_w and phase lags ϕ_w

$$\Theta_w\left(\bar{x}\right) = A_w\left(\bar{x}\right)\exp\left[-i\phi_w\left(\bar{x}\right)\right], \quad A_w\left(\bar{x}\right) = \left\{\left[\mathrm{Re}\,\Theta_w\left(\bar{x}\right)\right]^2 + \left[\mathrm{Im}\,\Theta_w\left(\bar{x}\right)\right]^2\right\}^{1/2},$$

$$\phi_w\left(\bar{x}\right) = \tan^{-1}\frac{\mathrm{Im}\,\Theta_w\left(\bar{x}\right)}{\mathrm{Re}\,\Theta_w\left(\bar{x}\right)} \tag{10.5}$$

Numerical results show the effects of the parameters: conjugation (Lu), wall axial conduction (b), and Biot number (Bi). For the smallest values of $\mathrm{Lu} \approx 5\times10^{-4}$, the oscillations in the fluid temperature are damped within a short distance from the duct inlet because of the larger thermal capacitances of the wall. For the larger values of $\mathrm{Lu} \approx 5\times10^{-3}$, the thermal wave penetrates further downstream due to the smaller thermal capacitance of the wall, and, consequently, requires a longer length for the same amount of heat to be stored at the wall. The amplitudes of the wall temperature flattened when parameter b increases. This occurs due to improved heat diffusion along the wall, especially very close to inlet where the thermal gradients are large. The amplitudes decay slower when the Reynolds number increases from 10^4 to 10^5. In the case of $b = 0$, the wall temperature amplitudes decay faster as the value of Lu decreases. The effect of Lu on the amplitude of the bulk fluid temperature is similar to that for the wall temperature, while the effect of axial wall conduction on bulk fluid temperature turns out to be little. Similar trends are observed with higher Biot number because an increasing Biot number decreases the wall amplitudes since the external thermal resistance becomes smaller. The smaller is the value of Lu, the larger the heat flux amplitudes. This occurs due to significant attenuation in the wall temperature. On the other hand, with the larger Lu, the axial wall conduction leads to increasing of the heat flux amplitudes. This effect is more pronounced at the inlet and is negligible for small values of Lu.

These results are obtained in the case of fixed dimensionless frequency $\bar{\omega} = 0.1$ and indicate, in particular, that for systems of gases flowing inside metal walls, the effect of conjugation cannot be neglected in regions close to inlet.

Example 10.3n Horizontal channel heated from below by $q_w = const.$ full developed laminar flow [3] (recall: this Dirihlet problem we considered in Example 2.8).

The system of equation for incompressible flow consists of two-dimensional Navier–Stokes and energy equations. The Navier–Stokes equations are written in the form with sources, which is used in the software. In this case, the pressure gradient in the first equation for u is considered as the source $q_x(x, y) = -dP/dx$, and the source $q_y(x, y) = dP/dy + (Gr/\mathrm{Re}^2)\theta$ is used in the second equation for velocity v, where the second term $[\theta = (T - T_\infty)\lambda/q_wH, H$ is channel height] takes into account the buoyancy effects (Sec. 2.8). For the walls, the Laplace equation is solved. The conjugate and the boundary conditions are different for the heated and for both insulated parts of the wall at the inlet and outlet of the channel. The total heat transfer coefficient, that takes into account the resistances of the walls from the plexiglass and from the fiberglass at the insulations, is used by formulation of the conjugate and boundary conditions.

It is assumed that the flow enters with parabolic velocity profile and with ambient temperature. This assumption is verified by measurements. The existing conditions at the ends of the channel are quite complex and depend basically on recirculation effects. According to some publication, it is assumed that $\partial u/\partial x = v = 0, \theta = 0$ if $u \leq 0$ (inflow) and $\partial\theta/\partial x = 0$ if $u \geq 0$ (outflow). A finite-volume technique with different grid size for body and flow is employed. Successive over relaxation (SOR) code is used for solving the pressure equation and tridiagonal matrix algorithm (TDMA) for the solution of nonlinear coupled system of momentum, energy, and continuity equations. The grid density is increased until two successive solutions for flow and thermal fields differ by less than one percent.

Comment 10.1 TDMA and SOR method are approaches for solving linear systems of algebraic equations. The first is based on a tridiagonal matrix that has only three nonzero elements in main diagonal and in two first diagonals above and below it and the second is a variant of the Gauss–Seidel method (see *Advanced Engineering Mathematics*).

The numerical results are in good agreement with author's experiments. The basic conclusions are as follows:

- In the case of low Reynolds number (Re = 9.48), the buoyancy causes two rolls with axis of rotation perpendicular to the flow direction. The upstream roll produces a recirculation zone, whereas the downstream roll entrains flow from outside. As the Reynolds number increases to Re = 29.7, these rolls become smaller and flow entrainment is reduced. It is observed that the intensity of these and transverse rolls as well as oscillatory and turbulence depend on the channel cross-section sizes and heating rates. The study shows that the effect of wall conductivity on the rolls' location is small.

- The longitudinal asymmetry of temperature profiles is caused by channel flow. A heated region temperature is higher than that of insulation, which results in generation of thermal plume above the surface. In the case of aluminum-heated region, the temperature distribution is highly uniform, which differs significantly from the case of the ceramic-heated region when the temperature uniformity is reduced due to poor thermal material diffusivity and to an increased heat transfer to the insulation.
- The comparison of the results obtained for conjugate and nonconjugante approaches demonstrates a significance of the conjugate modeling. This comparison is made for different heated region conductivities in the range corresponding to the materials such as plexiglass, ceramics, stainless steel, and aluminum. In the case of uniform heated surface, the nonconjugate model predicts a highly nonuniform temperature profile. In contrast, the conjugate model gives the highly uniform temperature profile. This occurs due to not only the redistributing the thermal energy by itself, but also to the increase of the thermal energy loss to the insulation. These results are confirmed by analysis of the effect of the thermal conductivity of the heated and insulation regions and of the wall thickness on the ratio of average heated region temperature predicted by both nonconjugate and conjugate models.
- The numerical and experimental results show that the conjugate effects are significant and usually should be taken into account, except two cases: (i) for the thin walls, low insulation, and/or high thermal conductivity of the heated regions, when the surface can be considered as an isothermal, and (ii) at the low heated region thermal conductivity when the heat transfer may be modeled as the $q_w = const.$ process.

Example 10.4n Heat transfer between the forced convective flow inside and the natural convective flow on the outside of the vertical pipe [4].

The flow enters the pipe with the temperature T_e and fully developed velocity profile. It is assumed that $T_e > T_\infty$ so there is the energy exchange between the internal flow and environment. For the internal flow only the energy equation is solved, whereas for the external flow, the full system of equations is considered (the first line):

$$\left(1-r^2\right)\frac{\partial\theta}{\partial x} = \frac{1}{r}\frac{\partial}{\partial r}\left(r\frac{\partial\theta}{\partial r}\right), \quad x=0, \theta=1, r=0, \quad u\frac{\partial u}{\partial x}+v\frac{\partial u}{\partial r} = \frac{(\text{Ra/Pe})\theta}{8\,\text{Pr}_0}+\frac{1}{r}\frac{\partial}{\partial r}\left(r\frac{\partial u}{\partial r}\right) \quad (10.6)$$

$$u\frac{\partial\theta}{\partial x}+v\frac{\partial\theta}{\partial r} = \frac{1}{r}\frac{\partial}{\partial r}\left(r\frac{\partial\theta}{\partial r}\right), \quad \text{Ra} = \frac{g\beta_o\left(T_e-T_\infty\right)D^3\,\text{Pr}_0}{v_0^2} \quad r\to\infty, x\to 0, \theta, u\to 0$$

The conjugate conditions are as usual $\theta_i = \theta_0, \left(\partial\theta/\partial r\right)_i = \left(\lambda_0/\lambda_i\right)\left(\partial\theta/\partial r\right)$, where $\theta = \left(T-T_\infty\right)./\left(T_e-T_\infty\right)$. Dimensionless variables, applied here, are scaled using R for r, $R\text{Pe}$ for x, $R\text{Pe}/v_0$ for u, R/v_0 for v, and subscripts $0, i$ denote outer and inner flows, respectively. Since the wall is considered as thin with no thermal resistance, the variable r is less than one ($r \leq 1$) and more than one ($r \geq 1$) for inside and outside of the pipe, respectively.

Numerical solutions of both the energy equation inside flow and the full system of equations for outside flow are performed using the Patankar–Spalding approach [5], which is similar to SIMPLE discussed above (Sec. 8.3.1). The iterative procedure is applied to conjugate both the inside and outside solutions. To improve the convergence of iterations, the local heat transfer coefficient distribution on the interface is used. The procedure starts from solving external problem using the boundary condition of uniform wall temperature. These results are used as a boundary condition for the inside pipe flow. Then, two equations (θ_{bl} is the balk temperature; j is the number of iteration)

$$\left(-\frac{\partial \theta}{\partial r}\right)_0^j = \frac{Nu_i^j \lambda_i \left(\theta_{\mathrm{bl}}^{j-1} - \theta_w^j\right)}{2\left(1 + K^j\right)\lambda_0}, \quad T_{\mathrm{bl}}^j = \frac{T_{\mathrm{bl}}^{j-1} + K^j T_w^j}{1 + K^j}, \quad K^j = 2\Delta x^j Nu_i^j, \tag{10.7}$$

which follows from the heat balance and conjugate conditions, along with equations (10.6) are employed to solve the outer problem. These data and relation for inner flow similar to the last one $\left(-\partial \theta / \partial r\right) = \left(Nu_0 / 2\right)\left(\lambda_0 / \lambda_i\right)\theta_{wi}$ are used to solve the inner problem in so on until the converge of the temperature and heat flux on the interface is achieved. Due to using heat transfer coefficient distribution instead of usually applying the temperature or heat flux distribution for the first interaction, the convergence is achieved during three to five iterations.

The following results are obtained:

- The Nusselt number for the forced convection of inside flow is insensitive to the thermal boundary conditions. Data of conjugate values of the Nusselt number for different $\lambda_0 / \lambda_i, \mathrm{Pr}_0$, and Ra/Pe are bounded between those for uniform wall temperature and uniform heat flux. Near the inlet, the Nusselt number values tend to be closer to uniform wall temperature, whereas for the larger distance from inlet, the Nusselt number values become closer to uniform heat flux.

- The Nusselt number for the external flow is compared with the results for isothermal vertical cylinder Nu_{cy}. Analyzing of the ratio cylinder Nu_0 / Nu_{cy} shows that the values obtained in the conjugate problem are always lower than that in the case of uniform temperature. Increasing λ_0 / λ_i and Ra/Pe leads to rapid decreasing of the temperature difference $\left(T_w - T_\infty\right)$ along the pipe and finally results in decreasing of the Nusselt number. At the fixed values of these parameters, the variations of Pr_0 from 0.7 (air) to 5 (water) do not practically affect the values of the Nusselt number.

- The wall temperature decreases with an increase in the downstream distance along the pipe. This effect intensifies at large values of λ_0 / λ_i and Ra/Pe. A special case is the large values of λ_0 / λ_i, which may be regarded as corresponding to internal air flow and external water flow. It is clear that such a situation shows that the external thermal resistance is very small in comparison with that of internal flow, and this results in the significant drop of the wall temperature upstream at the small distances from the entrance. A comparison of $\mathrm{Pr}_0 = 0.7$ and $\mathrm{Pr}_0 = 5$ indicates that this parameter does not play a significant role.

The bulk temperature ratio $\left(T_{bl} - T_\infty\right) / \left(T_e - T_\infty\right)$ determines the heat transfer effectiveness because it compares the heat transfer rate for the length of pipe between entrance and certain location to that for infinitely long pipe in the case of the same mass flow. The results obtained show that bulk temperature decreases with x as heat is transferred from the inside flow to external flow. Higher values of λ_0 / λ_i and Ra/Pe increase the heat transfer rate, which results in more rapid decrease in the bulk temperature. Analyzing data shows that $T_w < T_{bl}$ and also that the difference between these values is greater for larger values of λ_0 / λ_i. This occurs because the internal resistance becomes greater when the ratio λ_0 / λ_i increases, which leads to increase in the temperature difference. The comparison of the data for $\mathrm{Pr}_0 = 0.7$ and $\mathrm{Pr}_0 = 5$ reveals that the bulk temperature is affected slightly by the Prandtl number Pr_0.

10.1.2 Heat Exchangers and Finned Surfaces

Example 10.5a Heat transfer between two fluids separated by a thin wall flowing concurrently or countercurrently in the double pipe: Conjugate Graetz problem [6].

The flow velocities are fully developed. One fluid (1) is flowing in a tubular space of the double pipe, whereas the other one (2) flows in annual space exchanging the energy with the surrounding. The axial conductions in a fluid and in a wall are taken into account.

The energy equation for both fluids and the boundary conditions are

$$-\frac{1}{r}\frac{\partial}{\partial r}\left[\lambda r \frac{\partial T}{\partial r}\right] - \lambda \frac{\partial^2 T}{\partial r^2} + \rho c_p u_x (r) \frac{\partial T}{\partial x} = 0, \quad -\lambda_1 \frac{\partial T}{\partial r}\bigg|_{r_1} = -\lambda_2 \frac{r_1 + \Delta}{r_1}\frac{\partial T}{\partial r}\bigg|_{r_1 + \Delta}, \quad \frac{\partial T}{\partial r}\bigg|_{r=0} = 0$$

$$u_{x1} = U_1\left(1 - \frac{r^2}{r_1^2}\right), \quad u_{x2} = \pm U_2\left[1 - \left(\frac{r}{R}\right)^2 + \left(\bar{\Delta}^2 - 1\right)\frac{\ln(r/R)}{\ln \bar{\Delta}}\right] \tag{10.8}$$

$$-\lambda_2 \frac{\partial T}{\partial r}\bigg|_{r=R} = h_e\left[T(x, R) - T_0\right], \quad -\lambda_1 \frac{\partial T}{\partial r}\bigg|_{r_1} = h\left[T(x, r_1) - T(x, r_1 + \Delta)\right]$$

Here, U is the characteristic velocity, h and h_e are heat transfer coefficients for the resistance of separated and outer walls, respectively; $\bar{\Delta} = (r_1 + \Delta)/R$, r_1, R, r_2 are the radiuses of the inner and outer tubes and width of the annulus of the double pipe. The plus at u_{x2} is for the concurrent and the minus is for the countercurrent problems.

To decompose energy equation (to get separated equations for each fluid), two functions $\hat{S}$ are used resulting in solution of the same problem for both fluids

$$\hat{S}_1 = \int_0^r \left(-\lambda_1 \frac{\partial T}{\partial x} + \rho_1 c_{p1} u_{x1} T\right) 2\pi \zeta \, d\zeta, \quad \hat{S}_2 = \hat{S}_1 + \int_{r_1 + \Delta}^r \left(-\lambda_2 \frac{\partial T}{\partial z} + \rho c_p u_{x2} T\right) 2\pi \zeta \, d\zeta \tag{10.9}$$

$$\frac{\partial \hat{S}}{\partial x} = -2\pi r \left(-\lambda \frac{\partial T}{\partial r}\right), \quad \frac{\partial \hat{S}}{\partial r} = 2\pi r \left(-\lambda \frac{\partial T}{\partial x} + \rho c_p u_x T\right), \quad \hat{S}_1 \ (0 < r < r_1), \hat{S}_2 (r_1 + \Delta < r < R)$$

The last two equations (10.9) are modified applying dimensionless variables (see [6]). The numerical results are obtained for the flow in an annual space around a solid cylinder with a uniform heat source. Two versions are considered: with heat transfer through the outer walls and with insulated outer walls. Solutions are presented in the usual Graetz type form of asymptotic series of the eigenfunctions (Sec. 1.5.3). It is known that such series are efficient for high abscissas but converge slowly for small values of x. Authors conclude that in their examples the series converge unusually fast at Pe = 5 but did not point out the reason of this. They found that unlike the single-stream Graetz problem the effect of axial heat conduction in the fluid cannot be ignored, even for the Peclet numbers larger than 40–50, and only for the Peclet numbers significantly higher than 100, the axial conduction in the fluid becomes insignificant.

Example 10.6a/n Laminar flow in a double-pipe heat exchanger [7].

This problem is similar to the previous one; however, the solution and studied aspects of the problem are different. As well two flows in steady state with the fully developed velocity profiles inside a double-pipe are considered. The double pipe consists of an inner central tube and an outer annual channel. Concurrent and countercurrent cases are studied, but in contrast, the thermal conduction in the fluids is neglected. The governing system includes energy equations for the inner and outer streams (indices 1 and 2) and for the separating wall, boundary and conjugate conditions in the form similar to system in Example 10.1:

$$u_1 \frac{\partial \theta_1}{\partial x} = 4 \left(\frac{1}{r} \frac{\partial \theta_1}{\partial r} + \frac{\partial^2 \theta_1}{\partial r^2} \right) \quad \theta_1(0, r) = \theta_{01}, \quad \frac{\partial \theta_1}{\partial r}(x, 0) = 0$$

$$u_2 \frac{\partial \theta_2}{\partial x} = 4 \frac{\lambda_2}{(mc_p)_2} \left[R_i^2 - (1+\Delta)^2 \right] \left(\frac{1}{r} \frac{\partial \theta_2}{\partial r} + \frac{\partial^2 \theta_2}{\partial r^2} \right),$$

$$\theta_2(0, r) = \theta_{20}, \theta_2(L, r) = \theta_{20}, \frac{\partial \theta_2}{\partial r}(x, R_i) = 0 \tag{10.10}$$

$$\text{Pe}_1^2 \left(\frac{1}{r} \frac{\partial \theta_s}{\partial r} + \frac{\partial^2 \theta_s}{\partial r^2} \right) + \frac{\partial^2 \theta_s}{\partial x^2} = 0 \quad \frac{\partial \theta_s}{\partial x}(0, r) = \frac{\partial \theta_s}{\partial r}(L, r) = 0 \quad \theta_1(x, 1) = \theta_w(x, 1)$$

$$\theta_2(x, 1+\Delta) = \theta_w(x, 1+\Delta), \frac{\partial \theta_2}{\partial r}(x, 1+\Delta) = \frac{\lambda_w}{\lambda} \frac{\partial \theta_w}{\partial r}(x, 1+\Delta), \frac{\partial \theta_1}{\partial r}(x, 1) = \lambda_w \frac{\partial \theta_w}{\partial r}(x, 1)$$

The dimensionless variables are scaled as follows: linear sizes by internal radius of the inner duct R_i, velocities by mean axial velocity U, and temperatures by inlet temperature of the inner fluid $T_{0,1}$, thermal conductivity λ_2 and capacity $(mc_p)_2$ by corresponding values of the inner flow. The first and the second conditions in the second equation (10.10) pertain to concurrent and countercurrent cases, respectively.

The only concurrent case is investigated in details. The superposition method is used to solve the energy equations for fluids. In this case, the dependence between heat flux and temperature on the solid surface is given by Duhamel's integral (Sec. 1.4):

$$\theta_w - \theta_{b1} = \frac{2q_1(0)}{\text{Nu}_q} + \int_0^x \frac{2}{\text{Nu}_q(x-\xi)} \frac{dq_1}{d\xi} d\xi \tag{10.11}$$

This expression for both flows along with the energy equation for a body governed the solution of the problem. The energy equation for the wall is solved numerically applying the finite element method (Sec. 8.2). The iterative procedure starts with guessing of the distributions of the bulk temperature and the Nusselt number, which are used as a boundary condition for the wall energy equation. As a result, new distributions of the wall temperature and the Nusselt number are obtained. Then, the updated values of the Nusselt numbers and the bulk temperatures are calculated from conjugate conditions. Finally, the refined wall temperature distribution is obtained using equation (10.11). This and corresponding values of Nusselt number allow to start a new run. The convergence is achieved in less than 14 iterations.

The numerical results are obtained for following dimensionless parameters:

$L = 10$ and 100; $\Delta = 5$ and 2; $R_i = 3$ and 6; $\lambda_w = 1, 10, 100, 1000$, and $10,000$; $\lambda_2 = 0.1, 1$, and 5; $\text{Pe}_1 = 500, 1000$, and $10,000$; $(mc_p)_2 = 0.5, 1$, and 2.

The following results are formulated:

- The conjugate effect of the two streams is studied by comparing results obtained with and without the thermal wall conduction. In contrast to the latter case, two isothermal zones are created at the interface, and the wall temperatures do not coincide with inlet fluid temperatures. The length of these zones, as well as the wall-to-fluid temperature difference, increases due to account for axial wall conduction. With increasing wall conductivity, the wall temperature becomes more uniform, and the outlet temperature of the internal stream decreases, whereas the outlet temperature of the external stream increases correspondingly. For the relatively small wall conductivities up to $\lambda_w = 100$, the two streams are uncoupled and a central zone with uniform and equal heat fluxes for both sides exist. This is similar to the case when the wall conduction effect is neglected. For high wall conductivities, the situation completely changes, and the heat fluxes monotonically decreases from the inlet to the outlet with one crossing point. Near the inlet, the Nusselt number values coincide with that for the case of an isothermal condition. Downstream, both the Nusselt number and heat flux reach the asymptotic values corresponding to isothermal or to uniform heat flux conditions.
- Distribution of the local entropy production in the wall and streams calculated as suggested in [8] are sensitive to wall conductivity. For $\lambda_w = 10,000$, the entropy production monotonically decreases downstream from the maximum at the inlet. For reducing wall conductivities, the entropy production decreases progressively at the inlet, increasing in the outlet regions. Analyzing the distribution of entropy generation rate in the fluids indicates that the maxima in entropy production in streams correspond to high values of heat flux or wall-to-fluid temperature difference. Distribution of entropy production in the wall shows that the wall radial thermal resistance is dominant when the wall conductivity is low. For $\lambda_w = 1$, the entropy production distribution is similar to that for the wall with zero conductivity. For the high conductivities, the situation is completely revised, and the maximum is found at the middle of the wall instead of minimum existence here in the case of low conductivities.
- Accounting the wall conduction changes the character of dependence of the heat exchanger effectiveness on the wall conductivity. Instead of monotonically increased effectiveness with growing wall conductivity, for the intermediate λ_w, the maximum effectiveness exist at any given values of Pe_1 and capacity ratio $(mc_p)_2$. For low wall conductivities, the entropy production is concentrated in the wall. In the short device, the increasing λ_w leads to monotonically decreasing of the wall contribution. In a long exchanger, the minimum of the entropy production is observed. The reduction of effectiveness due to the wall conduction effect increases for increasing wall thickness, but it is slightly affected by variation of the pipe diameter ratio. The effectiveness also reduces with the increasing Peclet number. Increasing the fluids conductivity ratio has strong and positive effect on effectiveness.
- The proper choice of the wall material is needed for optimization. For example, in the case of countercurrent water streams separated by copper wall, the wall conductivity yields a small reduction in effectiveness. The optimum can be achieved by using a steel wall with $\lambda_w \approx 100$. For the glass wall, the order of magnitude drops to 1, and the effectiveness decreases due to the high radial resistance. For gaseous fluids, the wall conduction effect is more pronounced indicting, for instance, that corresponding value of λ_w becomes higher than 10,000 for two streams of air separated by copper wall.

The same problem for countercurrent flows is considered in [9]. The system of governing equations is solved numerically using Galerkin's method (Sec. 8.2). The results are compared with one-dimensional approach when the overall heat transfer coefficient is used and with two-dimensional solutions based on plug approximation (Sec. 2.7.3) for streams. It is shown that errors of such approximation approaches depend on the Biot number and are applicable only for small Biot numbers, $Bi < 0.1$.

Example 10.7n Microchannel heat sink as an element of heat exchanger [10].

The microchannel heat sink studied in this work is considered as an element of the microchannel heat exchanger. The heat transfer characteristics of such an exchanger should significantly increase due to the great reduction of the thickness of a thermal boundary layer and overall notable capacity based on the large surface/volume ratio. The studied model is a rectangular silicon channel with hydraulic diameter $D_h \approx 100\,\mu m$, which is a basic element of experimentally investigated model in [11].

Comment 10.2 The surface/volume ratio is an important characteristic for small objects because it determines the value of surface per unit volume. This ratio increases as the body size decreases since the surface is proportional to second power of size, while the volume changes as the third power of a size. Therefore, for sphere, this parameter is inversely proportional to radius. In many cases, the value of surface/volume ratio helps to understand the basic features of phenomenon. In particular, it is clear that heat exchanger efficiency should significantly grow with the size decreasing.

The following assumptions are used: (i) A range of the Knudsen number $Kn = l/D_h$ (Sec. 2.3.2) lies in continuum flow regime, and hence the Navier–Stokes equations are appropriate. (ii) The flow is incompressible, laminar, and steady state. The thermophysical properties are temperature dependent. (iii) The largest temperature gradients and thermal stresses are expected to occur at the inlet of the channel. Therefore, the development of the flow and temperature at the inlet should be carefully resolved. (iv) Thermal radiation is negligible since the typical operation temperature is below 100°C.

The three-dimensional Navier–Stokes and energy equations for fluid with temperature-dependent properties and energy equation for walls are employed:

$$\nabla(\rho \mathbf{V}) = 0, \quad \mathbf{V} \cdot \nabla(\rho \mathbf{V}) = -\nabla p + \nabla \cdot (\mu \nabla \mathbf{V}),$$
$$\mathbf{V} \cdot \nabla(\rho c_p T) = \nabla \cdot (\lambda \nabla T), \quad \nabla(\lambda_w \nabla T_S) = 0 \tag{10.12}$$

A uniform heat flux is imposed at one of the channel walls, whereas the others are isolated. The entering flow velocity and temperature are given, and gradients of velocity and temperature at the exit are taken to be zero. The no-slip condition at the walls and conjugate conditions at the fluid–solid interface are prescribed.

Patankar's technique of discretization and SIMPLER algorithm (Chapter 8) are used to solve the system of governing equations. The predicted and experimental data from [11] are in good agreement. The following results and conclusions are deduced:

• The local temperature distribution shows that the walls are isothermal, but the temperature field in fluid is essentially nonuniform. Initially, the high temperature gradient zone forms at the inlet of the channel and then increasing fluid core temperature is observed.

Three basic conclusions can be stated: (i) the maximum heat fluxes occur at the inlet of channel; (ii) the heat flux imposed at the wall is spread out by conduction within walls and finally is transferred to fluid; and (iii) due to the effect of conjugation, the thermal development occupies the entire channel. The temperature distributions at the inner and outer wall surfaces show very complex pattern, resulting from convective heat transfer and three-dimensional conductions.

•   The distribution of the local heat fluxes on the inner walls, which are the fluid–wall interface, confirms the observation deduced by the temperature distribution analysis in the first conclusion. In particular, the local heat fluxes are the greatest at the inlet of the channel where the temperature gradients are high. The local heat flux inside the channel is distributed high nonuniformly so that magnitude variation in the heat fluxes ranges several orders. The reason of this is the difference in spacing between channel walls. The channel cross-section is a stretched rectangular such that the distance between two walls in one direction is about three times smaller than that between two other walls. As a result, the boundary layer between small spacing walls is much thinner and consequently the convective heat transfer is much larger. The complicated heat flow structure is observed in the corners where the heat flux is a product of interaction between the boundary layers developed along both adjacent walls. The resulting configuration of heat flow in corners shows the negative heat fluxes directed from the fluid to walls. In such a case, the heat transfer coefficients are also negative (Sec. 5.4) when traditional methods cannot be used, and only conjugate formulation of the problem can give the realistic results (Chapter 5).

•   The average characteristics in general conform to local distribution quantities. The average wall temperature increases significantly in the entering portion due to high local temperatures in this area. In contrast to this, the fluid bulk temperature grows gradually along the channel approaching the wall temperature at the exit. Large temperature gradients in the inlet channel portion may result in significant thermal stresses, which is important to take into account during design. The average heat fluxes and average heat transfer coefficient gradually decrease along the channel. The average heat flux of all walls of the channel is smaller than initially imposed heat flux everywhere except inlet portion, where the average heat flux is greater than imposed one. This occurs because the area where the heat flux imposed is much smaller than that of the surfaces of the walls inside channel.

Heat transfer in micro-channels is investigated also in [12]–[14].

Example 10.8 * n Horizontal fin array [15].

The two adjacent internal long fins on a base are considered. The constant temperature of the base $T_{w,0}$ is higher than ambient air temperature T_∞. Heat is transferred from the fins to ambient air by convection and radiation. The problem is formulated as for a closure formed by two vertical fins and a horizontal base and is governed by the two-dimensional mass, momentum, and energy equations for the fluid and one-dimensional conduction equation for the fins.

The radiation heat transfer part is included in energy equation as a heat source, which is calculated as follows. The radiation exchange occurs between the left (subscript 1) and right (2) fins surfaces, the base (3), and the walls of the room through the open top (4), front side (5), and rear side (6) of closure. The open top and sides are considered as imaginary surfaces.

The black body irradiations of the fins and base are $E_{b1} = E_{b2} = \sigma T_w^4$ and $E_{b3} = \sigma T_{w,0}^4$, while $E_{b4} = E_{b5} = E_{b6} = J_4 = J_5 = J_6 = \sigma T_\infty^4$, where J is the radiosity. The radiation heat fluxes from fin and base are present as

$$q_{R1} = \frac{\varepsilon_1}{1-\varepsilon_1}[S_{13}(J_1 - J_3) + (S_{14} + 2S_{15})(J_1 - E_{b4})], \quad J_1 = \frac{a_{22}b_1 + S_{11}b_2}{a_{11}a_{22} - 2S_{13}S_{34}} \tag{10.13}$$

$$q_{R3} = \frac{\varepsilon_3}{1-\varepsilon_3}[2S_{31}(J_3 - J_1) + (S_{14} + 2S_{15})(J_3 - E_{b4})], \quad \frac{2S_{31}b_1 + a_{11}b_2}{a_{11}a_{22} - 2S_{13}S_{31}} \tag{10.14}$$

$$S_{ij} = \frac{1-\varepsilon_i}{\varepsilon_i}F_{ij}, a_{11} = 1 + S_{13} + S_{14}, a_{22} = 1 + 2S_{31} + S_{34}, b_1 = E_{b1} + S_{14}E_{b4}, b_2 = E_{b3} + S_{34}E_{b4}$$

Here, ε is the emissivity, S is the spacing between adjacent fins, $i = 1, 3$, and $j = 1-6$. The shape factors F_{ij} are calculated by formulae given in [16].

The governing system consists of energy equation, the momentum equation in vorticity-stream ($\omega - \psi$) form (Secs. 2.2.2.3 and 2.2.2.4), and two equations for fins

$$\frac{\partial T}{\partial t} + u\frac{\partial T}{\partial x} + v\frac{\partial T}{\partial y} = \frac{1}{Pr}\left(\frac{1}{Gr^{1/2}}\frac{\partial^2 T}{\partial x^2} + \frac{\partial^2 T}{\partial y^2}\right) + \frac{1}{Pr\,Gr^{1/4}}\left(2q_{R1} + \frac{q_{R3}}{A_R}\right) \tag{10.15}$$

$$\frac{\partial \omega}{\partial t} + u\frac{\partial \omega}{\partial x} + v\frac{\partial \omega}{\partial y} = \frac{\partial T}{\partial y} + \frac{1}{Gr^{1/2}}\frac{\partial^2 \omega}{\partial x^2} + \frac{\partial^2 \omega}{\partial y^2} \quad \omega = \frac{1}{Gr^{1/2}}\frac{\partial^2 \psi}{\partial x^2} + \frac{\partial^2 \psi}{\partial y^2} \tag{10.16}$$

$$\frac{Pr\,Gr^{1/2}}{\alpha}\frac{\partial T_w}{\partial t} = \frac{\partial^2 T_w}{\partial x^2} + \Lambda\frac{\partial T}{\partial y}\Big|_{y=0} \quad \frac{Pr\,Gr^{1/2}}{\alpha}\frac{\partial T_w}{\partial t} = \frac{\partial^2 T_w}{\partial x^2} - \Lambda\frac{\partial T}{\partial y}\Big|_{y=Gr^{1/4}} \tag{10.17}$$

The radiation heat fluxes q_{R1} and q_{R3} in energy equation are given by equations (10.13) and (10.14). The following scales are used for dimensionless variables in final equations (10.14)–(10.16): $S, S/Gr^{1/4}, S^2/v\,Gr^{1/2}, v\,Gr^{1/2}/S, v\,Gr^{1/4}/S, T_{w0} - T_\infty, v\,Gr^{1/4}, v\,Gr^{3/4}/S^2, \alpha$, $\lambda(T_{w,0} - T_\infty)Gr^{1/4}/S$ and σT_∞^4 for variables $x, y, t, u, v, T - T_\infty, \psi, \omega, \alpha_w, q_R, E_b$, respectively, and the Grashof number is defined as $Gr = g\beta(T_{w,0} - T_\infty)S^3/v$ (see [15]). The solution results are presented depending on the conduction–convection parameter, the radiation parameter, aspect ratio, and temperature ratio:

$$\Lambda = \frac{\lambda PS}{\lambda_w A_w}, \quad N_R = \frac{S\sigma T_\infty^4}{\lambda Gr^{1/4}(T_{w,0} - T_\infty)}, \quad A_R = \frac{H}{S}, \quad A_w = \frac{\Delta W}{2}, \quad P = \Delta + W, \tag{10.18}$$

where P, A_w, H, and W are half perimeter, half section area, and height and width of the fin. The boundary conditions include $T = T_{w,0}$, no-slip condition for the fins, and the base and $\partial T/\partial x = 0$ for the fin tips. For open top, the following boundary conditions are applied:

$$v = \partial u/\partial x = \partial \psi/\partial x = \omega = 0 \text{ and } \partial T/\partial x = 0.$$

An alternating direction method (ADI) is used to solve numerically dimensionless system (10.15)–(10.17). The temperature distributions are obtained in the fluid from the first and in the fins from the fourth and fifth equations, respectively. The vorticity is calculated using equation (10.16), then the stream function is computed from the third equation, and finally velocity components are obtained knowing the stream function. The procedure is continued for step-by-step time until the steady-state field is obtained.

Comment 10.3 ADI is a finite-difference approach for solving partial differential equations. The idea of the ADI method is to split the finite differential equation into two others with the x-derivative and y-derivative implicitly. These simple equation may be solved by the tridiagonal matrix algorithm (see the ADI method in Wikipedia).

The numerical results are obtained for varies fin arrays with different thermal characteristics and are compared with available experimental data.

- Calculation of average Nusselt numbers for a four-fin array for low and high emissivities agree with known experimental data. Analyses of these results show that the contributions of the fins, base, and end fins to total heat transfer are 36, 13.5, and 50.5%, respectively. The effect of fins spacing on heat fluxes is studied for arrays with different number of fins over a fixed base. As the number of fins increases from 4 to 16 and the value of spacing S decreases from 20 to 2.8 mm, the heat fluxes from the fin and from the base decrease from 149 to 44 W/m^2 and from 379 to 148 W/m^2. Despite increased numbers of fins, the heat transfer rate and effectiveness remain almost the same, but the average heat transfer coefficient decreases remarkably from 5.29 to 1.48 W/m^2 K. The effect of the base temperature is studied for the case investigated experimentally. Both results numerical and experimental are in reasonable agreement and indicate that the total heat transfer rate increases as the base temperature grows for any studied values of spacing and heights. The effectiveness increases as well for all heights, but it is found that for small values of S, effectiveness decreases as the base temperature grows. The results obtained for different fins thicknesses indicate that in the case of low heights and high thermal conductivities, the heat flux from fin practically does not depend on thickness. It is observed that the decreasing thermal conductivity leads to the reduction of the fin heat flux, and the increase in emissivity yields growing heat flux due to increasing radiation.
- The temperature profiles obtained for two different spacing show that the temperature far away from the fins is lower for higher spacing than that for the smaller one. At the same time, the velocity profiles indicate that there is a greater recirculation at larger S, resulting in higher velocities near the wall and lower velocities at the distance $S/2$. The isotherms and streamlines for the same two enclosures indicate that the air temperature is high in the middle of the enclosure with smaller spacing, whereas in the other enclosure with the larger spacing, the heating is confined to the air near the fins and the base. It can be seen from the streamline distribution that the streamlines travel upwards along the fins, where the temperature is high compared to that of the air in the enclosure.

Example 10.9a Flat finned surface in a transverse flow [17].

An incompressible fluid flows along a finned surface transversely to the fins. Since the flow is normal to the fins, the eddy forms in the each interfin space (Fig. 10.1). On side 1 of the fin, the temperature gradient increases, whereas on side 2, it decreases in the direction of the flow. On base 3, the temperature gradient may be assumed to be constant to a first approximation. It is assumed that at intersections, the surfaces are rounded, owing to which stagnant zones or secondary eddy flows do not form in the corners. It is also assumed that the boundary layer formed on the finned surface makes the main contribution to the thermal resistance to heat transfer between the surface and the potential eddy flow. This boundary layer develops from the end face to the base on the front (in the direction of the flow) fin surface 1 and increases from the base to the end on the back fin surface 2 (Fig. 10.1). These presentations of the dynamic boundary layer on the wall of the cavern under the conditions of eddy flow inside it are developed by Batchelor [18].

Assuming that the conditions in all cells on the surface are identical, the model of the problem of heat transfer in fins is formulated as bilateral flow over the body schematically shown in Figure 10.1. Here, the body surface represents the surface of the fin and of two adjacent cells. The upper body surface represents the right cell surface; the end of the model body corresponds to the ends of the two adjacent fins, and the lower body surface represents the surface of the left cell. The numbers in the model correspond to these in the scheme of the finned surface. Thus, numbers 1, 2, and 3 correspond to sections with increasing, decreasing, and constant temperature gradients, respectively, and the model represents the case of countercurrent flows with complicated velocity and temperature distributions.

For thermally thin fin, the governing equation and boundary condition are

$$\lambda_w \Delta \frac{d^2 T_w}{dx^2} = q_1(x) + q_2(x), \quad \frac{dT}{dx}\bigg|_{x=0} = 0, \quad T\big|_{x=H} = T_0, \tag{10.19}$$

where $q_1(x)$ and $q_2(x)$ are heat fluxes from surfaces of fin, H is the height of fin, and T_w and T_0 are the fin and the base temperatures, respectively. The heat fluxes are calculate by equation (4.14)

$$q_2(x) = h_*(x)(2H + H_B - x)$$

$$\times \left[T_E - T_D + \int_{2H+H_B}^{H+H_B} f\left(\frac{2H + H_B - \xi}{2H + H_B - x}\right) \frac{dT_w}{dx} d\xi \int_{H_B}^{x} f\left(\frac{2H + H_B - \xi}{2H + H_B - x}\right) \frac{dT_w}{d\xi} d\xi \right] q_1(x)$$

$$= h_*(x) \left[T_E - T_D + \int_0^x f\left(\frac{\xi}{x}\right) \frac{dT_w}{d\xi} d\xi \right] \tag{10.20}$$

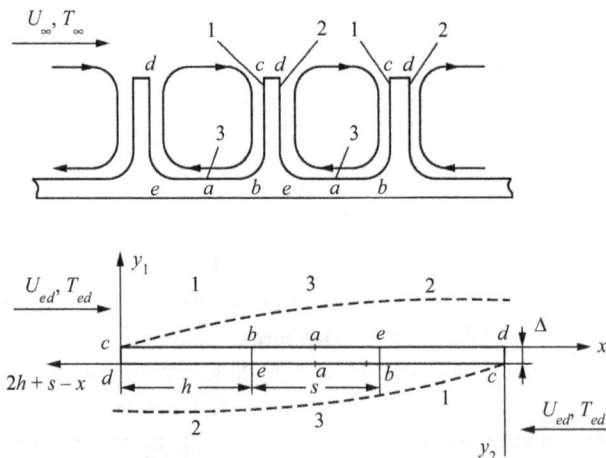

Figure 10.1. Scheme of a transverse flow over a finned surface.

Here, T_E and T_D are temperatures of end of the fin and inside the eddy flow (the last one plays the role of the temperature of the external flow for the boundary layer on the fin); H_B is the distance between the fins, and the influence function is given by equation (2.129).

The expression for heat flux $q_2(x)$ takes into account that the boundary layer on the back surface of the fin develops starting from the end point c (Fig. 10.1) of the front surface. Therefore, the heat transfer on the back surface depends on the temperature distribution on the front surface (section bc) and the length H_B of the interfin section. This fact is taken into account by the first integral in equation for $q_2(x)$, whereas the second integral determines the effect of the section (ex) on the back surface.

Substituting the expressions for the heat fluxes into equations (10.19) yields the intego-differential equation determining the temperature distribution over the height of the fin. In dimensionless variables, this equation and the boundary conditions are as follows:

$$\frac{d^2\theta_w}{d\eta^2} = N^2 \left\{ \varphi(\eta) \left[\theta_E + \int_0^\eta f\left(\frac{\zeta}{\eta}\right) \frac{d\theta_w}{d\zeta} d\zeta \right] + \varphi(\eta_0 - \eta) \left[\theta_E + \int_0^1 f\left(\frac{\zeta}{\eta_0 - \eta}\right) \frac{d\theta_w}{d\zeta} d\zeta \right. \right.$$

$$\left. \left. - \int_\eta^1 f\left(\frac{\zeta}{\eta_0 - \eta}\right) \frac{d\theta_w}{d\zeta} d\zeta \right] \right\} \quad \frac{d\theta}{d\eta}\bigg|_{\eta=0} = 0, \ \theta|_{\eta=1} = 1 \qquad (10.21)$$

$$\theta_w = \frac{T_w - T_D}{T_0 - T_D}, \quad N^2 = \frac{\bar{h}_* H^2}{\lambda_w \Delta} \quad \varphi(\eta) = \frac{h_*(\eta)}{\bar{h}_*} \quad \bar{h}_* = \frac{1}{H} \int_0^{H_B} \left[h_*(\zeta) + h_*(2H + H_B - \zeta) \right] d\zeta$$

Here, $\eta = x/H, \eta_0 = 2 + H_B/H, \bar{h}_*$ is the average heat transfer coefficient of an isothermal fin, and the parameter N^2 in fact is the Biot number (note that H^2/Δ is the linear characteristic) determining the conjugation effect (Sec. 5.2).

Equation (10.21) is solved using the method of reduction of the conjugate problem to conduction problem (Sec. 4.4.3). According to this approach, equation (10.21) is reduced to the system of ordinary differential equation and equation for error:

$$\frac{d^2\theta_w}{d\eta^2} = N^2 \left\{ \varphi(\eta) \left[\theta_w + g_1 \eta \frac{d\theta_w}{d\eta} + g_2 \eta^2 \frac{d^2\theta_w}{d\eta^2} \right] + \varphi(\eta_0 - \eta) \left[\theta_w - g_1(\eta_0 - \eta) \frac{d\theta_w}{d\eta} \right. \right.$$

$$\left. \left. + g_2(\eta_0 - \eta)^2 \frac{d^2\theta_w}{d\eta^2} \right] \right\} + \varepsilon(\eta) \quad \varepsilon(\eta) = \theta_w^{\text{int}} - \theta_w^{\text{diff}} \qquad (10.22)$$

The first approximation is obtained assuming $\varepsilon(\eta) = 0$ and solving the ordinary differential equation. Using the first approximation results for $\theta_w(\eta)$, the next approximation for $\theta_w(\eta)$ is obtained by integrating the right-hand side of the intego-differential equation (10.21) and calculating the error $\varepsilon(\eta)$. Then, solving equation (10.21) refined by including $\varepsilon(\eta)$ from relation (10.22), the next approximation is obtained (Sec. 4.4.3). The procedure is continued until a desired accuracy is achieved.

Numerical results are obtained for laminar flow: $C_1 = 3/4$, $C_2 = 1/3$, $g_1 = 0.63$, and $g_2 = -0.14$ (Chapter 4). The following conclusions are formulated:

- The solutions of the conjugate problem are compared with the results of approximate calculations using local and average heat transfer coefficients of the isothermal surface. The comparison shows that in the range $0 \leq N^2 \leq 2$, the results obtained using the conjugate model and both simplified methods are in good agreement. For $N^2 > 2$, the difference between the results for both fin characteristics becomes more significant. The values of the fin efficiency and the total heat flux removed by fin obtained in the simplified methods are too low, and the error increases as conjugate parameter N^2 increases reaching for large N^2 the values of 60–70%. The local heat transfer coefficients and heat fluxes obtained in the conjugate model are substantially nonuniform over the height of the fin reaching maximum values near the end face on the front side of the fin.
- On the back side of the fin, for $N^2 \geq 1.9$, the heat flux inversion is observed when the heat flux is directed toward the fin despite that the temperature head is still positive (Sec. 5.4). As the value of N^2 increases, the absolute magnitude of the inversed heat flux and the length of the heated section (ed) (Fig. 10.1) increase. Since the inversion effect cannot be obtained with the use of the simplified methods, neglecting the conjugation of the problem in this case yields not only quantitative errors, but also leads to qualitative incorrect results. The reason of this is that on the back side of the fin, the temperature head decreases in the flow direction (Chapter 5).
- It is found that for large values of conjugate parameter $N^2 > 2$, the heat flux removed by fin is maximum for fins with $1 < H_B/H < 1.5$. This result is a consequence of nonisothermicity of the finned surface and cannot be obtained using simplified methods.

10.2 Cooling Systems

10.2.1 Electronic Packages

Example 10.10n Horizontal channel with protruding heat sources [19].

The model consists of a horizontal channel with four volumetric sources mounted on the bottom wall. The heat transfer occurs by mixed convection and radiation. The air is considered as a cooling agent. The flow is assumed to be incompressible, laminar, and hydrodynamically and thermally developing. Two-dimensional Navier–Stokes equations (2.1)–(2.3) and energy equation (2.5) (without dissipation) are used for fluid. To take into account natural convection, the term $(Gr/Re)\theta$ for buoyancy force (Sec. 2.8) is adding in equation (2.3). Two-dimensional Laplace equations are used for walls and sources:

$$\frac{Pe}{Lu_w} \frac{\partial \theta}{\partial t} = \frac{\lambda_w}{\lambda} \left(\frac{\partial^2 \theta}{\partial x^2} + \frac{\partial^2 \theta}{\partial y^2} \right), \qquad \frac{Pe}{Lu_h} \frac{\partial \theta}{\partial t} = \frac{\lambda_h}{\lambda} \left(\frac{\partial^2 \theta}{\partial x^2} + \frac{\partial^2 \theta}{\partial y^2} \right) + \frac{R^2}{L\Delta} \qquad (10.23)$$

An additional energy equation is applied to determine the temperature of sources that are streamlined. The variables are dimensionless: $Re = UR/v$, $Pe = UR/\alpha$, and subscript h denotes the heat source. Scales are: R for x and y, U for u and v and ρU^2 for p. R is the channel height, Δ and L are height and width of sources, $\text{Lu}_w = \rho c/(\rho c)_w$, $\text{Lu}_h = \rho c/(\rho c)_h$ are Luikov numbers for channel wall and for heat sources; temperature and the Grashof number are defined in terms of the volumetric heat generation q_v as $\theta = (T - T_\infty)/\Delta T_{\text{ref}}$ and $\text{Gr} = g\beta q_v L\Delta R^3/v^2$, $\Delta T_{\text{ref}} = q_v L\Delta/\lambda$.

The uniform inlet conditions for fluid and no-slip boundary conditions at the walls are used. The outer wall surfaces are assumed to be adiabatic (Sec. 2.4.6.2). At the outlet of the channel, an extended domain is used to avoid the influence of large recirculation, which occurs at the last heat source. It is assumed that at the outlet and on the walls of extended domain, the following boundary conditions are appropriate: $\partial u/\partial y = v = \partial\theta/\partial y = 0$ and $\partial^2 u/\partial x^2 = \partial^2 v/\partial x^2 = \partial^2\theta/\partial x^2 = 0$.

Comment 10.4 The extended domain is used to get a uniform velocity profile at outlet and then to apply those conditions for Dirichlet problem formulation (Sec. 2.3.1).

Radiative heat transfer is calculated using the radiosity/irradiation approach (see, for example, [16]). The surfaces are considered as opaque, diffuse, and gray, and the inlet and outlet of the channel are treated as black surfaces at ambient temperature. The radiative heat flux q_{Ri} from the discrete surface and radiosity $J_i/\sigma T_\infty^4$ are determined as

$$q_{Ri} = \frac{\varepsilon_i}{1-\varepsilon_i}\left[\left(\frac{T_i}{T_\infty}\right)^4 - \frac{J_i}{\sigma T_\infty^4}\right] \qquad \frac{J_i}{\sigma T_\infty^4} = \varepsilon_i\left(\frac{T_i}{T_\infty}\right)^4 + (1-\varepsilon_i)\sum_{j=1}^{n}\frac{J_i}{\sigma T_\infty^4}F_{ij}, \qquad (10.24)$$

where ε and F_{ij} are emissivity and shape factor, respectively. Temperatures at the channel walls–fluid interface and at heat source–fluid interface are determined from energy balance equations

$$-\frac{\lambda_w}{\lambda}\left(\frac{\partial\theta}{\partial y}\right)_w = -\frac{\partial\theta}{\partial y} + q_R N_{RC} \qquad N_{RC} = \frac{\sigma T_\infty^4}{\lambda}\frac{R}{\Delta T_{\text{ref}}} \qquad \frac{\lambda_h}{\lambda}\left(\frac{\partial\theta}{\partial x}\right)_w \Delta y - \frac{\lambda_h}{\lambda}\left(\frac{\partial\theta}{\partial y}\right)_w$$

$$\Delta x + \left(\frac{\partial\theta}{\partial y}\right)_n \Delta x + \frac{\lambda_h}{\lambda}\left(\frac{\partial\theta}{\partial x}\right)_e \Delta x + \frac{R^2}{L\Delta}\Delta x\Delta y - q_R N_{RC}\Delta x = \text{Pe}\frac{\partial\theta}{\partial t}\Delta x\Delta y, \quad (10.25)$$

where Δx and Δy are dimensionless width and height (scaled by R) of an element taken into account in the balance, and subscript e refers to the extended domain.

The SIMPLE algorithm (Sec. 8.3.1) and the point-by-point Gauss–Siedel iteration method are used to solve the governing equations for velocity component, pressure, temperature, and radiosity. The following basic results and conclusions are obtained.

- It is observed that the increasing Reynolds number leads to shafting the centre of circulation to the right-side wall of the cavities. Beyond the last heat source, the intensity of circulation increases as Re grows. The temperature of the first chip is lower than others due to contact with incoming fresh air. The maximum temperature of the last chip is lower than that of the preceding one due to high recirculation at the last chip. The first and the last chips are exposed to the atmosphere. As a result, the radiation heat transfer from these chips is higher than that from the interior chips. The radiation effect and the maximum temperature become smaller as the Reynolds number grows.

- To study the buoyancy effect, different Grashof numbers are obtained by changing the values of the volumetric heat generation q_v. The results show that with the increasing Grashof number, the dimensionless temperature decreases linearly from which follows that the effect of buoyancy is negligible for the range of parameters studied. Despite that the dimensionless temperature decreases, the actual dimension temperature increases as the Grashof number increases as it is expected.
- The dimensionless temperature decreases as the emissivity of heat sources and of the walls increase at all Reynolds number analyzed. The comparison shows that the effect of the wall emissivity is more significant than that of the heat sources. The temperature decreases also when the emissivity of the substrate grows. As the emissivity of substrate changes from 0.02 to 0.85, the maximum temperature decreases in 11°C, while the same change of emissivity of heat sources gives only 4°C drops in maximum temperature. When the emissivity of the top wall increases, the convective heat transfer from the wall also increases due to radiation interaction increasing.
- The dimensionless temperature decreases as the thermal conductivities of the heat sources and substrate increase. At $\lambda_h/\lambda = 500$, the heat sources become isothermal. When the thermal conductivity ratio of heat sources and fluid λ_h/λ changes from 50 to 500, the calculation shows a 20% drop in the maximum temperature. Similar behavior of the dimensionless temperature is observed when the wall–fluid conductivity ratio is varied.
- As the emissivity of substrate and top wall increases, the contribution by radiation decreases, while the convective contribution increases. Increasing of the heat sources emissivity leads to increasing radiation contribution. In particular, at Re = 250, the radiation contribution increases from 10.5% at emissivity of heat sources $\varepsilon_h = 0.02$ to 19% at $\varepsilon_h = 0.85$. As Reynolds number increases, the radiation fraction of heat transfer becomes less.
- The correlation expressions for maximum temperature are obtained in the form $\left(A/A_{ref} \right)^m$ for $A = $ Re, Gr, λ_h/λ, λ_w/λ, ε_h and ε with $A_{ref} = 500, 8.65 \cdot 10^5, 100, 50, 0.55, 0.55$ and $m = 0.2, 1, 0.6, 0.6, 0.65, 0.56$. That means: the maximum temperature at $Re_{ref} = 500$ changes according to relation $\left(Re/500 \right)^{0.2}$ and similarly for other parameters.

Example 10.11n Elements and units of electronic systems [20].

It is considered an unsteady heat transfer in a small hermetically sealed unit of an electronic system with a heat source that has a core of constant power. The case of a transient regime when the housing walls are thermally insulated is studied. A computing domain consists of four walls formed an air-filled cavity with a heat-emitting element (for example, a transistor), which generates a constant heat flux. The flow is laminar; the buoyancy force is taken into account, like in the previous example, but the radiative effect is negligible. The mathematical model consists of two-dimensional unsteady Navier–Stokes equations (2.20) and (2.17) in $\omega - \psi$ variables with an additional term for natural convection, energy equation (2.5) for temperature, and conduction equation with source of generation for walls. Relations $\psi = \omega = 0, T = T_0$ are initial conditions; no-slip and thermal insulation expressions are conditions for walls, and the conjugate conditions are used in the form of equalities of temperatures and heat fluxes between gas and walls and between heat sources and gas. The problem is solved by the finite-difference method. The calculations are performed for square steel and fiberglass boxes with 0.1 m of length and 0.005 m thickness containing the heat source of 0.02 m length and 0.015 m height with volumetric heat generation in interval $10^5 - 10^7$ W/m^3. Other model characteristics are given in [20]. The following conclusions are formulated.

- For heat generation $q_v = 10^5$ W/m^3, in the cavity two convective cells are formed with smaller secondary flows in the corner regions. Immediately above the heat source, there is a domain with ascending flow, whereas the descending flows occupy the regions near the walls. The energy is transmitted from the source to the gas, giving rise to a thermal plume determined by the ascending flows. In this case, the conductivity heat transfer plays a dominant role, while the convective heat transfer contribution is small.
- As the power of volumetric generation increases, the flow velocities grow but the size of vortices decreases. The size of secondary vortices and the gas temperature increase, leading to more clear shape of thermal plume. These trends become more pronounced as the power of generation increases.
- Similar flow investigations for different wall material and various values of heat–source generation power show that (i) increase in the thermal diffusivity of the wall material leads to significant decreasing in the temperature of the air inside the cavity and of the heat–source temperature itself and (ii) the power of heat–source plays a critical role in formation of the thermo–hydrodynamic regimes. The data also demonstrated that the transient effect is important and should be taken into account.

The heat transfer in packages was also studied in articles [21] and [22].

10.2.2 Turbine Blades and Rocket

Film cooling is used to protect turbine blades and vanes of the first and second rows from direct contact with stream of hot gas. Injected cold air covers the surface of the blade or vane, producing a layer of cold air between protected object and hot gas. This type of the film-cooling system is investigated in [23]*. Here, we review a study of another type of cooling system for turbine blades.

Example 10.12n Turbine blades with radial cooling channels [24].
This type of cooling system is performed using radial channels with flowing cold air. Such channels are made though blade in the radial (vertical) direction perpendicular to blade cross-section. The cold air flowing though these channels cools the blade or vane.

The set of governing equations describing a model of such a cooling system consists of the two-dimensional Navier–Stokes equation for the fluid domain and the Laplace equation for computing heat conduction in solid. The finite-volume method (Sec. 8.2) is applied for the numerical solution of both equations. The turbulence is taken into account by one of turbulence models (Chapter 3). To calculate the heat fluxes from the blade to cooling air in the radial channels, the heat transfer coefficient is prescribed.

The iteration procedure is employed for conjugating the solutions of different domains. The Dirichlet problem is considered for fluid, while the Neumann problem formulation (Sec. 1.5.4) is used for heat conduction equation. It is noted that such an approach gives the stable solutions, but vice versa version, when the heat fluxes are considered as the boundary condition for the Navier–Stokes equation and the surface temperatures are employed to solve the Laplace equation, leads to unstable results.

Three blade configurations with different size and location of cooling channels at two exit Mach numbers $M = 0.59$ and $M = 0.95$ have been investigated. Parameter distributions around different blades are calculated showing that:

- Cooling duct configurations have a little effect on the pressure distribution around the blade, but the temperature distribution strongly depends on the size and location of cooling ducts.
- The blades with small channels and relatively uniform positions show better cooling effects and smaller mass flow rate. This effect is more pronounced on the pressure blade side.
- On both blade sides, the temperature decreases approaching the minimum close to stagnation point and then rapidly increases to its maximum.
- Mach number gradually increases on the suction side of the blade, while on the pressure side, Mach number increases almost until the exit and then goes down. For higher Mach number at the exit ($M = 0.95$), this behavior is more pronounced and leads to the supersonic values at the maximum.
- Accordingly, the gage pressure on the suction side starting from stagnation point is almost constant except small area close to the exit where it drops to zero. At the same time, on the other side in conformity with Mach number behavior, the pressure decreases, approaches the negative values of gage pressure at the minimum, and then grows to zero at the exit.

Example 10.13n Solid propellant rocket [25].

A charring material exposed to high temperature is a process of decomposition and loss of surface material by ablating to absorbing the heat. Such processes are used, for instance, for internal thermal protection of rocket combustion chambers or for thermal shield of reentry vehicles.

This paper studies the charring material process using the three-dimensional model composed of three zones: the virgin zone, the decomposition zone, and the char zone. These zones are disposed one over another so that on the top, the chair zone appears along with the working fluid flow. In the first zone, the material changes are negligible; in the second one, the material undergoes chemical and physical changes and energy is absorbed by decomposition; and in the third zone, composed mainly of char, the heat is transferred by conduction and convection. The changes in the material proceed by two ways: by the free material surface recession and by its decomposition, when the surface does not move, but the material properties are changed. During this process due to heating, the material releases pyrolysis gas (pyrolysis is a process of decomposition of material at elevated temperatures without oxygen), which passes through the solid into fluid which flows along the upper char zone.

To simplify the mathematical description of the problem, the two basic assumptions are used: (i) the pyrolysis gas velocity is approximately orthogonal to the receding surface and (ii) the surface regression is locally uniform and occurs along the normal direction to the surface. The governing equations in the fluid region are the conservation of mass, momentum, and energy equations:

$$\frac{D\rho}{Dt} + \rho \nabla \cdot \mathbf{v} = 0, \quad \rho \frac{D\mathbf{v}}{Dt} = \rho g - \nabla p + \mu \nabla^2 \mathbf{v} + \frac{\mu}{3} \nabla (\nabla \cdot \mathbf{v}),$$

$$\rho \frac{DT}{Dt} = \rho \frac{Dp}{Dt} + \nabla \cdot (\lambda \nabla T) + \mu S \tag{10.26}$$

The energy equation for decomposition charring material and its derivation on the base of the Arrhenius decomposition law are given in the [25]:

$$\rho c_p \left(\frac{\partial T}{\partial t} \right)_{\xi} = \nabla \cdot \left(\lambda \nabla T \right) + \rho c_p \mathbf{v} \cdot \nabla T + \left(J_g - \hat{J} \right) \left(\frac{\partial \rho}{\partial t} \right)_x + \rho_g \mathbf{v}_g \cdot \nabla J_g \qquad (10.27)$$

$$\hat{J} = \frac{\rho_v J_v - \rho_c J_c}{\rho_v - \rho_c}, \quad J_c(T) = J_c(T_r) + \int_{T_r}^{T} c_{pc}(t)\,dt, \quad J_v(T) = J_v(T_r) + \int_{T_r}^{T} c_{pv}(t)\,dt,$$

$$\lambda = x\lambda_v + (1-x)\lambda_c, \quad c_p = x c_{pv} + (1-x)c_{pc}, \quad x = \frac{\rho_v}{\rho_v - \rho_c}\left(1 - \frac{\rho_c}{\rho}\right)$$

In equations (10.26) and (10.27) $\mathbf{v}$ is the thermal protection recession velocity vector, S is the dissipation function [equation (2.5)], J is the solid enthalpy, x is a fraction of virgin, and the subscripts g, c, v denote pyrolysis gas, charred, and virgin values, while the subscripts x and ξ refer to derivatives at constant x in a fixed frame and at constant ξ in a moving frame.

Comment 10.5 Arrhenius equation determines the dependence of the chemical reaction constant k on the temperature and activating energy $k = A\exp(-E_a/RT)$, where A is so-called prefactor and R is the gas constant.

In equation (10.27), the term on the left-hand side is the sensible energy accumulation; the first term on the right-hand side is the conduction term; the second one is the energy convected by the motion of the reference frame; the third term is the difference between the energy convected away by pyrolysis gas and the chemical energy accumulation; and the fourth one is the energy convected by pyrolysis gases passing through the solid. The turbulence is taken into account by the $k - \omega$ model (Sec. 3.4.3.1).

The governing equations are solved numerically applying three-dimensional code Phoenics. This software uses a finite volume element approach and as well as other similar programs, like SIMPLE or SIMPLER, can be used for solving the energy equation of both the fluid and solid by accounting corresponding boundary and velocity conditions (Sec. 8.4). The other details of numerical performance are given in [25].

The method developed was used to predict results of several heat transfer problems for comparing to known analytical, experimental, or numerical solutions.

- Results of simulation of heat transfer in a blast tube with thermal protection are compared with data obtained in [26]. The initial protection temperature is 300 K. The laminar flow of combustion gas in the tube is at temperature 3,600 K. The temperature and density profiles in the fluid and in solid obtained in both studies are in agreement.
- The process in Material Test Motor (MTM) for testing new ablative materials with protection is simulated for the charring material ES59A with a low-density thermal protection ESA-ESTEC developed for space rockets [27]. The dimensions of the model cross-section have been extrapolated from the test section in MTM, and the suitable curvilinear mesh is used. The velocity and pressure as well as turbulent viscosity and conductivity show low numerical errors, which indicates that these are computed correctly. The results indicate that assumptions of negligible propellant reduction and constant

pressure in the chamber during burning are possible. The temperature profiles obtained in this case are similar to these in the blast tube with a typical sudden derivative variation arising due to passage through the different computation domains. The density also changes suddenly between charred and virgin zones at approximately one half of the solid thickness. The comparison of material affected depths (MAD) obtained experimentally and predicted by three different approaches shows a reasonable agreement indicating that after 25.3 s burning time, the MAD is about 4 mm. The mass flow rate and heat fluxes predicted by conjugate heat transfer approach are slightly underestimated, while the MAD predictions are basically overestimated.

• The heat exchange analysis of igniter of solid propellant rocket during the turbulent combustion is performed. Some approximations are employed to simplify the problem. The pressure at rocket exit is assumed to be constant; the external thermal protection of the igniter is considered as nondecomposing, and the igniter switching time is assumed to be negligible. Two thermal protections are considered: one from aramidic fiber and another by using Silica Phenolic with reduction erosion rate. The adiabatic combustion gas temperature 3,424 K is uniformly kept everywhere except the solid parts which are at 300 K. The velocities in the chamber are slow after a short transient period, and therefore, the buoyancy convection is taken into account, while the radiation is assumed to be negligible. The MAD and the thermal fields in gas and solid are calculated. The results indicate that the steel interface temperature is between 350 and 400 K under the initial temperature of 300 K. The temperature profiles in the internal thermal protection of the igniter are calculated as well as are the profiles of the solid density in the thermal protection.

10.2.3 Nuclear Reactor

Example 10.14a Emergency loss of coolant in the nuclear reactor [28].

The rewetting process existing between hot surfaces and adjacent flowing liquid film is one of the ways used for cooling nuclear fuel during emergency situation. The considered model in contrast to other knowing solution of this problem takes into account the transient character of the real rewetting process. The essential difference of these two approaches is that instead of usual assumption of infinite long surface, the semi-infinite plate is considered in the reviewing model. As a result, the length of the wet portion covered by film becomes time dependent as well as the surface temperature at the moving film front becomes to be an unknown function of time.

The following assumptions are used: (i) the lower plate surface is adiabatic, (ii) the heat transfer coefficient between the plate and the film is constant and known, (iii) the film is supplied at constant velocity U, and (iv) heat losses to surrounding are negligible. The one-dimensional conduction equation in the moving frame with the origin at the film front and initial and boundary conditions for wet and dry plate portions are as follows:

$$\lambda_w \frac{\partial^2 T}{\partial x^2} - \frac{h}{\Delta}\left(T - T_f\right) = \rho_w c_w \frac{\partial T}{\partial t} \tag{10.28}$$

$$wet \ T\left(x, t = 0\right) = T_i, \ T\left(x = 0, t > 0\right) = T_w\left(t\right), \ \frac{\partial T}{\partial x}\left(x = -U_w t, t > 0\right) = 0$$

dry $T(x, t = 0) = T_i,\ T(x = 0, t > 0) = T_w(t),\ T(x \to \infty, t > 0) = T_i.$

Here, T, T_f, T_i are plate, film, and initial plate temperatures, respectively, $U_w t$ is the length of wet part of the plate at time t, where U_w is the front velocity which due to evaporation is less than constant U. To convert the time-dependent length to a constant value and reduce the number of parameters, the dimensionless variables for the wet portion are introduced:

$$\eta = \frac{x}{Ut}, \quad z = \frac{h}{c\rho\Delta}, \quad \theta = \frac{T - T_f}{T_i - T_f}, \quad \theta_w = \frac{T_w - T_f}{T_i - T_f} \tag{10.29}$$

Applying these variables to the governing equation and boundary conditions yields the system depending on only one parameter $\mathrm{Bi/Pe}^2$. This ratio determines the rate of the transient cooling process. The greater this ratio, the shorter is the dimensionless time required to cool the plate to a given dimensionless temperature. Since cooling by a moving thin film proceeds in the boiling transitional state, it was suggested to name the ratio $\mathrm{Bi/Pe}^2$ as the Leidehfrost number, Ls, similar to the Leidenfrost point on the boiling curve [28].

The governing equation for wet part in dimensionless variables becomes

$$\mathrm{Ls}\frac{\partial^2\theta}{\partial\eta^2} + z\eta\frac{\partial\theta}{\partial\eta} - z^2\frac{\partial\theta}{\partial z} - z^2\theta = 0, \quad \mathrm{Ls} = \frac{\mathrm{Bi}}{\mathrm{Pe}^2} = \frac{\lambda_w h}{\rho_w^2 c_w^2 U_w^2 \Delta} \tag{10.30}$$

$$\theta(\eta, z = 0) = 1, \quad \theta(\eta = 0, z > 0) = \theta_w(z), \quad \frac{\partial\theta}{\partial\eta}(\eta = -1, z > 0) = 0$$

For the dry region, it is assumed that $h = 0$, and the governing equation is used in another form:

$$\frac{\partial^2\vartheta}{\partial\xi^2} - \frac{\partial\vartheta}{\partial z} = 0, \quad \xi = x\sqrt{\frac{h}{\lambda_w\Delta}}, \quad \vartheta = \frac{T - T_i}{T_f - T_i}, \quad \vartheta_w = \frac{T_w - T_i}{T_f - T_i} \tag{10.31}$$

$$\vartheta(\xi, z = 0) = 0, \quad \vartheta(\xi = 0, z > 0) = \vartheta_w(z)\ \vartheta(\xi \to \infty, z > 0) = 0$$

To determine the unknown temperature at the moving film front, the solution of equations (10.30) and (10.31) should be presented for the arbitrary temperature distribution. Using universal functions of type (1.23) in the form of series for this purpose and substituting results for temperatures θ and ϑ (with coefficients G and H, respectively) in (10.30) and (10.31), one gets two infinite systems of differential equations with constant initial and boundary conditions:

$$\theta = \sum_{n=1}^{\infty} G_n(\eta, z)\frac{\partial^n\theta_w}{\partial z^n} \qquad \vartheta = \sum_{n=1}^{\infty} H_n(\eta, z)\frac{\partial^n\vartheta_w}{\partial z^n}$$

$$\mathrm{Ls}\frac{\partial^2 G}{\partial\eta^2} + \eta z\frac{\partial G_n}{\partial\eta} - z^2\frac{\partial G_n}{\partial z} - z^2 G_n - z^2 G_{n-1} = 0, \quad \frac{\partial^2 H_n}{\partial\xi^2} - \frac{\partial H_n}{\partial z} - H_{n-1} = 0 \tag{10.32}$$

$$G_{-1} = 0, \quad z = 0\ G_0 = 1, G_n = 0; \quad \eta = 0\ G_0 = 1, G_n = 0; \quad \eta = -1\ \frac{\partial G_n}{\partial\eta} = 0$$

$$H_{-1} = 0; \quad z = 0\ H_n = 0; \quad \xi = 0\ H_0 = 1, H_n = 0; \quad \xi \to \infty\ H_n = 0$$

Solutions for function H_n are given by the error functions (Sec. 1.3). The first two are

$$H_0(\xi, z) = 1 - erf\frac{\xi}{2\sqrt{z}}, \quad H_1(\xi, z) = \frac{\xi}{2}erfc\frac{\xi}{2\sqrt{z}} - \xi\sqrt{\frac{z}{\pi}}\exp\left(-\frac{\xi^2}{4z}\right) \tag{10.33}$$

Equations (10.32) for G_n with variable coefficients are solved by approximate method of moments (Sec. 8.2) obtaining the following results:

$$G_0(\eta, z) = 1 + \left(\frac{\eta}{2} + \eta\right)A_0(z) + \left(\frac{\eta^3}{3} + \eta\right)B_0(z)$$

$$G_n(\eta, z) = \left(\frac{\eta}{2} + \eta\right)A_n(z) + \left(\frac{\eta^3}{3} + \eta\right)B_n(z) \tag{10.34}$$

Function A_n and B_n are defined by ordinary differential equations. For $n = 0$, these are

$$\frac{dA_0}{dz} = A_0(z)\left(-\frac{28Ls}{z^2} + \frac{14}{3z} - 1\right) + B_0(z)\frac{16}{3}\left(\frac{Ls}{z^2} - \frac{1}{z}\right) + 28 \quad A_0(0) = 0 \tag{10.35}$$

$$\frac{dB_0}{dz} = A_0(z)10\left(-\frac{2Ls}{z^2} + \frac{1}{3z}\right) + B_0(z)\left(\frac{20Ls}{z^2} + \frac{11}{3z} + 1\right) + 20 \quad B_0(0) = 0$$

The temperature fields in the wet and dry plate portions should be conjugated at the moving film front. The conjugate conditions for that are derived using the energy balance at the moving front. The heat $q^+(t)$ conducted from the dry region is slightly absorbed by evaporation and sputtering $q_w(t)$ at the moving film front, while the majority of the heat $q^-(t)$ is transferred to the wet region. Thus, the balance equation is

$$x = \eta = 0 \quad q^+(t) = q^-(t) + q_w(t,), \quad q_w(t) = h_w(T_w - T_0), \tag{10.36}$$

where h_w is the heat transfer coefficient of evaporation and sputtering. Heat fluxes are found using equations (10.32) for temperatures. Substituting heat fluxes (the first line) into (10.36) gives an ordinary differential equation (the second line), determining the plate temperature at the moving front:

$$q^-(z) = -\frac{\lambda_w(T_i - T_f)}{U_w t}\sum_{n=1}^{\infty}\left(\frac{\partial G_n}{\partial \eta}\right)_{\eta=0}\frac{\partial^n \theta_w}{\partial z^n},$$

$$q^+(z) = \lambda_w(T_f - T_i)\sqrt{\frac{h}{\lambda\Delta}}\sum_{n=1}^{\infty}\left(\frac{\partial G_n}{\partial \xi}\right)_{\xi=0}\frac{\partial^n \vartheta_w}{\partial z^n} \tag{10.37}$$

$$\left[z + (g_1 - g_0)\sqrt{z\pi Ls}\right]\frac{d\theta_w}{dz} + \theta_w\left[1 + g_0\sqrt{\frac{\pi Ls}{z}} + \frac{h_w}{h}\sqrt{z\pi Bi}\right] - \left[1 + \frac{h_w}{h}\sqrt{z\pi Bi}\theta_0\right] = 0, \theta_w(0) = 1,$$

where $g_0(z) = A_0(z) - B_0(z)$ and $g_1(z) = [A_1(z) - B_1(z)]/z$. If the first term in this equation is relatively small, the first approximation solution for θ_w can be obtained:

$$\theta_w = \frac{1 + (h_w/h)\sqrt{z\pi \mathrm{Bi}}\theta_0}{1 + g_0\sqrt{\pi\,\mathrm{Ls}/z} + (h_w/h)\sqrt{z\pi \mathrm{Bi}}}, \qquad \theta_0 = \frac{1}{1 + g_0\sqrt{\pi\,\mathrm{Ls}/z_0}} \qquad (10.38)$$

Here, z_0 is the onset time, which defines the dimensionless time required to cool the plate at the film front to the rewetting temperature. Comparing two equations (10.38), one sees that these relations coincide if $h_w = 0$. It follows from this that the onset time depends on the rewetting temperature and on the Leidenfrost number, but does not depend, at least in the first approximation, on heat absorbed by evaporation and sputtering.

The numerical results indicate that:

- The dimensionless temperature sharply decreases at the beginning of the cooling process. As time passes, the rate of cooling decreases and finally becomes zero at the minimum temperature. For the case of negligible heat absorbed at the front, the character of function $\theta_w(z)$ depends only on the Leidenfrost number. As the Leidenfrost number increases, the plate cools faster, the achieved minimum temperature is smaller, and the time z_{min} required reaching it decreases.
- The model used here describes only the transient part of the wetting process when the plate temperature at the moving front is higher than the wetting temperature. Although the model does not describe the second part of the process with practically constant rewetting temperature, the model consist of another dry plate portion with initial high temperature $T \leq T_i$. Therefore, in this case, the plate temperature at the moving front starts to increase after it reaches the minimum, and in conformity with boundary condition (10.27) ultimately as $t \to \infty$ becomes equal to the initial plate temperature.
- Although the onset time does not depend on heat absorbed by evaporation and sputtering, it follows from equations (10.38) that this amount of heat significantly affects the dependence $\theta_w(z, \mathrm{Ls}, \gamma)$, where $\gamma = (h_w/h)\sqrt{\mathrm{Bi}}$. The calculations show that despite all the curves $\theta_w(z)$ for different values of γ have the same initial and end points, the dependence $\theta_w(z)$ varies remarkable with γ. For small values of γ, the curves $\theta_w(z)$ are close to linear, but as the absorbed heat grows, the dependence $\theta_w(z)$ changes so that the temperature sharply decreases at the beginning of the process and then continues decreasing with time with gradually reducing cooling rate.

The same problem in traditional formulation considering the infinite plate or rod is analyzed in many publications, for example, consider articles [29] and [30].

10.3 Energy Systems

Example 10.15n Diesel engine piston temperature field [31].

Prediction of the heat transfer characteristics of an engine piston is a complicated task due to high values of temperature, heat fluxes, and velocity gradients of transient turbulent flow near the cylinder wall. In most studies, this problem is solved using the wall functions in the well-known form $u^+ = f(y^+)$ (Sec. 3.3.2). Such an approach is reasonable since the gap between cylinder wall and piston is small. However, wall functions for incompressible flow might be not appropriate for this case with variable gas density.

In reviewed article on the base of usual near the wall assumptions, the wall function (WF) are formulated taking into account gas compressibility by integrating appropriate energy equation. The detailed derivation of these functions is given in [31]. Here are presented only some final equations and the results of comparing the data obtained in conjugate heat transfer problem using the suggested and known wall functions. These results give the heat transfer parameters for piston and, at the same time, show the necessity of new type of the wall functions.

The final expressions for heat flux and heat transfer coefficient in the new variables are presented as follows:

$$q_w = \frac{c_{p,p}\rho_p u_{\tau,p}\left(T_p - T_w\right)}{T_p^*}, \quad h = \frac{c_{p,p}\rho_p u_{\tau,p}}{T_p^*}, \quad d\hat{T} = \left(\frac{1}{\mathrm{Pr}} + \frac{\mathrm{Pr}\hat{v}^2}{0.7 + 0.85\,\mathrm{Pr}\hat{v}}\right)^{-1} d\hat{y}, \quad (10.39)$$

where $\hat{y} = y\rho_p u_{\tau,p}/\mu$, $\hat{v} = \gamma^4/(\gamma^3 + 382.5)$, $\gamma = \kappa\hat{y}\sqrt{T_p/T}$ and the parameters with subscript p are taken at the centre of the cell located near the wall. The value $\hat{T}$ is obtained by integration of the last equation (10.39) using formulae for $\hat{y}, \hat{v}$ and iterations. The conjugate heat transfer problem for piston was carried employing a simplified model. The procedure used the same program for both fluid and piston domains, taking into account appropriate physical characteristics for each domain, as is described in Section 8.4. The details of simplified model and computation grid are given in [31].

Several dependences for piston heat transfer characteristics are simulated and plotted applying standard (SWF), modified (MWF), and hybrid (HWF) (partly with new) wall functions.

- The maximum surface temperature predicted by MWF is higher than that given by two others. This result would be invaluable in designing pistons with extreme loads. The same results are obtained for heat transfer from the charge to the piston.
- The cylinder pressure predicted by MWF is also marginally higher than that obtained with SWF. Although the difference is small, this is surprising because according to the previous result, the more heat is flowing through the piston in the case of MWF.
- The difference of cumulative energy (defined as a reminder between heat release and heat loss through boundaries) predicted by modified and standard functions is positive maximum between angles 369° and 375°. In this angle interval, the maximum cylinder pressure predicted by MWF is also higher.
- The computation time for modified wall function is about 20% greater.
- The suggested variable density wall functions gave reasonable predictions like existed incompressible functions, but comparison with experimental data is required to make a conclusion of usefulness of these functions.

Example 10.16n Solar energy storage unit [32].

The study of flow and heat transfer in energy storage units is stimulated by growing use of the solar systems due to its periodic nature. The model under consideration consists of parallel rectangular plates forming the channels with flowing heat transfer fluid (HTF), which exchanges heat with the phase-change material (PCM) located on the other side of the plates.

The following assumptions are used: (i) the density and specific heat of liquid and solid phases of the PCM are the same; (ii) heat conduction in x and z directions in the PCM are negligible in comparison with that in the y direction; (iii) heat conduction in the channels are negligible compared to convective heat transfer; (iv) the natural convection is negligible; (v) the flow is fully developed, and dissipation is negligible.

The system of governing equations and initial, boundary, and conjugate conditions for the HTF, container walls, and the PCM are, respectively

$$\frac{\partial \theta}{\partial t} + u\frac{\partial \theta}{\partial x} = \frac{\partial}{\partial y}\left(\alpha \frac{\partial \theta}{\partial y}\right),\ \frac{\partial \theta_w}{\partial t} = \frac{\partial}{\partial y}\left(\alpha_w \frac{\partial \theta_w}{\partial y}\right),\ \frac{\partial \theta_m}{\partial t} = \frac{\partial}{\partial y}\left(\alpha_m \frac{\partial \theta_m}{\partial y}\right) - \frac{1}{Ste}\frac{\partial f}{\partial t}$$

$$t = 0,\ \theta = \theta_w = \theta_m = 0,\ \ x = 0\ \theta = 1,\ y = 0,\ \frac{\partial \theta}{\partial y} = 0,\ y = \frac{1}{2},\ \lambda \frac{\partial \theta}{\partial y} = \lambda_w \frac{\partial \theta_w}{\partial y}$$

$$y = \frac{1}{2},\ \lambda_w \frac{\partial \theta_w}{\partial y} = \lambda \frac{\partial \theta}{\partial y},\ y = \frac{1}{2} + \frac{\Delta_w}{\Delta},\ \lambda_w \frac{\partial \theta_w}{\partial y} = \lambda_m \frac{\partial \theta_m}{\partial y}\ \ \theta = \frac{T - T_{melt}}{T_{in} - T_{melt}} \tag{10.40}$$

$$y = \frac{1}{2} + \Delta_w,\ \lambda_m \frac{\partial \theta_m}{\partial y} = \lambda_w \frac{\partial \theta_w}{\partial y},\ y = \frac{1}{2} + \Delta_w + \frac{\Delta_m}{2},\ \frac{\partial \theta_w}{\partial y} = 0$$

$$u = \frac{1}{2}\alpha \Pr \operatorname{Re}\left[1 - (2y)^n\right]\left[1 - (2az)^m\right]\left(1 + \frac{1}{n}\right)\left(1 + \frac{1}{m}\right)(1 + a)$$

$$\alpha_m = f + (1 - f)\alpha_s,\ n = 2\left[a \le (1/3)\right] \text{ or } 2 + 0.3a - 0.1,\ m = 1.7 + 0.5a^{-1.4}$$

Here, variables are dimensionless and are scaled by channel thickness Δ for coordinates and other dimensions, Δ^2/α_L for t, α_L and λ_L for other thermal diffusivities and thermal conductivities, respectively. $\operatorname{Re} = \bar{u}D_h/\nu$, $Ste = c_m (T_{in} - T_{melt})/\Lambda$ are the Reynolds and Stephan numbers, Λ is the latent heat, $a = \Delta/W$ is the channel aspect ratio, W is the width of channel, Δ, Δ_w, Δ_m are the thicknesses of channel, channel walls, and PCM, respectively, f denotes liquid fraction, n and m are exponents in velocity profiles. Subscripts w, m, L, in, s and $melt$ refer to plate, PCM, liquid, initial, solid and melting qualities, respectively.

The numerical solution of this set of equations is obtained using an enthalpy finite difference technique detailed outlined in [33]. The value of the liquid fraction changes from 1 in the fully liquid region to 0 in the fully solid region. At each grid point, the value of this fraction is determined using the numerical solutions for PCM and is updated following the condition $f^{k+1} = 0$ if $f^k < 1$ or $f^{k+1} = 1$ if $f^k > 1$. The set of algebraic equations is solved iteratively by TDMA (Comment 10.1). The details of the grid, step sizes, conservation criterion, and results of validation are given in [32].

Comment 10.6 The enthalpy method is an approach applying for the problem with phase changes, when it is necessary taking into account the latent heat, like, for example, process of solidification (Example 11.1). The adventure of enthalpy formulation is that this approach allows satisfying the energy balance at the phase front, explicitly tracking the required interface position.

The numerical results are obtained for n-octadecane as a PCM ($\rho = 776.5\,\text{kg/m}^3$, $c = 2165$ J/kg K, $\lambda_L = 0.148$, $\lambda_S = 0.358$ W/m K, $\Lambda = 243.5 \times 10^3$ J/kg, $T_{melt} = 300$ K), for water as HTF and for walls with dimensions $L = W = 0.5\,\mathrm{m}$, $\Delta = 0.01\,\mathrm{m}$ and properties $\rho = 8933\,\mathrm{kg/m^3}$, $c = 385$ J/kg K, $\lambda_w = 401$W/m K. The following conclusions are stated:

- At the inlet, the temperature is uniform across channel thickness for all time due to the high thermal conductivity of the walls. At small times $(t = 0.1)$, the temperature of PCM decreases along y and reaches zero. Then, it increases with time reaching almost the inlet HTF temperature. With time, the thermal boundary layer develops, which leads to increasing HTF temperature.
- Nusselt number is high at small x decreasing along the channel and increasing with time. At small distances from inlet, the local Nusselt numbers are close to the values corresponding to the constant heat flux boundary condition. For larger distances, the Nusselt number becomes close to the isothermal value at small time $(t = 0.1)$ and in time $(t = 0.8)$ reaches values corresponding to constant heat flux on the surface. However, for the full length of the channel for all time, the local Nusselt numbers differ significantly from those obtained at constant temperature or heat flux on the surface, so that using such values for prediction leads to remarkable errors.
- The outlet HTF temperature increases with time during the heat charge period approaching the values close to the inlet temperature. During this period, regardless of the Reynolds number value, the outlet temperature first rapidly changes due to noticeable heat storage. Then, the outlet temperature becomes to change slowly due to latent heat storage. Finally, as the melting is completed, the data show the rapid change of the outlet temperature, the reason of which is the sensible heat storage. On the whole, increasing Reynolds number yields the heat transfer rate growing, which results in rapid melting of PCM. This is confirmed by variation of liquid fraction in time. As a result, the system rapidly reaches the steady state, and the temperature of HTF reaches its inlet value.
- The melting front penetrates more quickly at smaller x for any Reynolds number and time since close to inlet, the Nusselt number and temperature of HTF are higher than at larger distances from the starting channel section.

The same problem of behavior of energy storage unit is considered in [34]. In this study, the hybrid scheme combining analytical and numerical techniques is used. The heat transfer in the PCM is studied by the source-and-sink method, employing a heat source at the freezing front and a heat sink at the melting front. This work contains a review of methods studying the phase-change processes (Stephan problems).

REFERENCES

[1] Barozzi, G. S., & Pagliarini, G. (1985). A method to solve conjugate heat transfer problems—the case of fully-developed laminar flow in a pipe. *ASME Journal of Heat Transfer 107*, 77–83. doi: http://dx.doi.org/10.1115/1.3247406

[2] Guedes, R. O. C., Ozisik, M. N., & Cotta, R. M. (1994). Conjugated periodic turbulent forced convection in parallel plate channel. *ASME Journal of Heat Transfer 116*, 40–46. doi: http://dx.doi.org/10.1115/1.2910881

[3] Chiu, W. K. S., Richards, C, J., & Jaluria, Y. (2001). Experimental and numerical study of conjugate heat transfer in a horizontal channel heated from below. *ASME Journal of Heat Transfer 123*, 688–697. doi: http://dx.doi.org/10.1115/1.1372316

[4] Sparrow, E. M., & Faghri, M. (1980). Fluid-to- fluid conjugate heat transfer for a vertical pipe-internal forced convection and external natural convection. *ASME Journal of Heat Transfer 102*, 402–407. doi: http://dx.doi.org/10.1115/1.3244313

[5] Patankar, S. V., & Spalding, D. B. (1972). A calculation procedure for heat, mass and momentum transfer in three-dimensional parabolic flows. *International Journal of Heat Mass Transfer 15*, 1787–1806. doi: http://dx.doi.org/10.1016/0017-9310(72)90054-3

[6] Papoutsakis, E., & Ramkrishna, D. (1981). Conjugated Graetz problems. *Chemical Engineering Science 36*, 381–391.

[7] Pagliarini, G, & Barozzi, G. S. (1991) Thermal coupling in laminar flow double-pipe exchangers. *ASME Journal of Heat Transfer 113*, 526–534. doi: http://dx.doi.org/10.1115/1.2910595

[8] Bejan, A. (1982). Second law analysis in heat transfer and thermal design. In T. F. Irvine and J. P. Hartnett (Eds.), *Advances heat transfer,* vol. 15 (pp. 1–58). Elsevier. doi: http://dx.doi.org/10.1016/S0065-2717(08)70172-2

[9] Song, W., & Li, B. Q. (2002). Finite element solution of conjugate heat transfer problems with and without the use of gap elements. *International Journal of Numerical Methods for Heat and Fluid Flow 12*, 81–99. doi: http://dx.doi.org/10.1108/09615530210413181

[10] Fedorov, A. G. & Viskanta, R. (2000). Three-dimensional conjugate heat transfer in the microchannel heat sinkfor electronic packaging. *International Journal of Heat and Mass Transfer 43*, 399–415. doi: http://dx.doi.org/10.1016/S0017-9310(99)00151-9

[11] Kawano, K., Minakami, K., Iwasaki, H., & Ishizuka, M. (1998). Development of microchannels heat exchanging. In R. A. Nelson, Jr, L. W. Swanson, M. V. A. Bianchi, & C. Camci (Eds.), *Application of heat transfer in equipment, systems and education,* HTD-vol. 361-3/PID- vol. 3 (pp. 173–180). New York: ASME.

[12] Joyce, G., & Soliman, H. M. (2009). Analysis of the transient single-phase thermal performance of micro-channel heat sinks. *Heat Transfer Engineering 30*, 1058–1067. doi: http://dx.doi.org/10.1080/01457630902921386

[13] *Mushtaq, I., Hasan, A., Rageb, A., Yaghoubi, M., & Homayoni, H. (2009). Influence of channel geometry on the performance of a counter flow microchannel heat exchanger. *International Journal of Thermal Sciences 45*, 1607–1618.

[14] Rosa, P., Karayiannis, T. G., & Collins, M. W. (2009). Single-phase heat transfer in microchannels: The importance of scaling effects (Review). *Applied Thermal Engineering 29*, 3447–3468. doi: http://dx.doi.org/10.1016/j.applthermaleng.2009.05.015

[15] Rao, V. D., Naidu, S, V., Rao, B. G, & Sharma, K. V. (2006). Heat transfer from a horizontal fin array by natural convection and radiation—A conjugate analysis. *International Journal of Heat and Mass Transfer 49*, 3379–3391. doi: http://dx.doi.org/10.1016/j.ijheatmasstransfer.2006.03.010

[16] Incropera, F. P., & Dewitt, D. P. (1996) *Fundamental Heat and Mass Transfer*. Fourth ed. New York: Wiley.

[17] Grechannyi, O. A., Dorfman, A. S., & Gorobets, V. G. (1986) Coupled heat transfer and effectiveness of flat finned surface in a transverse flow. *High Temperature 24*, 678–683.

[18] Chang, K. (1970) *Separation of flow*. Oxford: Pergamon Press.

[19] Premachandran, B., & Balaji, C. (2006). Conjugate mixed convection with surface radiation from a horizontal channel with protruding heat sources. *International Journal of Heat and Mass Transfer 49*, 3568–3582. doi: http://dx.doi.org/10.1016/j.ijheatmasstransfer.2006.02.044

[20] Kuznetsov, G. V., & Sheremet, M. A. (2009). Numerical modeling of temperature fields in the elements and units of electronic systems. *Microelectronica 38*, 344–352.

[21] *Wang, Q., & Jaluria, Y. (2004). Three-dimensional conjugate heat transfer in a horizontal channel with discrete heating. *Journal of Heat and Mass Transfer 126*, 642–647.

[22] Yoshinoa, H., Fujii, M., Zhang, X., Takeuchia, T., & Toyomasua, T. Conjugate heat transfer from an electronic module package cooled by air in a rectangular duct, http://www.google.com/#hl=en&source=hp&q (on+&btnG=Google+Search&aq=f&aqi=&oq=&fp=aa7ac58 34e645580).

[23] Kassab, A., Divo, E., Heidmann, J., Steinthorsson, E., & Rodriguez, F. (2003). BEM/FVM conjugate heat transfer analysis of a three-dimensional film cooling turbine blade. *International Journal of Numerical Methods for Heat and Fluid Flow 13*, 582–610.

[24] Croce, G. (2001). A conjugate heat transfer procedure for gas turbine blades. *Annals New York Academy Science 934*, 273–280.

[25] Baiocco, P., & Bellomi, P. (1996). A coupled thermo-ablative and fluid dynamic analysis for numerical application to solid propellant rockets. AIAA 96-1811, 31st. *AIAA Thermophysics Conference*, June 17–20, New Orleans, LA.

[26] Aerotherm Corporation Ed. (1992). User's manual non proprietary aerotherm charring material response and ablation program. CMA925, Mountain View, California.

[27] BPD Ed. (1994). EBM motor, low density liner characterization. RE-EBM-7101. Ed. 3.

[28] Dorfman A. S. (2004). Transient heat transfer between a semi-infinite hot plate and a flowing cooling liquid film. *ASME Journal of Heat Transfer 126*, 149–154. doi: http://dx.doi.org/10.1115/1.1650389

[29] Olek. S. Zvirin, Y., & Elias, E. (1988). Rewetting of hot surfaces by falling liquid films as a conjugate heat transfer problem. *International Journal of Multiphase Flow 14*, 13–33. doi: http://dx.doi.org/10.1016/0301-9322(88)90031-6

[30] Yeh, H. C. (1980). An analytical solution to fuel-and-cladding model of the rewetting of a nuclear fuel rod. *Nuclear Engineering and Design 61*, 101–112. doi: http://dx.doi.org/10.1016/0029-5493(80)90081-3

[31] Nuutinen, M., Kaario, O., & Larmi. (2008) Conjugate heat transfer in CI engine CFD simulations, *SAE World Congress & Exhibition*, Detroit, MI, USA, Session: Multi-Dimensional Engine Modeling (Part 4).

[32] El Qarnia, H. (2004). Theoretical study of transient response of a rectangular latent heat thermal energy storage system with conjugate forced convection. *Energy Conversion and Management 45*, 1537–1551. doi: http://dx.doi.org/10.1016/j.enconman.2003.08.024

[33] Voller, V. R. (1990) Fast implicit finite-difference method for the analysis of phase change problems. *Numerical Heat Transfer 17B*, 155–169.

[34] Li, H., Hsieh, C. K., & Goswami, D. Y. (1996). Conjugate heat transfer analysis of fluid flow in a phase-change energy storage unit. *International Journal of Numerical Methods for Heat and Fluid Flow 6*, 77–90. doi: http://dx.doi.org/10.1108/EUM0000000004105

CHAPTER 11

HEAT AND MASS TRANSFER IN TECHNOLOGY PROCESSES

11.1   Multiphase and Phase-Changing Processes

Problems containing the phase-changing processes are complex tasks, which requires to take into account moving boundary between two or more phases, the exchange of the thermal energy between phases, their different thermophysical properties, and latent heat of melting or solidification process. Modeling processes of this type has a practical interest in metallurgy, purification of metals, crystal grows, casting, welding, and many other technologies. Three examples of studying conjugate heat transfer in phase-changing processes demonstrate the modern methods employing in such cases.

Example 11.1n Solidification in the enclosed region [1].

The studied model consists of a mold with a liquid lump undergoing solidification and the solid around it. The process of phase change is described using the conservation equations with enthalpy as a basic variable. The system of governing equations and boundary conditions take into account the thermal buoyancy as a driving force:

$$\frac{Du}{Dt} = \Pr \nabla^2 u - \frac{\partial p}{\partial x} + q_u, \quad \frac{Dv}{Dt} = \Pr \nabla^2 v - \frac{\partial p}{\partial y} + q_v + \mathrm{Ra} \Pr \theta, \quad \frac{D\rho J}{Dt} = \nabla \cdot \left(\frac{\lambda}{c_p} \nabla J \right) - q_h \quad (11.1)$$

$$\theta_m = \theta_s, \quad \nabla \theta_s = \lambda_m \cdot \nabla \theta_m \cdot \mathbf{n}, \quad \nabla \theta_m \cdot \mathbf{n} = -\mathrm{Bi}\theta_m, \quad \theta = \frac{T - T_a}{T_i - T_a} \quad \mathrm{Ra} = \frac{g\beta (T_i - T_a) W^3}{\nu \alpha}$$

The first two momentum equations are supplemented by usual continuity equation and are written in dimensionless variables as well as the boundary conditions. The third energy equation is written in dimension variables in terms of enthalpy J determined as integrated differential $c_p dT$. The dimensionless variables are scaled by mold width W for coordinates, α_L/W for velocities, λ_S for thermal conductivity, and W^2/α_L for time. The first two boundary conditions are the conjugate equalities at the inner surfaces of the mold, while the third is the heat transfer condition at the outer surfaces of the mold. The others (not shown here) are no-slip condition and thermal conditions at the boundaries given in [2]. The subscripts L, S, m, i, and a denote liquid, solid, mold, initial, and ambient, respectively.

The phase-change control volume is considered using the model of porous medium with a liquid fraction f, depending on the latent heat content. The source terms q_u and q_v incorporated in the momentum equations are expressed applying the Darcy law according to which the velocity is proportional to pressure gradient. The change of phase is regarded by source term q_h included in the energy equation. For more details see [2].

The numerical solution is performed using the SIMPLE algorithm. The marker and cell scheme, which is similar to the weighted residual method and staggered grid, are employed for discretization (Chapter 8).

The following results and conclusions are obtained:

- One of the reasons of using the enthalpy method is that in this approach a simple examination of numerical solution shows the location of grid points involved in the phase-change process (Comment 10.6). It is known that in the enthalpy formulation the change-phase front takes up several grid points [3]. In this case, it occupies two points which can be correspond to one control volume if a grid is enough fine. Extrapolation or interpolation of these results leads to unrealistic stepwise behavior of the moving boundary and the time history of the temperature.
- The Nusselt number is calculated for all four walls at the solid–mold interface.

The influence of mold material, width, and aspect ratio on the solidification process is studied.

The following parameters are used for the reference computation: aspect ratio $a = 2$, Stefan number Ste=2.4, Pr = 0.011 (the tin), Re = 10^4. The average Nusselt number at the right wall is computed for the tin as a cast metal in copper, steel, and sand molds. The cooper mold show the largest and the sand mold the smallest rates of heat transfer. Similar results are obtained for the top wall. On the curves, $\mathrm{Nu}(t)$ obtained for metal molds waviness can be seen, which is less for the sand mold due to low conductivity. At the same time, high thermal resistance of the sand mold results in the higher vertical velocity and highest circulation. The smallest velocity is observed in the copper mold. The interface movement is less in the sand mold than that in copper and steel molds, which yields more effective dependence $\mathrm{Ra}(t)$ and higher velocity.

- Temperature–time history obtained for a point at $x = 3$, $y = 5$ shows that due to high heat transfer in the copper mold, the process of solidification is completed about 75% when the temperature at this point achieves the freezing level. In the steel mold at the same time, the solidification is less completed, and in the sand mold, only about 10% is solidified. In addition, a highly conducting mold heats up very gradually. Temperature profiles across the section indicate that temperature gradients in the liquid are lower than these in the solid. In the cooper mold, the liquid temperatures are isothermal across the section, while the profiles in two other molds show temperature variation across the section.
- The effect of three values 0.5, 2, and 4 of aspect ratio is investigated in a steel mold. The interface is basically flat for low aspect ratio (width to height) and curved at the

top and bottom due to high heat transfer from the liquid–solid region to the mold. This confirms the usefulness of an one-dimensional model for molds with small aspect ratio. At the same time, the one-dimensional approach in the case of large aspect ratio yields significant errors. The solidification proceeds faster in cavities with small aspect ratio because under adiabatic top and bottom only the side walls conduct the heat. The temperature profiles are observed to be parabolic in the solid and almost uniform in liquid. The streamlines and isotherms in the cavity with small aspect ratio show two tall counter-rotating cells in the melt region, size of which decreases as the solidification progresses. The isotherms also show high-temperature gradients near the solid–liquid interface and approximately constant in time temperature gradients in the mold.

- The data obtained for three mold widths indicate that as width increases, the heat transfer through the mold grows due to increased energy storage. The circulation in thinner cavities is higher because the heat transfer in those is less. The isotherms are curved at the top and bottom at early times and later become to be flat.

- It is concluded that the problem of solidification must be considered in conjugate formulation to obtain the accurate results.

Example 11.2 *n Concrete at evaluated temperatures [4, 5].

The solid skeleton of the hardened cement paste is porous hygroscopic material comprising various chemical compounds and pores filled with liquid and vapor water and dry air. On the condition of high temperature, heat changes the chemical compounds and fluid content resulting in changes of physical structure, which affect the mechanical and other properties of the concrete. The articles in question consider the heat and moisture transfer in concrete exposed to high temperature using comprehensive 3D models. Both works took into account the basic phenomena, but additional effects of capillary pressure and adsorbed water are studied in [5].

The basic assumptions are [4]: (i) there is a thermal equilibrium between phases; (ii) vapor and air behave as ideal gases; and (iii) the temperature dependence of mass fluxes is negligible. The conservation of mass and energy equations for dry air and moisture are

$$\frac{\partial \varepsilon_G \tilde{\rho}_A}{\partial t} = \nabla \cdot \mathbf{J}_A \quad \frac{\partial \varepsilon_G \tilde{\rho}_V}{\partial t} + \frac{\partial \varepsilon_{\mathrm{FW}} \rho_L}{\partial t} - \frac{\partial \varepsilon_D \rho_L}{\partial t} = -\nabla \cdot \left(\mathbf{J}_V + \mathbf{J}_{\mathrm{FW}} \right) \tag{11.2}$$

$$\rho c \frac{\partial T}{\partial t} - \Lambda_E \frac{\partial \varepsilon_{\mathrm{FW}} \rho_L}{\partial t} + \left(\Lambda_E + \Lambda_D \right) \frac{\partial \varepsilon_D \rho_L}{\partial t} = \nabla \cdot \left(\lambda \nabla T \right) + \Lambda_E \nabla \cdot \mathbf{J}_{\mathrm{FW}} - \left(\rho c v \right) \cdot \nabla T$$

Here, ε and $\mathbf{J}$ are the volume fraction and mass flux of phase, respectively, subscripts A, D, G, L, S, V, and FW denote dry air, chemically bound water released by dehydration, gas, liquid, solid, vapor, and free water (combined liquid and adsorbed), Λ_E and Λ_D are specific heat of evaporation and dehydration. The mass fluxes of dry air, vapor, and free water are defined using Darcy's and Fick's laws ignoring the diffusion of the adsorbed water on the surface:

$$\mathbf{J}_A = \varepsilon_G \tilde{\rho}_A \left[\mathbf{v}_G - D_{\mathrm{AV}} \nabla \left(\tilde{\rho}_A / \tilde{\rho}_G \right) \right], \mathbf{J}_V = \varepsilon_G \tilde{\rho}_A \left[\mathbf{v}_G - D_{\mathrm{VA}} \nabla \left(\tilde{\rho}_F / \tilde{\rho}_G \right) \right],$$

$$\mathbf{J}_{\mathrm{FW}} = \varepsilon_{\mathrm{FW}} \rho_L \mathbf{v}_G \tag{11.3}$$

$$\mathbf{v}_G = \left(KK_G / \mu_G \right) \nabla p_G, \quad \mathbf{v}_L = \left(KK_L / \mu_L \right) \nabla p_G, \quad \varepsilon_{\mathrm{FW}} = \left(\varepsilon_{\mathrm{CM}} / \rho_{\mathrm{CM}} \right) f \left[\left(p_V / p_{\mathrm{ST}} \right), T \right]$$

In these relations, $D_{AV} \approx D_{VA}$ are diffusion coefficients of dry air and water vapor, $\tilde{\rho}$ is the mass of a phase per unit volume, $K, K_G,$ and K_L are the permeability of dry concrete and the relative permeabilities of the gas and liquid. The pressure of air p_A and of vapor p_V is determining using the state equation of ideal gas $p = R\tilde{\rho}T$ as well as the pressure in a gas and water $p_L \approx p_G = p_A + p_V$. The volume fraction of free water ε_{FW} is defined from the equation of sorption isotherms, which relate the free water content to the cement content $\varepsilon_{CM}/\rho_{CM}$, temperature and relative humidity p_V/p_{ST}, where p_{ST} is the saturation pressure of the vapor (Comment 11.4). For temperatures above the critical point of water $(374.14°C)$, the free water content is zero and the gas volume fraction can be defined from $\phi = \varepsilon_{FW} + \varepsilon_G$, where ϕ is the concrete porosity.

Comment 11.1 Darcy's law describes the flow of fluid through a porous medium according to which the flux is proportional to porosity and pressure gradient.

Using these basic relations, the system of differential equations is formulated, which consists of ten equations given in [4] and additional six equations given in [5]. The boundary conditions that are the same in both models include the following expressions:

$$\frac{\partial T}{\partial n} = \frac{h_{GR}}{\lambda}(T_\infty - T), \quad \mathbf{J} \cdot \mathbf{n} - \gamma(\tilde{\rho}_V - \tilde{\rho}_{V,\infty}),$$

$$\frac{\partial \tilde{\rho}_V}{\partial n} + \frac{K_{VT}h_{GR}}{K_{VV}\lambda}(T_\infty - T) + \frac{\gamma}{K_{VV}}(\tilde{\rho}_{V,\infty} - \tilde{\rho}_V), \tag{11.4}$$

where h_{GR} and γ are coefficients of heat transfer and of vapor mass transfer on the boundary. The first and the second equations are the energy and gaseous mixture mass balances, while the third expression determines the vapor content gradient on the boundary.

The basic results may be formulated as follows:

- In both studies, a steep drying front is observed. The vapor and liquid water content of the phase mixture changes from high and low on the hot side to low and high on the cold side. The water content increases ahead of drying front due to the recondensation in the cooler zone, which is named "moisture clog zone." The maximum of peaks in a gas pressure and vapor content obtained in the modified model [5] are lower than that for the initial model [4]. This may be significant in analyzing potential spalling because the internal pore pressure is considered as the cause of it. The modified model also predicts more extensive moisture clog zones in which the liquid pressure gradient drives water away from the face exposed to fire. These results show that ignoring the adsorbed water flux can significantly affect the predicted values of free water flux, vapor content, and gas pressure.
- The values of liquid pressure predicted by two models are considerably different. The capillary pressure is zero according the initial model [4], while it increases rapidly with decreasing of relative humidity according to the modified model [5]. However, the overall results given by both models are very similar, showing that the capillary pressure has little or no effect on the transport in concrete under intense heating.
- If both capillary pressure and adsorbed water are taken into account, the gas pressure and vapor content are higher than those shown by initial model or by modified model, including only capillary pressure. In this case, the results are physically realistic in contrast to unrealistic behavior of capillary pressure as the adsorbed water is ignored.

Example 11.3 *n Czochralski crystal growth process [6–8].

The Czochralski process is used for growing a crystal to get semiconductors (e.g., silicon, germanium, and gallium arsenide) and crystals of metals and salts. It is named after Polish scientist Jan Czochralski, who discovered this process in 1916. The process is performed in apparatus with cylindrical crucible heated in the furnace above melting temperature of the melt. The crystal and the crucible are rotated in opposite directions. The resulting cylindrical crystal is vertically pulled from the crucible.

Since the temperature in this process is high, the radiative heat transfer in addition to conduction and convection should be taken into account. The 3D axisymmetric heat transfer model is studied in this article. Due to axial symmetry, only cross-section containing liquid, solid, and filling gas areas is considered. The following additional to usual assumptions are used [8]: (i) the diameter of the crystal pulled from the crucible is constant; (ii) the studied region with liquid, solid, and gas is an enclosure; (iii) the effect of rotation is negligible; (iv) the crystal and liquid have the same radiative properties.

The 3D steady-state momentum (with buoyancy), energy (with radiative term),

$$\rho(\mathbf{v}\cdot\nabla)\mathbf{v} = -\nabla p + \mu\nabla^2\mathbf{v} + \rho\mathbf{g}\beta(T - T_\infty), \quad \nabla T(\mathbf{r}) = (1/\alpha)\mathbf{v}\cdot\nabla + (1/\lambda)q(\mathbf{r}), \quad (11.5)$$

and continuity $\nabla \cdot \mathbf{v} = 0$ equations governed the problem [7]. The radiative heat source is calculated considering the liquid and solid phases as semitransparent medium with constant absorption coefficients k_L and k_S, while the surrounding gas is considered as transparent medium with $k = 0$. The radiative source q_r is defined as a radiant energy absorbed by infinitesimal surface within infinitesimal time. In 3D case, this amount of energy is determined by the following coupled integral equations [7, 8]:

$$q_r(\mathbf{p}) + \varepsilon(\mathbf{p})e_b\big[T(\mathbf{p})\big] = \varepsilon(\mathbf{p})\int_S\left\{e_b\big[T(\mathbf{r})\big] + \frac{1-\varepsilon(\mathbf{r})}{\varepsilon(\mathbf{r})}q_r(\mathbf{r})\right\}\tau(\mathbf{r},\mathbf{p})\mathbf{K}(\mathbf{r},\mathbf{p})dS(\mathbf{r})$$

$$+\varepsilon(\mathbf{p})\int_S\left\{\int_{L_{rp}} k(\mathbf{r}')e_b\big[T^m(\mathbf{r}')\tau(\mathbf{r}',\mathbf{p})\big]dL(\mathbf{r}')\right\}\mathbf{K}(\mathbf{r},\mathbf{p})dS(\mathbf{r}) \quad (11.6)$$

$$q_{rv}(\mathbf{p}) + 4k(\mathbf{p})e_b\big[T^m(\mathbf{p})\big] = k(\mathbf{p})\int_S\left\{e_b\big[T(\mathbf{r})\big] + \frac{1-\varepsilon(\mathbf{r})}{\varepsilon(\mathbf{r})}q_r(\mathbf{r})\right\}\tau(\mathbf{r},\mathbf{p})\mathbf{K}(\mathbf{r},\mathbf{p})dS(\mathbf{r})$$

$$+ k(p)\int_S\left\{\int_{L_{rp}} k(\mathbf{r}')e_b\big[T^m(\mathbf{r}')\tau(\mathbf{r}',\mathbf{p})\big]dL(\mathbf{r}')\right\}\mathbf{K}_\mathbf{r}(\mathbf{r}',\mathbf{p})dS(\mathbf{r}) \quad (11.7)$$

Here, $\mathbf{r}$ and $\mathbf{p}$ refer to current and observed points, respectively, whereas point $\mathbf{r}'$ is located on the line L_{rp} connecting points $\mathbf{r}$ and $\mathbf{p}$, the integration is performed over the surface of the computational domain S and along the line L_{rp}, ε is the emissivity, and $e_b(T)$ and $e_b(T^m)$ are the black-body emissivities of the computational domain surface and of the semitransparent medium, respectively. The kernel functions and transmissivity are

$$\mathbf{K}(\mathbf{r},\mathbf{p}) = \frac{\cos\varphi_r\cos\varphi_p}{\pi|\mathbf{r}-\mathbf{p}|^2}, \quad \mathbf{K}_\mathbf{r}(\mathbf{r},\mathbf{p}) = \frac{\cos\varphi_p}{\pi|\mathbf{r}-\mathbf{p}|^2} \quad \tau(\mathbf{r},\mathbf{p}) = \exp\left[-\int_{L_{rp}} k(\mathbf{r}')dL_{rp}(\mathbf{r}')\right] \quad (11.8)$$

Derivation of these equations and other details one may find in [8].

Usual boundary conditions should be satisfied on the external surfaces of studied enclosure. The conjugate condition on the phase-change front involves continuity of melting temperature, a jump in heat flux, and no-slip condition for the melt velocity:

$$T_L(\mathbf{r}) = T_s(\mathbf{r}) = T_{ph} \quad -\lambda_L \frac{\partial T_L}{\partial n_{ph}} + \lambda_s \frac{\partial T_s}{\partial n_{ph}} = -\Lambda_{ph}\rho_s \left(v \cdot \mathbf{n}_{ph} \right), \quad \mathbf{v}_L = \mathbf{v}_x \qquad (11.9)$$

On the solid–gas and liquid–gas interfaces equalities of temperatures and heat fluxes including the radiative component q_r should be satisfied:

$$T_L(\mathbf{r}) = T_G(\mathbf{r}), \quad \lambda_L \frac{\partial T_L}{\partial n} = \lambda_G \frac{\partial T_G}{\partial n} + q_{rL}, \quad T_s(r) = T_G(r), \quad \lambda_S \frac{\partial T_s}{\partial n} = \lambda_G \frac{\partial T_G}{\partial n} + q_{rs} \qquad (11.10)$$

In these conditions T_{ph}, $\mathbf{n}_{ph}$, and Λ_{ph} are the phase-change temperature, vector normal to phase-change surface, and the latent heat.

Numerical solution is performed using commercial package FLUENT, which is based on the finite-volume approach, and another software based on the boundary element method (Chapter 8). The FLUENT is applied for determining velocity, pressure, and temperature fields under known distribution of the radiative heat source. The computational domain is divided in liquid, solid, and gas subdomains that are numerically analyzed separately. To calculate the radiative heat fluxes and source, the BEM code is used. The iterative procedure is employed to couple both numerical results. More details are given in [8]. As an example, the velocity and temperature profiles in the liquid phase and radiative heat source in a typical 3D Czochralski process are given in [7].

The solution of the other problems with phase-changing phenomena can be seen in [9–11].

11.2 Manufacturing Processes Simulation

Example 11.4n Continuous wires casting [12].
The modern continuous casting directly from melt has special benefits as follows:

(i)   the near-net shape formation of metallic materials accompanied by reductions of energy, time, and labor;
(ii) the development of new functional properties caused by structural modification, such as the formation of nonequilibrium phases;
(iii) increased control over the process through automation;
(iv) the improvement of mechanical properties caused by decreased segregation and the refinement of grain size.

The model involves a casting channel formed by a static component and a rotating copper wheel. Liquid metal is fed into the cavity at the top of a shoe and drawn into a gap of decreasing cross-section as the wheel rotates. It is intended that the solid–liquid interface be contained within the die cavity giving the ability to predict and manipulate its position through variation of geometry and processing. Three main assumptions are made:

(i) steady state is achieved,
(ii) fluid motion within the melt is Newtonian, incompressible, and laminar;
(iii) equilibrium solidification applies; and
(iv) instantaneous and complete filling of channel cavity.

The problem is governed by set of the following equations:

$$f_i = \frac{\Delta I}{\Lambda} = \begin{cases} 1 & T > T_l \\ (T - T_s)/(T_l - T_s) & T_s \le T \le T_l, \\ 0 & T < T_s \end{cases} \quad \frac{\partial \rho J}{\partial t} + \frac{\partial \rho \Delta I}{\partial t} + \frac{\partial \rho u_i J}{\partial x_i} = \frac{\partial \lambda \nabla T}{\partial x_i} + q \quad (11.11)$$

$$q_a = h_a(T_a - T_w) + \varepsilon_a \sigma(T_a^4 - T_w^4), \quad \Delta I^{n+1} = \Delta I^n + a_H c_p \left[T - (T_l - T_s)f_l - T_s \right] \quad (11.12)$$

Here, ΔI is the latent heat content, Λ is the latent heat of solidification, T_l is the liquidus temperature, T_s is the solidus temperatures, J is the enthalpy, q is a source term, ε is the emissivity, a_H is the under-relaxation factor (Comment 10.1), and subscript a denotes external values.

The conjugate problem is solved using software FLUENT that is based on the enthalpy–porosity approach (Comment 10.6). The liquid fraction value f_i for each control volume lying between zero and unity is defined by the first equation (11.11). The phase change is calculated employing the second equation (11.11) and lever rule. Both values are obtained iterative applying the second equation (11.12). The heat flux through an external wall as a result of radiation and convection to an external free air stream is computed by first relation (11.12).

Comment 11.2 Liquidus is the temperature above which a substance is liquid, and solidus specifies the temperature below which a substance is solid. The lever rule is used to determine the percent weight of liquid and solid phases in binary composition at the temperature that is between the liquidus and solidus (see Wikipedia).

The basic results and conclusions are as follows:

- Three parameters are most influential on temperature, liquid fraction, and velocity profiles in the casting channel: (i) initial inlet melt velocity, (ii) initial melt temperature at inlet, and (iii) rotation speed of the wheel.
- The cast material must be sufficiently solidified on reaching the exit of the casting channel. However, if the temperature is too low near the exit, the wire will fail due to a high deformation resistance.
- The relationship between wheel speed and melt depth is the main factor determining strip thickness in metal melt spinning. The maximum achievable mass flow rate at the inlet is 23 gs^{-1} (corresponding to a line speed of 0.78 ms^{-1}). The inlet melt velocities below this value cause excessive solidification throughout the channel so the wheel speed is fixed at 9.85 ms^{-1}.
- The thermal contact between the shoe and the wheel also caused complete solidification of the melt within the channel and would require large amounts of superheat to achieve solidification at a satisfactory distance from the channel exit.

- The model presented here is only an approximate because it does not take into account nonequilibrium solidification. The model also assumes that the solidified melt is a very viscous fluid neglecting the effects on the velocity and "turbulence" of the melt due to solidification. Nevertheless, the heat transfer model itself is reasonably accurate.

Example 11.5 *n Optical fiber coating process [13].

A study of the coating process is very important because improper coating, in particular, nonuniform or missing coating, leads to microbending and finally affects the transmission loss. In the typical optical fiber system, a bare fiber is drown from a furnace where its temperature reaches over 1600°C. After cooling to a proper wetting temperature, fiber moves through a coating applicator with a coating fluid, laser micrometer for controlling the diameter, and finally gets through ultra-violet (UV) curing oven to a take-up a spool.

In this study, an axisymmetric two-dimensional process is considered in cylindrical coordinates with the radial distance r measured from a centre of the fiber and the axial distance z measured upward from a die exit. The coating material is a UV-curable acrylate, viscosity of which highly depends on the temperature. This yields the coupled momentum and energy equations with nonlinear diffusion terms.

The system of governing equations is used in general variables ϕ, I (Sec. 8.4):

$$\frac{1}{r}\frac{\partial(\rho r u \phi)}{\partial r} + \frac{\partial(\rho r u \phi)}{\partial z} = \frac{1}{r}\frac{\partial}{\partial r}\left(r\mu\frac{\partial\phi}{\partial r}\right) + \frac{\partial}{\partial z}\left(\mu\frac{\partial\phi}{\partial z}\right) + I$$

$$\phi = \begin{pmatrix} 1 \\ u \\ v \end{pmatrix} \quad I = \begin{pmatrix} 0 \\ -\frac{\partial p}{\partial r} + \frac{1}{r}\frac{\partial}{\partial r}\left(r\mu\frac{\partial u}{\partial r}\right) + \frac{\partial}{\partial r}\left(\mu\frac{\partial u}{\partial r}\right) - 2\mu\frac{u}{r} \\ -\frac{\partial p}{\partial z} + \frac{1}{r}\frac{\partial}{\partial r}\left(r\mu\frac{\partial v}{\partial z}\right) + \frac{\partial}{\partial r}\left(\mu\frac{\partial v}{\partial z}\right) + \rho g_z \end{pmatrix}, \quad \rho_w c_{pw} v_w \frac{\partial T}{\partial z} = \frac{2\lambda_{in}}{r_w}\frac{\partial T}{\partial r} \qquad (11.13)$$

$$\frac{1}{r}\frac{\partial(\rho r u T)}{\partial r} + \frac{\partial(\rho r u T)}{\partial z} = \frac{1}{r}\frac{\partial}{\partial r}(r\lambda) + \frac{\partial}{\partial z}\left(\lambda\frac{\partial T}{\partial z}\right) + 2\mu\left[\left(\frac{\partial u}{\partial r}\right)^2 + \left(\frac{v}{r}\right)^2 + \left(\frac{\partial u}{\partial z}\right)^2 + \left(\frac{\partial u}{\partial z} + \frac{\partial v}{\partial r}\right)\right]$$

Here, the first equation is valid for continuity (the first line in the second equation) and for momentum equations for fluid for both velocity components. The source term I arises from the variable viscosity. The energy equation for fiber [the last equation (11.13)] is simplified [see equation (2.5)] taking into account that (i) the Biot number Bi $= 2hr_w/\lambda_w$ is small due to small fiber diameter ($2r_w = 125\,\mu m$) allowing considering the radially averaged fiber temperature, (ii) the axial conduction may be neglected because it is small comparatively to energy carried from fiber by convection, and (iii) the buoyancy effect are neglected since the Gr/Re$^2 \ll 1$. The equation (11.13) for T (which contains indices w) defines the fiber surface temperature. Here, λ_{in} is the harmonic thermal conductivity at the fiber–fluid interface (a special mean value that mitigate the influence of large outliers and increase the influence of small values, see Wikipedia).

The boundary conditions are as follows: (i) a shear free and adiabatic conditions at the top surface in the pressured applicator; (ii) at the fiber surface no slip radial condition, a given axial speed, and conjugate thermal conditions; (iii) no-slip condition at all walls, isothermal condition at die wall, an adiabatic condition or a given heat transfer coefficient at the applicator wall;

(iv) uniform speed and temperature 298.15 K of coating material at the feed inlet; and (v) at the die exit, a fully developed flow and thermal conditions and the specified meniscus. Details of free surface modeling are given in [14].

The primitive variables, that is, velocity, pressure, and temperature, after Landau transformation $\xi = (r - r_i)/(r_0 - r_i)$ and $\eta = z/L$, where r_i is the inner radius [15], are used to solve the governing equations. Highly clustered grids are employed in the regions with large velocity or/and temperature gradients. The second-order upwind scheme (Sec. 8.3.1.2) is applied. The algorithm, like SIMPLE (Sec. 8.3), for pressure is repeated until the convergence is achieved. For validation, the final coating thicknesses at variable fiber speed in laminar flow in a circular duct are calculated to compare with known data

The basic results are as follows:

- The moving fiber creates the thermal field in which the cooler fluid removes the energy from the fiber. Nevertheless, the fiber temperature increases with speed. The reason of this is the higher temperatures of the fluid and the wall, which occurs due to increased dissipation effects with the growing speed. As the moving fiber first meets the fluid, it loses energy to the cooler fluid due to high Nusselt numbers. As the fluid heats up, the temperature gradient becomes smaller. When the speed increases, the Nusselt number becomes negative and the fluid heats the fiber. It follows from this fact that the smallest gap between die wall and the fiber is responsible for the rise of the Nusselt number and the fall of the fiber temperature near the die exit.

- The viscous dissipation increases considerably as the speed grows. Due to this, the fiber temperature is higher than that at the die entrance. The temperature gradient at the fiber is very sharp, and temperature increasing is the highest in the die. This very large temperature occurring due the dissipation at the high speed should be avoided since the polymer may start to crosslink and degrade. As the fiber moves away from the dynamic contact point to the die exit, the fiber temperature decreases due to removal heat at the fiber surface. However, when the fiber speed increases, the temperature also increases, and the Nusselt number reaches an asymptotic value.

- The die exit diameter is one of the critical variables determining the quality of the coating process. The numerical data show that the coating thickness increases linearly with the die exit diameter at the different fixed speed values. The ratio of thermal conductivities of the fiber and coating fluid also affect the process significantly. At the fixed material properties, the rise of coating fluid conductivity leads to increasing the coating thickness. This effect becomes unnoticeable when the fiber speed grows. If the fluid thermal conductivity is high, the Nusselt number variation is strong as well. This and increasing the coating thickness with the growing fluid conductivity is due to enhanced thermal diffusion in coating fluid.

- The effect of entrance temperature is high. As entrance temperature increases the coating thickness decreases linearly. The fiber speed affects this dependence only slightly. At the lower entrance temperature, the viscous heating becomes a major factor. The condition on the outer wall of applicator does not significantly affect the resulting coating thickness so that practically the conditions at the applicator wall can be flexible.

- To understand the overall coating process, this work should be used along with the other results as isothermal flow and the exit meniscus modeling, which are important for improving the quality of the coating, its uniformity, and the production rate.

Comment 11.3 The meniscus formed across the interface of two fluids due to surface tension. Its creation is described by Young–Laplace equation that relates the pressure difference to the shape of the surface wall (see Wikipedia).

Example 11.6n Twin-screw extruder simulation [16].

The actual complicated flow domain as a circular region is divided in two subdomains named translation and intermeshing regions. The translation domain is modeled similar to single screw extruder as a channel with cross-section $H/W \ll 1$, where H and W are height and width of the channel, respectively. For steady developing, the creeping approximation is used for two-dimensional flow at low Reynolds number. In contrast to the boundary layer approximation, in this case the inertia terms are neglected, while the viscose terms are important (Sec. 2.5.1).

The momentum conservation equations are considered in $y-z$ cross-section, where z is directed along the screw helix, y is normal to this direction, and x is normal to $y-z$ cross-section. The heat convection equation is written by taking into account the temperature dependence of physical properties and dissipation since the treatment materials are strong viscous. The governing system for translating domain is

$$\frac{\partial p}{\partial x} = \frac{\partial}{\partial y}\left(\mu\frac{\partial u}{\partial y}\right), \quad \frac{\partial p}{\partial z} = \frac{\partial}{\partial y}\left(\mu\frac{\partial w}{\partial y}\right), \quad \frac{\partial p}{\partial y} = 0,$$

$$\rho c_p w\frac{\partial T}{\partial z} = \frac{\partial}{\partial y}\left(\lambda\frac{\partial T}{\partial y}\right) + \mu\left(\frac{\partial u}{\partial y}\right)^2 + \mu\left(\frac{\partial w}{\partial y}\right)^2 \tag{11.14}$$

The rheology power law for non-Newtonian fluid is used to describe the dependence $\mu(T,\dot{\gamma})$, where $\dot{\gamma}$ is the strain rate and m in the first relation of equation (11.15) is the temperature coefficient:

$$\mu = \mu_0\left(\frac{\dot{\gamma}}{\dot{\gamma}_0}\right)^{n-1}\exp\left[-m(T-T_0)\right], \quad \dot{\gamma} = \left[\left(\frac{\partial u}{\partial y}\right)^2 + \left(\frac{\partial w}{\partial y}\right)^2\right]^{1/2} \tag{11.15}$$

These equations for adiabatic screw under usual no-slip boundary conditions and specified barrel temperature are solved using the finite difference method.

The flow in intermeshing domain is also simplified using steady creeping flow equations for the case of completely filled with a fluid screw channel. Since the axial length of most extruders is much larger than other sizes, the velocity gradients in the z direction are neglected so that the governing equations in x, y directions are as follows:

$$\frac{\partial u}{\partial x} + \frac{\partial v}{\partial y} = 0 \quad \frac{\partial p}{\partial x} = \frac{\partial \tau_{xx}}{\partial x} + \frac{\partial \tau_{xy}}{\partial y}, \quad \frac{\partial p}{\partial y} = \frac{\partial \tau_{xy}}{\partial x} + \frac{\partial \tau_{yy}}{\partial y}, \quad \frac{\partial p}{\partial z} = \frac{\partial \tau_{xz}}{\partial x} + \frac{\partial \tau_{yz}}{\partial y}, \tag{11.16}$$

where stresses in Einstein's notations are $\tau_{ij} = \mu\left(u_{i,j} + u_{j,i}\right)$. The same rheology power law (11.15) for non-Newtonian fluid is applied, but expression for strain rate as well as energy equation in this case are more complicated due to dissipation function S:

$$\frac{\partial}{\partial x}\left(\lambda\frac{\partial T}{\partial x}\right) + \frac{\partial}{\partial y}\left(\lambda\frac{\partial T}{\partial y}\right) + \rho c_p\left(u\frac{\partial T}{\partial x} + v\frac{\partial T}{\partial y} + w\frac{\partial T}{\partial z}\right) + \mu S = 0 \tag{11.17}$$

$$S = 2\left(\frac{\partial u}{\partial x}\right)^2 + \left(\frac{\partial u}{\partial y} + \frac{\partial v}{\partial x}\right)^2 + 2\left(\frac{\partial v}{\partial y}\right)^2 + \left(\frac{\partial w}{\partial x} + \frac{\partial w}{\partial y}\right)^2, \quad \dot{\gamma} = S^{1/2}$$

In this case, the velocity component w is directed along the extruder axis. The same as in the previous domain no-slip boundary conditions, assumption of adiabatic screw, and specified barrel temperature are employed. The second and the third equation (11.16) in the x and y directions are solved by finite element method, while the last one in the z direction is solved using the Galerkin method (Chapter 8). The axial velocity component w is perpendicular to the $x - y$ plane, where the u, v components are located. These components are coupled in the case of the non-Newtonian fluids but are uncoupled for the Newtonian fluids. Therefore, in the case of non-Newtonian fluid the equations for velocity components, pressure and temperature should be solved iteratively.

The calculations are performed for concrete screw with a diameter of 23.3 mm, a channel depth of 4.77 mm, and a barrel diameter of 30.84 mm:

- It is found numerically by varying the location of the interface between two separated domains that the intermeshing domain occupies two-third of the total region.
- The results for translating region show that the flow temperature increases above the barrel temperature, which is $275°C$. The reason of this increasing is the viscous dissipation. As a result of such a temperature difference, the heat goes from fluid to barrel. The temperature variations slightly affect the velocity field that is defined mainly by flow rate. Knowing the velocity and temperature distributions, one can calculate other characteristics of the screw, such as stress or strain distributions.
- The flow and pressure are screw-symmetric in the intermeshing domain. At the centre of this region, a significant pressure change can be seen. The linear velocity profile at the outlet yields a nonlinear pressure change along the annulus. The temperature at the root of the screw is lower than that at the barrel, and the temperature rise in this domain is small in comparison with that in translating domain. It is found that the portion of total flow in the screw channel goes into the other channel, while the remaining flow is retained in the same channel. The ratio of flow portion leaving the channel to the total flow is defined as the division ratio. It is shown that this ratio decreases as the index n in the power law (11.15) increases, but it increases when the throughput increases or the depth of the screw channel decreases.
- The complete flow is obtained by coupling both translating and intermeshing models. The velocity, temperature, shear, and pressure fields indicate that the region where the both flows interact is relatively small. The pressure patterns show the different regions with low, high, and uniform pressure. The pressure rise in the intermeshing region is small as it is compared with that in the translating region, and the same conclusion is valid for the bulk temperature. At the smaller die openings when the throughputs are smaller also, the bulk temperature is higher because in this case the velocity profiles are much steeper giving higher viscous dissipation.

Example 11.7a/n Films and fibers production [17].

In such a production system, a continuous sheet or a rod of material extruded from a die at temperature T_0 is pulled through a surrounding coolant with constant velocity U_w molding plastic or other type of film or a rod.

It is shown in [18] that the conjugate solution of the problem in question should be solved in two parts. The first part from $x = 0$ to the point $x = x_{cr}$ (Fig. 4.14) where the initial boundary layers fill the cross-section up is a conjugate problem of heat transfer between a moving semi-infinite slab and the surrounding medium. For this case, the thermal boundary layer is formed analogous to that on the plate in an external medium, and hence, the equation for the plate in variables x/L and $\varphi = y\sqrt{U_w/2\nu x}$ coincides with the first equation (4.2) without right-hand side for the $\text{Pr} \to 0$. Consequently, the solution for the plate is given by (4.1) with coefficients (4.10) for g_k:

$$2x\frac{\partial\theta}{\partial x} + \varphi\frac{\partial\theta}{\partial\varphi} = \frac{1}{\text{Pr}}\frac{\partial^2\theta}{\partial\varphi^2}, \quad q_w = \frac{\lambda_w\sqrt{\text{Re}}}{\theta_w}\sqrt{\frac{\text{Pr}}{x\pi}}\left(T_w - T_0 + \sum_{k=1}^{\infty}g_k x^k \frac{d^k T_w}{dx^k}\right) \tag{11.18}$$

where $g_{k\infty} = (-1)^{k+1}/k!(2k-1)$. Substituting heat fluxes from the plate and from fluid sides in the conjugate condition yields the expression whence it follows that the plate temperature at the die is constant determining from general solution at $\partial^k T_w/\partial x^k = 0$,

$$K(T_w - T_\infty) + (T_w - T_0) + \sum_{k=1}^{\infty}(Kg_k + g_{k\infty}) = 0 \, x^k \frac{\partial^k T_w}{\partial x^k}, \quad \frac{T_w - T_\infty}{T_0 - T_\infty} = \frac{1}{1+K}, \tag{11.19}$$

where $K = g_k \, \text{Pr}^{1/2}\sqrt{\pi/b}$ and $b = c_p\rho\lambda/(c_p\rho\lambda)_w = (\lambda/\lambda_w)\text{Lu}$. Thus, the temperature profiles in the fluid and in the solid are similar. The constancy of the surface temperature is a consequence of the proportional growth of the thermal boundary layer in the moving solid and in viscous flow entrained by it. The jump of the surface temperature at the point $x = 0$ is attributable to a general property of the equations of a parabolic type describing transport processes involving the propagation of the disturbances with infinite speed.

For the second part of the plate for $x > x_{cr}$, where the boundary layer on the inner surfaces of the solid interact, the conjugate problem of heat transfer between a moving continuous flat plate and surrounding is solved numerically by reduction to an equivalent conduction problem outlined in Section 4.4.3. Equation for the plate and combined boundary condition (4.44) are transformed to variables $\tilde{x} = x\alpha_w/\Delta^2 U_w$ and $\tilde{y} = y/\Delta$

$$\frac{\partial\theta}{\partial\tilde{x}} = \frac{\partial^2\theta}{\partial\tilde{y}^2}, \quad q_w = h_*\left[\theta_w(\tilde{x}) + g_1\tilde{x}\frac{d\theta_w}{d\tilde{x}} + \varepsilon(\tilde{x})\right],$$

$$\varepsilon(\tilde{x}) = \frac{q_w^{\text{int}}}{h_*} + \theta_w(0) - \theta_w(\tilde{x}) - g_1\tilde{x}\frac{d\theta_w}{d\tilde{x}} \tag{11.20}$$

where q_w^{int} is an integral expression for heat flux (4.41). This equation with boundary condition applied at $\tilde{y} = 1$ and $\tilde{y} = 0$ is solved numerically using implicit differential scheme, inscribed on a four-point stencil. In the first iteration, the correction $\varepsilon(\tilde{x})$ is taken as zero. After the first iteration, this correction is calculated and computation is repeated. This process is continued until the wall temperature is practically the same in two successive iterations. The solution is obtained in the third iteration. More details are given in [17].

The calculations are performed for symmetrical flow around the plate and different values of ratio $\phi = U_\infty/U_w$, where U_∞ is the free stream velocity and $b = c_p\rho\lambda/(c_p\rho\lambda)_w = (\lambda/\lambda_w)\text{Lu}$. The results are plotted in Figure 11.1.

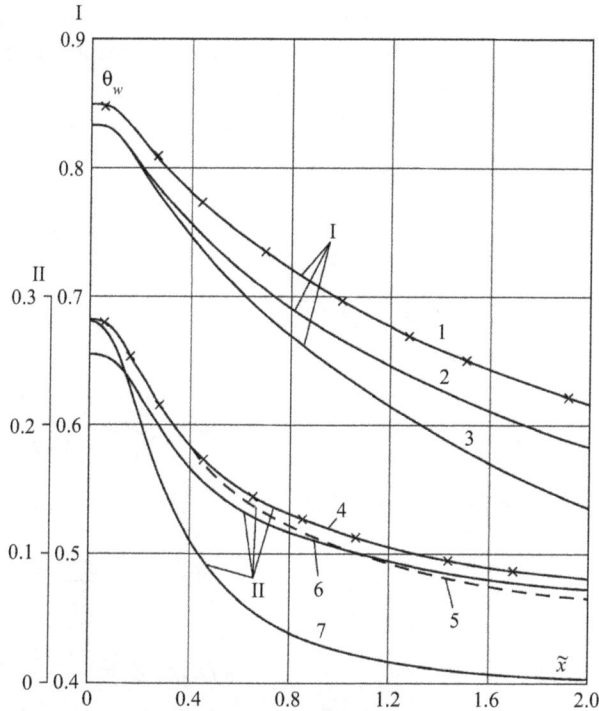

Figure 11.1. Temperature of the plate surface in symmetrical flow: I) $b = 0.042$, $\mathrm{Pr} = 5.5$; 1) $\phi = 0$. 2) $\phi = 0.8$, 3) $\phi = 0$, $g_1 = 0$, $\varepsilon = 0$, II) $b = 8.51$, $\mathrm{Pr} = 6.1$; 4) $\phi = 0$, 5) $\phi = 0$, $\varepsilon = 0$, 6) $\phi = 0.8$, 7) $\phi = 0$, $\varepsilon = 0$, $g_1 = 0$.

It is evident from the second equation (11.20) that for $\varepsilon = 0$ and $g_1 = 0$, the usual conduction problem with boundary condition of the third kind is obtained. The result of its solution is represented by curve 7 in Figure 11.1. The dashed curve in this figure corresponds to the first iteration when only the first derivative is taken into account and $\varepsilon = 0$. The crosses on the curves given the temperature distribution indicate the experimental data from [18]. It is seen that the result obtained after correction (curves 1 and 4) well agrees with experimental data, while the significant deviation of the temperature distribution calculated using the boundary condition of the third kind implies that the problem of heat transfer between a moving continuous plate and the surrounding medium must be solved strictly as a conjugate problem.

The same problem is solved in [18] numerically.

11.3 Draing Technology

Example 11.8 *n A brick drying [19].

A water-saturated brick of aspect ratio 2 is placed in air flow with the same initial temperature, and the concentration difference between the brick surface and ambient causes the evaporation of water from a brick. The reduction of brick temperature caused by evaporation is continued until brick reaches the wet bulb temperature at which it remains for a long period of time. Finally, the air flow heats the brick.

To study this drying process, the two-dimensional Navier–Stokes equations, including buoyancy terms, are solved for the flow domain because the boundary layer approximation is not applicable for a short brick. This solution is conjugated with solutions of energy and moisture transport equations. The governing system includes continuity, momentum, energy, and concentration equations for flow:

$$\frac{\partial u}{\partial x}+\frac{\partial v}{\partial y}=0 \quad \frac{\partial u}{\partial t}+u\frac{\partial u}{\partial x}+v\frac{\partial u}{\partial y}=-\frac{1}{\rho}\frac{\partial p}{\partial x}+\frac{\mu}{\rho}\left[\frac{\partial^2 u}{\partial x^2}+\frac{\partial^2 u}{\partial y^2}\right]$$

$$\frac{\partial v}{\partial t}+u\frac{\partial v}{\partial x}+v\frac{\partial v}{\partial y}=-\frac{1}{\rho}\frac{\partial p}{\partial y}+\frac{\mu}{\rho}\left[\frac{\partial^2 v}{\partial x^2}+\frac{\partial^2 v}{\partial y^2}\right]+g\beta\left(T-T_\infty\right)+g\beta\left(C-C_\infty\right) \qquad (11.21)$$

$$\frac{\partial T}{\partial t}+u\frac{\partial T}{\partial x}+v\frac{\partial T}{\partial y}=\frac{1}{\rho c_p}\left[\frac{\partial^2 T}{\partial x^2}+\frac{\partial^2 T}{\partial y^2}\right], \quad \frac{\partial C}{\partial t}+u\frac{\partial C}{\partial x}+v\frac{\partial C}{\partial y}=D\left[\frac{\partial^2 C}{\partial x^2}+\frac{\partial^2 C}{\partial y^2}\right],$$

and energy and conservation for total (liquid and vapor) moisture equations:

$$\rho_0\frac{\partial}{\partial t}\left(\Sigma M_j I_j\right)=\frac{\partial}{\partial x}\left(\lambda\frac{\partial T}{\partial x}\right)+\frac{\partial}{\partial y}\left(\lambda\frac{\partial T}{\partial y}\right)-\frac{\partial}{\partial x}\left(\Sigma J_j I_j\right)-\frac{\partial}{\partial y}\left(\Sigma J_j I_j\right) \qquad (11.22)$$

$$\frac{\partial M}{\partial t}=\frac{\partial}{\partial x}\left[\left(D_{M1}+D_{M2}\right)\frac{\partial M}{\partial x}+\left(D_{T1}+D_{T2}\right)\frac{\partial T}{\partial x}\right]+\frac{\partial}{\partial y}\left[\left(D_{M1}+D_{M2}\right)\frac{\partial M}{\partial y}+\left(D_{T1}+D_{T2}\right)\frac{\partial T}{\partial y}\right]$$

Here, D_T and D_M are isothermal and nonisothermal diffusion coefficients, I and J are enthalpy and mass flux, and j is 0, 1, 2 representing dry solid, liquid and vapor, respectively. The last equation (11.22) is obtained applying Darcy's law for capillary liquid mass flow and Fick's law for diffusive mass flux.

The boundary conditions on the interface consist of no slip of velocities condition, continuity of temperature and moisture conditions, and heat and species balance (f refers to the flow domain). At the inlet and far away from solid, the air flow has the ambient parameters (index a):

$$u=v=0, T_w=T_f, C_f=C\left(T,M\right)_w, \quad \left[\lambda+\left(I_2-I_1\right)D_{T2}\right]\frac{\partial T}{\partial n}+\left(I_2-I_1\right)D_{M2}\frac{\partial M}{\partial n}=\lambda_f\frac{\partial T_f}{\partial n}$$

$$+\left(I_2-I_1\right)D\frac{\partial C_f}{\partial n}, \quad D_{T2}\frac{\partial T}{\partial n}+D_{M2}\frac{\partial M}{\partial n}=D\frac{\partial C_f}{\partial n}, \quad u=U, v=0, T=T_a, C=C_a \qquad (11.23)$$

At outflow, a smooth extrapolation is used, and the bottom is adiabatic.

The finite element Galerkin's method (Sec. 8.2) is applied to solve the set of governing equations. First, equations (11.22) are simplified using some assumptions [19]. Since the characteristic time of the flow field is much smaller than that of the heat and moisture transfer, the quasi-steady state approximation is applicable (Sec. 4.4.4). Due to this, solutions of energy and concentration equations (11.21) are obtained using steady-state velocities. To calculate fluxes at the interface from equations (11.23), the concentration of the fluid at the interface is needed. It is determined knowing local relative humidity from the equation of desorption isotherm [20] $\phi=1-\exp\left(-17M^{0.6}\right)$.

Comment 11.4 Desorption (opposite of sorption) is a process of a substance releasing from or through a surface. Desorption (sorption) isotherm gives in a graphical or analytical form the equilibrium relation between water content and relative humidity of the material at a constant temperature.

The fluxes on the interface are used as boundary conditions for solving equations (11.22) for the brick. With known temperature and concentration distribution on the surface of the solid, the equations (11.21) for air are solved for the next time step. This calculation procedure of drying is continued until final parameters of the brick are achieved. The full drying time of 36 h takes 252 h calculation time.

The following basic results and conclusions are obtained:

- Different examples are computed to compare with available literature data. In particular, the same problem as shown in [21] (next Example) was simulated. The comparison shows that the differences in the temperature distributions between results obtained by Navier–Stokes and by boundary layer equations (as in [21]) are small, while the differences in moisture distributions are more considerable. These differences are seen to increase away from the leading edge. The studying of conjugation effects shows that the use of constant heat and mass transfer coefficients overpredicts the moisture removal.
- The conjugate drying result for mixed convection indicates that in the initial period, the surface temperature decreases since the heat for vaporization is absorbed from the brick. In time, the convective heat flux from ambient goes into solid. However, the amount of this heat is still not enough for vaporization and the brick temperature decreases finally reaching the uniform wet bulb temperature.

Comment 11.5 The wet bulb temperature is determined by wet bulb thermometer showing the lowest temperature that may be reached by evaporation of water.

The distribution of the moisture content along the brick surface shows that the evaporation is higher at the leading edge than that in other regions. This occurs due to a thin concentration boundary layer developed at the leading edge. With time, the rate of evaporation decreases but finally, when the solid attains uniform temperature, the significant moisture gradients still exist due to the smaller mass transfer diffusion coefficient.

- The result for mixed drying is compared with that for forced drying obtained by ignoring the buoyancy terms in equations (11.21). The comparison shows that the temperature drop in the case of a mixed regime is more than that for the only forced convection. As a result, the wet bulb temperature is reached much earlier in the case of mixed convection. The same qualitatively result is observed for moisture content.
- Variations of average Nusselt and Sherwood numbers in time are studied. Two definitions of Nusselt and Sherwood numbers are considered: using instantaneous temperature and concentration differences between the ambient and the surface of the brick and the similar differences between the ambient and wet bulb conditions. It is shown that the Nusselt and Sherwood numbers of the second type asymptotically approach the values of Nusselt and Sherwood numbers of the first type. The values of the Nusselt number increases with time until the wet bulb conditions are reached. The Sherwood number is constant for most of drying period or decreases due to diminishing moisture potential. It is observed that the analogy between heat and mass transfer is applicable only after wet bulb conditions are reached.

Example 11.9n Drying a rectangular wood board [21].

The problem is simplified using several assumptions: (i) there are equilibrium at each point and time; (ii) the board is unsaturated and homogeneous; (iii) gravity effects are negligible; (iv) the characteristic length scales of drying medium are much smaller than those of the external fluid; (v) the thickness of the interfacial surface is negligible.

The governing system includes continuity, momentum, energy, and conservation of species boundary layer equations and boundary and conjugate conditions:

$$\nabla \cdot \mathbf{u} = 0, \quad \mathbf{u} \cdot \nabla \mathbf{u} = -\nabla p + \frac{1}{\mathrm{Re}} \nabla^2 \mathbf{u}, \quad \mathbf{u} \cdot \nabla T = \frac{1}{\mathrm{Re\,Pr}} \nabla^2 T, \quad u \cdot \nabla \omega = \frac{1}{\mathrm{Re\,Sc}} \nabla^2 \omega$$

$$\lambda_a \nabla T \cdot \mathbf{n} + \Lambda \rho_a D \nabla \omega \cdot \mathbf{n} = \left(\lambda_a + \varepsilon \Lambda \alpha_m \gamma \right) \nabla T \cdot \mathbf{n} + \varepsilon \Lambda \alpha_m \nabla M \cdot \mathbf{n} \qquad (11.24)$$

$$T = T_w, \quad \omega = \phi\left(T_w, M_w\right), \qquad \rho_a D \nabla \omega \cdot \mathbf{n} = \alpha_m \gamma \nabla T \cdot \mathbf{n} + \alpha_m \nabla M \cdot \mathbf{n}$$

where ω is the water vapor mass fraction and γ is the relative thermal diffusion coefficient. All variables are dimensionless scaled by $L, u_\infty, \rho u_\infty^2, T_\infty, L/u_\infty$, and ω_∞ for the coordinates, velocity, pressure, temperature, time and water vapor mass fraction, and subscripts a and m denote air and moisture, respectively. Function $\phi\left(T_w, M_w\right)$ relates absolute humidity of the air to equilibrium values of temperature and moisture content on the solid (Comment 11.4), which in this steady is [22]

$$\phi = 1 - \exp\left(AM^B\right), \quad A = -\frac{0.13}{T}\left(1 - \frac{T}{T_c}\right)^{-6.46},$$

$$B = 110 T^{-0.75}, \quad \omega = 0.62198 \frac{\phi p_w}{p - \phi p_w} \qquad (11.25)$$

where $T_c = 647;14$ K is the critical temperature of water. The last relation gives the vapor mass fraction, where p is the total pressure of the humid air and p_w is the saturation pressure. The finite-element method is used to solve the set of governing equations. Two examples of streamlined suspended rectangular board are considered. Initial conditions of the stream of air and solid are: $T_\infty = 60°C$, $\omega_\infty = 0.116$ kg/kg, Re = 200, Sc = 0.6 and $T_0 = 25°C$, $M_0 = 222°M(m.c.\,40\%)$, respectively.

- In the first example, the drying of only upper surface of the board is studied. Both vertical surfaces are adiabatic and impermeable. The solid temperature and moisture potential distributions at 3 h time drying show that both the heating and the drying of the board are nonuniform. The leading part of the board heats much faster than the rest of it. As the air flows along the surface, the temperature gradient between the flow and solid decreases, and due to this, the heat fluxes decrease also in the flow direction. For the longer drying time due to evaporation, the heat fluxes increase in the flow direction. A dry zone appears close to the leading edge, while the remainder remains still wet. As air flows, it becomes more saturated, and both the humidity gradients and moisture fluxes of the surface decrease in the flow direction.
- In the second example, drying of the entire board under the same conditions is investigated. In this case, the results are close to those outlined for the first example. Only the temperature and moisture potential distribution are different due to the fact that the

vertical surfaces are also involved in the process. Both heat and mass transfer are less intensive on the vertical surfaces because the air velocities at the leading and aft edges are significantly reduced. The drying front near the aft edge is less penetrated than that in the area close to the leading edge. The average moisture potential variations show that board drying in this example occurs faster than that in the first case. Despite the difference in final moisture content is only about 1%, the difference in moisture distribution is considerable, which can be important for the using the wood board.

- These results are obtained under assumption of a homogeneous wood board and are pertinent only to this case. For longer drying times, the observed difference between the final of average moisture content in two cases is expected to increase.

Example 11.10n Drying of porous materials [23].

The external air flow is flowing parallel to porous material, which is assumed to be unsaturated, homogeneous, and nondeformable. This relatively simple model is used to simulate various situations studying the behavior of heat and mass transfer coefficients at the interface during the drying process.

The system of governing equations and boundary conditions are almost the same as in the previous example. The system for solid in both studies is based on the Luikov theory [24]. The finite element method is used to calculate the heat and mass transfer coefficients. The first (the moisture content is above the maximum sorption value) and the second periods are considered.

The main results are as follows:

- For the first period, the initial moisture content is assumed to be 8%. The results show the thermal and mass leading edge effect, which leads to high heat and mass fluxes at the interface. Intensive evaporation and mass transfer toward the interface occur. For the second period, the initial moisture content is assumed to be 3%. The obtained temperature and moisture fields show a dry or sorption zone linked to recession of an evaporation front. In the dry zone, the basic moisture transport occurs by vapor. In the sorption zone, the bound water exists. The leading edge effect is also seen in this case also.

- The heat and mass transfer analogy is observed in the first drying period. This result can be expected since the specific humidity at the interface depends only on the temperature. However, depending on situation, this analogy may or may not be valid. In the second period, the specific humidity on the interface depends on the temperature and moisture content. Therefore, the boundary conditions at the interface for heat and mass transfer may be different. Thus, in this case, the analogy between heat and mass transfer may not be valid.

- Since the temperature and the specific humidity at the interface are not uniform, the obtained heat and mass transfer coefficients differ from standard values corresponding to the case of plate with constant temperature and moisture content. However, the variations from the reference values in this study are only about 10%. This result is in conflict with huge variations from standard values obtained by many authors using one-dimensional simulation. It is shown by comparing two- and one-dimensional solutions of the same problem that the one-dimensional approach cannot generally give realistic results because the one-dimensional simulation ignores the thermal and mass boundary layer leading edge effects.

- The length scales of external flow are usually large compared to microscopic scales of the porous medium. Due to this fact, the porous medium is assumed to be homogeneous.

To estimate the effect of macroscopic heterogeneities of moisture content on the mass transfer coefficient, the moisture content distribution obtained in the problem in question is used. The same problem is solved applying this nonuniform moisture content distribution and mass transfer coefficient. The average mass flux at the interface obtained by using this mass transfer coefficient and nonuniform moisture distribution differs from standard one of about 10%.

Example 11.11a/n Drying of pulled continuous materials in the initial period [25].

The model schematically is the same as that in Section 4.2.7 and Example 11.7 (Fig. 4.14). The thermophysical properties of the heat transfer agent (air, subscript 0), vapor (1), liquid (2), and dry material (3) are known as well as the relative concentration of vapor in air $\rho_{10,\infty}$, and the initial temperature $T(0)$ and moisture content $M(0)$ of the material. The problem is governed in the flow domain by boundary layer equations for velocity, temperature, and vapor concentration and corresponding boundary conditions:

$$\frac{\partial u}{\partial x}+\frac{\partial v}{\partial y}=0 \quad u\frac{\partial u}{\partial x}+v\frac{\partial u}{\partial y}=v\frac{\partial^2 u}{\partial y^2} \quad u\frac{\partial T}{\partial x}+v\frac{\partial T}{\partial y}=\alpha\frac{\partial^2 T}{\partial y^2}$$

$$u\frac{\partial \rho_{10}}{\partial x}+v\frac{\partial \rho_{10}}{\partial y}=D\frac{\partial^2 \rho_{10}}{\partial y^2} \tag{11.26}$$

$$y=\Delta/2, u=U_x, v=U_y, \rho_{10}=\rho_{10,w}^+(x), T=T_w^+(x), y\to\infty, u=v\to 0, \rho_{10}\to\rho_{10,\infty}, T\to T_\infty$$

Here, D is a vapor diffusion coefficient, Δ is the thickness of material that is pulled with the speed U_x, and the superscripts plus and minus denote the values calculated at the interface from the agent side ($y>\Delta/2$) and from the material side ($y<\Delta/2$).

The heat and mass transfer in the capillary-porous body is defined by Luikov equations and the initial and symmetry conditions [24]:

$$c_M\rho_3\frac{\partial T}{\partial t}=\frac{\partial}{\partial x}\left(\lambda_M\frac{\partial T}{\partial x}\right)+\frac{\partial}{\partial y}\left(\lambda_M\frac{\partial T}{\partial y}\right)+\rho_3\varepsilon\Lambda\frac{\partial M}{\partial t}, U_y=-\frac{D}{1-\rho_{10,w}}\left.\frac{\partial \rho_{10}}{\partial y}\right|_w \tag{11.27}$$

$$\rho_3\frac{\partial M}{\partial t}=\frac{\partial}{\partial x}\left[\rho_3\alpha_M\left(\frac{\partial M}{\partial x}+\gamma\frac{\partial T}{\partial x}\right)\right]+\frac{\partial}{\partial y}\left[\rho_3\alpha_M\left(\frac{\partial M}{\partial y}+\gamma\frac{\partial T}{\partial y}\right)\right]$$

$$x=0, T=T(0), M=M(0), \quad y=0, \partial T/\partial y=0, \partial M/\partial y=0,$$

where subscript M refers to moist material, γ is the relative thermal diffusion coefficient, ε is a phase change coefficient defining the part of a vapor in the moisture content, Λ is a heat of evaporation, and U_y is the transverse velocity on the surface.

This system of equations is simplified and transformed using the following:

(i) the terms in (11.27) determining the heat and mass transfer in the x direction are relatively small and may be neglected;

(ii) four conjugate conditions should be satisfied:

$$T_w^+(x)=T_w^-(x), \quad \rho_{1,w}^+(x)=\rho_{1,w}^-(x), \quad I_w^+(x)=I_w^-(x),$$

$$-q_w^+ +q_w^- =(1-\varepsilon_w)I_w^-\Lambda \tag{11.28}$$

Three of those are the equalities of temperatures, vapor densities, and mass fluxes $I(x)$ from both material sides. The fourth is the balance of heat on the material surface: the difference between heats incoming from coolant and absorbing by material is utilized for evaporation;

(iii) the vapor density at the material surface from coolant side $\rho_{1,w}^{+}$ is defined through vapor concentration $\rho_{10,w}$, considering the vapor and air as ideal gases, while the vapor density from the body side $\rho_{1,w}^{-}$ is determined by the desorption isotherm equation $\rho_{1,w}^{-}/\rho_{st}(T_w) = \phi(T_w, M_w)$ knowing the saturated vapor density $\rho_{st}(T_w)$:

$$\rho_{1,w}^{+} = \frac{p_{\infty}}{R_0 T_0 \left[\left(1/\rho_{10,w}^{+} \right) - 1 + R_1/R_0 \right]},$$

$$\frac{1}{\phi(T_w, M_w)} = \frac{p_{st}(T_w)}{p_{\infty}} \left[1 + \frac{R_0}{R_1} \left(\frac{1}{\rho_{10,w}} - 1 \right) \right] \tag{11.29}$$

(iv) The system (11.26) is solved using the universal function in differential (4.1) and integral (4.14) forms for energy equation and similar relations for diffusion equation:

$$q_w = h_* \left[T_w - T_\infty + \sum_{k=1}^{\infty} g_k x^k \frac{d^k T_w}{dx^k} \right]$$

$$= h_* \left\{ T_w(0) - T_\infty + \int_0^x \left[1 - \left(\frac{\xi}{x} \right)^{C_1} \right]^{-C_2} \frac{dT_w}{d\xi} d\xi \right\} \tag{11.30}$$

$$j_w = h_{m*} \left[\rho_{10,w} - \rho_{10,\infty} + \sum_{k=1}^{\infty} g_k x^k \frac{d^k \rho_{10,w}}{dx^k} \right]_w$$

$$= h_{m*} \left\{ \rho_{10,w}(0) - \rho_\infty + \int_0^x \left[1 - \left(\frac{\xi}{x} \right)^{C_1} \right]^{-C_2} \frac{d\rho_{10,w}}{d\xi} d\xi \right\}$$

Here, $h_* = g_0(\mathrm{Pr})\lambda\sqrt{U_x/\nu x}, h_{m*} = (h_*/c_p)\sqrt{\mathrm{Le}}$ with the Lewis number for air $\mathrm{Le} \approx 1$, the values of g_k and C are given in Section 4.2.7, depending on the Prandtl number; for air, these are: $g_0 = 0.351$, $g_1 = 1.35$, $g_2 = -0.18$, $g_3 = 0.03$, $C_1 = 1.2$, and $C_2 = 0.57$; in the second equation (11.30), the coefficients and exponents depend on the Schmidt number, Sc; for air, $Sc \approx \mathrm{Pr}$, and hence the same values of g_k and C may be used in this case; (v) for the initial period when the moisture content of the material exceeds the maximum sorptive moisture content $M_{m,s}$, the partial pressure of the vapor above the surface equals the saturation pressure; in view of this $\varphi = 1$; in addition when $M > M_{m,s}$, the phase change coefficient $\varepsilon = 0$.

Applying the listed considerations, one gets the transformed starting system of equations and boundary conditions (11.26) and (11.27) in dimensionless variables:

$$\frac{c_M}{c_3} \frac{\partial \theta}{\partial Fo} = \frac{\partial}{\partial \eta} \left(\frac{\lambda_M}{\lambda_3} \frac{\partial \theta}{\partial \eta} \right), \quad \frac{\partial \vartheta}{\partial Fo} = \frac{\partial}{\partial \eta} \left[\frac{\alpha_M}{\alpha_3} \left(\frac{\partial \vartheta}{\partial \eta} + \gamma \frac{T(0) - T_\infty}{M(0) - M_{m,s}} \frac{\partial \theta}{\partial \eta} \right) \right] \tag{11.31}$$

$$\theta(0)=1, \ \ \vartheta(0)=1, \ \ \frac{\partial\theta}{\partial\eta}=0, \ \ \frac{\partial\vartheta}{\partial\eta}=0$$

$$\text{Fo}=\frac{x\lambda_3}{c_3\rho_3 U_x\Delta^2}=\frac{t\alpha_3}{\Delta^2}, \eta=\frac{y}{\Delta}, \theta=\frac{T-T_\infty}{T(0)-T_\infty}, \vartheta=\frac{M-M_{m,s}}{M(0)-M_{m,s}}.$$

The solution of this system should also satisfy the conjugate condition (11.28), two last of which after transforming to dimensionless variables (11.31) become the form

$$-\frac{\partial\theta}{\partial\eta}\bigg|_w=\frac{g_0(\text{Pr})\lambda_3}{\lambda_M}\left(\frac{\lambda\text{Lu}}{\lambda_3\,\text{Pr Fo}}\right)^{1/2}\left\{1+\int_0^{\text{Fo}}\left[1-\left(\frac{\hat{\text{Fo}}}{\text{Fo}}\right)^{C_1}\right]^{-C_2}\frac{d\theta_w}{d\hat{\text{Fo}}}\,d\hat{\text{Fo}}\right\}+\Lambda\Big\{\big(1-\rho_{10,w}\big)$$

$$\times\Big[c_{p1}\rho_{10,w}+c_{p0}\big(1-\rho_{10,w}\big)\Big]\Big\}^{-1},$$

$$\left\{\frac{\rho_{10,w}(0)-\rho_{10,\infty}}{T(0)-T_\infty}+\int_0^{\text{Fo}}\left[1-\left(\frac{\hat{\text{Fo}}}{\text{Fo}}\right)^{C_1}\right]^{-C_2}\frac{d\rho_{10,w}}{dT_w}\frac{d\theta_w}{d\hat{\text{Fo}}}\,d\hat{\text{Fo}}\right\} \tag{11.32}$$

$$\frac{\partial\vartheta}{\partial\eta}\bigg|_w=\frac{g_0(\text{Pr})c_3}{(\alpha_M/\alpha_3)\big(1-\rho_{10,w}\big)\Big[c_{p1}\rho_{10,w}+c_{p0}\big(1-\rho_{10,w}\big)\Big]}\left(\frac{\lambda\text{Lu}}{\lambda_3\,\text{Pr Fo}}\right)^{1/2}\left\{\frac{\rho_{10,w}(0)-\rho_{10,\infty}}{M(0)-M_{m,s}}\right.$$

$$\left.+\frac{T(0)-T_\infty}{M(0)-M_{m,s}}\int_0^{\text{Fo}}\left[1-\left(\frac{\hat{\text{Fo}}}{\text{Fo}}\right)^{C_1}\right]^{-C_2}\frac{d\rho_{10,w}}{dT_w}\frac{d\theta_w}{d\hat{\text{Fo}}}\,d\hat{\text{Fo}}\right\}+\gamma\frac{T(0)-T_\infty}{M(0)-M_{m,s}}\frac{\partial\theta}{\partial\eta}\bigg|_w \tag{11.33}$$

Here, physical properties of moist air are calculated additively $c_{p1}\rho_{10,w}+c_{p0}(1-\rho_{10,w})$, $\text{Lu}=c_p\rho/(c\rho)_3$ is the Luikov number (Sec. 4.4.4), and the relative vapor concentration is determined by the second equation (11.29) that for an initial period when $\phi=1$ becomes $\rho_{10,w}=\left\{1+(R_1/R_0)\left[(p_\infty/p_{st})-1\right]\right\}^{-1}$.

Analyzing system (11.31)–(11.33) enables some general conclusions that are valid for the first period of drying:

- The considered conjugate problem for the fixed material and heat-transfer agent is governed by four parameters $\rho_{10,\infty}$, T_∞, $T(0)$ and $M(0)$.
- When the dependence of the material properties on temperature and moisture content is neglected and calculations are confined to the mean values of thermophysical coefficients, the number of parameters governed the problem at the fixed material and heat-transfer agent is reduced to two

$$\frac{T(0)-T_\infty}{M(0)-M_{m,s}}, \frac{\rho_{10,w}-\rho_{10,\infty}}{M(0)-M_{m,s}}. \tag{11.34}$$

- The body thickness is incorporated only in the Fourier number. Therefore, the time needed for the material to reach the definite state characterized by certain values of temperature and moisture content and the drying duration are proportional to material thickness squared.

- The velocity at which the material is pulled through the coolant does not affect the drying rate and time. It only determines the distance from the die over which the material reaches a certain state. The explanation is that the heat and mass fluxes in the laminar flow regime are inversely proportional to the ratio x/U_x, that is, to the time of the material pulling as it follows from equations (11.30) containing coefficients h_* and $h_{m,*}$. It can readily be demonstrated that in the turbulent flow regime, the drying duration depends on the pulling velocity, since coefficients h_* and $h_{m,*}$ are proportional to $U_x^{0.6}$.

Equations (11.31) under initial and boundary conditions and conjugate conditions (11.32) and (11.33) are solved numerically by differential technique using the tridiagonal matrix algorithm (Comment 10.1) and implicit difference scheme. The first equation (11.31) is solved with the first condition (11.32), and thereafter, the second equation (11.31) with the second condition (11.32). To calculate the coefficients that depend on the functions sought, iterations are carried out.

Calculations are performed for the following conditions: $T_\infty = 90°C, \rho_{10,\infty} = 0.125$ and $M(0) = 0.25$. Two cases are considered: with the initial temperature of the material equal to $T(0) = 70$ and $50°C$. The first of them is higher, and the second is lower than the dew point temperature corresponding to the assigned relative vapor concentration in the heat-transfer agent $\rho_{10,\infty} = 0.125$. Therefore, in the first case, drying proceeds from the very beginning, whereas in the second case the material is first moistened, and drying begins after sometime interval.

The calculations are made for the thermophysical properties characteristics of the paper-type material: $\rho_3 = 800 \, kg/m^3$, $c_3 = 1500 \, J/kg \, K$, $c_M = c_3 + c_{H_2O}M$, the maximum sorptive moisture content at the dew point temperature is $M_{m,s} = 0.19$, and the thermal conductivity of the moist material is taken to be constant $\lambda_M = 0.4 \, W/m \, K$. The conjugate parameter and dimensionless coefficient of moisture diffusion calculated using these value are: $c_p \rho \lambda /(c\rho)_3 \lambda_M = (\lambda/\lambda_M)$ Lu $= 6 \cdot 10^{-5}$ and $\alpha_M (c\rho)_3/\lambda_M = 0.125$. Accordingly with the foregoing, the thickness and the pulling velocity are not specified.

Prediction allows one to draw the following conclusions:

- The temperature and moisture content varies little across the material, and their values on the surface do not actually differ from those mean integrals over the thickness.
- The rate of heat and mass transfer predicted with conjugation effects is lower than that resulting from the calculation with the third kind of boundary conditions and the heat and mass transfer coefficients for constant temperature and concentration heads. In the case of drying, the material moisture content is higher and in the case of moistening lower than the corresponding values predicted by the third kind of boundary conditions.

The reason of this is that in general when the head grows, either in the direction of heat transfer agent flow or in time, the heat and mass transfer coefficients appear to be higher, while in the reverse case, they turn out to be lower than the corresponding coefficients obtained for the constant heads (Chapter 5). In the considered drying and moistening processes, the concentration heads diminish. The temperature head increases in the drying process and decreases during the moistening process. In conformity with this, the mass transfer coefficients in both cases are smaller than $h_{m*}(\chi_p < 1)$. The heat transfer coefficient is smaller than h_* for the moistening ($\chi_t < 1$) and larger than h_* for the drying ($\chi_t > 1$). Despite in the considered drying process $h > h_*$, the resulting heat transfer rate decreases. This occurs due to the fact that the mass transfer coefficient decreases much more than the heat transfer coefficient increases.

Comment 11.6 Here $\chi_p = h_m / h_{m*}$ is the coefficient of nonisobaricity showing how much the mass transfer coefficient obtained in the conjugate problem differs from that at constant concentration (or partial vapor pressure), which is similar to nonisotermicity coefficient $\chi_t = h/h_*$ defined in Section 4.2.2, regarding the heat transfer.

- The observed decrease of heat transfer rate takes place in any convective drying or moistening process because those proceed, as well as the one in considered example, under conditions of falling concentration head. The degree of such decrease is determined in each case by the specific conditions, but the qualitative results will be the same.
- The analogy between heat and mass coefficients, frequently employed in predictions, is not observed. This is because for such an analogy, the coincidence is required not only of differential equations, but also of the appropriate boundary conditions determined by the distribution of the temperature and concentration heads along the surface or in time. Computing data show that these distributions substantially differ resulting in considerable different heat and mass transfer coefficients. While the heat transfer coefficients are close to the isothermal coefficients h_*, the mass transfer coefficients differ drastically from the isobaric one h_{m*}, especially in the moistening process when the mass transfer coefficient reduces to zero, resulting in the phenomenon similar to well-known heat flow inversion (Sec. 5.4). In this case, the mass flux reduces to zero much earlier than the concentration head does, which is in contrast to results of nonconjugated problem showing that both points coincide. In the conjugate problem, at a point, where the mass flux reduces to zero, the concentration head is finite and, therefore, the mass transfer coefficient becomes zero. After this point, the mass flux changes its sign, and the drying process begins even though the direction of the concentration head remains the same $\left(\rho_{10,w} < \rho_{10,\infty} \right)$. Thus, the mass transfer coefficient is negative in this section. Such a pattern remains up to the point at which the concentration head vanishes. Since at this point the mass flux is finite, the vanishing concentration head results in the mass transfer coefficient tending to infinity losing virtually the meaning.

Physically, the inversion phenomenon is explained by the inertia properties of the coolant, due to which the change in concentration near the wall is manifested much earlier in its immediate vicinity than far from the wall. More detailed explanation of the inverse phenomenon is given in Section 5.4.

11.4 Food Processing

Example 11.12n Freeze drying of slab-shaped food [26].

Unlike to the conventional drying process, which is based on capillary motion and evaporation of water, the freeze drying process used sublimation of ice to dry the object. The low temperature and pressure below the triple point in freeze drying provide high quality of dry product. Despite high quality of final product, the conventional drying methods are basically used in food production due to long drying time and high cost of the freeze drying process. There are many investigations of drying, but most of them studied the alternative conventional drying methods.

The model consists of planar and slab-shaped products that are divided by a sublimation interface into the dried and frozen parts. The moisture or ice in the products sublimate under vacuum pressure and developed vapor m_V diffuses through pores to exit. The energy for sublimation comes from the bottom by conduction and by radiation from the upper heating plate. As a result, the uniform sublimation interface below the top forms in the planar product. In the slab-shaped product, the additional radiation energy comes through the lateral surface opened to the drying chamber. Due to this, the formed below the top and beside the lateral surface sublimation interfaces are nonuniform. Such a drying process in the slab-shaped product in contract to that in the planar product requires multidimensional analysis.

The problem is governed by the mass and energy conservation equations, derivation of which is given in details in [26]:

$$\varepsilon \frac{(1-S)\rho_V -\left(1-S^0\right)\rho_V^0}{\Delta t}\Delta V + \sum_{j=E,W,N,S} (\mathbf{m}_V)_j \cdot \mathbf{n}_j = -\varepsilon\rho_I \frac{S-S^0}{\Delta t}\Delta V \qquad (11.35)$$

$$\left(\rho c_p\right)^0 \frac{T-T^0}{\Delta t}\Delta V + \sum_{j=E,W,N,S} (-\lambda\nabla T)_j \cdot \mathbf{n}_j = -\varepsilon\rho_I\Lambda_S \frac{S-S^0}{\Delta t}$$

Here, ΔV is the volume of grid cell, ε and ε_I denote the porosity and the fraction of the ice volume to the total volume, $S = \varepsilon_I/\varepsilon$ is the ice saturation, Λ_S is the heat of sublimation of ice, $\mathbf{n}_j$ is the outward normal vector, the subscript j denotes the control surface of a cell with East, West, North, and South faces, and superscript 0 refers to initial values. The vapor flux is determined by summing the diffusion and flow through porous products:

$$m_V = -(1-S)\frac{m_M}{RT}\left[D + p_V \frac{K}{\mu_V}\right]\nabla p_V, \qquad (11.36)$$

where m_M, D, and K are molecular mass, the effective diffusivity, and Darcy permeability, respectively.

The first term of the mass conservation in the first equation (11.35) represents the change in the vapor containing in a particular cell, and the second term determines the vapor flow out of this cell. These changes yield the reduction of the rate of ice saturation S defined by right term of the mass conservation equation. Similarly, the first and second terms of the energy conservation second equation (11.35) represent the change of energy inside the cell and heat flux through the control surface. The sum of these two terms is equal to the last term of energy equation, which defines the latent heat arises due to the sublimation of ice.

The temperature boundary conditions for the top, bottom, and lateral surface opened to the drying chamber are

$$q_T = \sigma F_T \left(T_H^4 - T_T^4\right) \quad q_B = h\left(T_H - T_B\right) \quad q_S = \sigma F_S \left(T_H^4 - T_S^4\right), \qquad (11.37)$$

where subscripts T, B, S and H denote top, bottom, side surfaces of product and heating plates, respectively, F is the radiation shape factor, and h is the overall heat transfer coefficient for the bottom. The pressure boundary condition for surfaces opened to drying chamber (subscript C) is $p_V = p_{VC}$. As an initial condition at $t = 0$, the uniform temperature, pressure, and rate of ice saturation are used.

The numerical procedure starts from computing temperature distribution by solving the second equation (11.35). Then, the distribution of vapor pressure is obtained from the first equation only for dried cells with $S = 0$. To calculate the pressure in the frozen and sublimation cells, those are treated as Dirichlet boundaries assuming the local thermodynamic equilibrium when the vapor is saturated, and its pressure is defined as $p_V = p_S(T)$. The evolution of ice saturation S in the frozen and sublimation cells is calculated by solving the mass conservation equation [the first equation (11.35)]. Iterative calculations are repeated until desired accuracy is obtained.

The numerical results are obtained using the beef as a product

- The average sublimation temperature of the slab-shaped product is 5–10% lower than that of the planar product. The reason of this is the curved sublimation interfaces in the slab-shaped product caused by literal surface opened to the drying chamber. As a result, the diffusion length is decreased, and the interfacial area is increased. These two effects enabled the shorter drying time with a lower sublimation temperature and the primary direction of drying changes from vertical from top to bottom in the planar product to radial in the slab-shaped product from the lateral surface to the inner core.
- The distribution of ice saturation, temperature, and vapor pressure in the planar product shows the existence of the second sublimation interface near the bottom. The vapor from the secondary interface is transported out of a product by diffusion or is deposited in the frozen region as ice. The maximum temperature and vapor pressure increase in time. The ice saturation in the slab-shaped product in the frozen region remains relatively constant about 0.7. The distribution of the temperature and vapor pressure in the slab-shaped product shows that the heat and mass transfer during the frozen drying in this case is a fully multidimensional process. Despite that the spatial temperature gradients in the dried area are larger than that in the frozen region, the heat transfer through this area is small due to small thermal conductivity. The spatial vapor pressure gradients in the frozen region caused by the temperature gradients according to dependence for saturated pressure $p_S(T)$ are also relatively small because of a small pore space in the frozen area.
- The main source of energy is the conduction from bottom, while the radiation from the top and lateral surface supplies also about 40% of total energy. Since a dried region is developed near the bottom, the heat flow from the bottom decreases from the initial 2.86 to the final 0.93 J/s. The initial vapor flow through the lateral surfaces is about two times larger than that through the top surface. At the end of the drying process, almost 80% of vapor flows through the lateral surface.
- The results obtained for product of heights 5, 10, 15, and 20 mm indicate the relatively constant drying rate for the planar product. The slab-shaped product shows more non-linear behavior. The high drying rate at the beginning changes to lower drying rate in the latter parts of the drying process. The reason of this is the insulating dried region surrounded the frozen area. For the height 5 mm, both products planar and slab-shaped exhibit almost the same drying time, but the difference increases with time because the

drying time of the slab-shaped product is less sensitive to the product height than that of the planar product. The configuration of the sublimation interfaces of the slab-shaped product with different heights indicates that the primary drying direction is the radial from the lateral surface to the inner core.

- The lateral surface of the slab-shaped product is favorable for the reduction of both the drying time and sublimation temperature by increasing the vapor diffusion and the interfacial area for sublimation.

Example 11.13n Food and polymer flow through extrusion dies [27].

A flow of non-Newtonian fluid with significant viscous dissipation and temperature-dependent viscosity is considered. The considered fluid flows through a circular pipe with a contraction angle of 45° Although such a problem simulates wide applications of food and polymer melts flowing through the extrusion dies, very few studies were performed especially in conjugate formulation. One of the reasons of this is the complication of such a problem, which results not only due to the above-mentioned features, but also because the involving materials are chemically reactive and the viscosity is a function of the temperature and moisture content.

The full system of governing equations in cylindrical two-dimensional coordinates with temperature-dependent viscosity is complicated:

$$\frac{\partial u}{\partial x}+\frac{1}{r}\frac{\partial rv}{\partial r}=0, \quad u\frac{\partial u}{\partial x}+v\frac{\partial u}{\partial r}=-\frac{\partial p}{\partial x}+\frac{1}{Re}\left\{\left[\frac{1}{r}\frac{\partial}{\partial r}\left(\mu r\frac{\partial u}{\partial r}\right)+\frac{\partial}{\partial x}\left(\mu\frac{\partial u}{\partial x}\right)\right]+\left[\frac{\partial\mu}{\partial r}\frac{\partial v}{\partial x}+\frac{\partial\mu}{\partial x}\frac{\partial u}{\partial x}\right]\right\}$$

$$u\frac{\partial v}{\partial x}+v\frac{\partial v}{\partial r}=-\frac{\partial p}{\partial r}+\frac{1}{Re}\left\{\left[\frac{1}{r}\frac{\partial}{\partial r}\left(\mu r\frac{\partial v}{\partial r}\right)+\frac{\partial}{\partial x}\left(\mu\frac{\partial v}{\partial x}\right)\right]+\left[\frac{\partial\mu}{\partial r}\frac{\partial v}{\partial r}+\frac{\partial\mu}{\partial x}\frac{\partial u}{\partial r}+\mu\frac{v}{r^2}\right]\right\} \quad (11.38)$$

$$u\frac{\partial\theta}{\partial x}+v\frac{\partial\theta}{\partial r}=+\frac{1}{Pe}\left[\frac{\partial}{\partial x}\left(\lambda\frac{\partial\theta}{\partial x}\right)+\frac{1}{r}\frac{\partial}{\partial r}\left(\lambda r\frac{\partial\theta}{\partial r}\right)\right]+\frac{Ec}{Re}\mu S$$

Variable are scaled by radius of the channel R for x and r, average velocity U for u and v, ρU^2 for p, and reference values μ_0 and λ_0 for μ and λ. The rheology power law (11.15) is used for polymers viscosity with strain rate $\dot{\gamma}=S^{1/2}$. For food materials, such as a corn meal and other reactive polymers, more complex law is used:

$$\mu=\mu_0\left(\frac{\dot{\gamma}}{\dot{\gamma}_0}\right)^{n-1}\exp\left[-m(T-T_0)\right]\exp\left(-m_cC\right) \text{ or}$$

$$\mu=0.49\dot{\gamma}^{-0.32}\exp\left(3969/T\right)\exp\left(-0.03C\right), \quad (11.39)$$

where C is the concentration of moisture on dry-weight in percent. The second expression is used for simulation of the flow of corn meal to compare the prediction with experimental data.

The following boundary conditions are employed: (i) at the inlet for flow, the fully developed velocity and uniform temperature profiles; (ii) at the outlet for flow: developed condition: $\partial^2 u/\partial x^2=\partial v/\partial x=\partial^2\theta/\partial x^2=0$; (iii) at inlet and outlet for walls: adiabatic condition; (iv) at the outer surface: isothermal conditions or convection from ambient with given heat transfer coefficient $Bi(\theta_0-\theta_a)=-(\partial\theta/\partial r)_0$, where the subscripts 0 and a refer to outer surface and ambient values. The solution method is similar to the SIMPLER, and zero velocity at walls is achieved by setting viscosity to a large value as 10^{20} (Sec. 8.4). The constricted part of the pipe is approximated using rectangular steps. Dicredization and other details may be found in authors' article [28].

The results are presented in four separated parts:

- *Unconstricted tubes.* The flow is studied in the case with specified temperature θ_0 at the outer surface. The balk temperature decreases along with decreasing θ_0. This occurs because the heat from the wall decreases and viscosity becomes higher, which results in more rapidly drop of pressure. The effect of outer surface temperature increases as the conductivity ratio λ_w/λ increases because the thermal resistance is smaller when the conductivity is greater. Thus, as λ_w/λ increases, the fluid and internal surface temperatures as well as the Nusselt number become larger. At high Peclet numbers, the convective heat transfer dominates, and as a result, the role of conductivity ratio decreases. Therefore, the effect of small polymer conductivity is relatively small. As expected, the data show that the temperatures inside the tube and Nusselt number are smaller for greater walls thicknesses.

- *Constricted tubes with specified temperature at the outer surface.* Due to restructured flow at the inlet to the constricted region, the largest variation of the fluid temperature is observed near the internal surface. The fluid temperature increases downstream so that at the exit the fluid temperature achieved the highest value. This occurs in both cases of cold and hot walls because here is the largest viscous dissipation. Although the pressure drops much faster in the constricted part, decreasing pressure in unconstricted part of the tube is also significant. There are peaks of heat fluxes at the inlet to the constricted part where the temperature gradients are large. Downstream at the exit, the fluid loses heat to the wall due to high temperature, which arises from the dissipation.

- *Constricted tubes with a specified Biot number at the outer surface.* Although the results show that for high (100) Biot numbers, the temperature gradients are greater than that for low one (10), in both cases the maximum fluid temperatures are observed at the exit close to the internal surface due to the effect of dissipation as it was indicated in situations considered above. The larger Biot number causes the larger heat loses from fluid, inlet temperature of which is higher than that of ambient. Due to that, the temperature of the internal surface is lower for the higher Bi. This leads to a higher fluid velocity at the centerline in comparison with the fluid velocity at the internal surface where the temperature is lower and hence the viscosity is larger. Despite that the fluid temperature is lower at the internal surface, the bulk temperature downstream at the centerline is higher when the Biot number is higher. The reason of this is the high effect of dissipation, which affects the flow far from walls much stronger than it does the outer boundary condition.

- *Food extrusion.* The calculation is made for actual food extruder as a long straight tube with die of three narrow steps. The food is the corn meal, ambient temperature is 20°C, and the heat transfer coefficient at the outer surface is 10 W/m²K as these quantities were in an experiment. The comparison of both data shows quite good agreement. Using results obtained for different values of parameters, the correlation is found between pressure drop in MPa and the mass flow rate $\dot{m}$ in kg/h, moisture content m_c in percent and temperature

$$\Delta p = KT^a m^b m_c^c, \; K = \exp(55.129), \; a = -8.723, \; b = 0.432, \; c = -1.0276 \qquad (11.40)$$

Detailed calculation performed for data, $m_c = 21\%$, $T_{in} = 382$ K, $\dot{m} = 51.8$ kg/h, shows an almost isothermal temperature field near the centerline and the largest temperature gradients near the boundary. Due to the high viscosity, no recirculation arises. The most pressure drop takes place at the narrow exit region. The heat transfer coefficient strongly affects the pressure and temperature distribution. As the mass flow rate increases, the required pressure drop grows, while the bulk temperature decreasing becomes smaller.

• The data presented in this study indicate that wall conduction significantly influences the flow and heat transfer, and hence the process in an extruder should be investigated using conjugate formulation of the problem.

REFERENCES

[1] Viswanath, R., & Jaluria, Y. (1995). Numerical study of conjugate transient solidification in an enclosed region. *Numerical Heat transfer Part A 27*, 519–536. doi: http://dx.doi.org/10.108 0/10407789508913716

[2] Viswanath, R., & Jaluria, Y. (1993). Comparison of different solution methodologies for melting and solidification problems in enclosures. *Numerical Heat transfer Part A 24*, 77–105. doi: http: //dx.doi.org/10.1080/10407799308955883

[3] Li, H., Hsieh, C. K., & Goswami, D. Y. (1996). Conjugate heat transfer analysis of fluid flow in a phase-change energy storage unit. *International Journal of Numerical Methods for Heat & Fluid Flow 6*, 77–90. doi: http://dx.doi.org/10.1108/EUM0000000004105

[4] Tenchev, R. T., Li, L. Y., & Purkiss, J. A. (2001). Finite element analysis of coupled heat and moisture transfer in concrete subjected to fire. *Numerical Heat Transfer Part A 39*, 685–710. doi: http://dx.doi.org/10.1080/10407780152032839

[5] Davie, C. T., Pearce, C. J., & Bicanic, N. (2006). Coupled heat and moisture transport in concrete at evaluated temperatures—effects of capillary pressure and adsorbed water. *Numerical Heat Transfer Part A 49*, 733–763. doi: http://dx.doi.org/10.1080/10407780500503854

[6] Nowak, A. J., Biaecki, R. A., Fic, A., Wecei, G., Wrobel, L.C., & Sarier, B. (2002). Coupling of conductive, convective and radiative heat transfer in Czochralski crystal growth process. *Computational Materials Science 25*, 570–576. doi: http://dx.doi.org/10.1016/S0927-0256(02)00336-1

[7] Nowak, A. J., Biaecki, R. A., Fic, A., & Wecei, G. (2003). Analysis of fluid and energy transport in Czochralski's process. *Computers & Fluids 32*, 85–95. doi: http://dx.doi.org/10.1016/S004 -7930(01)00101-3

[8] Wecel, G. (2006). BEM-FVM solution of the conjugate radiative and convective heat transfer problems. *Archives of Computational Methods in Engineering 13*, 171–248. doi: http://dx.doi.org /10.1007/BF02980230

[9] Trp, A. (2005). An experimental and numerical investigation of heat transfer during technical grade paraffin melting and solidification in a shell-and-tube latent thermal energy storage unit. *Solar Energy 79*, 648–660. doi: http://dx.doi.org/10.1016/j.solener.2005.03.006

[10] *Wan, Z. M., Liu, W., Tu, Z. K., & Nakayama, A. (2009). Conjúgate numéricas analysis of flow and heat transfer with phase change in a miniature flat plate CPL evaporator. *International Journal of Heat and Mass Transfer 52*, 422–430. doi: http://dx.doi.org/10.1016/j.ijheatmasstransfer.2008.06.019

[11] Choi, C. Y., & Hsieh, C. K. (1992). Solution of Stefan problems imposed with cyclic temperature and flux boundary conditions. *International Journal of Heat and Mass Transfer 35*, 1181–1195. doi: http://dx.doi.org/10.1016/0017-9310(92)90178-U

[12] Blasé, T. A., Guo, Z. X., Shi, Z., Long, Z. K., & Hopkins, W. G. (2004). A 3D conjugate heat transfer model for continuous wire casting. *Materials Science and Engineering: A 365*, 318–324. doi: http://dx.doi.org/10.1016/j.msea.2003.09.042

[13] Yoo, S. Y., & Jaluria, Y. (2007). Conjugate heat transfer in an optical fiber process. *Numerical Heat Transfer Part A 51*, 109–127. doi: http://dx.doi.org/10.1080/10407780600710342

[14] Yoo, S. Y., & Jaluria, Y. (2008). Numerical simulation of the meniscus in nonisothermal free surface flow at the exit of a coating die. *Numerical Heat Transfer Part A 53*, 111–131. doi: http://dx.doi.org/10.1080/10407780701454089

[15] Landau, H. G. (1950). Heat conduction in a melting solid. *Quarterly of Applied Mathematics 8*, 81–94.

[16] Sastrohartono, T., Jaluria, Y., & Karwe, M. V. (1994). Numerical coupling of multiple-region simulations to study transport in a twin-screw extruder. *Numerical Heat Transfer Part A 25*, 541–557. doi: http://dx.doi.org/10.1080/10407789408955965

[17] Dorfman, A. S., Grechannyy, O. A., & Novikov, V. G. (1981). Conjugate heat transfer problem for a moving continuous plate, in a fluid flow. *High Temperature 19*, 706–714. doi: http://dx.doi.org/10.1016/0017-9310(76)90158-7

[18] Chida, K., & Katto, Y. (1976). Conjugate heat transfer of continuously moving surfaces. *International Journal of Heat and Mass Transfer 19*, 461–470.

[19] Murugesan, K., Suresh, H. N., Seetharamu, K. N., Narayana, P. A. A., & Sundararajan, T. (2001). A theoretical model of brick drying as a conjugate problem. *International Journal of Heat and Mass Transfer 44*, 4075–4086. doi: http://dx.doi.org/10.1016/S0017-9310(01)00065-5

[20] Fortes. M., & Okos, M. R. (1981) Heat and mass transfer in hygroscopic capillary extruded products. *AIChE Journal 27*, 255–262. doi: http://dx.doi.org/10.1002/aic.690270212

[21] Oliveira, L. S., & Haghighi, K. (1998). Conjugate heat and mass transfer in convective drying of porous media. *Numerical Heat Transfer Part A 34*, 105–117. doi: http://dx.doi.org/10.1080/10407789808913980

[22] Zuriz, C., Singh, R. P., Moini, S. M., & Henderson, S. M. (1979). Desorption isotherms of rough rice from 10°C to 40°C. *Transactions of the ASME 22*, 433–440.

[23] Masmoudi, W., & Prat, M. (1991). Heat and mass transfer between a porous medium and parallel external flow. *International Journal of Heat and Mass Transfer 34*, 1975–1989. doi: http://dx.doi.org/10.1016/0017-9310(91)90209-W

[24] Luikov, A. V. (1966). *Heat and mass transfer in capillary porous bodies*. Oxford: Pergamon Press.

[25] Dolinskiy, A. A., Dorfman, A. S., & Davidenko, B. V. (1991). Conjugate heat and mass transfer in continuous processes of convective drying. *International Journal of Heat and Mass Transfer 34*, 2883–2889. doi: http://dx.doi.org/10.1016/0017-9310(91)90248-D

[26] Nam, J. H., & Song, C. S. (2007). Numerical simulation of conjugate heat and mass transfer during multi-dimensional freeze drying of slab-shaped food products. *International Journal of Heat and Mass Transfer 50*, 4891–4900. doi: http://dx.doi.org/10.1016/j.ijheatmasstransfer.2007.08.004

[27] Lin, P., & Jaluria, Y. (1997). Conjugate transport in polymer melt flow through extrusion dies. *Polymer Engineering & Science 37*, 1582–1596. doi: http://dx.doi.org/10.1002/pen.11806

[28] Lin, P., & Jaluria, Y. (1996). Numerical approach to model heat transfer in polymer melts flowing in constricted channels. *Numerical Heat Transfer 30*, 103–123. doi: http://dx.doi.org/10.1080/10407789608913831

FLUID FLOW AND HEAT TRANSFER IN BIOLOGY AND CLINICAL MEDICINE

12.1 Blood Flow in Normal and Pathologic Vessels

Example 12.1n Arterial stenoses modeling [1].

The arterial stenosis is a constriction of blood vessel cross-section that occurs due to formation plaques in an arterial wall, leading to restriction of blood flow and to separation of flow once the stenosis becomes large. The flow disturbed by stenosis leads to abnormal blood circulation, resulting in vascular disorders such as post-stenoic dilatation, losses in pressure, abnormally high shear stresses, that may provoke blood problems (red cells and platelets).

The model consists of a tube with constriction (stenosis) of the length $2D$ and the form $r(z)/D = 0.5 - A(1 + \cos \pi z/D)$, where A is a constant, r and z are coordinates with origin $z = 0$ at the throat of the stenosis, and D is the unobstructed tube diameter. The stenosis is located at $15\,D$ from the inflow and $20\,D$ prior the outflow sections. It is assumed that: (i) the blood is a homogeneous, incompressible Newtonian fluid with constant kinematic viscosity $v = 0.035$ cm^2/s and density $\rho = 1.06$ g/cm^3; (ii) the flow is steady laminar current with parabolic profile at the inflow that remains laminar proximal to the stenosis; (iii) the disturbed blood flow after stenosis is turbulent and may be simulated using a turbulence model for the moderate Reynolds number ranging from 400 (human carotid artery) to 1,500 (human ascending aorta).

The system describing disturbed flow includes the steady-state Navier–Stokes equation, and two transport equations (3.38) and (3.39) determined k and ω in the $k - \omega$ turbulence model considered in Section 3.4.3. The boundary conditions are as follows:

(i) at the wall, the usual no-slip conditions for the velocity and $k = 0$ and $\omega = 6\mu/\beta_\omega \Delta^2$ for $k - \omega$ model, where $\beta_\omega = 0.8333$ is a constant in equation (3.39) for ω and Δ is the height of the first node above the wall;

(ii) the symmetry conditions at the central axis;

(iii) at the channel inlet for laminar flow, very small values are taken $k = 0.0001$, $\omega = 0.45$, which corresponds to $\varepsilon = 0.000045$ [see equation (3.39)];

(iv) at the outlet, the normal and tangential stresses are zero, the flow is fully developed, and properties are no longer varied with distance along the channel so that $\partial k/\partial z = \partial \omega/\partial z = 0$.

The above-mentioned system of three equations, including boundary conditions was solved by the finite element software Fluid Dynamic Analysis Package (FIDAP). Three stenosis models with the lumen area reduction, 50% $(A = 0.073)$, 75% $(A = 0.125)$, and 86% $(A = 0.1565)$ (defined as $1-(r/R)^2$), are studied showing the following results.

- The stenosis leads to separation of flow and to formation of a zone with vortex; the vortex length increases almost linearly with the Reynolds number until it reaches the maximum value at critical Reynolds number, indicating that flow distal to the stenosis is laminar, and it becomes transitional or/and turbulent when Re is greater than critical value; according to calculation, the critical Reynolds numbers are: 1,100, 400, 230 for the first, second, and third models, respectively; comparison with data obtained by laminar flow modeling shows that in the laminar flow range, both results are in agreement, but for Reynolds numbers larger than the critical value, the laminar flow model overestimates the vortex length.
- The wall static pressure distribution obtained for the second model at Re = 2,000 well agrees with experimental data and corresponds to streamline patterns showing sharp pressure dropping in the stenosis throat; comparison with data given by $k - \varepsilon$ model at Re = 15,000 indicates that the $k - \varepsilon$ model predicts much higher pressure post stenosis.
- The wall shear stress distribution reveals that the highest value takes place at the throat and the lowest negative pressure appears at the reattachment point of the vortex. In the vortex region, the wall share stresses are negative, indicating that they are acting in opposite directions upstream and downstream of the reattachment point; in the case of laminar flow, the wall share stresses in the area downstream of the reattachment point approaches, but never exceeds the corresponding value for fully developed Poiseille flow. In contrast, when the flow is transient or turbulent in this area, the shear stresses are higher than the fully developed Poiseille flow value farther decreasing to the laminar flow rate, which shows that relaminarization takes place.
- The turbulence intensity along the central line are computed for the second model at Re = 2,000, and the results are compared with experimental measurements and data obtained by $k - \varepsilon$ model at Re = 15,000; the computed results as well as experimental data reveal that the turbulent intensity increases after the throat and reaches the peak at the vortex center in the flow separation zone, but the computed prediction underestimates the turbulence intensity giving only qualitatively correct results, while the $k - \varepsilon$ model prediction differs much more from the experimental data.
- The developed model is suitable for blood flow studies in the areas of the arterial tree where both laminar and transitional/turbulent flows coexist.

Example 12.2n Blood flow though series stenoses [2].

Multiple stenoses in a blood vessel may occur since the primary stenosis resulting in downstream circulation in time forms a secondary stenosis. This may result in a third one, and so on. This paper presents a detailed analysis on the flow dynamics with double bell-shaped stenoses at the relatively low, realistic Reynolds numbers ranging from 100 to 4,000. The dynamic characteristics including separation, reattachment, the formation of recirculation eddy, and the distribution of the kinetic energy are investigated for the cases of one and two stenoses at different distances between those.

The model consists of a tube in cylindrical coordinates z, r with two constrictions modeling the stenoses. The walls along the constriction have the bell-shaped profile,

$$f(z) = 1 - c_i \exp\left[-c_s (z - s_i)^2\right] \tag{12.1}$$

which in curvilinear coordinates, used in this study, transforms in the rectangular domain. The problem is governed by two-dimensional Reynolds averaged Navier–Stokes equation (RANS) with turbulent characteristics determined by $k - \omega$ turbulence model (Chapter 3). Both systems of equation for RANS and for the $k - \omega$ model are transformed in the curvilinear coordinates applying corresponding formulae given in the paper. Using curvilinear coordinates has some adventures; particularly, in this case, the equations are simpler at the walls, and therefore, it is easier to integrate those through the viscous sublayer without additional damping functions. The boundary conditions at the inflow are specified as $u = 1 - r, k = 1.5 I_{tb}^2 u^2, \omega = \sqrt{k}/C_\mu^{1/4} l, l = \min(\kappa y_w, 0.1 \text{Re})$. At the outflow, the velocities are extrapolated from interior and a constant static pressure is imposed. The gradients of k and ω at exit are assumed to be zero $\partial k/\partial z = \partial \omega/\partial z = 0$. The same zero gradients for k and ω with respect to r are applied along the axis of symmetry. At the walls, the usual no-slip condition for velocities and zero pressure gradient are employed taking also $k = 0$ and $\omega = 6\nu/\beta_\omega l$. In the expressions above, c_s is the shape constant, $c_i = (D - d_i)/D$ is the constriction ratio, and s_i is the distance of stenosis from the inlet section.

Other variables are dimensionless scaled by r_0, u_0 (for lengths and velocities), ρu_0^2 (for p), u_0^2 (for k), u_0/r_0 (for ω), $I_{tb} \approx 1\%$ is the turbulence intensity, $C_\mu = 0.09, \beta_\omega = 0.075$, Karman constant $\kappa = 0.41$, y_w is a normal distance from the wall, and l is a scale of length.

The solution procedure is based on artificially modifying governing elliptical equations to make them hyperbolic. This is achieved by adding artificial unsteady terms. These modified unsteady equations unlike the elliptical equations may be solved straightforward starting from the initial condition using some implicit numerical method (in this case the Gauss–Seidel implicit algorithm was employed) to obtain required steady solution as a limit at $t \to \infty$. Two tests were performed to verify the accuracy of established program: (i) the fully developed steady channel flow was calculated and compared with data obtained by direct numerical simulation and (ii) computed steady turbulent flow inside a circular tube with a constriction was compared with known experimental data.

The following results are obtained in this study.

- The flow-through stenosis is complicated because the laminar, transitional, and turbulent regimes coexist there. Since the type of the flow is unknown in advance, it is important to have a model capable to simulate at least laminar and turbulent flows. The developed approach, including the $k - \omega$ turbulence model has such an ability.
- The streamline patterns for turbulent flow through stenosis shows the circulation zone divided in two parts: the circulation flow behind the stenosis and the main flow near the center of the tube with relatively straight and parallel streamlines; the length of the vortex increases with Reynolds number increasing in laminar flow until it reaches a critical Reynolds number, whose value is about 300, and then when the flow becomes transitional or turbulent, it decreases as the Reynolds number grows;

- The dimensionless vorticity at the wall, which related to wall shear stress, increases rapidly as flow approaches stenosis and reaches a peak value slightly upstream of the maximum stenosed area; then downstream it, the value decreases and becomes negative where the separation on the wall occurs. The value of the vorticity peak increases and tends to shift upstream as the Reynolds number increases. The negative magnitude of the wall vorticity in recirculation zone also increases as the Reynolds number grows.

- The wall pressure and centerline velocity distributions for Re = 100 and 300 in comparison with data obtained for laminar flow in the same problem [3] indicate that (i) for Re = 100, the $k - \omega$ turbulence model gives the same results as those obtained for laminar flow and (ii) for Re = 300, the centerline velocity giving by laminar flow modeling is much slower because the flow distal to the stenosis becomes transitional in this case, and the laminar model cannot take into account such a property.

- In the case of two stenoses, the recirculation eddies are formed downstream of each stenosis; there is a separation streamline dividing the flow in recirculation region distal to each stenosis and region with main flow near the center line. When the relative distance between stenoses S/D is less than 3, a recirculation zone fills the valley region between the two stenoses and the reattachment point is located on the front of the second stenosis; the recirculation zone distal to the second stenosis is reduced as S/D increases so that for $S/D = 4$, it appears to be much smaller than that for the first stenosis.

- The two peaks exist in the wall vorticity distribution such that the wall vorticity peak generated by second stenosis is smaller than that generated by the first stenosis; as the distance between stenoses increases the second vorticity peak grows, and at $S/D = 4$, they become almost equal. The negative wall vorticity peak occurs proximal to the second stenosis when S/D is less than 3, while it does not occur when S/D is 3 or 4; the maximum centerline velocity occurs slightly downstream of the stenosis because the formation of a recirculation zone behind each stenosis reduces the cross-sectional area.

- The maximum of centerline disturbance intensity, which characterized the turbulence intensity, near the second stenosis is higher than that near the first stenosis for all the spacing ratios; at the same time, the downstream peak value of wall vorticity increases with spacing ratio increasing until the spacing ratio S/D reaches 4. Thus, it may be deduced that the double stenoses have the strongest effect on the distribution of turbulence intensity in the downstream region when the spacing ratio is $S/D = 4$.

- As the Reynolds number increases: (i) the value of two wall vorticity peaks grows as well as the value of their difference, (ii) the value of centerline velocity decreases, and (iii) the value of the peak of centerline disturbance intensity goes up, while the distance between its peak location and stenosis decreases.

- The data obtained for different values of ratio c_i of the stenosis profile (12.1) show that $c_i = 0.5$ is a critical value of stenosis, which blocks the vessel leading to abrupt changes in the flow properties; this result is in line with the clinic practice usually based on the treatment of the artery disease, depending on whether the artery is more than 75% stenotic which corresponds to inequality $c_i > 0.5$.

Example 12.3a Blood flow in artery with multi-stenosis under magnetic field [4].

This paper presents the results of studying the effects of arterial wall parameters on the blood flow through an elastic artery with multi-stenosis under effect of an external magnetic field in the porous media. Such an analysis is of interest, for example, in understanding blood flow properties during the magnetic resonance angiography (MRA) or magnetic resonance imaging (MRI).

The model is constructed as an anisotropically elastic cylindrical tube with multi-stenosis filled with a viscous incompressible electrically conducting fluid modeling the blood. The geometry of the multi-stenosis is given by following relations:

$$R(z) = \begin{cases} a\left\{1 - a_1\left[s_l^{m-1}(z - d_l) - (z - d_l)^m\right]\right\}, & d_l \le z \le d_l + s_l \\ a & \text{otherwise} \end{cases}, \quad a_1 = \frac{\delta_l m^{m/(m-1)}}{a s_l^m (m-1)}, \quad (12.2)$$

where $m \ge 2$ is the shape parameter, a is the constant radius of the normal artery, s_l is the length of the stenosis, and d_l measures the location of stenosis, where $l = 1, 2, 3...,\delta_l$ is the maximum height of the stenosis located at $z = a + s_l / m^{1/(m-1)}$, and the ratio δ/a of the height of the stenosis to the radius of normal artery is much less than unity.

The governing equations for fluid are two-dimensional unsteady Navier–Stokes equations written in cylindrical coordinates with additional terms taking into account the effects of magnetic field and permeability $\left(\sigma B^2 + \mu/k\right)u$ and $\left(\sigma B^2 + \mu/k\right)w$ (like in Example 12.5), where u and w are the axial and radial velocity components, respectively.

The significant distinction of this study is that the given below governing equations take into account (i) the motion of the arterial wall subjected to inertial forces, (ii) wall surface forces and (iii) constraint forces representing the reactions of the surrounding connective tissues

$$(T_t - T_\theta)\frac{dR}{dz} + R\frac{\partial T_t}{\partial z} - R\left[M_0\frac{\partial^2 \xi}{\partial t^2} + C_1\frac{\partial \xi}{\partial t} + K_1\xi + \left(M_0\frac{\partial^2 \eta}{\partial t^2} + C_r\frac{\partial \eta}{\partial t} + K_r\eta\right)\frac{dR}{dz}\right]$$

$$+ R\left[1 + (dR/dz)^2\right]^{-1/2}\left\{\frac{dR}{dz}(T_{zz} - T_{rr}) + \left[\left(\frac{dR}{dz}\right)^2 - 1\right]T_{rz}\right\}_{R-h/2} = 0 \quad (12.3)$$

$$\frac{T_\theta}{R}\left[1 + (dR/dz)^2\right]^{-1/2} - \frac{d^2R}{dz^2}\frac{T_t}{\left[1 + (dR/dz)^2\right]^{3/2}} - \left[1 + (dR/dz)^2\right]^{-1/2}\frac{dR}{dz}\left(M_0\frac{\partial^2 \xi}{\partial t^2} + C_1\frac{\partial \xi}{\partial t} + K_1\xi\right)$$

$$- \left(M_0\frac{\partial^2 \eta}{\partial t^2} + C_r\frac{\partial \eta}{\partial t} + K_r\eta\right) - \left[1 + (dR/dz)^2\right]\left[2\frac{dR}{dz}T_{rz} - T_{rr} - \left(\frac{dR}{dz}\right)^2 T_{zz}\right]_{R-h/2} = 0 \quad (12.4)$$

Here, $M_0 = \rho_0 h + M_a$, where ρ_0 and h are the density and thickness of the arterial wall, ξ and η are the displacement components of the vessel wall along the axial and radial directions, T_t and T_θ are the viscoelastic stress components acting along the longitudinal and the circumferential directions, K_1, C_1 and M_a represent (per unit area) the spring coefficient, the frictional coefficient of the dashpot, and the additional mass of the mechanical model in longitudinal tethering, respectively, and K_r and C_r are those in the radial direction.

The boundary conditions are determined as the velocities on the arterial wall and zero normal velocity and zero gradient axial velocity on the symmetry axis:

$$u(r, z, t) = \frac{\partial \eta}{\partial t}, \quad w(r, z, t) = \frac{\partial \xi}{\partial t} \text{ on } r = R(z) \quad u(r, z, t) = 0,$$

$$\frac{\partial w(r, z, t)}{\partial r} = 0 \text{ on } r = 0 \quad (12.5)$$

The equations governing both the fluid motion and the motion of the arterial wall are nonlinear. Solution of this complicated system is found considering the first term of perturbation series of order $\varepsilon = \delta/a$, which is expended in Taylor series at $r = R(z) - h/2$ presenting solution for any function in the form

$$f(r, z, t)\big|_{R(z)-h/2} = f_0(a, z, t) + \varepsilon\left[f_1(a, z, t) + R_1(z)\frac{\partial f_0(a, z, t)}{\partial r} \right] + O(\varepsilon^2), \qquad (12.6)$$

where $f(a, z, t)$ are solutions for unperturbed artery without stenosis. A tedious procedure of solution is given in the reviewed paper in details. Calculation was performed for the case of three stenosis and experimental data of parameters indicated in paper as well.

The following basic results are observed.

- The resistance to flow (resistive impedances) at the wall surface is influenced by the unsteady behavior of the flowing blood as well as by the vessel wall distensibility, the Hartman number ($Ha = BL\sqrt{\sigma/\mu}$), permeability constant k, the maximum height of stenosis δ, and shape parameter m.
- In the first stenosis region ($0.375 \le z \le 0.875$), the resistance impedance increases at the onset of the stenosis until its maximum at the throat, then it decreases steeply to reach the end point of the constriction. This observation holds true for the second stenosis region ($1.25 \le z \le 1.75$) and for the third one ($2.125 \le z \le 2.625$); the resistance increases also with Hartman number and the maximum height of stenosis, while it decreases with permeability k and shape parameter m, attaining its maximum for symmetrical stenosis.
- The wall shear stress τ_{rz} ($t = 0.1s$) in the first stenosis area decreases at the onset until it maximum at the throat, then it decreases steeply to reach the end point of the constriction; the observation is similar to that in the second and the third stenosis. Stress τ_{rz} increases with increasing viscoelastic stress along the longitudinal direction and anisotropy, while it decreases with increasing the viscoelastic stress along the circumferential direction, the total mass of the vessel and surrounding tissues and the contributions of the viscous and elastic constraints to the total tethering; the magnitude of the wall shear stress decreases in the converging zone in the stenosic area as shape parameter increases, while it increases in the diverging zone in a similar situation; for any given stenosis shape, the wall shear stress steeply decreases in the upstream from its approached values to the peak value at the throat, then increases in the downstream of the throat to reach its approached magnitude at the end point of the constriction profiles; the transmission of wall shear stress distribution through a tethered tube is substantially lower than that through the free tube, while the shearing stress distribution at the stenosis throat have the inverse character through totally tethered and free tubes.
- The stream lines of the fluid at the arterial surface wall decrease its value gradually by increasing the number of stenosis at the arterial wall.
- The radial velocity at the surface wall decreases with increasing anisotropy, the contribution elastic constraints of the total tethering K, while it increases with increasing the initial stresses components in the longitudinal and circumferential directions; at the same time, the effects of the contribution viscous constraints of the total tethering C and total mass of the vessel and surrounding tissues of the radial velocity are negligibly small. Similarly, the axial velocity at the surface wall decreases with increasing the

contributions of the viscous and elastic constraints to the total tethering and the total mass of the vessel and the surrounding tissues, while it increases with increasing anisotropy and the initial components of the longitudinal and circumferential directions.

• The bolus defined as a volume of fluid bounded by closet streamlines in the wave frame is transported at the wave; the trapped bolus at the central line increases in the size as the permeability increases, while it decreases in size by increasing the Hartman number. The trapped bolus appear gradually in the non-symmetric stenosis but they disappear in the case of symmetric stenosis; they also are smaller in the free tube than in the tethered tube.

Example 12.4a Simulation of blood flow in small vessels [5].

In this study, the blood flow in small vessels is modeled by considering blood as non-Newtonian fluid with Herschel–Bulkley rheological behavior flowing in varying cross-section channels. It is assumed that the progressive sinusoidal wave [first equation (12.7)] imposed on such a channel propagates along it with a constant speed c. The problem is governed by Navier–Stokes equations with the Herschel–Bulkley model taking into account the combined effect of Bingham plastic and power-law behavior of fluid according to the second expression (12.7):

$$H = H_0 + \Lambda x + a \sin\left[(2\pi/\lambda)(x - ct) \right] \qquad \mu = \frac{\tau_0 + \alpha\left[\dot{\gamma}^n - (\tau_0/\mu_0)^n \right]}{\dot{\gamma}} \qquad (12.7)$$

Here, H_0 is the half-width at the inlet (for small vessel, its values are $10-60\,\mu\text{m}$), Λ is a constant defining the channel form, which is chosen here so that (i)for converging tube (arterioles), the width of the outlet of one wavelength is 25% less than that of the inlet, and (ii) for divergent tube (venules), the width of the outlet of one wavelength is 25% more than that of the inlet and a is an amplitude. According to expression for μ, at low strain rate $\dot{\gamma} < (\tau_0/\mu_0)$, the substance behaves as viscose fluid with constant viscosity μ_0, while when the strain rate grows and reaches threshold, τ_0, the fluid behavior is described by the power law (12.7), where α is the consistency factor and n is the power law index determining the thinning ($n < 1$) or thickening ($n > 1$) fluid. The other typical for small blood vessels parameter values used in this study are amplitude ratio $\phi = a/H_0 = 0.1-0.9$, $H_0/\lambda = 0.01-0.02$, $\Delta p - 300 - 50$, $\tau = 0-0.2$, $n = 1/3-2$.

Under usual assumptions of small Reynolds number and long wavelength, the governing equations and boundary conditions in the fixed frame simplify and become

$$\frac{\partial p}{\partial x} = \frac{\partial \tau_{yx}}{\partial y}, \ \frac{\partial p}{\partial y} = 0, \ \tau_{yx} = \left(\tau_0 + \left| \frac{\partial u}{\partial y} \right|^n \right) \text{sgn}\left(\frac{\partial u}{\partial y} \right),$$

$$\psi = 0, \ \frac{\partial u}{\partial y} = 0, \ y = 0, \tau_{yx} = 0, \ y = H, u = 0$$

$$[(n_1 + 1)P_1]^{-1}\left[(P_1 H - \tau_0)^{n_1 + 1} - (P_1 y - \tau_0)^{n_1 + 1} \right],$$

$$[(n_1 + 1)P_1]^{-1}\left[(-P_1 H - \tau_0)^{n_1 + 1} - (P_1 y - \tau_0)^{n_1 + 1} \right] \qquad (12.8)$$

The two last expressions (12.8) determine the velocity $u(x, y, t)$ as a result of solution of the first equation (12.8) satisfying other relations and boundary conditions (12.8). Here, the first solution is valid if $y \geq 0$ and the second if $y < 0$, $P_1 = -\partial p/\partial x$ and $n_1 = 1/n$. All variables are dimensionless and are scaled: $x, y, H, u, \psi, p,$ and τ_{yx} by $\lambda, H_0, H_0, c, cH_0, \mu c^n \lambda/H_0^{n+1},$ and $\mu(cH_0)^n$, respectively.

The following results are obtained:

- The data for velocity distribution in the case of Newtonian fluid $(n = 1)$ agree with results obtained before by Takabatake and Ayukawa ([4] Chapter 7); the effect of non-Newtonian rheology leads to disturbed parabolic profile in which disturbance increases as the index n grows; for a converging channel, the magnitude of velocity is greater than that for the uniform channel and for a diverging channel, the result is altogether different. The influence of the pressure on the velocity distribution for $n < 1/n > 1$ fluids show that in the case of $\Delta P = -1$, for a shear thinning fluid, flow reversal is totally absent in the uniform/divergent channel; for a converging channel, although there is a reduction in the region of flow reversal, it does not vanish altogether, irrespective of whether the fluid is of shear thinning or of shear thickening type.
- The relationship between the pressure difference and the mean flow rate is nonlinear for converging and diverging channels for both values of the power index $n \neq 1$, while it becomes linear for Newtonian fluid at $n = 1$, which is in line with before obtained results; the mean flow rate increases as the pressure difference decreases. The pumping region $(\Delta P > 0)$ increases with the growing of amplitude ratio ϕ for both shear thinning and shear thickening fluids as well as with increasing the power index n; in the co-pumping region $(\Delta P < 0)$, the pressure rise decreases when Q exceeds a certain value. Pumping region also increases with τ increasing, and this effect is greater in the case of shear thickening fluids.
- The data for wall shear stress distribution obtained for one wave period divided in four parts indicate that in each of these time instants, there exist two peaks: a negative τ_{min} and a maximum τ_{max}; it is observed that the transition from τ_{min} to τ_{max} takes place between the maximal and the minimal channel heights. Since at the point of occlusion, the pressure to the left of the point with τ_{max} is maximum as well, the authors show that such a situation may be responsible for a number of consequences, for example, in the case of high shear at the crest, a dissolving wavy wall may have a tendency to level out, or some specific chemical reactions are likely to occur. Shear stress increases as ϕ increases and also as power index grows; the last effect is observed for all types of channels.

On the basis of these detailed results and streamline patterns, the general conclusions are made.

- At any instant time, there is a retrograde flow region for both Newtonian and non-Newtonian fluids, when $\Delta P = 0$ and also for some negative values of ΔP.
- With an increase in the value of n and ϕ, the regions of forward/retrograde flow advance at a faster rate.
- The parabolic nature of the velocity profiles are significantly affected by the rheological power index n.

- In the case of a shear thinning fluid, the flow reversal does not occur in the uniform channel; it transforms to the forward flow in the diverging channel. For a shear thickening fluid, flow reversal reduces, but it does not vanish totally when ΔP changes from 0 to -1.
- Non-uniform geometry affects quite significantly the distribution of velocity and wall shear stress as well as the pumping phenomena and other flow characteristics.
- Peristaltic pumping characteristics as well as the distribution of velocity and the shear stress are strongly influenced by the amplitude ratio ϕ and the rheological index n.

Example 12.5a/n Simulation of the blood flow during electromagnetic hyperthermia [6].

The electromagnetic hyperthermia is a procedure for cancer treatment that is based on the experimental fact that at the temperature higher than $41°C$, the transformed malignant cells are more sensitive than the normal cells. The procedure consists of injecting the magnetic fluid into an artery supplying the malignant tissues or directly into tumor and then subjecting the system to an alternating current magnetic field. The elevating temperature of $45–47°C$ generated in the injected magnetic fluid owing to the imposed magnetic field results in destroying the cancer cells. This method is applicable for treatment of some tumors sites, including brain, soft tissues, liver, abdominal, pancreatic cancer, and head/neck tumors.

To simulate the blood flow under the action of the magnetic field, the model considers a two-dimensional channel with a flow of a biomagnetic non-Newtonian viscoelastic fluid whose rheology behavior is described by second-grade equation. To model the arteries sides, the channel walls are assumed to be porous and stretchable with velocity proportional to the longitudinal distance from the origin. Due to the no-slip condition, such walls induced the similar fluid flow. The external magnetic field is generated by dipole located at the distance b above the channel.

Comment 12.1 Except Newtonian fluid whose viscosity depends only on temperature, there are non-Newtonian fluids whose viscosity depends on spatial or/and time deformation. With two non-Newtonian fluids, we encountered the power law fluids (Sec. 4.2.8) and couple-stress fluid (Example 7.6). Other examples of non-Newtonian fluids one may see reading article "rheology" on Wikipedia. Such many non-Newtonian fluid models exist because each substance or a group of it shows individual rheological behavior. Due to that in practice different complexity functions are used for describing various rheological laws. The power function is one of the simples. Couple-stress and second-grade laws using in studying blood flows are examples of more complicated models. Note that this is an example when two different rheological laws are employed for studying flow of the same substance and that also typically in using non-Newtonian fluid. Moreover, even for the same non-Newtonian fluid, sometimes different equations are used. For example, in reviewing study, relations $\mu \geq 0$, $\alpha_1 \geq \alpha_2$, $\alpha_1 + \alpha_2 = 0$ (α_1 and α_2 are normal stress moduli) are used for the second-grade fluid, whereas others are applied as well in other cases [7]. These examples are considered here to give a reader an understanding how complicated the non-Newtonian fluid area is.

The problem is governed by two-dimensional Navier–Stokes and energy equations with the following additional terms accounting for the viscoelasticity, permeability, and magnetic effects:

$$-k_0\left[u\left(\frac{\partial^3 u}{\partial x^3}+\frac{\partial^3 u}{\partial x \partial y^2}\right)+v\left(\frac{\partial^3 u}{\partial x^2 \partial y}+\frac{\partial^3 u}{\partial y^3}\right)-\frac{\partial u}{\partial x}\frac{\partial^2 u}{\partial y^2}-\frac{\partial u}{\partial y}\frac{\partial^2 v}{\partial y^2}-2\left(\frac{\partial u}{\partial x}\frac{\partial^2 u}{\partial x^2}+\frac{\partial v}{\partial y}\frac{\partial^2 u}{\partial y^2}\right)\right]$$
$$+\mu_0 M\frac{\partial \Omega}{\partial x}-\frac{\mu u}{\tilde{k}}$$

$$-k_0\left[u\left(\frac{\partial^3 v}{\partial x^3}+\frac{\partial^3 v}{\partial x\partial y^2}\right)+v\left(\frac{\partial^3 v}{\partial x^2\partial y}+\frac{\partial^3 v}{\partial y^3}\right)-\frac{\partial v}{\partial x}\frac{\partial^2 u}{\partial y^2}-\frac{\partial v}{\partial y}\frac{\partial^2 v}{\partial y^2}-2\left(\frac{\partial u}{\partial x}\frac{\partial^2 v}{\partial x^2}+\frac{\partial v}{\partial y}\frac{\partial^2 v}{\partial y^2}\right)\right]$$

$$+\mu_0 M\frac{\partial\Omega}{\partial y}-\frac{\mu v}{\tilde{k}}-k_0\frac{\partial u}{\partial y}\frac{\partial}{\partial y}\left(u\frac{\partial u}{\partial x}+v\frac{\partial u}{\partial y}\right)+\mu_0 T\frac{\partial M}{\partial T}\left(u\frac{\partial\Omega}{\partial x}+v\frac{\partial\Omega}{\partial y}\right). \tag{12.9}$$

The boundary conditions consist of symmetry conditions on the parallel to the walls' central axis x and $u=\tilde{c}x$, $v=0$, $T=T_w$, $p_0=p+(\rho/2)\left(u^2+v^2\right)$ with constant coefficient of proportionality $\tilde{c}$, temperature T_w, and total pressure p_0 at the walls. In the expressions (12.9), k_0, $\tilde{k}$, μ_0, M, and Ω are, respectively, coefficients of permeability and viscoelasticity, magnetic permeability, magnetic moment, and magnetic field strength. Expressions containing k_0, μ_0 and $\tilde{k}$ determine viscoelasticity and magnetic effects.

System (12.9) is reduced to ordinary differential equations for five dimensionless functions: $f(\eta)=\psi/\tilde{c}H^2\xi$, $P_1(\eta)+\xi P_2(\eta)=-p/\rho\tilde{c}^2 H^2$, $\theta_1(\eta)+\xi^2\theta_2(\eta)=T/T_w$, depending on variables $\xi=x/H$ and $\eta=y/H$ and on seven dimensionless quantities: $\text{Re}=\rho\tilde{c}H^2/\mu$, $\text{Pr}=\mu c_p/\lambda$, distance from dipole, $\tilde{b}=b/H+\eta$ and viscoelastic $K=k_0\tilde{c}/\mu$, permeability $\tilde{K}=\tilde{k}/H^2$, viscous dissipation $K_T=\tilde{c}\mu^2/\rho k_{pm}T_w$, ferromagnetic interaction $B=\gamma\mu_0 k_{pm}T_w\rho/2\pi\mu^2$ parameters, where $\tilde{k}$, k_{pm}, and γ are permeability, pyromagnetic, and magnetic strength coefficients, respectively, while other notations are defined above. As one may see five functions f, P_1, P_2, θ_1, θ_2 depending on variable η determine velocity components, pressure, and temperature. The system of five ordinary differential equations defined these functions was solved applying perturbation series in small viscoelastic parameter K and numerically using the finite difference method.

The effect of various parameters on the velocity components, pressure, skin friction coefficient, and temperature of the arteries blood during electromagnetic hyperthermia was investigated numerically for the following ranges of parameters: $\text{Re}=1-2$, $\text{Pr}=5-8$, $\tilde{b}=2.5-4$, $K=0.05-0.1$, $\tilde{K}=0.08-2$, $B=0-10$. Analysis of 12 graphs presenting detailed results gives several basic conclusions.

- The axial velocity decreases up to central line reaching negative minimum close to it and then increases; such a distribution is obtained for all studied values of parameters, showing that the back flow occurs near the central line. It develops due to the stretching walls and may be considerably reduced by applying strong external magnetic field; separation close to the lower wall is observed for large Reynolds and Prandtl numbers, for near located magnetic dipole (large $\tilde{b}$) and small porosity permeability ($\tilde{K}$).
- For all studied cases, the transverse velocity increases with distance from lower wall reaching maximum at $\eta\approx 0.7$ and then monotonically decreases; it decreases as the Reynolds and Prandtl numbers increases. For large Prandtl numbers, the reversal flow take place in the vicinity of the central line. The same tendency is observed for magnetic dipole located far from lower wall and small porosity permeability.
- The data obtained for all values of parameters for function $\theta_1(\eta)$ (θ_2 was not studied) indicate that temperature monotonically decreases across the channel section and grows as the Reynolds or Prandtl number decreases as well as ferromagnetic interaction or permeability parameter.

- In considered cases, the distribution of function $P_2(\eta)\,P_1(\eta)$ was not studied) shows that pressure decreases across the channel section, reaches minimum close to the upper wall $(\eta \approx 0.8)$, and then remains almost constant. The pressure increases as the Reynolds or Prandtl number increases.

- Skin friction decreases for any values of Pr, K, $\tilde{K}$, and $\tilde{b}$ as the ferromagnetic interaction parameter B grows, except two cases: increasing the Reynolds number when the skin friction coefficient increases and large Prandtl numbers when it grows for the values of B between one and two gradually decreasing as B becomes larger.

12.2 Peristaltic Flow in Disordered Human Organs

Example 12.6a Particle motion in peristaltic flow with application to the ureter [8].

Studying the interaction of small particle with peristaltic flow is important for many applications, like for the cells in blood or stones and/or bacterium in ureteral flow, microorganisms in a solution, and for hydraulic transport of particles.

In the reviewed article, the interaction between small particle and incompressible Newtonian two-dimensional peristaltic flow is considered. The flow in the channel is described by two-dimensional steady Navier–Stokes equations in the moving frame (Sec. 7.2). The sinusoidal waves are imposed on the walls that are assumed to be flexible in the transverse direction. The Navier–Stokes equation for the stream function is used in the form (7.5) (without time Laplacian), as in many other studies. The solution of this equation is obtained by perturbation series up to ε^2, similar to Fung and Yih approach (Examples 7.2 and 7.6).

The momentum equation for a particle suspended in a moving fluid is formulated applying the Basset–Boussinesq–Oseen (BBO) approach in the dimensionless form. This equation is applicable to small spherical Stokes type particle in fluid at a low Reynolds number and takes into account specific forces significant in this case:

$$\frac{d\bar{u}_p}{dt} = \frac{2\tilde{\rho}(\bar{u}-\bar{u}_p)}{\mathrm{St}(2\tilde{\rho}+1)} + \frac{3}{2\tilde{\rho}+1}\frac{d\bar{u}}{dt} + \frac{r^2}{40(2\tilde{\rho}+1)}\frac{d}{dt}\nabla^2\bar{u} + \frac{\tilde{\rho}r^2}{12\mathrm{St}(2\tilde{\rho}+1)}\nabla^2\bar{u}$$

$$\times \sqrt{\frac{9}{2\pi\tilde{\rho}\mathrm{St}}}\,\frac{2\tilde{\rho}}{2\tilde{\rho}+1}\left[\int_0^t \frac{d(\bar{u}-\bar{u}_p)}{dt}\frac{dt'}{\sqrt{t-t'}} + \frac{\bar{u}_0-\bar{u}_{p0}}{\sqrt{t}}\right] + \left[\frac{2(\tilde{\rho}-1)}{2\tilde{\rho}+1}\right]\frac{g_0\tau_c}{c}\bar{g}. \quad (12.10)$$

Here, $\bar{u}\{u,v\}$, $\bar{u}_p\{u_p,v_p\}$, and $\bar{g}$ are velocity and gravity vectors, $\tilde{\rho}=\rho_p/\rho$, r is the particle radius, $\mathrm{St}=\tau_p/\tau_c$ is the Stokes number, $\tau_p=4\rho_p r^2/18\mu$ is the particle relaxation time, c is the wave speed, $\tau_c=\lambda/\pi c$, $\nabla^2=\varepsilon^2\partial^2/X^2+\partial^2/Y^2$, $\varepsilon=\pi R/\lambda$, R is the half of channel height, and subscript p refers to a particle. Variables are scaled: coordinate X in the fixed frame by λ/π, Y and r by R, velocities by wave speed c, and time t by τ_c. The terms of sum on the right-hand side of the BBO equation, which equals the particle acceleration $d\bar{u}/dt$, represent steady-state Stokes drags force, virtual mass forces (second and third terms), Faxen, Basset, and gravity forces.

Comment 12.2 (i) The Basset force arises due to temporary delay of the boundary layer development when a body accelerates in the Stokes flow. The corresponding term in the dynamic equation is known as history or the Boussinesq–Basset term, who proposed this force independently at the end of 19 century. (ii) The Faxen force, named after the physicist who suggested it in 1922, is a correction to the Stokes law for a sphere (Sec. 2.5.1.1) that is valid when the body is moving close to the wall. (iii) The correction to the creeping flow developed by Oseen in 1910 (Sec. 2.5.1.2) takes into account neglected at the low Reynolds numbers inertia effect. (iv) The virtual mass is some mass of fluid added to moving body to take into account the inertia effect arising at changing the body velocity.

Analysis of the solutions for peristaltic flow and for submerged particle (equation 12.10) leads to the following basic results and conclusions.

- The streamlines and velocity profiles indicate that flow is nonuniform near the wave trough in which the flow enters; below the wave crest, the downward–upward region grows with the flow rate increasing. The pressure distribution shows an adverse pressure gradient in opposite to the peristaltic flow direction in the upper half plane and the favorable pressure gradient in the flow direction in the lower half plane.

- The satisfactory agreement of these results obtained from solution up to order ε^2 with experimental data [9] is observed for $\varepsilon \, \mathrm{Re} = \omega R^2 / v \leq 10$, where $\omega = \pi c / \lambda$ is the frequency. This is better than it is expected since the Reynolds number is assumed to be of order unity. The explanation is the sufficient convergence of perturbation series.

- Two real situations are simulated and analyzed: peristaltic transport of a fluid with particles modeling the flow of mixture along the ureter with (i) stones or their small pieces after acoustic breakup of the stones and (ii) urine with bacteria that usually is a result of the stones or of a treatment. The numerical integration of system of equations for velocities components obtained from BBO equation (12.10) gives trajectories of simulated stones and bacteria using particles of corresponding sizes and characteristics. Analyzing of trajectories gives an insight into individual particles behavior, showing the domains with normal particles motion and reflux.

- The trajectories of both types of particles in the case of retrograde motion near the longitudinal axis are similar; for zero pumping regime, the particles near the longitudinal axis have positive displacement, while near the wall the net displacement is negative. This corresponds to the reflux situation reported in the early studies (Example 7.1) showing that there is a possibility of transport of bacteria to the upper urinary tract.

- The behavior of a group of particles is investigated using simplified equation (12.10). It is shown by comparative calculations that the most significant contribution in the BBO equation are made by the Stokes drag, gravity, and Basset forces; it is also assumed that: (i) each particle acts independently of the others and (ii) only flow affects the particle behavior but the back effect may be neglected which is the same assumption as in the whole study.

- The particles in the group initially distributed uniformly near the center move forward, whereas those near the wall are delayed; particles in groups of stones and bacteria behave similar: some particles get closer to the wall and others participate in formation of a bolus creating satellite structures from which after a while the bolus detaches; the observed particles tend to move to the wall is a possible explanation for the failure of calculi after successful ESWL (extracorporeal shock wave lithotripsy) for acoustic breakup of the stones.

Example 12.7a Simulation of chyme flow during gastrointestinal endoscopy [10].

The endoscope is a powerful means in diagnosis and management of intestinal illnesses. The interaction between intestinal, chime flow, and endoscope during the gastrointestinal endoscopy is studied in this article. The pressure, pressure drop, velocity, and forces acting by endoscope and the intestine on chime flow are calculated. The model consists of cylindrical annulus bounded at the outer boundary by small intestine and at the inner boundary by the inserted endoscope. The sinusoidal wave is imposed on the outer tube-small intestine. The two-dimensional Navier–Stokes equations in cylindrical coordinates governed the problem. In dimensionless variables in the frame moving with wave at speed c, these equations and boundary conditions are

$$\partial(ru)/\partial r + \partial(rw)/\partial z = 0, \ \varepsilon^3 \operatorname{Re}\left(u\frac{\partial u}{\partial r} + w\frac{\partial u}{\partial z} \right) = -\frac{\partial p}{\partial r} + \varepsilon^2 \left(\frac{1}{r}\frac{\partial}{\partial r}\left(r\frac{\partial u}{\partial r} \right) - \frac{u}{r^2} + \varepsilon^2 \frac{\partial^2 u}{\partial z^2} \right) \quad (12.11)$$

$$\varepsilon \operatorname{Re}\left(u\frac{\partial w}{\partial r} + w\frac{\partial w}{\partial z} \right) = -\frac{\partial p}{\partial z} + \varepsilon^2 \left(\frac{1}{r}\frac{\partial}{\partial r}\left(r\frac{\partial w}{\partial r} \right) - \frac{u}{r^2} + \varepsilon^2 \frac{\partial^2 w}{\partial z^2} \right) \quad \begin{array}{l} u = 0, w = -1 \text{ at } r = \Delta \\ w = -1, \text{ at } r = 1 + \phi\sin 2\pi z \end{array}$$

The variables are scaled by the radius of the outer tube r_0, wavelength $\lambda, \lambda/r_0 c, c, \lambda/c, \lambda c\mu/r_0$ for radial and axial distances, radial and axial velocities, time and pressure, respectively, $\operatorname{Re} = \rho cr_0/\mu, \varepsilon = r_0/\lambda, \Delta = r_e/r_0, \phi = a/r_0$, where r_e is the endoscope radius and a is an amplitude of sinusoidal wave $\eta = r_0 + a\sin(2\pi/\lambda)(z - ct)$.

Under assumptions of small Reynolds number and long wavelength that are relevant in the case studied (intestine is small as compared to wavelength, $\varepsilon < 1$), equations (12.11) simplify showing that the pressure is independent of the radial variable. Then, the last equation (12.11) may be integrated twice with respect to r giving

$$w = -1 - 0.25\frac{dp}{dz}\left[\Delta^2 - r^2 + \left(\Delta^2 - \eta^2 \right)\frac{\ln(r/\Delta)}{\ln(\Delta/\eta)} \right],$$

$$q = \pi\left(\Delta^2 - \eta^2 \right)\left\{ 1 + 0.125\frac{dp}{dz}\left[\Delta^2 + \eta^2 - \frac{\Delta^2\eta^2}{\ln(\Delta/\eta)} \right] \right\} \quad (12.12)$$

Here, q is the dimensionless volume flow rate in the wave frame that is found by integration of the expression for w across the annulus section. Solving equation (12.12) defining q for dp/dz and knowing that $q = Q - \pi\left(1 + \phi^2/2 - \Delta^2 \right)$, one finds three basic characteristics: pressure drop Δp across one wavelength and the frictional forces F_e and F_0 over one wavelength acting on endoscope and intestine (Q is the mean flow rate):

$$\Delta p = \int_0^1 G(z)\,dz, \quad F_e = \pi\int_0^1 G(z)\left[\Delta^2 - \frac{\Delta^2 - \eta^2}{2\ln(\Delta/\eta)} \right]dz,$$

$$F_0 = \pi\int_0^1 G(z)\left[\eta^2 - \frac{\Delta^2 - \eta^2}{2\ln(\Delta/\eta)} \right]dz \quad (12.13)$$

$$G(z) = -\left\{ 8\left[Q/\pi + \left(\eta^2 - 1 - \phi^2/2 \right) \right] / \left[\Delta^4 - \eta^4 - \left(\Delta^2 - \eta^2 \right)^2 / \ln(\Delta/\eta) \right] \right\}$$

These relatively simple expressions give an exact solution of the problem in question in the case of small Reynolds number and long wavelength. Performing integration according to relations (12.13) leads to the following conclusions.

- The pressure drop is generally positive (which means that the pressure in the chime flow decreases with increasing the axial coordinate) when the flow rate is not sufficiently small, and it is negative otherwise; thus, there is the pressure gradient acting on the flow, if the flow rate is not too small, while this pressure gradient acts in the opposite direction in the case of too small flow rates. The magnitude of the pressure drop increases as the wave amplitude or endoscope diameter increases.

- The inner friction force is generally negative (the force act by the chyme flow on the endoscope) provided that the flow rate is not sufficiently small being otherwise the positive one; in any regime, the magnitude of inner force increases with either the wave amplitude or endoscope cross-section size growing; thus, for small rates, the force acts by the endoscope on chime flow and may become strong in the case of large amplitude or endoscope size , while at large flow rate this force acts in the opposite direction so that the chyme flow affect the endoscope.

- The outer friction force is generally positive (the force act by the chime flow on the intestine) provided that the flow rate is not too small and is negative otherwise; for sufficiently large flow rate, this force is larger for larger amplitude or aspect ratio; thus, for the large flow rate, the chime exerts on the intestine, but at the small flow rate, the intestine acts on the chime flow increasing this effect as amplitude or endoscope grows.

Example 12.8a Simulation of bile flow in a duct with stones [11].

The article presents a study of the motion of a mixture of the fluid (simulate the bile) and the solid particles (simulate the stones) that forms a dense porous mass. The model consists of a channel with sinusoidal wave propagating along it walls. The problem is governed by Brinkman equations, which are Navier–Stokes equations with different right-hand part taking into account porosity (e) and permeability (k) effects (Example 12.9). In this study, the Brinkman equation is considered in streamline form obtained after pressure eliminating (like for Navier–Stokes equation in Sec. 2.2.2.4) subjected to the Saffman slip and sinusoidal conditions on the walls

$$R\left(\psi_{tyy} + \psi_{txx} + \psi_y\psi_{yyx} - \psi_x\psi_{yyy} + \psi_y\psi_{xxx} - \psi_x\psi_{xxy}\right)$$

$$= \frac{1}{e}\left(\psi_{yyyy} + \psi_{xxyy} + \psi_{xxxx}\right) - \frac{1}{k}\left(\psi_{yy} + \psi_{xx}\right) \tag{12.14}$$

$$\psi_y = \mp s\psi_{yy} \text{ at } y = \pm H \pm \eta, \quad s = b/\sqrt{k}, \quad -\psi_x = \pm\frac{2\pi ac}{\lambda}\sin\left[\frac{2\pi}{\lambda}(x - ct)\right] \text{ at } \pm H \pm \eta$$

Here, all variables are dimensionless and are scaled: $x, y, u, v, \psi, \eta, p, t, k$ by H, c, cH, H, $\rho c^2, H/c, H^2$, respectively, s is the slip parameter in the Soffman slip boundary condition, and $\varepsilon = a/H$ is the dimensionless amplitude.

Comment 12.3 Saffman slip condition is used for smooth or permeable boundaries and has the form $\partial u/\partial y = (b/\sqrt{k})\, u$, where b is a constant depending only on the properties of a porous material and k is permeability coefficient [12].

The problem (12.14) is solved using perturbation series up to ε^2 in standard form

$$\psi = \psi_0 + \varepsilon\psi_1 + \varepsilon^2\psi_2 + \cdots, \quad \frac{\partial p}{\partial x} = \left(\frac{\partial p}{\partial x}\right)_0 + \varepsilon\left(\frac{\partial p}{\partial x}\right)_1 + \varepsilon^2\left(\frac{\partial p}{\partial x}\right)_2 + \cdots, \tag{12.15}$$

where the first term in the last equation corresponds to the imposed pressure gradient and the others are arising due to peristaltic motion. Detailed solution procedure is given in [11]. In general case, the solution is reduced to the forth-order differential equation, which may be solved numerically. In a special case, for initially stagnation fluid without traveling peristaltic wave, a close solution is obtained. It is observed that in such a case, the maximum pressure gradient that can be created by one wave of small amplitude is of the order ε^2. Using this solution, the analytical relations are derived for velocity, time-averaged velocity, and for critical pressure of reflux. Computation and analyzing results yield the following conclusions.

- The parabolic mean velocity perturbation term (i) decreases as the Reynolds number grows, (ii) not very much changes for the considered here range of the wave number and slip parameter, (iii) increases when the Darcy number k and porosity parameter e increase.
- The time-averaged mean axial velocity increases as the Reynolds number grows; the reflux occurs in the central region when pressure gradient attains a certain critical value 0.220966. The bile velocity strongly depends on the amplitude ratio increasing as the amplitude ratio increases and resulting in reversal flow near the boundaries at high values of amplitude ratio; these results are in conformity with known experimental data.
- The velocity in the central region reduces with the decrease in the Darcy number and the velocity profiles in this case are quite different from the familiar Poiseuille profile; this indicates that the bile velocity decreases as the number of stones increases. At the same time, the velocity increases as the porosity parameter grows under other fixed parameters in the case when the Darcy number exceeds the value 0.05, though its parabolic nature changes for small Darcy numbers showing that in the case of the absence of stones, the bile velocity is the greatest; it is also observed that the velocity strongly depends on the slip parameter in the consider case of porous medium, while the wave number affects the velocity more prominently in the vicinity of the boundary.

The two general conclusions are formulated:

- Bile velocity decreases as the number of stones increases.
- When bile contains a very large number of stones, reflux occurs at the quite very small critical pressure.

12.3   Biologic Transport Processes

Example 12.9a Modeling transport processes in the cerebral perivascular space [13].

Perivascular space (PVS) within the brain is an important pathway for interstitial fluid (ISF) and solute transport in the cerebral cortex that has significant impacts on physiology. In this paper, a model of fluid flow in the cerebral perivascular space induced by peristaltic motion of the blood vessel is developed, and effects of various physiological parameters on perivasclar fluid transport are studied. The model is a thin annual fluid-filled porous medium, surrounding a blood vessel. The outer wall is fixed at the distance $r = R_2$ from the axis z of cylindrical coordinates. The inner wall that corresponds to the blood vessel wall is assumed to oscillate sinusoidal as $\eta = R_1 + a\sin(2\pi/\lambda)(z - ct)$.

The problem is governed by the Brinkman equation that is the Navier–Stokes equation with different right-hand side containing term $(\varepsilon\mu/k)\mathbf{v}$ instead of $\mu\nabla^2\mathbf{v}$, where ε and k are porosity and permeability characteristics, respectively. This term takes into account the effects important in porous media according to generalized Darcy's law (Comment 11.1). It is shown in [13] that in moving frame under usual assumptions of small Reynolds number and long wavelength, the governing system reduces to two equations leading to the following solutions for pressure, axial velocity, and the volumetric flow rate:

$$\frac{dp}{dr} = 0, \quad \frac{dp}{dz} = -\frac{\varepsilon\mu}{k}u - \frac{\varepsilon\mu c}{k}, \quad p = p(z), \quad u = -\frac{k}{\varepsilon\mu}\frac{dp}{dz} - c, \quad Q = \pi\left[R_2^2 - \eta^2(z)\right]u \quad (12.16)$$

Eliminating u from last two relations and solving the resulting equation for dp/dz gives the expression that after integration yields pressure drop on one wavelength:

$$\frac{dp}{dz} = -\frac{\varepsilon\mu}{k}\left\{\frac{Q}{\pi\left[R_2^2 - \eta^2(z)\right]} + c\right\}, \quad \Delta p_\lambda = \frac{\varepsilon c\mu\lambda}{k}\left[-\frac{Q}{4\pi cR_2^2}\left(\frac{1}{\sqrt{R_+}} + \frac{1}{\sqrt{R_-}}\right) - 1\right], \quad (12.17)$$

where $R_\pm = \left[1 \pm (R_1/R_2)\right]^2 + (a/R_2)^2$. Analysis yields the following conclusions.

- The total volumetric flow rate is the sum of contributions from the pressure gradient and the peristaltic movement of the boundary, which are coupled in solution; the first part is the flow rate for pressure-driven flow through an annulus filled with the porous medium, while the second part comes from peristaltic wave.
- The time-averaged displacement of a tracer particle is always positive regardless of the initial position of the particle, which means that there is no reverse transport anywhere in the PVS; the explanation of the reverse perivascular transport proposed by Schley et al. [14] is based on the other model simulating flow in non-porous space using Navier–Stokes equation so that there are no contradictions between two different results.
- Convection-enhanced delivery (CED) is the method in which drags are infused directly into the brain tissue through a needle or catheter; it is shown that interaction between the peristaltic motion of the blood vessel under investigation and CED infusion pressure gradient depends on the orientation of the blood vessel and results in the maximum effect in the case when the blood vessel is oriented in the radial direction when the peristaltic wave travels in the outward radial direction as well.
- The comparative calculations of the effects of CED therapy in the presence and in the absence of peristaltic blood vessel show that whether fluid transport in the PVS is predominantly driven by CED or by the peristaltic wave depends on the distance from the needle and on the values of permeability k and amplitude a; at sufficiently large distances, the main contribution to fluid transport comes from the peristaltic wave, while near the infusion source, the importance of peristaltic wave depends on the values of k and a. In some cases, the peristaltic wave contribution is as large as 80%; in general, it seems reasonable that peristaltic contribution should be greater for the lower values of k and higher values of a.

Example 12.10a/n Simulation of macromolecules transport in tumors [15].

The therapeutic efficiency of various genetically engineered macromolecules or monoclonal antibodies depends on their delivery and distribution in tumors. The mathematical model developed in the considered paper describes the transport of fluid and macromolecules in solid tumors, presenting physical characteristics of this process.

Comment 12.4 The term "genetically engineered" stand for subjects changed by genetic engineering methods. Monoclonal antibodies are products developed following idea of a "magic bullet" coined by Paul Ehrlich at the beginning of the 20th century: if a compound selectively target against a disease-causing organism, then a toxin for that organism could be delivered along with this compound. In 1970s, this idea was realized by production of monoclonal antibodies that now widely used in biochemistry, molecular biology, and medicine. One possible treatment for cancer involves monoclonal antibodies that bind only cancer cell-specific antigens and induce an immunological response against the target cancer cell.

The model consists of a cylindrical region surrounding an individual blood vessel of radius r_b and intercapillary half-distance L that is streamlined by streamlined by ISF with velocity u_∞. The pressure and velocity profiles around an infinite cylinder are estimated. In contrast to previous studies, the nonuniform filtration and convection resulting from a heterogeneous pressure distribution are taken into account in the simplified model with only radial pressure gradients for an isolated capillary. The role of various physiological parameters in interstitial transport of fluid and macromolecules in tumors are studied. The macromolecular concentration profiles are described using a diffusion equation with convection and binding. The flow of fluid in the interstitial space is modeled using Darcy's law for flow through a porous medium. For a cylindrical region surrounding an individual blood vessel, the equation and boundary conditions are as follows:

$$\mathbf{u} = -K\nabla p = -K\left(\frac{\partial p}{\partial r}\mathbf{i}_r + \frac{1}{r}\frac{\partial p}{\partial \theta}\mathbf{i}_\theta\right), \quad u_r\big|_{r=r_b} = -K\frac{\partial p}{\partial r}\bigg|_{r=r_b}, \quad \begin{aligned}u_r\big|_{r\to\infty} &= u_\infty\cos\theta\\ u_\theta\big|_{r\to\infty} &= u_\infty\sin\theta\end{aligned}, \tag{12.18}$$

where K is the hydraulic conductivity of the interstitium and r is the distance from the center of the blood vessel.

Comment 12.5 The diffusion equation with convection and binding is a sum of a passive diffusion equation and irreversible binding of the antibodies with binding sites on the cell surface. Such an equation is applied in this study.

The solution of the problem (12.18) for the pressure and velocity has the form [16]

$$p = p_i^0 - \frac{L_p r_b}{K}(p_e - p_i^0)\ln\frac{r}{r_b} + \frac{u_\infty}{K}r\cos\theta\left[1 + \frac{r_b^2}{r^2}\left(\frac{1 - L_p r_b/K}{1 + L_p r_b/K}\right)\right] \tag{12.19}$$

$$u_r = -u_\infty\cos\theta\left[1 - \frac{r_b^2}{r^2}\left(\frac{1 - L_p r_b/K}{1 + L_p r_b/K}\right)\right] + L_p(p_e - p_i^0)\frac{r_b}{r}, \quad u_\theta = u_\infty\sin\theta\left[1 + \frac{r_b^2}{r^2}\left(\frac{1 - L_p r_b/K}{1 + L_p r_b/K}\right)\right]$$

Here, L_p is the hydraulic conductivity of the vessel wall, $p_e = p_v - \sigma(\pi_v - \pi_i)$ is the effective vascular pressure, p_v is the vascular pressure of vessel, $(\pi_v - \pi_i)$ is the difference in osmotic pressure of plasma and ISF, σ is the osmotic coefficient, and p_i^0 is the interstitial pressure at $r = r_b$ and $\theta = \pi/2$. The interstitial concentration is determined employing a convection-diffusion equation for the space surrounding a vessel with boundary conditions: (i) at the vessel wall for transcapillary solute flux and (ii) far away from the blood vessel for no-flux condition at one-half of the intercapillary distance L

$$\frac{\partial C_i}{\partial t} = \nabla \cdot \left(D \nabla C_i \right) - \mathbf{u} \cdot \nabla C_i - k_f C_i \left(B_{max} - B_i \right) + k_r B_i,$$

$$\frac{\partial B_i}{\partial t} = k_f C_i \left(B_{max} - B_i \right) - (k_r + k_e) B_i \qquad (12.20)$$

$$\left(-D \frac{\partial C_i}{\partial r} + u_r C_i \right)_{r=r_b} = u_r \big|_{r=r_b} (1 - \sigma) C_p + P(C_p - C_i) \left(\frac{Pe}{ePe_{-1}} \right), \quad \left(-D \frac{\partial C_i}{\partial r} + u C_i \right)_{r=L} = 0$$

Here, C_i and B_i are the interstitial free solute and the bound solute concentrations, B_{max} is the concentration of binding sites available, D is the interstitial diffusion coefficient, k_f is the forward binding rate constant, k_r is the reverse (dissociation) rate constant, k_e is the elimination (or metabolism) rate constant, and Pe is the transcapillary Peclet number defined as a ratio of convective to diffusive fluxes. The boundary condition (i) for transcapillary solute flux [the third relation (12.20)] is formulated using Starling's law. The boundary condition (ii) [the last relation (12.20)] is approximately one ignoring interaction with other blood vessels and valid only at early times when $t < L/u$. The problem (12.20) was solved numerically using the finite element approach described in [17].

Comment 12.6 The Starling equation was formulated in 1896 by the British physiologist Ernest Starling, also known as the Frank–Starling law of the heart. The Starling equation states that net filtration (or net fluid movement) is proportional to net driving force determined in using here notations as $p_v - p_i - \sigma (\pi_v - \pi_i)$ (see Wikipedia).

Calculations were performed using typical for tumors values of parameters obtaining the following results and conclusions.

- The interstitial pressure profile for baseline parameters corresponding to those in a lymphatic tumor surrounded by normal tissue is not symmetric due to the filtration from the blood vessel; this leads to a stagnation point upstream to the vessel, but none downstream. When the far-away velocity increases by factor 100, the profile resembles a classical solution of inviscid flow around the cylinder; the interstitial velocity streamlines are perpendicular to the pressure isobars, that is, parallel to the pressure gradients.
- These results confirm two assumptions used in macroscopic models; the first, although the pressure is not symmetric, the extravasation is a weak function of θ, and the second, the contribution of convection to interstitial transport and an effect of nonuniform velocities are limited on the macroscopic scale. Binding reduces the diffusive and convective transport rates by the same amount, keeping their relative contributions equal.
- Unlike the distorted pressure profiles, the concentration profiles for nonbinding macromolecule at the vessel wall corresponding to the same baseline parameters of a lymphatic tumor are relatively unaffected. This is because the Peclet number (ratio of convective and diffusion fluxes) is small. Since the influence of convection (small Pe) is small, the angular dependence on profiles is weak, and as a result, the one-dimensional model may be used for sensitivity analysis, which, in particular, have shown that for early times here considered, there is insufficient material to saturate the binding sites; therefore increasing the plasma concentration moderately has no effect on the dimensionless concentration profile. Only significant increasing in plasma concentration approaching a level close to

saturation will result in profile change; conversely, if the forward rate constant and the binding affinity (k_f/k_r) are increased, then penetration into regions far away from the blood vessel is diminished, and larger concentrations are found outside of the blood vessel. This reduction in the capability to penetrate is known as the "binding site barrier."

- The major limitation in the current microscopic analysis is that interactions between blood vessels are neglected so that pressure and velocity fields are computed for a single vessel in an infinite medium; axial variations in pressure, permeability, or vessel diameter are a secondary limitation since there is known to be gradients of these characteristics in the microcirculation; however, for the physiological parameters chosen, the perturbation of the uniform velocity field is limited essentially to a couple of vessel radii and is nearly uniform at a distance $r = L$ as well as the smallest capillaries are $100\mu m$ long, which is an order of magnitude greater than the radius of the vessel.

Example 12.11n Simulation of embryo transport [8].

Within 3 days after fertilization, the embryo is transported along the uterine in the upper part of the uterus. Unlike sperm, the embryo does not have a self-propelling mechanism, and therefore it is transported by intrauterine fluid flow patterns to its final site of implantation during the early process of human reproduction. In the outlined paper, the transport characteristics of this peristaltic motion, which serves as a vehicle for embryo, are investigated. The model simulated the uterine cavity as a two-dimensional channel, which is closed at the rigid end and open toward the cervix. The fluid motion is induced by two trains of sinusoidal waves propagating along the channel walls. The sinusoidal wall motility gradually decreases toward the rigid fundal end $(x=L)$ according to the expression for upper (index 1 and +) and lower $(2$ and $-)$ walls, respectively:

$$\eta_{1,2} = \pm H + a\cos\left\{2\pi\left[(x/\lambda)-(t/T)\right]\pm(\varphi/2)\right\}\tanh\left[\phi(L-x)\right] \qquad (12.21)$$

Here, $2H$ is the unperturbed channel width, T and φ are the period and the phase difference between upper and lower walls, respectively, and other notations are as given in the previous examples. The hyperbolic tangent is introduced to enforce the conditions of anchored boundary conditions near the fundal end that is assumed to be a simple rigid circular curve. The angle ϕ controls the slope of the hyperbolic function. As in other analogous models, it is assumed that walls are moving only in the y-direction but are also extensible.

Comment 12.7 Although this model is similar to the others with sinusoidal moving walls, it differs having a closed end. Such a model more realistic describes the uterine cavity geometry providing possible trajectories by which the embryo is transported.

The problem is governed by two-dimensional Navier–Stokes equations and no-slip and no-penetration boundary conditions on the walls. It is assumed that at the open inlet, fluid may flow into and out of the channel. Computations were performed using the finite volume package FLUENT for moving boundaries according to equations (12.21). The mesh was composed using 20,000 triangle cells. Other details of computational process may found in [18]. The velocity components as functions of time at different positions are obtained and analyzed, leading to following general conclusions.

- The present study reproduces the previous results obtained for uterus modeled as an open channel [19]; as before, it is found that the velocity profiles are dependent on the wall motility, level of asymmetry, and frequency of peristalsis. At the same time, the present study shows that flow characteristics in a closed model are affected by the closed end indicating, in particular, that the magnitude of the axial velocity is increasing toward the open end of the channel.

- The trajectories of particles revealed periodic motions in small moving loops; particles initially separated by one wavelength are transported in almost identical pattern. While this outcome is profound for the simulation with a small wavelength, the particles initially located at full wavelength ($x = n\lambda$) experience small velocities so that their displacement is negligible.

- Particles within the channel recirculate around their initial location along and across the channel; trajectories pattern illustrate the overall transport of embryo after its entering the uterine cavity, where in the idealized conditions, the embryo should recirculate around its initial location until it will be ready for implantation. However, results show that the real embryo may never reach the fundal end being implanted in the anterior or posterior walls. These finding support the observations that implantation of the embryo occurs in the area where it was placed, either naturally or artificially.

- Peristaltic motions due to uterine contractions toward the fundus practically "lock the embryos" within a small area around the location where they were deposited during the transport process; this result is in conformity with observation that standing after transport procedure does not affect the final position of the embryo.

Example 12.12a Modeling the bioheat transfer in human tissues [20].

The bioheat processes are important since heat transfer in human body determines the performance of thermoregulation system and the efficiency of various procedures in thermotherapy. Heat transfer in human body is complicated involving different processes, like conduction in human tissues, perfusion of the arterial-venous blood through the tissue pores, metabolic heat generation, and interaction of blood flow with external magnetohydrodynamic (MHD) and electromagnetic fields.

To study these effects, the relevant model is used that consists of electrically conducting fluid filling the porous space in an asymmetrical channel in the presence of transversely directed magnetic field B. The induced magnetic field is neglected. Two asymmetric sinusoidal waves are imposed on the upper (η_1) and lower (η_2) walls:

$$\eta_1 = d_1 + a_1 \cos\left[\left(2\pi/\lambda\right)\left(x - ct\right)\right] \qquad \eta_2 = -d_2 + a_2 \cos\left[\left(2\pi/\lambda\right)\left(x - ct\right) + \phi\right] \qquad (12.22)$$

where d_1 and d_2 are distances from the x axis to upper and lower walls. The governing system consists of Navier–Stokes and energy equations with additional terms $\mu u/\tilde{k}$, $\mu v/\tilde{k}$, and $\sigma B^2 u$ accounting for permeability and magnetic effects similar to that in equations (12.9), where σ is the electrical conductivity, $\tilde{k} = k_0/d_1^2$, and k_0 is the permeability characteristic. Under the same as in the previous example, assumptions of small Reynolds number and long wavelength, the governing system reduces in three basic equations:

$$\frac{\partial^4 \psi}{\partial y^4} - \left(\frac{1}{\tilde{k}} + M^2\right)\frac{\partial^2 \psi}{\partial y^2} = 0, \quad \frac{1}{Pr}\frac{\partial^2 \theta}{\partial y^2} + E\frac{\partial^2 \psi}{\partial y^2} = 0, \quad \frac{dp}{dx} = \frac{\partial^3 \psi}{\partial y^3} - \left(\frac{1}{\tilde{k}} + M^2\right)\left(\frac{\partial \psi}{\partial y} - 1\right) \qquad (12.23)$$

Here variables are dimensionless scaled by cd_1, λ, d_1, c, $\lambda\mu c/d_1^2$ for stream function, longitudinal and transversal coordinates, axial velocity, and pressure, respectively, $\theta = \left(T - T_1\right)/\left(T_2 - T_1\right)$,

$\mathrm{E} = c^2 / c_v \left(T_2 - T_1 \right)$, and $\mathrm{M} = \sigma B^2 d_1^2 / \mu$ are Eckert and Hartman numbers. Solutions of equations (12.23) has the form

$$\psi = C_1 y + C_2 + C_3 \cosh \left(\sqrt{\frac{1}{\tilde{k}} + \mathrm{M}^2} \right) y + C_4 \sinh \left(\sqrt{\frac{1}{\tilde{k}} + \mathrm{M}^2} \right) y,$$

$$\frac{dp}{dx} = -\left(\frac{1}{\tilde{k}} + \mathrm{M}^2 \right)\left(C_1 + 1 \right) \tag{12.24}$$

Constants are determined using boundary conditions. Results as well as the expression for temperature are awkward [20]. The following basic conclusions are formulated.

- The variation of the pressure rise per wavelength against flux $\Delta p_\lambda (Q)$ shows three regions: peristaltic pumping ($\Delta p_\lambda > 0$ and $Q > 0$), free pumping ($\Delta p_\lambda = 0$), and augmented pumping ($\Delta p_\lambda < 0$ and $Q > 0$). In the first region, the pumping rate increases as M increases, reaches the maximum critical value at $Q = 0.6$, and decreases as M grows further; similar behavior is observed in both the other regions.
- The effect of permeability in all regions is opposite to that of M with the same critical value of Q; an increase in distances d leads to decreasing in Δp_λ in peristaltic and free pumping regions; in augmented pumping region, the reverse effect is observed.
- The velocity profile is parabolic at the inlet; for large permeability, the increase in M results in the increased axial velocities in the neighborhood of the walls and their decreasing close to the channel center.
- Significant variations in temperature profiles occur near the lower wall and in the channel center, where a reduction in θ is noticed with increasing M; an increase in permeability affects the temperature profile in a way opposite to that of M: the porous medium resists the heat flow, and this resistance increases as permeability decreases. An increase in the Brinkman number Br = EcPr leads to increase in the heat transfer rate.
- In the wave frame, the stream lines split to trap a bolus; in a symmetrical channel, the bolus is symmetrical about centerline, and the bolus size is reduced as M increases. In asymmetric case, the bolus tend to shift to the left side of the channel; size of bolus increases if permeability grows in both symmetric and asymmetric cases.
- The heat transfer coefficient for MHD flow is greater when compared with that for the hydrodynamic flow; heat transfer coefficient increases as M increases and when permeability or Brinkman number decreases.

REFERNCES

[1] Ghalichi, F., Deng, X., De Champlain, A., van Douville, Y., King, M., & Guidoin, R. (1998). Low Reynolds number turbulence modeling of blood flow in arterial stenoses. *Biorheology 35*, 281–294. doi: http://dx.doi.org/10.1016/S0006-355X(99)80011-0

[2] Lee, T. S., Liao, W., & Low, H. T. (2003). Numerical simulation of turbulent flow through series stenoses. *International Journal for Numerical Methods in Fluids 42*, 717–740. doi: http://dx.doi .org/10.1002/fld.550

[3] Lee, T. S., Liao, W., & Low, H. T. (2001). Development of an artificial compressibility methodology with implicit LU-SGS method. *International Journal of Computational Fluid Dynamics 15*, 197–208. doi: http://dx.doi.org/10.1080/10618560108970029

[4] Mekheimer, Kh. S., Haroun, M. H., & Elkot, M. A. (2011). Effects of magnetic field, porosity and wall properties for anisotropically elastic multi-stenosis arteries on blood flow characteristics. *Applied Mathematics and Mechanics (English Edition) 32*, 1047–1064. doi: http://dx.doi .org/10.1007/s10483-011-1480-7

[5] Misra, J. C., & Maiti, S. (2012). Peristaltic pumping of blood through small vessels of varying cross-section. *Journal of Applied Mechanics–Transactions of the Asme* arXiv1006.0176v2 [physics. flu-dyn] doi: http://dx.doi.org/10.1115/1.4006635

[6] Misra, J. C., Sinha, A., & Shit, G. C. (2010). Flow of a biomagnetic viscoelastic fluid: application to estimation of blood flow in arteries during electromagnetic hyperthermia, a therapeutic procedure for cancer treatment. *Applied Mathematics and Mechanics (English Edition) 31*, 1405–1420. doi: http: //dx.doi.org/10.1007/s10483-010-1371-6

[7] Rotational flow of generalized second grade fluid between two circular cylinders. prr.hec.gov.pk /Chapters/436S-2.pdf

[8] Jimenez-Lozano, J., Sen, M., & Dunn, P. F. (2009). Particle motion in unsteady two-dimensional peristaltic flow with application to the ureter. *Physical Review* E 79 041901.

[9] Weinberg, S. L., Eckstein, E. C., & Shapiro, A. H. (1971). An experimentally study of peristaltic pumping. *Journal of Fluid Mechanics 49*, 461–479. doi: http://dx.doi.org/10.1017/S0022112071002209

[10] Roy, R., Rios, F., & Riahi, D. N. (2011). Mathematical models for flow of chime during gastrointestinal endoscopy. *Applied Mathematics 2*, 600-607. doi: http://dx.doi.org/10.4236/am.2011.25080

[11] Maiti, S., & Misra, J. C. (2011). Peristaltic flow of a fluid in a porous channel: a study having relevance to flow of bile within duct in pathological state. *International Journal of Hydrogen Energy 49*, 950–966.

[12] Bhatt, B. S., & Sacheti, N. C. (1979). On the analogy in slip flows. *Indian Journal of Pure and Applied Mathematics 10*, 303–306.

[13] Wang, P., & Olbricht, W. L. (2011). Fluid mechanics in the perivascular space. *The Journal of Theoretical Biology 274*, 52–57. doi: http://dx.doi.org/10.1016/j.jtbi.2011.01.014

[14] Schley, D., Carare-Nnadi, R., Please, C. P., Perry, V. H., & Weller, R. O. (2006). Mechanisms to explain the reverse perivascular transport of solutes out of the brain. *The Journal of Theoretical Biology 238*, 962–974. doi: http://dx.doi.org/10.1016/j.jtbi.2005.07.005

[15] Baxter, L. T., & Jain, R. K. (1990). Transport of fluid and macromolecules in tumors IV. A microscopic model of the perivascular distribution. *Microvascular Research 41*, 252–272.

[16] Baxter, L. T. (1990). Transport of fluid and macromolecules in normal and neoplastic tissue. Ph.D. Thesis. Carnegie Mellon University, Pittsburgh, PA.

[17] Becker, E. B., Carey, G. F., & Oden, J. T. (1981). *Finite elements: an introduction.* Eaglewood Cliffs, NJ: Prentice- Hall. doi: http://dx.doi.org/10.1016/0026-2862(91)90026-8

[18] Yaniv, S., Jaffa, A. J., Eytan, O., & Elad, D. (2009). Simulation of embryo transport in a closed uterine cavity model. *European Journal of Obstetrics & Gynecology and Reproductive Biology 144S*, S50–S60. doi: http://dx.doi.org/10.1016/j.ejogrb.2009.02.019

[19] Eytan, O., & Elad, D. (1999). Analysis of intra-uterine fluid motion induced by uterine contractions. *Bulletin of Mathematical Biology 61*, 221–238. doi: http://dx.doi.org/10.1006/bulm.1998.0069

[20] Hayat, T., Qureshi, M. U., & Hussain, Q. (2009). Effect of heat transfer on the peristaltic flow of an electrically conducting fluid in porous space. *Applied Mathematical Modelling 33*, 1862–1873. doi: http://dx.doi.org/10.1016/j.apm.2008.03.024

CONCLUSION

HOW COMPLICATED A MATHEMATICAL MODEL SHOULD BE?

Contemporary analytical and numerical methods outlined in this book are powerful tools for investigating natural phenomena and engineering systems. The basic feature of these methods that distinguish them from other studying methods is the representation of physical understanding of real object in mathematical terms. This is achieved by creating the mathematical models and results in quantitative information in the form of the functional dependencies, describing the behavior of studying phenomena.

In fact, the level of a quality of employed models represents the progress in the corresponding area of knowledge. For example, the first turbulence model created by Reynolds has just the form of Reynolds number. The second model that was a great achievement presented the first turbulent flow equations (RANS). The next step was the development of the $k - \varepsilon$ and the $k - \omega$ turbulence models, and finally, when the computer resources allowed, the direct simulation approach (DNS) and other methods of this type (Chapter 9) were created.

The models improving one may also see comparing the boundary conditions of the third and fourth kinds (Chapters 4–6). In this case, the first model in the form of the simple Newton law of heat flux and temperature head proportionality used in boundary conditions of the third kind is now substituted by the modern approach based on the strong boundary condition of fourth kind (conjugate). At the same time, it is important to keep in mind that despite a high accuracy of some contemporary models, there is often a reason in using the more simple old models. In particular, such a situation exists in both considered cases. In the first one, the turbulence $k - \varepsilon$ and the $k - \omega$ models are still the basic practical tools since the direct solution of the Navier–Stokes equation is restricted to relatively small Reynolds numbers and even the recently developed detached simulation method (DES) consists of, as a part, a solution of the average Navier–Stokes equation (RANS). In the second case, in heat transfer, it was shown that there are groups of problems, which do not demand the conjugate solutions (conclusion to Chapters 4–6). These should be solved by common, simple approach using boundary conditions of the third kind, satisfactorily accurate for these problems.

In the past, the almost entire field of knowledge (except the engineering part) was a descriptive science containing basically verbal and illustrative information and some experimental data. In time with computer advent, the situation has changed, and several sciences start to involve mathematics in the problem description. Development of this process led to a group of knowledge areas, such as biology and medicine, meteorology, economics, linguistics, and earth science, where the mathematical modeling was adapted and became a regular procedure. This changed, in principle, the quality of studying results, transforming these areas of knowledge from descriptive type to sciences describing the behavior of studying object in the mathematically grounded investigating data.

Examples presented in Chapter 12 of results obtained in biology and medicine in the last 50 years show how effective is the mathematical modeling. Besides the calculation of characteristics, such as blood pressure or urine flow rate, which can be measured as well, mathematical modeling provides determining parameters that could not be obtained experimentally, such as the stresses in blood vessels walls or pattern showing flow separation and reattachment after stenosis, or trajectories of particles in the ureter to estimate the possibility of reflux.

Creating a model is a challenge because this process consists of two essential conflicting parts: the system of equations describing the studied problem and the algorithm specifying the method of solution of this system. Usually, the more complete and, therefore, the more complex is the system of governing equations, the more involved is an algorithm for solving it. Although it is obvious that a complete model is desired, the realization of the complex model requires high computer memory and computing time and leads to excessive cost. Therefore, one of the important features of the applied model is a balance between the simplicity and usefulness of the model and problem solution.

The complexity of a model depends on the problem in question and on the details that we are supposed to get. These requirements together with corresponding algorithm form the model. For example, in the relatively simple case of laminar or turbulent flow at a high Reynolds number if we need only friction or/and heat and mass transfer coefficients, the model may be based on boundary layer equations. In such a case, there is no reason to create or apply existing software much easier to use known analytical relations (Chapter 2). The similar situation takes place if the studied flow is characterized by low Reynolds number when instead of Navier–Stokes and full-energy equation, the more simple creeping equations can be used. However, in the case of flow at a moderate Reynolds number, the Navier–Stokes, full-energy equations, and a corresponding numerical approach should be employed. The numerical method is also necessary to use in the majority of cases of boundary layer or creeping flows if the detailed characteristics, such as velocity or/and temperature and concentration profiles or contours of isolines (for example, isobars or isotherms), are required.

Another possible difficulty arising due to complexity is the error accumulation.

Since the computation process consists of many consecutive calculation steps, the small numerical errors in each step under some conditions might summarize, resulting in unacceptable inaccuracy. Especially, such an error accumulation produces problems in the process of iteration or in the case of recurrent formula when the value of a computing variable depends on the previous calculation results. With such recurrent dependency we encountered, in particular, in Chapter 4 considering equation (4.5). This inhomogeneous equation determines the function G_k through presenting previously obtained function G_{k-1} on the right-hand side. Due to that, an error of function G_1 affects the result for G_2 summarizing with its own error; then, that outcome affects the result for G_3 and so on, yielding rapidly growing inaccuracy. To overcome this trouble, in the considered case, the sum of functions (4.8) was applied transforming equation (4.5) in the homogeneous form. However, this is an artificial means, and there is no standard procedure for such a transformation of inhomogeneous equation.

Proceeding from these considerations, one concludes that the model should be as simple as possible while describing the essential physical properties of studied phenomenon with satisfactory accuracy. Therefore, it seems reasonable to begin with the simplest model, complicating it and see what modification works in a particular case. This can be done comparing the computation results with known analytical (Parts I and II) or/and numerical (Applications) solutions. In some cases, a simple preliminary estimation using, for example, the Van-Dyke formula for an error of the friction coefficient (Section 2.6.7) or an approximate evaluation of the boundary condition of the third kind accuracy (Section 4.4.2) may also be helpful. Indeed, such a comparing process is the mandatory step of the software validation, showing its applicability and accuracy (Section 8.1).

APPENDIX

Table A1. Error functions *erfz* and *erfcz*

z	*erf* z	*erfc* z
0	0	1·0
0.05	0.056372	0·943628
0·1	0·112463	0.887537
0·15	0.167996	0·832004
0.2	0.222703	0.777297
0·25	0.276326	0·723674
0·3	0·328627	0·671373
0·35	0·379382	0.620618
0·4	0·428392	0·571608
0.45	0·475482	0.524518
0·5	0·520500	0.479500
0.55	0·563323	0·436677
0.6	0.603856	0.396144
0·65	0·642029	0.357971
0·7	0·677801	0·322199
0·75	0.711156	0·288844
0·8	0·742101	0.257899
0.85	0·770668	0·229332
0·9	0·796908	0·203092
0.95	0·820891	0·179109
1·0	0.842701	0·157299
1·1	0.880205	0.119795
1·2	0.910314	0.089686

(Continued)

Table A1. Error functions *erfz* and *erfcz* (*Continued*)

z	erf z	erfc z
1·4	0·952285	0·047715
1·5	0.966105	0.033895
1·6	0.976348	0·023652
1.7	0.983790	0·016210
1.8	0·989091	0.010909
1.9	0·992790	0·007210
2·0	0·995322	0.004678
2·1	0·997021	0.002979
2·2	0.998137	0·001863
2·3	0.998857	0·001143
2·4	0·999311	0·000689
2·5	0.999593	0.000407
2.6	0.999764	0.000236
2·7	0.999866	0.000134
2.8	0·999925	0.000075
2·9	0.999959	0·000041
3.0	0·999978	0·000022

Table A2. Numerical solution for plane stagnation point flow (Howarth)

$\eta = \sqrt{\dfrac{C}{v}}\, y$	ϕ	$\dfrac{\mathrm{d}\phi}{\mathrm{d}\eta} = \dfrac{u}{U}$	$\dfrac{\mathrm{d}^2\phi}{\mathrm{d}\eta^2}$
0	0	0	1·2326
0·2	0·0233	0·2266	1·0345
0·4	0·0881	0·4145	0·8463
0·6	0·1867	0.5663	0·6752
0·8	0·3124	0.6859	0.5251
1·0	0·4592	0.7779	0·3980
1.2	0·6220	0·8467	0·2938
1.4	0·7967	0·8968	0·2110
1.6	0·9798	0·9323	0·1474
1·8	1.1689	0·9568	0'1000
2·0	1·3620	0·9732	0·0658

Table A2. (*Continued*)

$\eta = \sqrt{\dfrac{C}{\nu}}\,y$	ϕ	$\dfrac{d\phi}{d\eta} = \dfrac{u}{U}$	$\dfrac{d^2\phi}{d\eta^2}$
2·2	1·5578	0·9839	0·0420
2.4	1·7553	0·9905	0·0260
2·6	1·9538	0.9946	0·0156
2·8	2·1530	0·9970	0·0090
3·0	2·3526	0·9984	0.0051
3·2	2·5523	0·9992	0·0028
3·4	2·7522	0·9996	0.0014
3·6	2·9521	0·9998	0.0007
3·8	3·1521	0·9999	0.0004
4·0	3·3521	1·0000	0·0002
4·2	3·5521	1·0000	0·0001
4·4	3·7521	1·0000	0·0000
4.6	3·9521	1·0000	0·0000

Table A3. Numerical solution for boundary layer flow along the plate (Howarth)

$\eta = y\sqrt{\dfrac{U\infty}{\nu x}}$	f	$f' = \dfrac{u}{U\infty}$	f''
0	0	0	0·33206
0·2	0·00664	0·06641	0·33199
0·4	0·02656	0·13277	0·33147
0·6	0·05974	0·19894	0·33008
0·8	0·10611	0·26471	0·32739
1·0	0·16557	0·32979	0·32301
1·2	0·23795	0·39378	0·31659
1.4	0·32298	0·45627	0·30787
1·6	0·42032	0·51676	0·29667
1.8	0·52952	0·57477	0·28293
2·0	0·65003	0·62977	0·26675

(*Continued*)

Table A3. Numerical solution for boundary layer flow along the plate (Howarth) (*Continued*)

$\eta = y\sqrt{\dfrac{U\infty}{vx}}$	f	$f' = \dfrac{u}{U\infty}$	f''
2·2	0.78120	0·68132	0·24835
2·4	0·92230	0·72899	0·22809
2·6	1.07252	0·77246	0·20646
2·8	1·23099	0·81152	0·18401
3·0	1·39682	0·84605	0·16136
3·2	1·56911	0·87609	0·13913
3·4	1·74696	0·90177	0·11788
3·6	1·92954	0·92333	0·09809
3·8	2·11605	0·94112	0·08013
4·0	2·30576	0·95552	0·06424
4·2	2·49806	0·96696	0·05052
4·4	2·69238	0·97587	0·03897
4·6	2·88826	0·98269	0·02948
4·8	3·08534	0·98779	0·02187
5·0	3·28329	0·99155	0·01591
5·2	3.48189	0·99425	0·01134
5·4	3·68094	0·99616	0·00793
5·6	3·88031	0·99748	0·00543
5·8	4·07990	0·99838	0·00365
6·0	4·27964	0·99898	0·00240
6·2	4·47948	0·99937	0·00155
6·4	4·67938	0·99961	0·00098
6·6	4·87931	0·99977	0·00061
6·8	5·07928	0·99987	0·00037
7·0	5·27926	0·99992	0·00022
7·2	5·47925	0·99996	0·00013
7·4	5·67924	0·99998	0·00007
7·6	5·87924	0·99999	0·00004
7·8	6·07923	1·00000	0·00002
8·0	6·27923	1·00000	0·00001

Table A3. (*Continued*)

$\eta = y\sqrt{\dfrac{U\infty}{vx}}$	f	$f' = \dfrac{u}{U\infty}$	f''
8·2	6·47923	1·00000	0·00001
8·4	6·67923	1·00000	0·00000
8·6	6·87923	1·00000	0·00000
8·8	7·07923	1·00000	0·00000

Navier–Stokes equations in cylindrical coordinates

$$\rho\left(\frac{\partial v_r}{\partial t} + v_r\frac{\partial v_r}{\partial r} + \frac{v_\gamma}{r}\frac{\partial v_r}{\partial \gamma} + v_z\frac{\partial v_r}{\partial z} - \frac{v_\gamma^2}{r}\right) = -\frac{\partial p}{\partial r}$$

$$+ \mu\left[\frac{\partial}{\partial r}\left(\frac{1}{r}\frac{\partial}{\partial r}(rv_r)\right) + \frac{1}{r^2}\frac{\partial^2 v_r}{\partial \gamma^2} + \frac{\partial^2 v_r}{\partial z^2} - \frac{2}{r^2}\frac{\partial v_\gamma}{\partial \gamma}\right]$$

$$\rho\left(\frac{\partial v_\gamma}{\partial t} + v_r\frac{\partial v_\gamma}{\partial r} + \frac{v_\gamma}{r}\frac{\partial v_\gamma}{\partial \gamma} + v_z\frac{\partial v_\gamma}{\partial z} + \frac{v_r v_\gamma}{r}\right) = -\frac{\partial p}{\partial \gamma}$$

$$+ \mu\left[\frac{\partial}{\partial r}\left(\frac{1}{r}\frac{\partial}{\partial r}(rv_\gamma)\right) + \frac{1}{r^2}\frac{\partial^2 v_\gamma}{\partial \gamma^2} + \frac{\partial^2 v_\gamma}{\partial z^2} - \frac{2}{r^2}\frac{\partial v_r}{\partial \gamma}\right]$$

$$\rho\left(\frac{\partial v_z}{\partial t} + v_r\frac{\partial v_z}{\partial r} + \frac{v_\gamma}{r}\frac{\partial v_z}{\partial \gamma} + v_z\frac{\partial v_z}{\partial z}\right) = -\frac{\partial p}{\partial z} + \mu\left[\frac{1}{r}\frac{\partial}{\partial r}\left(r\frac{\partial v_z}{\partial r}\right) + \frac{1}{r^2}\frac{\partial^2 v_z}{\partial \gamma^2} + \frac{\partial^2 v_z}{\partial z^2}\right]$$

Navier–Stokes equations in spherical coordinates

$$\rho\left(\frac{\partial v_r}{\partial t} + v_r\frac{\partial v_r}{\partial r} + \frac{v_\gamma}{r}\frac{\partial v_r}{\partial \gamma} + \frac{v_\phi}{r\sin\gamma}\frac{\partial v_r}{\partial \phi} + \frac{v_\gamma^2 + v_\phi^2}{r}\right) = -\frac{\partial p}{\partial r}$$

$$+ \mu\left[\frac{1}{r}\frac{\partial^2}{\partial r^2}(r^2 v_r) + \frac{1}{r^2\sin\gamma}\frac{\partial}{\partial \gamma}\left(\sin\gamma\frac{\partial v_r}{\partial \gamma}\right) + \frac{1}{r^2\sin^2\gamma}\frac{\partial^2 v_r}{\partial \phi^2}\right]$$

$$\rho\left(\frac{\partial v_\gamma}{\partial t} + v_r\frac{\partial v_\gamma}{\partial r} + \frac{v_\gamma}{r}\frac{\partial v_\gamma}{\partial \gamma} + \frac{v_\phi}{r\sin\gamma}\frac{\partial v_\gamma}{\partial \phi} + \frac{v_r v_\gamma - v_\phi^2\cos\gamma}{r}\right) = -\frac{1}{r}\frac{\partial p}{\partial \gamma}$$

$$+\mu\left\{\frac{1}{r^2}\frac{\partial}{\partial r}\left(r^2\frac{\partial v_\gamma}{\partial r}\right) + \frac{1}{r^2}\frac{\partial}{\partial \gamma}\left[\frac{1}{\sin\gamma}\frac{\partial}{\partial \gamma}\left(v_\gamma\sin\gamma\right)\right] + \frac{1}{r^2\sin^2\gamma}\frac{\partial^2 v_\gamma}{\partial \phi^2} + \frac{2}{r^2}\frac{\partial v_r}{\partial \gamma} - \frac{2\cos\gamma}{r^2\sin\gamma}\frac{\partial v_\phi}{\partial \phi}\right\}$$

$$\rho\left(\frac{\partial v_\phi}{\partial t} + v_r\frac{\partial v_\phi}{\partial r} + \frac{v_\gamma}{r}\frac{\partial v_\phi}{\partial \gamma} + \frac{v_\phi}{r\sin\gamma}\frac{\partial v_\phi}{\partial \phi} + \frac{v_r v_\phi + v_\gamma v_\phi\cos\gamma}{r}\right) = -\frac{1}{r\sin\gamma}\frac{\partial p}{\partial \phi}$$

$$+\mu\left\{\frac{1}{r}\frac{\partial}{\partial r}\left(r^2\frac{\partial v_\phi}{\partial r}\right) + \frac{1}{r^2}\frac{\partial}{\partial \gamma}\left[\frac{1}{\sin\gamma}\frac{\partial}{\partial \gamma}\left(v_\phi\sin\gamma\right)\right] + \frac{1}{r^2\sin^2\gamma}\frac{\partial^2 v_\phi}{\partial \phi^2}\right.$$

$$\left.+\frac{2}{r^2\sin\gamma}\frac{\partial v_r}{\partial \phi} + \frac{2\cos\gamma}{r^2\sin\gamma}\frac{\partial v_\gamma}{\partial \phi}\right\}$$

Pioneers—Contributors

A
Allmarass, 135
Arrhenius, 299
Atwell, 135

B
Baldwin, 122, 125, 128, 129, 131, 135
Barth, 135
Basset, 347, 348
Batchelor, 108, 292
Bejan, 307
Benard, 106, 274
Bernoulli, 50, 61, 71
Bessel, 27, 28, 59
Bingham, 343
Biot, 3, 8, 9, 10, 159, 160, 187, 192, 193,
        206, 213, 217, 218, 222, 225, 226,
        233, 281, 282, 288, 293, 316, 334
Blasius, 72, 80–85, 86, 87, 88, 89, 90, 91, 92,
        102, 172
Boltzmann, 218
Boussinesq, 117, 118, 122, 135, 347, 348
Bradshaw, 135
Brebbia, 254, 267
Brinkman, 350, 351, 357
Brun, 193, 206
Bulkley, 343

C
Cauchy, 4, 49, 66
Cebeci, 122, 124, 128, 129, 131, 133, 144
Chang, 292
Chapman, 91, 171, 172, 197, 218
Chou, 138
Clauser, 119, 120, 122
Coles, 120, 121

Corrsin, 144
Couette, 59, 60, 62, 63, 83
Courant, 270
Crocco, 172
Czochralski, 313, 314

D
D'Alemberts, 41
Darcy, 310, 311, 312, 322, 331,
        351, 353
Dirac, 36, 256, 258, 266
Dirichlet, 28, 29, 57, 58, 63, 78, 155, 295,
        297, 332
Dorodnitsyn, 171, 172
Drake, 198
Duhamel, 15–18, 21, 23, 36, 38, 68, 91, 103,
        265, 286

E
Eckert, 75, 76, 109, 198
Ehrlich, 347, 353
Einstein, 47, 116, 272, 318
Erk, 197
Euler, 4, 24, 41, 66, 68, 238, 242, 243

F
Falkner, 85
Faxen, 348
Fermat, 137, 143
Ferriss, 135
Feynman, 274
Fick, 44, 311, 322
Fourier, 3, 6, 8, 14, 22, 23, 24, 25, 27,
        30, 31, 32, 33, 35, 37, 38, 44,
        68, 69, 75, 76, 96, 152, 174, 221,
        271, 328

Frank, 354
Frossling, 86
Fung, 242, 243, 244, 347

G
Galerkin, 255, 258, 319, 322
Gauss, 282, 295, 339
Gibson, 122, 129, 141, 178, 186
Gortler, 72–74, 99, 100, 128, 129,
          150, 151, 152, 155, 162, 165,
          174, 177, 190, 192,
          201, 265
Graetz, 4, 281, 285
Grashof, 4, 75, 76, 105, 290,
          295, 296
Green, 21, 36–38, 39
Grigull, 197
Grober, 197

H
Hadamard, 67
Hagen, 59, 60, 128
Hartman, 342, 343, 357
Hartree, 85, 101
Heaviside, 13, 19
Helmholtz, 21
Herschel, 343
Hiemenz, 61
Hook, 3
Howarth, 61, 81, 86, 87, 102

I
Illingworth, 171, 172
Imai, 91
Incropera, 290, 295

J
Jaffrin, 237, 242, 248
Jaluria, 58, 282, 297, 309, 310, 316, 317,
          318, 333, 334
Johonson, 131
Jou, 273

K
Kansa, 263
Karman, 93, 118, 120, 139, 221,
          222, 339
King, 131
Kistler, 144
Klebanoff, 144

Knudsen, 58, 288
Kolmogorov, 42, 136–137, 143, 269,
          270, 274
Kronecker, 47, 54
Kutateladse, 224–226
Kutta, 129, 152, 212

L
Lagrange, 238, 240, 241, 242
Lame, 6, 7, 51–53, 108
Laplace, 5, 10, 28, 29, 30, 32, 33–35, 38, 45,
          49, 53, 65, 66, 68, 149, 155, 185, 222,
          224, 264, 280, 282, 294, 297, 318
Laufer, 122, 270
Launder, 42, 138
Leibniz, 3, 31, 271
Leidenfrost, 301, 303
Levi-Civita, 47, 54, 55, 66
Lewis, 327
Lighthill, 104
Lilly, 272
Liouville, 24, 26, 27, 281
Lomax, 122, 125, 128, 129, 131
Luikov, 186, 193, 281, 295, 325, 326, 328

M
Mach, 139, 218, 235, 298
Mangler, 172, 176
Mellor, 122–124, 129, 141, 178
Mises, 72–74, 76, 99, 100, 129, 150, 152,
          172, 177, 210, 252
Mittra, 243
Moivre, 50
Monin, 124
Moretti, 181

N
Navier, 4, 41–55, 56–77, 64, 65, 67, 69, 80,
          84, 90, 91, 105, 106, 113, 114–117,
          129, 149, 150, 155, 236, 237, 238,
          240, 241, 242, 243, 245, 259, 263,
          264, 269, 270, 271, 272, 273, 274,
          282, 288, 294, 296, 297, 322, 323,
          339, 341, 343, 345, 347, 349, 351,
          352, 356
Neumann, 28, 29, 297
Newton 3, 8, 44, 69, 81, 147, 175, 176, 182,
          183, 190, 210, 233, 238, 244, 315,
          319, 337, 343, 344, 345, 347

Nusselt, 4, 75, 76, 84, 96, 155, 162, 174, 186, 217, 284, 286, 287, 291, 306, 310, 317, 323, 334

O
Orr, 113, 114, 239, 244
Oseen, 67–68, 347, 348

P
Panov, 251, 254
Patankar, 140, 264, 283, 288
Peclet, 4, 45, 64, 65, 68–69, 70, 75, 76, 186, 225, 261, 262, 279, 280, 285, 287, 334, 354
Perelman, 137, 143
Pohlhausen, 72, 80, 83, 93, 105, 139, 221, 222
Poincare, 137, 143
Poiseuille, 59, 60, 63, 90, 128, 238, 239, 240, 244, 246, 247, 351
Poisson, 4, 5, 10, 28, 56, 185, 245, 263, 264
Prandtl, 41, 42, 63, 68, 69, 70, 72–74, 75, 76, 83, 84, 85, 99, 100, 103, 104, 105, 112, 114, 117–118, 119, 122, 125, 128, 129, 130–131, 132, 134, 135, 141, 150, 152, 153, 154, 156, 157, 158, 159, 160, 161, 166, 167, 169, 172–178, 183, 189, 191, 192, 203, 210, 231, 234, 252, 272, 284, 327, 346, 347
Prasad, 243

R
Rayleigh, 4, 105, 106, 112, 114, 274
Reynolds, 4, 41, 42, 43, 45, 60, 64–77, 88, 90, 100, 105, 111, 112, 113, 114–117, 118, 119, 120, 125, 129, 130, 131, 132, 134, 135, 136, 137, 138, 140, 141, 149, 155, 161, 178, 181, 189, 191, 192, 218, 225, 233, 236, 237, 238, 239, 240, 242, 244, 245, 246, 247, 262, 263, 269, 270, 271, 272, 273, 274, 281, 282, 295, 296, 305, 306, 318, 347–349, 351, 352, 356
Rubesin, 91, 171, 172, 197, 218

Runge, 129, 152, 212
Rybczynski, 6

S
Saffman, 350
Saint-Venant, 4, 56
Sakiadis, 172
Schlichting, 84, 89, 90, 101, 102, 103, 111, 112, 114, 115, 119, 125, 126, 128, 130, 131, 140
Schmidt, 4, 327
Seidel, 282, 339
Shapiro, 237, 240, 242, 348
Sherwood, 323
Skan, 85
Smagorinsky, 271, 272, 273
Smith, 122, 124–125, 128, 129, 131
Sommerfeld, 113, 114, 239, 244
Spalart, 135, 273, 274
Spalding, 42, 154, 162, 283
Sparrow, 68, 283
Stanton, 4, 129, 130, 142
Starling, 354
Stefan, 218, 310
Stepanov, 172, 176
Stewartson, 171, 172
Stokes, 4, 41–55, 56–77, 64, 65, 67, 69, 80, 84, 90, 91, 105, 106, 113, 114–117, 129, 149, 150, 155, 236, 237, 238, 240, 241, 242, 243, 245, 259, 263, 264, 269, 270, 271, 272, 273, 274, 282, 288, 294, 296, 297, 322, 323, 339, 341, 343, 345, 347, 349, 351, 352, 355, 356
Strelets, 273
Strouhal, 242
Sturm, 24, 26, 27, 281

T
Taylor Brook, 35, 72, 73, 118, 209, 239, 252, 257, 261, 269, 342
Tollmien, 87, 88, 99

V
Van Dyke, 64, 76, 91, 225, 242
Van Drist, 196, 197, 210
Viskanta, 288

W
Walker, 254
Wilcox, 43, 120, 121, 125, 127, 128, 129,
        131, 132, 135, 136, 137, 138, 136
Wiles, 137, 143

Y
Yaglom, 124
Yih, 238, 239, 240, 242, 243, 244, 347
Young, 245, 318

AUTHOR INDEX

A

Abramzon, 68
Acharya, 258, 260, 262, 263
Allmaras, 273
Ayukawa, 240, 353

B

Baiocco, 298, 299
Balaji, 294
Baldwin, 125, 197, 202, 203, 207, 208
Baliga, 258, 262, 263
Barozzi, 279, 285
Barrett, 227
Batchelor, G. K., 50, 292
Bau, 242
Baxter, 352, 353
Becker, 354
Bejan, 287
Bellomi, 298, 299
Betz, 99
Bhatt, 350
Bialecki, 264, 265, 313, 314
Bicanic, 311–312
Bird, R. B., 52, 55
Blasé, 314
Borde, 68
Brebbia, 254, 257, 258

C

Carare-Nnadi, 352
Carey, 354
Carslaw, 6, 12–15, 35
Cebeci, 122, 124–125, 128–129, 131, 133
Chang, 292
Chapman, 91, 171, 172, 197, 219
Chaube, 243, 244
Chida, 320, 321
Chiu, W. K. S., 58, 282

Choi, 314
Clauser, 119, 120
Coles, 119, 121
Collins, 289
Corrsin, 124
Cotta, 280, 281
Croce, 297

D

Davidenko, 326
Davie, 311–312
De Champlain, 337
Deng, 337
Dewitt, 290, 295
Divo, 263, 264, 265, 297
Dolinskiy, 326
Dominguez, 254, 257, 258
Dorfman, 7, 68, 73, 74, 76, 100, 103, 104,
       129–132, 139, 140, 150, 165, 174,
       176, 209, 224, 291, 300, 301, 319,
       320, 326
Drake, 103, 198
Drew, T. B., 52
Dunn, 347, 355

E

Eckert, 103, 198
Eckstein, 348
Ede, 105
Elad, 355
Elias, 303
Elkot, 340, 344
El Qarnia, 304, 305
Erdélyi, 58
Erk, 197
Evans, 85
Eytan, 355

F
Faghri, 283
Fedorov, 288
Fedyayevskiy, 141
Fic, 313, 314
Fortes, 322
Fröpssing, 86
Fujii, 297
Fung, 237, 238, 239, 242, 243, 244, 347

G
Ghalichi, 337
Gibbings, 112
Gibson, 122–124, 129, 141, 178, 186
Ginevskii, 99, 141
Gorobets, 291
Gosman, 260
Goswami, 310
Grechannyy, 200, 203, 291, 319, 320
Greenberg, 120
Grigull, 197
Grober, 197
Guedes, 280, 281
Guidoin, 337
Guo, 314

H
Haghighi, 323, 324
Haroun, 340, 344
Hartree, 101
Hasan, 289
Hayat, 356, 357
Heidmann, 297
Henderson, 324
Hirst, 119, 121
Homayoni, 289
Hon, 263
Hopkins, 314
Howell, 219, 265
Hsieh, 310, 311
Hu, 242
Hussain, 356, 357

I
Incropera, 290, 295
Imai, 91
Ishizuka, 288
Issa, 260
Iwasaki, 288
Iyevlev, 142

J
Jaeger, 6, 11, 12, 13, 4, 15, 35
Jaffa, 355
Jaffrin, 237, 240, 242
Jain, 352
Jaluria, Y., 58, 282, 297, 309, 310, 316, 317, 318, 333, 334
Jansen, 272, 273, 274
Jimenez-Lozano, 347, 355
Jou, 273
Joyce, 289

K
Kaario, 303, 304
Kaneva, 58
Kansa, 263
Karayiannis, 289
Karki, 258, 262, 263
Karwe, 318
Kassab, 263, 264, 265, 297
Katto, 320, 321
Kawai, 240
Kawano, 288
Kays, 103, 176, 181
Kestin, 139
Kh, 340, 344
Kimura, M., 240
Kimura, S., 58
King, 337
Kistler, 124
Klebanoff, 124
Kolesnikov, 141
Konopliv, 68
Kutateladze, 141, 142
Kuznetsov, 296, 297

L
Landau, 317
Larmi, 303, 304
Laufer, 122, 270
Lee, 338, 340
Legros, 244, 245, 246
Leontiev, 141, 142
Levich, 67
Li, B. Q., 288
Li, H., 310
Li, L. Y., 311–312
Liao, 338, 340
Lightfoot, 52, 55
Lighthill, 104

Lin, 333, 334
Liu, 314
Lomax, 122, 125, 128, 129, 131
Long, 314
Love, 96
Low, 338, 340
Luikov, 186, 193, 281, 295, 325, 326, 328

M
Maiti, 343, 350
Masmoudi, 325
Mekheimer, 340, 344
Mellor, 122–124, 129, 141, 178, 186
Minakami, 288
Misra, 339, 343, 345, 350
Mittra, 243
Modi, 416
Moini, 324
Monin, 124
Moretti, 181
Moukalled, 260
Munson, B. R., 55
Murthy, 258, 262, 263
Murugesan, 321, 322
Mushtaq, 289

N
Nagolkina, 200, 203
Naidu, 106, 289, 290
Nakayama, 314
Nam, 326, 330, 331
Narayana, 321, 322
Norseman, T. N., 3
Novikiv, 174, 319
Nowak, 313, 314
Nuutinen, 303, 304

O
Oden, 354
Okiishi, T. H., 55
Okos, 322
Olbricht, 351, 352
Olek, 303
Oliveira, 323, 324
Ozisik, 121, 30, 36 280, 281

P
Pagliarini, 279, 285
Pan, 242, 253
Pandey, 243, 244

Panov, 251, 254
Papoutsakis, 285
Patankar, 139, 140, 251, 254, 258–262, 264, 283
Pearce, 311–312
Perry, 352
Piercey, V. I., 51, 52
Please, 352
Polyanin, 1, 42, 56
Pop, 58
Prakash, 258, 262, 263
Prasad, 243
Prat, 325
Premachandran, 294
Pun, 154, 162
Purkiss, 311–312

Q
Qureshi, 356, 357

R
Rageb, 289
Raithby, 260
Ramkrishna, 285
Rao, B. G., 106, 289, 290
Rao, V. D., 106, 289, 290
Riahi, 348
Richards, C. J., 58, 282
Richardson, 139
Rios, 348
Rodriguez, 297
Rosa, 289
Rotta, 119
Roy, 348
Rubesin, 91, 171, 197, 219

S
Sacheti, 350
Sakiadis, 172
Sarier, 313
Sastrohartono, 318
Schley, 352
Schlichting, H., 43, 55, 56, 60, 72, 76, 84,
        89–90, 111–112, 114–115, 119, 125,
        126, 128, 130, 131, 139, 140
Seetharamu, 321, 322
Selverov, 241
Sen, 347, 355
Senatos, 200, 203
Shapiro, 237, 240, 348
Sharma, 106, 289, 290

Sheremet, 296, 297
Shi, 314
Shit, 345
Shugan, 244, 245, 246
Singh, 324
Sinha, 345
Smirnov, 244, 245, 246
Smith, 122, 124–125, 128, 129, 131
Sohal, 219, 265
Soliman, 289
Song, C., 330, 331
Song, W., 288
Spalart, 135, 273, 274
Spalding, 139, 140, 154, 162, 283
Sparrow, 68, 283
Squires, 273, 274
Steinthorsson, 264, 265, 297
Stewart, W. E., 52, 55
Stone, 241
Strelets, 273
Sundararajan, 321, 322
Suresh, 321, 322

T
Takabatake, 240, 353
Takeuchia, 297
Tejeda-Martinez, 272, 273, 274
Tenchev, 311–312
Thomas, 140
Timoshenko, 371, 376
Townsend, 119
Toyomasua, 297
Trp, 314
Truckenbrodt, 223, 227
Tu, 314

V
Van Doormaal, 260
van Douville, 337

Van Driest, 124, 125, 133
Van Dyke, 64, 76, 91, 225, 242
Vanka, 258, 262, 263
Viskanta, 288
Viswanath, 309, 310
Voller, 305
Vynnycky, M., 58

W
Walker, 254, 257, 258
Wan, 314
Wang, P., 351, 352
Wang, Q., 297
Watkinc, 260
Wecel, 313, 314
Weinberg, 237, 240, 348
Weller, 352
Wilcox, 43, 115, 117–121, 125–132,
    135–138, 270–274, 408–412, 416
Wrobel, 313
Wylie, 143

Y
Yaghoubi, 289
Yaglom, 124
Yaniv, 355
Yeh, 303
Yi, 242
Yih, 237–240, 242, 243, 244, 347
Yoo, 316, 317
Yoshinoa, 297
Young, D. F., 55

Z
Zhang, 297
Zuriz, 324
Zvirin, 303

INDEX

A

ADI. *See* Alternating direction method
Adiabatic wall temperature, energy equation, 62–63
Adiabatic wall temperature estimation, 83–84
AFOSR Stanford Conference, 129, 136
Air Force Office of Scientific Research Conference, 129, 136
Algebraic equations, 70
Algebraic models
  1/2 equation model, 131–132
  applicability of, 132–133
  application of, 125–131
  Baldwin–Lomax model, 125
  boundary-layer flows, 129
  Cebeci–Smith model, 124–125
  far wake, 125–127
  flows, channel and pipe, 128
  isothermal surface, 129–130
  Mellor–Gibson model, 122–124
  Prandtl's mixing-length hypothesis, 117–119
  turbulent boundary layer, 118–121
  turbulent Prandtl number, 130–131
  two-dimensional jet, 127
  two parallel streams, 127–128
Alternating direction method (ADI), 290, 291
Analytical methods, 251–253
Approximate methods
  in boundary-layer theory, 93–99
  for solving differential equations, 253–258
Approximate methods in boundary-layer theory
  Karman-Pohlhausen integral method, 93–99
  limiting Prandtl numbers, thermal boundary-layer equations for, 103–104
  linearization, momentum boundary-layer equation, 99–103
Arbitrary pressure gradient flow, 154–155
Arrhenius equation, 299
Arterial stenoses modeling, 337–338
Asymmetric laminar-turbulent flow, 212
Axisymmetric body, 176

B

Baldwin-Lomax model, 125
Basset-Boussinesq-Oseen (BBO) approach, 347
Basset force, 348
BEM. *See* Boundary element method
Bernoulli equations, 50, 61, 71
Bessel function, 27
Bioheat transfer, in human tissues, 356–357
Biot number, 9, 159, 192–193
Biot numbers, 222
Blasius analytical solution, 82
Blasius and Pohlhausen solutions, 80–84
Blasius equation, 81
Blasius formula, 88, 91
Blasius variables, 89
Boundary element method (BEM), 254
Boundary-layer approximation
  derivation of, 69–72
  Euler equation, 68
  of higher order, 76–77
  Prandtl–Mises and Görtler transformations, 72–74
  theory of similarity and dimensionless numbers, 74–76
Boundary-layer equations, 72
  Blasius and Pohlhausen Solutions, 80–84
  flow in straight and convergent channels, 90
  fluid flows interaction, 87–90

power series form, 86–87
self-similar flows, 84–85
Boundary-layer partial differential
    equations, 84
Boundary-layer problems, 72
Boundary-layer structure, 105
Boundary-layer theory, approximate
    methods in, 93–104
Boundary-layer thickness estimation, 82
Boussinesq–Basset term, 347
Boussinesq formula, 118
Boussnesq and laminar shear stress
    formulae, 117
Buoyancy to viscous forces (Grashof), 75

C
Cartesian coordinates, 4–5, 51, 54
Cartesian units, 52
Cauchy–Riemann conditions, 49
Cebeci–Smith and Baldwin–Lomax
    models, 128, 129, 131
Cebeci–Smith model, 124–125
Cerebral perivascular space, modeling
    transport processes, 351–352
CFD. *See* Computational fluid dynamics
Chaos theory, 273–274
Chapman–Rubesin formula, 218
Charts, solving conjugate heat transfer
    problem
  applicability of, 221–225
  development, 212–215
  refining and estimating accuracy of, 221
  use, 215–221
Classical numerical methods, 251–265
Clauser law, 122
Clauser's relation, 120
Clinical medicine, fluid flow and heat
    transfer application
  arterial stenoses modeling, 337–338
  Bioheat transfer, in human tissues,
    356–357
  cerebral perivascular space, modeling
    transport processes, 351–352
  electromagnetic hyperthermia, simulation
    of blood, 345–347
  embryo transport, simulation of, 355–356
  gastrointestinal endoscopy, simulation of
    chyme flow, 348–350
  macromolecules transport in tumors,
    simulation of, 352–354

multi-stenosis under magnetic field, blood
    flow in artery, 340–343
  series stenoses, blood flow, 338–340
  small vessels, simulation of blood flow,
    343–345
  stones, simulation of bile flow, 350–351
Coles' composite wall-wake profile, 121
Coles' idea, 120
Compressible fluid flow, 171–172
Computational fluid dynamics (CFD),
    117, 262
Computing convection-diffusion terms,
    260–261
Computing flow, and heat transfer
    characteristics, 258–263
Computing pressure and velocity, 259–260
Conduction-convection parameter, 290
Conduction heat fluxes, 69
Conduction problem. *See* Conjugate
    heat transfer problem
Conduction to energy storage heat rates
    (Fourier), 75
Conjugate Graetz problem, 285
Conjugate heat transfer
  Biot number, 192–193
  in flow past plates, 225–231
  gradient analogy, 193–195
  heat flux inversion, 195–197
  optimizing heat transfer, 198–206
  solid-fluid interface, temperature
    singularities on, 209–212
  temperature head distribution effect,
    189–192
  zero heat transfer surfaces, 197–198
Conjugate heat transfer problem
  arbitrary pressure gradient flow, 154–155
  axisymmetric body, 176
  Biot number, 192–193
  charts, applicability of, 221–225
  charts for solving, 213–221
  combined boundary condition, 184–185
  compressible fluid flow, 171–172
  errors caused by boundary condition,
    182–184
  in flow past plates, 225–231
  formulation of, 149–150, 236–237
  general boundary condition, 182
  general properties, 189–206
  laminar fluid flow, 150–177
    (*See also* Laminar fluid flow)

moving continuous sheet, 172–174
one-dimensional approach, 221–225
peristaltic motion, 235–247
power-law non-Newtonian fluids, 174–175
pressure gradient flow, 156–158
recovery factor, 175–176
reducing, conjugate problem, 182–187
self-similar flow, 168–171
temperature head, 165–168
thermal boundary-layer equation, 150–154
turbulent fluid flow, 177–181
  (*See also* Turbulent fluid flow)
unsteady heat transfer, 185–186
Conjugate methods
  concept of, 145–146
  requirement, 146–147
Conjugate solutions, 243–247
Continuous wires casting, 314–315
Control-volume finite-difference
    method (CVFDM)
  computing convection-diffusion
      terms, 260–261
  computing pressure and velocity, 259–260
  false diffusion, 261–262
Control-volume finite-element method
    (CVFEM), 262–263
Convection to conduction heat rate
    (Peclet), 75
Convective flow,
  pipe, 283–284
Convective to conduction heat transfer rates
    across the interface (Nusselt), 76
Cooling systems
  electronic packages, 294–297
  turbine blades and rocket, 297–300
Couette flow, 59, 60
Couette flow, energy equation, 62
Couette motion, 59
Creeping approximation
  heat transfer from sphere, 68
  Laplace equations, 65
  Oseen's approximation, 67–68
  Stokes flow, 65–67
Curvilinear orthogonal coordinates, 51–55
CVFDM. *See* Control-volume finite-
    difference method
CVFEM. *See* Control-volume finite-element
    method
Cylindrical coordinates, 53
Czochralski crystal growth process, 313

D
D'Alamber paradox, 68
D'Alemberts paradox, 41
Darcy's law, 312
Derivation of boundary-layer equations,
    69–72
DES. *See* Detached eddy simulation
Detached eddy simulation (DES), 43, 117,
    272–273
Dihamel's integral evaluation, 21
Dimensionless continuity equation, 76
Dimensionless number, 74–76, 75
Dirac delta function, 36
Direct numerical simulation (DNS), 42,
    117, 270
Dirichlet problem, 57, 58, 63, 297
Dirihlet problem, 282
DNS. *See* Direct numerical simulation
Dorodnitsyn's variables, 171
Draing technology, 321–330
  brick drying, 321–322
  porous materials, 325–326
  of pulled continuous materials, 326–329
  rectangular wood board, 324–325
Duhamel integral derivation, 15–16
Duhamel's method, 91
  Duhamel integral derivation, 15–16
  time-dependent surface temperature, 17–20
Duhamel's principle, 36
Dynamic and thermal boundary layers, 84–85

E
Eddy-viscosity, 117, 122, 124, 125, 134, 135
Einstein and index notation, 47–48
Electromagnetic hyperthermia, simulation of
    blood, 345–347
Embryo transport, simulation of, 355–356
Energy, and mass transfer equations
  adiabatic wall temperature, 62–63
  analogy *vs.* transfer processes, 44–45
  couette flow, 62
  curvilinear orthogonal coordinates, 53–55
  Einstein and index notation, 47–48
  exact solutions of, 62–63
  irrotational inviscid two-dimensional
      flows, 49–50
  temperature distributions in, 63
  transport mechanism, 44–45
Energy systems, 303–306
  1/2 equation model, 131–132

Equilibrium boundary-layer flows, 118
Equilibrium boundary-layer solutions, 178
Equilibrium turbulent boundary layer, 121
Equivalent conduction problem. *See also*
　　Conjugate heat transfer problem
　with combined boundary condition,
　　184–185
　for unsteady heat transfer, 185–186
Error integral, solution for solid, 11–15
　infinite solid, 12–13
　semi-infinite body, 13–15
Euler equation, 68

F
False diffusion, 261–262
FDM. *See* Finite-difference methods
FEM. *See* Finite-element method
Fick's law, 44
Film-cooling system, 297
Films and fibers production, 319–321
Finite-difference methods (FDM), 251
Finite-element method (FEM), 254
Finite-volume method (FVM), 254
Flow, in straight and convergent channels, 90
Flow and heat transfer
　isothermal semi-infinite flat plate, 80–84
Flow past plates
　conjugate heat transfer in, 225–231
Flows, with pressure gradients, 97–99
Flow velocity, 115
Fluid flow
　biology and clinical medicine, 337–357
　classical numerical methods, 251–265
　clinical application, 337–357
Fluid flows interaction
　flow in wake of body, 87–88
　mixing layer of two parallel streams, 89–90
　two-dimensional jet, 88–89
Fluid Prandtl number, 68
Food processing
　constricted tubes, 334
　food and polymer flow, 333–335
　food extrusion, 334
　freeze drying, 330–333
　unconstricted tubes, 334
Fourier's law, 44, 152, 174
Fourier transform, 30–33, 37
Free convection
　and radiation from horizontal fin array, 106
　vertical plate, 105

Friction coefficient, 172
Friction velocity, 119, 120
Full-energy equations, 72, 76.
　　See also Navier–Stokes equations
Fung and Yih approach, 347

G
Gastrointestinal endoscopy, simulation of
　chyme flow, 348–350
General boundary condition, 182
Görtler transformations, 72–74
Görtler variables, 73, 151, 152, 190
Gradient analogy, 193–195
Grashof number, 105
Green's function method, 36–39

H
Hagen–Poiseuille laminar flow, 128
Hagen–Poiseuille profile, 59, 60
Hamilton operator, 45
Harmonic functions, 49
Heat conduction equation and problem
　　formulation
　boundary conditions, 8–10
　Cartesian coordinates, 4–6
　conjugate conditions, 8–10
　heat flux, universal function for, 7–8
　initial conditions, 8–10
　orthogonal curvilinear coordinates, 6–7
Heat exchange between two fluids, 158, 160
Heat exchangers and pipes
　and channels, 279–284
　and finned surfaces, 285–294
Heat flux, universal function for
　in arbitrary pressure gradient flow, 154–155
　for axisymmetric body, 176
　in compressible fluid flow, 171–172
　applications of, 158–162
　moving continuous sheet, 172–174
　pressure gradient flow, 156–158
　self-similar flow, 168–171
　thermal boundary-layer
　　equation, 150–154
　for turbulent fluid flow, 177–181
Heat fluxes estimation, 84
Heat flux inversion, 195–197
Heat transfer
　from fluid sphere, 68
　from nonisothermal plate, 96
Heat transfer fluid (HTF), 304

Heat transfer intensity, 189
Heat transfer methods
  applications, 277
  biology and clinical medicine,
    337–357
  Cartesian coordinates, 4–6
  classical numerical methods, 251–265
  cooling systems, 294–303
  Duhamel's method, 15–20
  energy systems, 303–306
  error integral, 11–15
  Green's function method, 36–39
  heat conduction equation, 4–11
  heat exchangers and pipes, 279–294
  historical notes, 3–4
  integral transforms, 30–35
  orthogonal curvilinear coordinates, 6–7
  problem formulation, 4–11
  separation variable method, 21–30
  in technology processes, 309–335
  Turbulent fluid flow methods, 111–143
  universal function, arbitrary
    nonisothermal surface, 7–11
Helmholtz equation, 21
Hiemenz flow, 61
Higher order, boundary-layer equations of,
    76–77
Homogeneous, 20–21
Homogeneous boundary conditions, 36
Horizontal fin array, 289
Howarth flow, 87
HTF. See Heat transfer fluid (HTF)
Hydrodynamic laws, 44
Hyperbolic equation, 135

I
Illingworth–Stewartson's variables, 171
Inertia to viscous forces (Reynolds), 75
Influence function, 103
Inhomogeneous problems, 20–21
Initial and boundary conditions
  energy equation, 57–58
  Navier–Stokes equations, 56–57
Integral methods, 139–142
Integral transform technique
  Fourier transform, 30–33
  Laplace transform, 33–35
Intermittency factor, 111
Irrotational inviscid two-dimensional
    flows, 49–51

Isothermal coefficient, 190
Isothermal heat transfer coefficient, 172, 211
Isothermal temperature, 75

K
Karman constant, 118, 120
Karman–Pohlhausen integral method, 139
  boundary-layer equation, 93
  flows with pressure gradients, 97–99
  friction and heat transfer, on
    flat plate, 95–97
  momentum integral equation, 94
Kinematic eddy-viscosity, 120
Kinetic energy to enthalpy (Eckert), 75
Knudsen number, 58, 288
Kolmogorov scale, 269
Kolmogorov's reasoning, 136

L
Lame coefficient, 52
Lame coefficients, 51, 53
Laminar boundary layer
  coefficients of universal
    function, 153, 166
  Prandtl number, 157
Laminar flow
  Dirihlet problem, 282
  in double-pipe heat exchanger, 285–288
  solid-fluid interface, 209
  stagnation point, 211–212
  uniform heat flux, 279–280
  universal functions, 215, 216
  zero-pressure gradient, 212
Laminar flow stability. See also Turbulent
    fluid flow methods
  characteristics, 111–112
  problem of, 112–114
Laminar fluid flow
  in arbitrary pressure gradient flow, 154–155
  for axisymmetric body, 176
  boundary-layer equations, 80–92
  boundary-layer theory, approximate
    methods in, 93–104
  in compressible fluid flow, 171–172
  conjugate heat transfer problem, 150–176
  energy, and mass transfer equations, 43–56
  heart flux, universal functions for, 158–162
  history, 41–42
  initial and boundary conditions, 56–58
  moving continuous sheet, 172–174

natural convection, 104–108
Navier–Stokes equations, 43–56
pressure gradient flow, 156–158
for recovery factor, 175–176
reynolds and peclet numbers, 64–80
self-similar flow, 169–172
for temperature head, 165–168
thermal boundary-layer
   equation, 150–154
universal function for, 150–176
Laminar gradient flow, power-law
   free-stream velocity, 212
Laplace equation, 49, 65, 66, 297
Laplace's and inhomogeneous Poisson's
   equations, 28
Laplace's and Poisson's equations, 5, 10
Laplace transform, 33–35
Large eddy simulation (LES), 43, 117,
   271–272
LES. *See* Large eddy simulation
Levi–Civita rules, 66
Linear energy equation, 80
Linearization, momentum boundary-layer
   equation
   Prandtl–Mises form, 99
   preseparation flow, 100
   skin friction coefficient, universal function
      for, 100–103
Linearized boundary-layer equation, 87
Linear potential velocity distribution, 102
Linear temperature head distribution, 161

M
Macromolecules transport in tumors,
   simulation of, 352–355
Manufacturing processes simulation,
   314–321
   continuous wires casting, 314–315
   films and fibers production, 319–321
   optical fiber coating process, 316–317
   twin-screw extruder simulation, 318–319
Mechanical energy dissipation, 175
Mellor–Gibson model, 122–124, 129, 178
MEMS. *See* Microelectromechnical system
Microchannel heat sink, heat exchanger, 288
Microelectromechanical system (MEMS), 241
Modern numerical methods
   Chaos theory, 273–274
   detached eddy simulation, 272–273
   direct numerical simulation, 270

large eddy simulation, 271–272
Momentum to thermal diffusivity
   (Prandtl), 75
Multiphase and phase–changing processes,
   309–314
   concrete, evaluated temperatures, 311
   Czochralski crystal growth process, 313
   Darcy's law, 312
   solidification, 309
Multi–stenosis under magnetic field, blood
   flow in artery, 340–343

N
Natural convection, 104–106
Navier–Stokes equations, 236, 238, 282
   Bernoulli equation, 50
   computing pressure and velocity, 259
   to curvilinear coordinates, 53
   Einstein and index notation, 47–48
   initial and boundary conditions, 56–57
   Reynolds averaging equations, 116
   stagnation point flow, 61–62
   steady flow, 59–60
   stream function form, 48–49
   three other unsteady problems, 59
   two stokes problems, 58–59
   vector form, 45–47
   velocity components, 43
   vorticity form, 48
Neumann problem, 29
Newtonian and non-Newtonian fluids, 238
Newton's viscosity law, 44
Nonisothermal and conjugate heat transfer, 155
   Biot number, 192–193
   gradient analogy, 193–195
   heat flux inversion, 195–197
   optimizing heat transfer, 198–206
   temperature head distribution effect,
      189–192
   zero heat transfer surfaces, 197–198
Nonisothermicity
   coefficient distribution, 161
   in turbulent and laminar flows, 191
Nonlinear approach, 238
Non-Newtonian fluids, 174–175, 238
Nuclear reactor, 300–303
Numerial methods, of conjugation, 263–265
Numerical methods, 251–253
Nusselt distribution, 162
Nusselt number, 284, 286

O
One-dimensional approach, 222–225
One-dimensional unsteady problems, 21–23
One-equation and two-equation models.
    See also Two-equation models
    applicability of, 138–139
    turbulence kinetic energy equation,
        134–135
Optical fiber coating process, 316–317
Optimizing heat transfer, 198–206
Orr–Sommerfeld equations, 114, 239
Orthogonal curvilinear coordinates
    cylindrical coordinates, 6–7
    elliptical cylindrical coordinates, 7
    spherical coordinates, 7
Oseen's approximation, 67–68

P
PCM. *See* Phase-change material
Peclet number, 279
Peclet numbers, 261
Peristaltic flow, particle motion in, 347–351
Peristaltic motion
    conjugate solutions, 243–247
    early works, 237–238
    formulation of conjugate problem,
        236–237
    overview, 235–236
    semi-conjugate solutions, 238–243
Phase-change material (PCM), 304
Plate
    heat transfer characterizes of, 217, 219
    nonisothermicity coefficient, 228
    turbulent flow, 229
Poiseuille profile, 90
Potential velocity, 87
Power function, 84
Power-law non-Newtonian fluids, 174–175,
    210
Power series form, solutions, 86–87
Prandtl approach, 69, 70, 80
Prandtl–Mises and Görtler transformations,
    72–74
Prandtl–Mises form, 252
Prandtl–Mises–Görtler form, 73, 129, 177
Prandtl–Mises–Görtler variables, 150, 151
Prandtl numbers, 68, 83, 154, 159, 174, 178,
    189, 192
Prandtl's assumption, 118
Prandtl's formulae, 119

Prandtl's idea, 118
Prandtl's mixing-length hypothesis, 117–119
Pressure gradient flow, 156–158

R
RANS equation. *See* Reynolds–Averaged
        Navier–Stokes equation
Rayleigh–Bénard flow, 274
Rayleigh equation, 114
Rayleigh number, 105
Recovery factor, universal functions for,
        175–176
Reynolds and Peclet numbers
    boundary-layer approximation, 68–77
    creeping approximation, 65–68
    Navier–Stokes equations, 64
Reynolds and Prandtl numbers, 161
Reynolds–Averaged Navier–Stokes
        (RANS) equation, 4, 273
    physical aspects, 114–115
    Reynolds averaging, 115–116
    Reynolds equations and Reynolds stresses,
        116–117
Reynolds–average models, 269
Reynolds averaging equations, 116
Reynolds numbers, 45, 60, 111, 112, 114,
        189, 270, 282
Reynolds procedure of statistical averaging
        Navier–Stokes equations (RANS), 41
Reynolds stress, 119, 125, 134, 135
Royal Aeronautical Society, 72
Runge–Kutta method, 152, 212

S
Second-order boundary-layer equations,
        76, 77
    solutions of, 91
Self-similar flows, 84–85, 101, 169–171
Semi-conjugate solutions, 238–243
Semi-Implicit Method for Pressure-Linked
        Equations (SIMPLE), 259
Separation variable method
    approach, 20–21
    eigenfunctions, orthogonality of, 24–28
    homogeneous, 20–21
    inhomogeneous problems, 20–21
    one-dimensional unsteady problems, 21–23
    two-dimensional steady problems, 28–30
Series stenoses, blood flow, 338–340
SGS. *See* Subgrid scales

Shear stresses, 69
Similarity theory, 74–76
SIMPLE. *See* Semi-Implicit Method for Pressure-Linked Equations
SIMPLER algorithm, 288–289
Six dimensionless numbers, 76
Skin friction, 81, 91
Small vessels, simulation of blood flow, 343–345
Solar energy storage unit, 304–305
Solid-fluid interface, temperature singularities
  asymmetric laminar-turbulent flow, 212
  basic equations, 210–211
  laminar flow, 209
  laminar flow, stagnation point, 211–212
  laminar flow, zero-pressure gradient, 212
  laminar gradient flow, power-law free-stream velocity, 212
  turbulent flow, zero-pressure gradient, 212
Solid propellant rocket, 298
SOR code. *See* Successive over relaxation code
Spalart–Allmarass model, 135
Sphere temperature, 68
Spherical coordinates, 55
Stability, of fluid between two horizontal plates, 106
Stagnation point flow, 61
Steady flow, 59–61
Steady-state and zero-pressure gradient boundary-layer equation, 89
Steady-state boundary-layer equation, 123
Stephan problems, 306
Stokes flow, 65–67
Stones, simulation of bile flow, 350–351
Stream function, 66
Stream function form, of Navier–Stokes equation, 48–49
Sturm–Liouville problem, 24, 26, 27
Subgrid scales (SGS), 271
Successive over relaxation (SOR) code, 282

T
Taylor expansion, 261
Taylor series, 73, 118
TDMA. *See* Tridiagonal matrix algorithm
Temperature boundary layer equations, 252
Temperature distributions, energy equation, 63

Temperature head derivatives, 189
Temperature head distribution effect, 189–192
  flow regime, 191–192
  pressure gradient, 192
  temperature head gradient, 190–191
Tensor order, 53
Theory of similarity, 74–76
Thermal boundary conditions, effect of, 85
Thermal boundary-layer equation, solutions of, 91
Thermal boundary-layer equations, 80, 150–154
  for limiting Prandtl numbers, 103–104
Thermal boundary-layer thickness estimation, 83
Thermally thin body assumption, applicability of, 221–222
Three-dimensional Navier–Stokes equation, 114
Time-dependent surface temperature, 17–20
Transverse coordinate, 73
Transverse flow, 102, 162, 168
  flat finned surface in, 291, 292
Tridiagonal matrix algorithm (TDMA), 282
Turbulence models, 270–274
Turbulence velocity, 115
Turbulent flow
  heat transfer characterizes, 218
  parallel plate, 280–282
  zero-pressure gradient, 212
Turbulent flow averaging, 116
Turbulent fluid flow
  conjugate heat transfer problem, 177–181
  universal function for, 177–181
Turbulent fluid flow methods
  algebraic models, 117–132
  integral methods, 139–142
  laminar flow stability problem, 112–114
  one-equation and two-equation models, 134–139
  Reynolds-averaged Navier–Stokes equation, 114–117
  transition from, 111–114
Twin-screw extruder simulation, 318–319
Two-dimensional channel, 237
Two-dimensional effects, 222–225
Two-dimensional flow, 49, 69
Two-dimensional jet, 88–89

Two-dimensional Navier–Stokes equations, 294
Two-dimensional steady problems, 28–30
Two-dimensional unsteady Navier–Stokes equations, 341
Two-equation models
  applicability of, 138–139
  k - ε model, 138
  k–kl models, 138
  k–τ models, 138
  k - ω model, 136–137
Two stokes problems, Navier–Stokes equation, 58–59

U
Universal Function for Heat Flux
Universal functions
  in arbitrary pressure gradient flow, 154–155
  for axisymmetric body, 176
  in compressible fluid flow, 171–172
  as general boundary condition, 182
  heart flux, universal functions for, 158–162
  for heat flux in self-similar flows, 150–154
  Laminar boundary layer, 153
  laminar flow, 215, 216
  for laminar fluid flow, 150–176
  moving continuous sheet, 172–174
  for power-law non-Newtonian fluids, 174–175
  pressure gradient flow, 156–158
  for recovery factor, 175–176
  self-similar flow, 169–171

skin friction coefficient, 101
for temperature head, 165–168
for turbulent, 215
for turbulent fluid flow, 177–181
Unsteady problems, 59
Unsteady thermal boundary-layer equation, 169–171
Unsteady three-dimensional Navier–Stokes equation, 269

V
Van Driest's formula, 124–125
Vector form
  diffusion equations, 45–46
  energy equations, 45–46
  Navier–Stokes equations, 45–46
Velocity distribution, 84
Velocity profile, turbulent boundary layer, 118–121
Viscosity coefficient, 117
Vorticity form, of Navier–Stokes equation, 48

W
Wall function (WF), 304
WF. See Wall function

Z
Zero-equation models, 117
Zero heat transfer surfaces, 197–198
Zero pressure gradient, 161, 170
Zero Reynolds number, 237
Zero velocity components, 56

Check Out the Other Mechanical Engineering We Have!

Announcing Digital Content Crafted by Librarians

Momentum Press offers digital content as authoritative treatments of advanced engineering topics, by leaders in their fields. Hosted on ebrary, MP provides practitioners, researchers, faculty and students in engineering, science and industry with innovative electronic content in sensors and controls engineering, advanced energy engineering, manufacturing, and materials science. **Momentum Press offers library-friendly terms:**

- perpetual access for a one-time fee
- no subscriptions or access fees required
- unlimited concurrent usage permitted
- downloadable PDFs provided
- free MARC records included
- free trials

The **Momentum Press** digital library is very affordable, with no obligation to buy in future years.

For more information, please visit **www.momentumpress.net/library** or to set up a trial in the US, please contact **Adam Chesler,** *adam.chesler@momentumpress.net.*